The Small Wars of the United States, 1899–2009

The Small Wars of the United States, 1899–2009 is the complete bibliography of works on U.S. military intervention and irregular warfare around the world, as well as efforts to quell insurgencies on behalf of American allies. The text covers conflicts from 1899 to present, with detailed annotations of selected sources. In this second edition, author Benjamin R. Beede revises his seminal work, bringing it completely up to date, including entries on the current conflicts in Iraq and Afghanistan. An invaluable research tool, *The Small Wars of the United States, 1899–2009* is a critical resource for students and scholars studying U.S. military history.

Routledge Research Guides to American Military Studies provide concise, annotated bibliographies to the major areas and events in American military history. With the inclusion of brief critical annotations after each entry, the student and researcher can easily assess the utility of each bibliographic source and evaluate the abundance of resources available with ease and efficiency. Comprehensive, concise, and current—**Routledge Research Guides to American Military Studies** are an essential research tool for any historian.

Benjamin R. Beede is a Librarian Emeritus of Rutgers–The State University of New Jersey, and author of numerous books and articles. He lives in North Brunswick, New Jersey.

Routledge Research Guides to American Military Studies

America and World War I
David R. Woodward

The War of 1812
John Grodzinski

The United States in the Vietnam War, 1954–1975
Louis A. Peake

The Western European and Mediterranean Theaters in World War II
Donal J. Sexton

The Korean War
Keith D. McFarland

The Small Wars of the United States, 1899–2009
Benjamin R. Beede

Coming Soon

Terrorism
Joshua Sinai

Women in the United States Military
Judith Bellafaire

The Spanish–American War and Philippines Insurgency, 1898–1902
Mark Barnes

The Small Wars of the United States, 1899–2009
An Annotated Bibliography

Revised Second Edition

Benjamin R. Beede

NEW YORK AND LONDON

First published 1985 by Garland Publishing
This edition published 2010 by Routledge
711 Third Avenue, New York, NY 10017

Simultaneously published in the U.K. by Routledge
2 Park Square, Milton Park, Abingdon, Oxfordshire OX14 4RN

First issued in paperback 2014

Routledge is an imprint of the Taylor and Francis Group, an informa business

© 2010 Benjamin R. Beede

Typeset in Times New Roman by
Taylor & Francis Books

All rights reserved. No part of this book may be reprinted or reproduced or utilized in any form or by any electronic, mechanical, or other means, now known or hereafter invented, including photocopying and recording, or in any information storage or retrieval system, without permission in writing from the publishers.

Trademark Notice: Product or corporate names may be trademarks or registered trademarks, and are used only for identification and explanation without intent to infringe.

Library of Congress Cataloging in Publication Data
Beede, Benjamin R.
  The small wars of the United States, 1899–2009 : an annotated bibliography / Benjamin R. Beede.
    p. cm. – (Routledge research guides to American military studies)
  Rev. ed. of: Intervention and counterinsurgency / Benjamin R. Beede.
  Includes bibliographical references and index.
  1. United States–History, Military–20th century–Bibliography. 2. Guerrilla warfare–United States–History–20th century–Bibliography. 3. Counterinsurgency–United States–History–20th century–Bibliography. 4. Low-intensity conflicts (Military science)–United States–History–20th century–Bibliography. 5. United States–History, Military–21st century–Bibliography. 6. Guerrilla warfare–United States–History–21st century–Bibliography. 7. Counterinsurgency–United States–History–21st century–Bibliography. 8. Low-intensity conflicts (Military science)–United States–History–21st century–Bibliography. I. Beede, Benjamin R. Intervention and counterinsurgency. II. Title.
  Z1249.M5B44 2010
  [E745]
  016.355–dc22
                                        2009043106

ISBN 978-0-415-98888-9 (hbk)
ISBN 978-1-138-86781-9 (pbk)
ISBN 978-0-203-85434-1 (ebk)

To my wife Anne Brugh

Contents

Acknowledgment		xv
Introduction		1
I	**Bibliographic sources**	10
	A. Catalogs of manuscripts	10
	B. Oral history	11
	C. Online periodical, newspaper, and report indexes	11
	D. Bibliographies	18
	E. Libraries and research centers	22
II	**Histories and general studies**	24
III	**Pre-World War II doctrines and general studies, 1899–1940**	33
	A. Doctrines	33
	B. Organizing for expeditionary service	34
	C. Counterinsurgency tactics	35
	D. The role of air power	37
	E. Military government/civic action	37
	F. Other topics	38
IV	**The Philippine–American War and the Moro campaigns, 1899–c.1918**	39
	A. General studies	39
	B. The Philippine–American War, 1899–c.1918	40
	C. The Moro campaigns, 1902–c.1913	66
V	**Boxer Uprising, 1898–1901**	76
	A. Published documentary collections and official reports	77
	B. General surveys	78
	C. The defense of the legations	81
	D. The Seymour Expedition	82
	E. The attack on Tianjin	82
	F. Subsequent fighting in the Tianjin area	83
	G. The advance to Beijing	83
	H. The role of the marines	84
	I. Medical aspects of the China Relief Expedition	87
	J. Military government	88

VI	**China, 1901–41**	89
	A. General studies	91
	B. Army service in China	92
	C. The Yangzi Patrol	94
	D. The *Panay* incident	97
	E. The end of the gunboat era	98
	F. Other naval operations in China	99
	G. Marine operations in China	99
VII	**Cuba, 1906–09**	106
	A. The Platt Amendment	107
	B. General studies	108
	C. Military studies	109
	D. Other activities	111
VIII	**Nicaragua, 1909–25**	112
	A. Interventions of 1909, 1910, and 1912	114
	B. The legation guard as an American presence in Nicaragua, 1912–25	117
IX	**The Mexican border, 1911–19**	119
	A. General treatments	122
	B. Taft and the intervention question	123
	C. Border skirmishes and the broader problem of intervention	124
	D. The attack on Columbus, New Mexico, 1916	127
	E. The Punitive Expedition: general studies	128
	F. The Punitive Expedition: major battles	132
	G. The Punitive Expedition: weapons and equipment	133
	H. The Punitive Expedition: logistics	134
	I. The Punitive Expedition: patrols and skirmishes	135
	J. The Punitive Expedition: commanders and others	136
	K. The Punitive Expedition: the role of aviation	136
	L. The Punitive Expedition: medical aspects	137
	M. The Punitive Expedition: the horses	137
	N. Mobilization of the National Guard for border service	137
X	**Veracruz, 1914**	141
	A. General studies	142
	B. Diplomatic aspects of the occupation	144
	C. The attack on Veracruz	146

	D. Military government/civic action	146
	E. Other topics	148

XI Haiti, 1915–34 — 149

A. Congressional hearings and official reports — 151
B. General studies — 151
C. The decision to intervene — 156
D. The campaigns — 156
E. Battle of Fort Rivière — 158
F. The role of air power — 159
G. Military advisory activities — 159
H. The occupation regime — 161
I. Civic action — 163
J. Haiti as a duty station — 164
K. The decision to withdraw — 165
L. The marine withdrawal — 166

XII Dominican Republic, 1916–24 — 167

A. Congressional hearings and official reports — 169
B. General studies — 169
C. Strategy and tactics — 172
D. The role of air power — 174
E. Military advisory activities — 174
F. Military government/civic action — 174
G. The decision to withdraw — 176
H. The marine withdrawal — 176
I. The aftermath — 177

XIII Cuba, 1917–22 — 178

A. Diplomatic background of the intervention — 179
B. Marine landings and the intervention — 179
C. Special studies — 180

XIV Russia, 1918–20 — 181

A. General studies — 181
B. Diplomatic and political studies — 184
C. Military analyses — 184

XV North Russia, 1918–19 — 186

A. Contemporary general studies — 187
B. Later analyses — 188
C. Combat narratives — 189

	D. Naval operations	190
	E. Political critiques	190
XVI	**Siberia, 1918–20**	**192**
	A. General studies	193
	B. Political and diplomatic studies	195
	C. Personal narratives	197
	D. Naval and marine operations	198
XVII	**Nicaragua, 1927–33**	**199**
	A. The decision to intervene	201
	B. General surveys	202
	C. Strategy and tactics	206
	D. Major battles and campaigns	210
	E. The role of air power	213
	F. Military advisory activities	216
	G. Other military topics	217
	H. Civic action	218
	I. Political developments during the intervention	220
	J. The decision to withdraw	222
	K. The evacuation	222
XVIII	**Post-World War II doctrines and general studies**	**224**
	A. Counterinsurgency in the 1950s	224
	B. Counterinsurgency doctrine	226
	C. Small wars and counterinsurgency: general studies	230
	D. Small wars, counterinsurgency, and the media	242
	E. The Kennedy counterinsurgency initiative	242
	F. Special Forces and related units	244
	G. The navy and counterinsurgency	246
	H. Counterinsurgency training in the 1960s	246
	I. Systems analysis and counterinsurgency	248
	J. Counterinsurgency tactics	249
	K. Counterinsurgency weapons	250
	L. Intelligence and counterinsurgency	250
	M. Other facets of counterinsurgency	251
	N. Civic action	252
XIX	**North China, 1945–49**	**255**
	A. General studies	257
	B. Diplomatic and political perspectives	260
	C. Repatriation of the Japanese in China	260

	D. Marines as railway guards	261
	E. Clashes with Chinese Communist forces	261
	F. Other topics	262

XX Lebanon, 1958 — 263

- A. Political surveys of the Lebanon crisis — 265
- B. General military studies — 268
- C. The landings — 269
- D. Political conflicts during the landings — 270
- E. The navy's role in Lebanon — 272
- F. The marines in Lebanon — 273
- G. Special military studies — 273
- H. The influence of the Lebanon operation on the armed forces — 274

XXI Dominican Republic, 1965–66 — 275

- A. General studies — 277
- B. General military studies — 283
- C. Naval operations — 284
- D. Marine operations — 284
- E. Army operations — 285
- F. Air force operations — 287

XXII From the Nixon Doctrine through the end of the Cold War, 1970–90 — 288

- A. Doctrines — 288
- B. Small wars: general studies — 289
- C. Small wars and the media — 291
- D. Counterinsurgency and low-intensity conflict theory after Vietnam — 293
- E. Counterinsurgency forces: general — 310
- F. Counterinsurgency forces: army and air force — 311
- G. Counterinsurgency forces: marines and navy — 312
- H. Evacuation operations — 314
- I. The legal aspects of counterinsurgency — 315
- J. Intelligence and counterinsurgency — 315
- K. Civil affairs/civic action — 316

XXIII Lebanon, 1982–84 — 319

- A. Congressional hearings and official reports — 320
- B. General studies — 321
- C. The withdrawal of the marines — 325
- D. Other aspects of the Lebanon intervention — 326

XXIV	**Grenada, 1983–85**	327
	A. Congressional hearings and official reports	328
	B. Documentary collections	329
	C. General studies	330
	D. The decision to intervene	331
	E. Military and naval operations	333
	F. Political and legal aspects	337
	G. The media and the Grenada invasion	338
	H. African-American perspectives on Grenada	339
XXV	**Panama, 1989**	340
	A. General studies	342
	B. Military studies: general	343
	C. Military studies: specialized	345
XXVI	**From the end of the Cold War to the War on Terrorism, 1991–2001**	347
	A. Small wars, military humanitarian efforts, and peacekeeping: general studies	347
	B. Counterinsurgency and low-intensity conflict theory	350
XXVII	**Persian Gulf War, 1990–91**	353
	A. Bibliography	356
	B. Dictionaries and encyclopedias	356
	C. Chronologies	357
	D. General studies	357
	E. Decision to go to war	362
	F. Air operations	364
	G. Army land operations	368
	H. Marine operations	370
	I. Naval operations	372
	J. Special military operations	373
	K. Special military studies	373
	L. Logistics	374
	M. Civil affairs/civic action	374
	N. Political aspects	375
	O. Legal aspects	376
	P. The Gulf War and the media and the role of public opinion	377

XXVIII	**Somalia, 1992–94**	381
	A. General studies	383
	B. Military studies: general	386
	C. Military studies: intelligence	388
	D. Military studies: logistics	389
	E. Other military studies	389
	F. The battle of Mogadishu	391
	G. Legal aspects	392
	H. The Somalia intervention and the media	393
XXIX	**Bosnia, 1992–96**	394
	A. General studies	396
	B. Air operations	399
	C. Other military studies	399
	D. The peace settlement	400
XXX	**Haiti, 1991–96**	401
	A. General studies	403
	B. Military studies: general	405
	C. Special military studies	406
XXXI	**Kosovo, 1999–**	408
	A. General studies and political analyses	410
	B. Military studies: general	411
	C. The air campaign	412
	D. The international force in Kosovo	413
XXXII	**Doctrines and general studies in the age of the War on Terrorism**	416
	A. Counterinsurgency doctrine	416
	B. Counterinsurgency doctrine: critiques	417
	C. Small wars: general studies	421
	D. Counterinsurgency: general studies	421
	E. Counterinsurgency: past lessons	428
	F. Counterinsurgency: tactics	432
	G. Counterinsurgency: logistics	435
	H. Intelligence and counterinsurgency	435
	I. Aviation and counterinsurgency	437
	J. Counterinsurgency and civic action	439
	K. Counterinsurgency: legal aspects	441

	L.	Counterinsurgency forces: general	442
	M.	Counterinsurgency forces: army and air force	442
	N.	Counterinsurgency forces: marines and navy	442
XXXIII	**Afghanistan, 2001–**		444
	A.	Encyclopedia	447
	B.	General studies	447
	C.	General military studies	450
	D.	Air operations	456
	E.	Ground combat	459
	F.	Special military operations	464
	G.	Intelligence	465
	H.	Military advisory activities	465
	I.	Logistics	466
	J.	Civil affairs/civic action	468
	K.	Psychological operations	470
	L.	Engineering operations	470
XXXIV	**Iraq War, 2003–**		472
	A.	General studies	475
	B.	The decision for war	478
	C.	General military studies	483
	D.	Air operations	486
	E.	Ground combat	487
	F.	Battles: Baghdad	493
	G.	Battles: Fallujah	494
	H.	Insurgency and the counterinsurgency campaign	494
	I.	Naval operations	498
	J.	Intelligence	499
	K.	Prisons and prisoners	499
	L.	Military advisory activities	501
	M.	Logistics	502
	N.	Civil affairs/civic action	503
	O.	Information programs/psychological operations	505
	P.	The media and the war	506
	Q.	Terminating the war	506

Author index   509

Subject index   524

Acknowledgments

This book owes a great deal to the guidance, good counsel, and effective editing of a number of people at Routledge. These members of the Routledge team include Matthew Kopel (Editorial Assistant), Kimberly Guinta (Acquisitions Editor for History), Ulrike Swientek (Senior eProduction Editor), Andy Soutter (copyediting), James Clark (proof-reading), and Tessa Hallam (typesetting).

The compilation and editing of a bibliography may appear to be a simple task, but it is not, especially when a manuscript grows beyond the length initially planned.

I am most grateful for the invitation to prepare a second edition of this bibliography, for the extreme care that was given to my manuscript, and for permission to expand the size of the project.

Preparation of a useable electronic manuscript would have been almost impossible for me without the considerable assistance of Dr. Thomas J. Kehoe, who knows vastly more about computers than do I. He also made helpful suggestions about the project generally.

The contributions of the Rutgers University Libraries and the Princeton University Library should be also be noted. Despite the difficult financial situation for higher education, these libraries continue to provide rich resources. Finally, I want to pay tribute to the Marine Corps Association which maintains indexes for and full-text coverage of two splendid magazines, the *Marine Corps Gazette and Leatherneck* for members and associate members of the association.

Introduction

The phrase "small war" should not be considered facetious; it is simply associated with the Spanish term *guerrilla*.

Many of the small wars included in this bibliography are unlikely to be familiar to the average person in the United States; some of them have been almost entirely forgotten, except by military historians. There is a difference between obscurity and insignificance. As Senator Barry Goldwater once remarked, the small wars should be examined "in the perspective of their times." In such a context, these campaigns and occupations often had more importance than seems apparent today.[1] The many interventions in Central America and the Caribbean left a legacy of bitterness in Latin America that impaired United States relations with that area for many years, and the United States involvements in North Russia and Siberia were remembered with anger by the Soviet Union. United States involvements in the Middle East are a newer example of the impact on the country's foreign relations of carrying out interventions.

Identifying the small wars of the United States is not easy. There are some useful lists,[2] but they do not answer every question. Certain of the episodes included in this bibliography may not seem like wars at all. A contemporary observer concluded that the second intervention in Cuba between 1906 and 1909 "was not war."[3] Nevertheless, some bloodless or relatively bloodless occupations have been brought into the bibliography, because they had significant political ramifications and because those rather large-scale military operations provided useful experience for the United States armed forces.

The classic definition of the term "small wars" was formulated by C.E. Callwell at the end of the nineteenth century:[4]

> campaigns undertaken to suppress rebellions and guerrilla warfare in all parts of the world where organized armies are struggling against opponents who will not meet them in the open field.

Thus, according to Callwell, the tactics employed are what define small wars as such. This definition is thoroughly appropriate for many of the conflicts included in this bibliography, that is, those that involve the use of guerrilla and counter-guerrilla tactics, but it is not applicable to all of them. Some of the small wars discussed, such as the Gulf War, were simply conventional wars on a scale smaller than the world wars. A complicating factor in the discussion of small wars is that some of them began as conventional conflicts and then, as conventional tactics failed, became guerrilla campaigns. This was the case with the Philippine–American War, for example. Vietnam was a small war in many respects, or rather, it was often a conventional conflict in some parts of the country at the same time it was an insurgency in other areas.

After decades of experience, the United States Marine Corps in 1940 formulated a detailed statement about small wars:[5]

> small wars are operations undertaken under executive authority, wherein military force is combined with diplomatic pressure in the internal or external affairs of another state whose government is unstable, inadequate, or unsatisfactory for the preservation of life and of such interests as are determined by the foreign policy of our nation.

This definition seems applicable to all of the episodes contained in this bibliography, except for the skirmishes along the Mexican–United States border in the early twentieth century. Some kinds of military and naval operations and certain specific conflicts and incidents have been omitted from the bibliography. Short-term rescue missions have not been dealt with. What might be termed "shows of force" have been ignored, moreover. Such use of the armed forces for diplomatic reasons has been defined as follows:[6]

> A political use of the armed forces occurs when physical actions are taken by one or more components of the uniformed military services as part of a deliberate attempt by the national authorities to influence, or to be prepared to influence, specific behavior of individuals in another nation without engaging in a continuing contest of violence.

Some of the episodes that Blechman and Kaplan examine in *Force Without War*, which was just cited, appear to be small wars in the sense that there was fighting between American forces and other elements. A great many of the incidents they mention are far from what one would generally think of as a war, however.[7]

Identifying the small wars is only part of the first phase of studying them. The question occasionally arises: "What should be the criteria used in dating unconventional wars?"[8] The writer was discussing the Philippine–American War, which lasted for years after its ostensible conclusion in 1902. Indeed, Congress once tried to change the duration of this war long after its end. In 1944, H.R. 4099 proposed inclusion of the major Moro campaigns in the Philippine–American War by moving the termination date from July 1, 1902, to December 31, 1913. President Franklin D. Roosevelt vetoed the bill.[9] The same kind of problem arises with the border unrest during the early years of the Mexican Revolution. United States forces clashed with Mexican regulars and irregulars sporadically for nearly a decade. Still another dating problem relates to the United States presence in China from the end of the Boxer Rebellion to Pearl Harbor. Once again, fighting occurred from time to time, not steadily. I have attempted to solve such problems of duration by employing the furthest parameters for dating the wars.

One of the major points at issue in the discussion of small wars is their general versus their specific characteristics. Can anyone who has a degree of responsibility for the conduct of a current or prospective small war gain anything from a study of the past? Obviously, many military and civilian writers have thought the answer is "yes," and their number seems to be growing if one may judge from the increasing number of published studies that seek to draw lessons from earlier wars.

The armed forces of the United States have rarely, except when under considerable pressure, as they were during the Kennedy and Johnson administrations, evinced much interest in fighting small wars or preparing for such conflicts. In the early decades of the twentieth century, the army and marines went at their tasks gallantly and gradually developed a valuable body of doctrine resulting from their experiences in Cuba, the Philippines, and other Third World countries. This process has been duplicated in several periods since then. At present, the armed forces are committed to two conflicts that can be considered small wars, and, thus, there is more concern with small war doctrine now than there has been since the Vietnam years.

Nevertheless, the primary commitment of the military services has necessarily been to prepare for a major war with another power or group of powers. Fighting small conventional wars takes special skills and makes demands not found in high-intensity conflicts. Waging counterinsurgency demands still other skills and is, in part, a police operation. This truth was recognized to some extent even before World War II, which accounts for United States sponsorship of constabularies in several Caribbean nations and in the Philippines. In the postwar era, many thousands of foreign police officers were trained by the United States, either in this country or in their homelands to assist in counterinsurgency campaigns, thus supplementing or perhaps replacing soldiers.

The aversion of the military services to small wars is doubtless justified. As the *Small Wars Manual, 1940*, developed by the marines in the 1930s, recognized, fighting guerrilla forces involves putting the civilian population in the middle, a distasteful assignment at best, and sometimes an overly demanding responsibility. A vigorous prosecution of counterinsurgency can breed hatred that is quite different in kind from that which develops in a conventional war, because of the uses both sides make of civilians and because of situations that almost invite atrocities as a normal part of the campaign. It is to the credit of the United States that relatively few accusations have been made about its conduct in small wars. The converse of that facet of the small war challenge is that a commander may be called upon to violate military doctrine in order to achieve nonmilitary goals. One of the leading marine experts in this field warned that:[10]

> measures justifiable in a regular war, tactically sound, and probably the most efficient available, must frequently be eliminated from the plan of campaign as not being in accord with public policy in the existing situation.

Sometimes their having to follow directives from political leaders in the United States places members of the armed forces in more danger than is really necessary.

During the 1960s the armed forces did not show much interest in earlier United States experiences with counterinsurgency, despite their possible relevance to post–World War II challenges, especially the Vietnam conflict. One critic of United States performance in Vietnam has taken issue with what seems to have been the prevailing attitude and has said that "a review of the Philippine Insurrection [the Philippine–American War] could have prevented many of our mistakes in Vietnam."[11] Another observer, also writing during the very last stages of United States commitment in Vietnam, asserted that such knowledge would have been of no real assistance to Vietnam planners.[12] He referred generally to the marine campaigns and the *Small Wars Manual, 1940*, which was issued by the marines, rather than to the Philippine–American War, but he may well have dismissed the lessons of that conflict, as well.

The failure to study earlier small wars was especially evident in the absence of any effort by the air force and the army to learn from the marines, who had had significant experience in counterinsurgency operations. Article after article appeared on the subject of Army Special Forces and the air force and navy counterparts in the 1960s. Almost nothing was said about marine involvement in Latin America. There may have been a conscious and quite understandable effort to put distance between pre–World War II interventions and post–World War II military involvements, because the former had often been criticized on the ground that military force was employed simply to protect United States investments abroad. An effort to learn from the marines' experience before World War II could have been accomplished without endorsing the national policies that brought the marines to several Caribbean nations, however.

The relatively successful pre–World War II campaigns in Latin America did provide lessons that might have improved the post–World War II interventions that have often proved so frustrating. Except for Haiti, the Caribbean countries were pacified without much loss of life. For the most part, the scale of warfare was kept to the patrol and company level. While aviation was used innovatively and while there were significant civilian casualties in Nicaragua, there was not the heavy bombing characteristic of World War II, Korea, and Vietnam. By avoiding widespread destruction, the United States probably avoided alienating even larger segments of the civilian population.

Another possible key to the relatively bloodless quality of the small wars before 1945, other than the Philippine–American War, was that the marines and soldiers involved were professionals. Fire discipline and measured responses to enemy attacks are essential for successful counterinsurgency. They are more likely to be attained with professional forces than with a mass, conscript army. "Civilian soldiers" are not necessarily more brutal than career personnel. They are likely to be less well trained and disciplined than career personnel, especially when there has been a rapid build-up, and they may be quite fearful. Fear may lead to cruelty and to wanton killing by members of the United States forces who may blame local civilians for their being in the armed forces. Career soldiers may be able to endure the frequent provocations of guerrilla fighters better than conscripts or what are in effect conscripts, that is, men who volunteered to avoid being drafted. Whatever the character of the troops employed in a small war, it remains true that "those who send in the soldiers, not the men themselves, are responsible for what occurs afterward."[13]

It is true that analysts in the 1960s had many examples of World War II guerrilla operations and counterinsurgency campaigns to draw upon. Certainly, there was much to learn from them, but it does seem to have been an error to ignore pre–World War II United States counterinsurgency, if only because episodes such as the marine war against Sandino in Nicaragua took place in a *United States context*, with all the problems and limitations inherent in that fact.

During the 1970s some observers argued that "the army is finished with counterinsurgency, although it lingers on as a potential responsibility."[14] Other analysts pointed to various restraints that seemed increasingly to be limiting the ability of the United States to wage small wars.[15] With the beginning of the 1980s there was a reaction to what has been termed the "Vietnam syndrome," that is, the determination not to intervene militarily in other countries. Some military and civilian planners and publicists began to argue that the United States would have to be ready to use limited military force to protect its interests in such areas as the Persian Gulf. Such voices had been heard during the Arab oil embargo of 1973–74, and as the memory of Vietnam faded, they have become more influential, especially after President Reagan's election.

Except for the fighting along the Mexican–United States border, the small wars included in this bibliography involved United States intervention in the affairs of other countries. Military intervention has occurred for many reasons. For Panama and Cuba before World War II, there were treaties permitting it. Ex post facto treaties were negotiated to legitimize the occupations of Haiti, the Dominican Republic, and Nicaragua. In some cases, an incumbent government asked for United States intervention. It would be a mistake to assume that the United States intervened only to preserve the status quo, however. There have been instances in which liberal elements in a country, such as Cuba, favored United States intervention to ensure an honest election.

United States policy-makers often feared that unless they acted, another power would. This was frequently the situation in Latin America before and during World War I and in Lebanon in 1958, to mention only a few cases. The interventions in Russia and North China were closely associated with United States participation in World Wars I and II, respectively. In 1918–20, the United States worked first with the Allies to re-establish an eastern front and later to limit Japanese expansion into Siberia. After World War II, marines landed in China to assist in the repatriation of Japanese personnel and to hold key coastal areas until nationalist forces could assume control.

The complexity of small wars, whether conventional or counterinsurgency, may add to their unpopularity among military specialists. The basic principles are relatively easy to postulate, but they are extremely difficult to implement. Despite the lip service given to the principle that counterinsurgency is as much a political as a military contest, time after time the concept has been neglected by decision-makers at home and by the forces in the field. Trying to integrate political and military approaches to a small war is a fundamental challenge. Decision-makers must determine who will make the policies: the diplomatic service, the intelligence service, or the armed forces. Policy compromises intended to placate political, intelligence, and military leaders may well produce compromised results and lost victories, although a careful decision-making process in which all stakeholders are represented can be a factor leading to success.

A major cause of complexity in small wars is the role of the local government, which must receive some degree of respect from the United States diplomats, intelligence officers, and military commanders on the scene. Working with the local government is often one of the major stumbling blocks in fighting the small war, however. Most students would agree with Pustay's assertion that "counterinsurgency operations must be overwhelmingly dominated by indigenous government elements, if they are to succeed."[16] The United States has had to violate this salutary rule on many occasions, because the local government proved incapable of protecting either its own interests or those of the United States or both. Greater United States intervention, in turn, underlines the weakness of the government and probably makes it still weaker in the long run. Moreover, as Pustay noted, putting pressure on "a friendly regime obviously appears less palatable than taking direct action against common enemies."[17]

The United States has often been profoundly disenchanted with its allies in the Third World. Many marine officers in Haiti, for example, believed that the Haitian elite exploited the black masses. During the 1960s the commitment to civic action programs often reflected a desire to help the people of Third World nations. Walt Rostow expressed his views in a remarkable speech at the Special Warfare School in 1961. He implied that United States advisers in Third World nations would ally themselves with modernizing elements who could and would confront both the Communist insurgents and the authoritarian governmental leaders who denied that there was any reason for fundamental social and economic change.[18] There are, of course, significant difficulties with Rostow's approach. Reaching the people behind or around the incumbent government seems to be a chimera. Certainly, roads can be built and medical care can be provided, but it is impossible to bring about basic political and other changes without intervention that denies the national sovereignty of the country involved.

Probably the best remedy for the problems of the Third World would be a mix of economic and social assistance through international agencies and private business, but on such a large scale that it would bear little resemblance to such efforts in the past. A profound commitment by developed countries would be required for such a scheme to have any chance of success. Initiating an aid program, even a massive one, after an insurgency has started, makes the success of the assistance effort problematic. Assistance would surely be more helpful before fighting started than once combat has begun.

Military personnel have been put into a number of countries, apparently without much thought having been given by United States leaders to the possible consequences. Moreover, they have sometimes seemed uncertain in their own minds about how national policy objectives would be promoted through the use of military power. Armed force can be a helpful expedient, especially in the short run, but it must be employed with extreme care. Domestic political considerations sometimes play a role by encouraging quick decisions to intervene, and such concerns tend to cloud the perspectives of decision-makers who should weigh competing policy alternatives more carefully before employing the armed forces. The hope of making domestic or even international political gains must always be balanced against the possibility that the objectives sought will not be reached at all, or at such high costs in resources that continuing an intervention will rebound to the disadvantage of the leaders who decided to use force.

Not surprisingly, given the frequent lack of appropriate analysis and planning, United States policy goals were not achieved by sending in the marines or the army. Constitutional democracy of a lasting character was never imposed on any country as the result of United States intervention. Communism was not stopped in North Russia, Siberia, or North China. Despite the assumption that the United States personnel were the primary actors in given situations, they sometimes became the pawns of local political and military elites. By intervening militarily, the United States sometimes lost rather than gained control. Once United States troops were fighting, the United States lost its chance to mediate. A point that was already clear from many Latin American interventions and the United States involvement in China immediately following World War II and other operations, is that once the United States is committed to one side in a civil war, it cannot be an impartial mediator.

Major powers will no doubt continue to intervene militarily in other nations as long as the national state system endures. For as long as the United States remains a world power, it will be concerned with the methods for fighting small wars and it may include special contingents in its defense establishment for conducting such wars, despite its often unhappy experiences with small wars in the past. Recognizing that the United States is likely to become involved in small wars should carry with it the realization that past interventions and counterinsurgency wars must be studied and re-studied for their possible lessons. Recent trends in the literature suggest that both members of the armed forces and civilian defense analysts are looking at the past more often than they once did and are actively seeking lessons for the more effective conduct of small wars.

The basic lesson of the small wars of the United States is that there are strict limits to the effective use of force. A military intervention is most likely to succeed when there has been careful political and military planning. Interventions, however, usually occur on short notice and in a crisis atmosphere that works against extended, effective planning. The situation may be governed by domestic political considerations, moreover, rather than by strategic and diplomatic objectives that have been precisely defined. All of these factors suggest that the cards are often stacked against a successful intervention.

Notes

1 U.S. Congress. Senate. Committee on Foreign Relations. *War Powers Legislation, Hearings on S. 731, S.J. Res. 18 and S.J. 59.* 92nd Cong., 1st sess., 1972, 355.
2 One of the most useful lists appeared initially as part of J.T. Emerson, "War Powers Legislation." *West Virginia Law Review* 74 (1972): 88–119. It was subsequently reprinted in several of the hearings associated with the war powers controversy.
3 Kernan, Francis J. "The Purpose of the Army." *Infantry Journal* 7 (March, 1911): 676.
4 Callwell, Charles Edward. *Small Wars: Their Principles and Practice.* 3rd ed. London: H.M. Stationery Office, 1906, 21.
5 U.S. Marine Corps. *Small Wars Manual, United States Marine Corps, 1940.* Washington, DC: Government Printing Office, 1940, 1.
6 Blechman, Barry M. and Stephen S. Kaplan, eds., with David K. Hall, William B. Quandt, Jerome N. Slater, Robert M. Slusser, and Philip Windsor, *Force Without War: U.S. Armed Forces as a Political Instrument.* Washington, DC: Brookings Institution Press, 1978, 12.
7 Ibid., 547–53.
8 Clayton James, D. "Commentary: The MacArthur Viewpoint," in *The American Military and the Far East: Proceedings of the Ninth Military History Symposium United States Air Force Academy 1–3 October 1980,* ed. Joe C. Dixon. Colorado Springs, CO: Headquarters, United States Air Force Academy and Office of Air Force History, 1980?, 116.
9 The veto message, dated December 8, 1944, was published as H.R. Doc. no. 804, 90th Cong.
10 Utley, Harold H. "An Introduction to the Tactics and Technique of Small Wars," *Marine Corps Gazette* 16, no. 5 (May, 1931): 52.
11 King, Edward L. *The Death of the Army: A Pre-Mortem.* New York: Saturday Review Press, 1972, 39.

12 Schaffer, Ronald "The 1940 Small Wars Manual and the 'Lessons of History.'" *Military Affairs* 36 (April, 1972): 46–51.
13 Knight, Melvin M. *The Americans in Santo Domingo*. New York: Vanguard Press, 1928, 82.
14 Grinter, Lawrence E. "Nation Building, Counterinsurgency, and Military Intervention," in *The Limits of Military Intervention*, ed. Ellen P. Stern. Beverly Hills, CA: Sage Publications, 1977, 251.
15 Etzold, Thomas H. *Defense or Delusion? America's Military in the 1980s*. New York: Harper & Row, 1982, 21. See also, generally, the essays in Stern, *Limits of Military Intervention*, cited in note 14.
16 Pustay, John S. *Counterinsurgency Warfare*. New York: Free Press, 1965, 157.
17 Ibid., 160.
18 "Guerrilla Warfare in the Underdeveloped Areas." *Department of State Bulletin* 45 (August 7, 1961): 233–38. The address was widely reprinted, appearing in a number of periodicals issued by military service associations.

I
Bibliographic sources

Numerous general and specialized publications and websites are identified in this section. Users of this bibliography are likely to have access to some of the commercial items and all of the governmental sources. This segment is divided into:

    A. Catalogs of manuscripts
    B. Oral history
    C. Online periodical, newspaper, and report indexes
    D. Bibliographies
    E. Libraries and research centers.

Because of space considerations, a large number of additional publications and websites had to be omitted.

A. Catalogs of manuscripts

1. *ArchiveGrid*. Dublin, OH: OCLC (Online Computer Library Center).

    Invaluable source of information about archival holdings in thousands of libraries and museums. Using the search terms, such as "Mexico Punitive Expedition," brings references to numerous items.

2. Machette, Robert B. and Jan Shelton Danis with Anne B. Eales and others, comp. *Guide to Federal Records in the National Archives of the United States*. Washington, DC: National Archives and Records Administration, 1995.

    Two volumes provide general descriptions of the record groups in the National Archives, and the third volume is a detailed subject index. This is an essential initial source for researchers that includes much material related to military affairs and foreign relations.

3. Paszek, Lawrence J., comp. *United States Air Force History: A Guide to Documentary Sources.* Washington, DC: Office of Air Force History, 1973.

   Identifies many collections relevant to the small wars. Paszek ranges far beyond air force archives. Materials are listed from the Library of Congress, state historical societies, and other institutions.

4. Russell, J. Thomas, comp. *Preliminary Guide to the Manuscript Collection of the U.S. Military Academy Library.* West Point, NY: U.S. Military Academy, West Point, Library, 1968.

   Includes a number of collections relevant to the Punitive Expedition and other conflicts. Positive identification of all appropriate materials is difficult, owing to the brevity of the annotations.

5. Sommers, Richard J., comp. *Manuscript Holdings of the Military History Research Collection: A Military History Research Collection Bibliography.* 2 volumes. Carlisle Barracks, Pennsylvania, 1972–75.

   Includes a great many collections of interest to the student of the small wars. Materials seem to be richest for the Moro campaigns and the Punitive Expedition.

6. U.S. Marine Corps. Headquarters. History and Museums Division. *Marine Corps Personal Collection Catalog.* Washington, DC: U.S. Marine Corps, Headquarters, History and Museums Division, 1980.

   Provides a wealth of material. Each item is annotated with a precise indication of the contents. A detailed subject index contributes importantly to the worth of this catalog.

B. Oral history

7. Barnard, Roy S., comp. *Oral History.* U.S. Army Military History Research Collection. Carlisle Barracks, Pennsylvania, 1976.

   Contains a number of oral histories with references to the small wars, ranging from pre-World War II service in China through the development of Special Forces in the 1950s and 1960s.

C. Online periodical, newspaper, and report indexes

Government publications

1. International

8. *Access UN.* Naples, FL: NewsBank Inc. Coverage varies.

   Current guide to literature produced by the United Nations. It also provides indexing for older materials from 1966 onward. Important because United States military operations are sometimes multi-national commitments, often under UN auspices.

9. *Official Document System of the United Nations (ODS)*. New York: United Nations.

   Chronological coverage varies. Documents other than resolutions date primarily from 1993 onward, although older items, such as United Nations resolutions from 1946, are being added.

2. United States

10. *CIS Index to Publications of the United States Congress*. Washington, DC: Congressional Information Service. 1970–.

    Although this source is not online, it needs to be included in this section because of its scope and the usefulness of the summaries of hearings that it provides. Indexing is excellent, which further enhances its value. There is a *CIS Annual* that cumulates earlier issues for a given year, and the texts of the materials indexed have been published in various formats.

11. *Congressional Research Reports*. Washington, DC: United States. Library of Congress. Congressional Research Service. 1914–.

    The reports are intended for the Congress and its components, but they are highly useful to researchers because of their careful preparation and non-partisan nature. Congressional policy has restricted the dissemination of the reports, but many have been made public. Collections in various formats have been prepared and continue to be prepared by private publishers. In addition, a considerable number can be accessed online without charge. *Open CRS Network* (http://opencrs.com/) is one of a number of possible sources.

12. *Declassified Documents Reference System*. Farmington Hills, MI: Gale Cengage Learning.

    Essential source of the texts of newly declassified United States documents, including papers of the Joint Chiefs of Staff and Central Intelligence Agency studies. An earlier version dates to 1975, but some of the items being released currently were prepared decades ago.

13. *GPO Access*. Washington, DC: United States Government Printing Office. Web address: www.gpoaccess.gov

    Offers full-text congressional hearings and reports, among many other items. Coverage of the *Catalog of United States Government Publications* dates to 1976. More and more current items are available as full text, and *GPO Access* citations include online access.

This is a vital source for anyone concerned with the involvement of the United States in small wars.

14. *National Technical Information Service.* Washington, DC: United States Department of Commerce, National Technical Information Service. Web address: www.ntis.gov. 1964–.

    Despite its title, this database provides large numbers of citations to publications from all elements of the federal government. Coverage of the Iraq and Afghanistan conflicts, for example, is extensive.

15. *Thomas.* Washington, DC: United States. Library of Congress. Coverage varies. Web address: http://thomas.loc.gov

    The service was initiated in January 1995, but some materials date back to the late 1980s. Provides the text of the *Congressional Record* and numerous other items, including committee reports. This is a highly accessible source regarding congressional activity.

Military

16. *Air University Index to Military Periodicals.* Maxwell Air Force Base, AL: Air University Library. 1949–99 (paper and other formats); 1987– (online).

    Unrivaled for many years, this index has covered many periodicals not indexed elsewhere. Other indexes are now available, but this remains a classic source. A considerable number of pertinent items in the *Air University Index to Military Periodicals* have been included in this compilation with annotations.

17. *Staff College Automated Military Periodicals Index (SCAMPI).* Norfolk, VA: Joint Forces Staff College. Ike Skelton Library. 1995–.

    Essential source for current and historical citations. Despite the title, this index usefully includes RAND and General Accounting Office reports. Brief summaries of articles are often provided.

Nongovernmental publications

General

18. *Academic Search Premier.* Ipswich, MA: EBSCO Publishing. Coverage varies.

    Primarily useful for identifying periodical articles, this source also provides references to government publications, conference proceedings, and other materials.

19. *Access World News*. Naples, FL: NewsBank Inc. Coverage varies.

   Extensive coverage of United States and foreign newspapers. Foreign language materials are not translated.

20. *Ethnic NewsWatch*. Bethesda, MD: ProQuest Information and Learning Company. 1960–.

   Broad, full-text coverage of publications, including newspapers, of special interest to ethnic and minority groups. A significant number of the hundreds of thousands of articles are in Spanish. This is a highly important source.

21. *Factiva*. Princeton, NJ: *Factiva*, a Dow Jones Company. Coverage varies.

   Huge source of full-text newspaper and magazine stories. Despite the service's emphasis on business, *Factiva* is important for most fields, including international affairs and military developments. Many foreign publications are included. *Factiva* is highly current, being updated every day.

22. *Ingenta*. Providence, RI: Ingenta, Inc. Dates vary, but the majority of publications have at least a 10-year back file.

   Vast index to tens of thousands of journal articles and reports. Important source of citations. Significant number of full-text articles.

23. *JSTOR*. Ann Arbor, MI: JSTOR Ann Arbor. Chronological coverage varies.

   In addition to being a sophisticated periodical index, the service provides full texts of articles for various fields of study. The latest issues of journals are not supplied, in order to prevent competition with publishers. This is a limited, but important, resource because many of the leading scholarly journals are covered.

24. *Left Index*. Baltimore, MD: National Information Services Corporation/Biblioline. 1982–99 (paper format); 1990– (online).

   Important source for criticisms of United States diplomatic and military moves. Coverage of the index is broad, embracing dissertations, book chapters, and journal articles, among many other types of publications.

25. *PAIS Lexis-Nexis Academic*. Albany, NY: Lexis/Nexis. Matthew Bender. 1973–.

   Indexes many kinds of materials, such as newspaper stories, law review articles, and court decisions. Chronological coverage varies considerably, depending upon the source.

Bibliographic sources 15

26. *PAIS International and Archive.* Bethesda, MD: ProQuest LLC. 1915–.

    Provides a wide variety of government publications, journal articles, and others for a long period, largely paralleling the years covered by this book. The *Archive* contains items between 1915 and 1976, and the current version deals with newer materials. This is strictly an indexing system; no full-text publications are included.

27. *Project Muse.* Baltimore, MD: Johns Hopkins University Press. 1990–, although the initial dates for indexing vary.

    Includes full texts of articles from a highly selective, exceptionally worthwhile group of journals.

28. *Readers' Guide Full Text.* Bronx, NY: The H.W. Wilson Company. 1983– (abstracts); 1994– (some journals as full text).

    Classic source of information about articles in a wide variety of popular and current news publications. Paper issues began in 1901, with a later supplement for the years 1890 to 1900, the *Nineteenth Century Readers' Guide.*

29. *TV-NewsSearch: The Database of the Vanderbilt Television News Archive.* Nashville, TN: Vanderbilt University Television News Archive. Coverage varies, earliest items date from 1968.

    Indexes material in the archive, and videotapes can be purchased from Vanderbilt. This is an important source for coverage of various wars, including the Gulf, Afghanistan, and Iraq wars.

30. *World News Connection.* Washington, DC: National Technical Information Service, Foreign Broadcast Information Service. 1996–.

    Provides current, full-text materials from a wide variety of sources, such as print publications, broadcasts, and reports. Foreign language materials are translated into English.

History

31. *America: History & Life.* Santa Barbara, CA: ABC-CLIO. 1964–.

    Extensive coverage of historical journals. Dissertations are also included. Currently, about 16,000 items are added to the database each year. This is a major reference tool for anyone interested in the small wars of the United States.

32. *Historical Abstracts. Part A: Modern History Abstracts, 1450–1914; Part B: Twentieth Century Abstracts, 1914 to the Present.* Santa Barbara, CA: ABC-CLIO. 1955–.

    Because it does not include references to the United States, this source is much less useful for studying the small wars in which this country has engaged than *America: History & Life*, but some of the materials cited may be relevant for background information.

Humanities

33. *Humanities Index Full Text*. Bronx, NY: The H.W. Wilson Company. 1984– (abstracts); 1994– (some titles as full text).

   Includes a number of major history journals. The index began as the *International Index* in 1907, and it has undergone various title changes.

34. *Philosophers Index*. Bowling Green, OH: Philosopher's Information Center. 1940–.

   Broad coverage of journals and books from a number of fields in many languages. Numerous items refer to current and past military topics, such as the United States invasion of Iraq.

Political science

35. *CIAO: Columbia International Affairs Online*. New York: Columbia University Press. 1991–.

   A rich source for the study of United States military involvements. Vast coverage of journals, research reports, policy papers, and many other types of materials. Much of the material referenced is available as full text.

36. *CQ Researcher*. Washington, DC: CQ Press. 1991–.

   Online version of a service that began in paper format in 1945. Although much of the content deals with domestic problems, the service is, nevertheless, useful for military research.

37. *CSA Worldwide Political Science Abstracts*. Bethesda, MD: ProQuest LLC. Coverage varies, some material from 1967 onward.

   Several databases were merged in 2000. Provides access not only to current material but also includes files of *Political Science Abstracts* (1975-2000) and *ABC POL SCI* (1984-2000). Broad coverage of relevant material from the social sciences.

Social sciences

38. *ASSIA: Applied Social Sciences Index and Abstracts*. Bethesda, MD: ProQuest LLC. 1987–.

   Includes some references to websites, as well as scholarly journals. Hundreds of thousands of items are already included.

Bibliographic sources 17

39. *Index to Legal Periodicals*. Bronx, NY: The H.W. Wilson Company. 1982– (abstracts); 1994– (some journals as full text).

   Includes some social science, as well as legal journals. This is an important source, given the relevance of both domestic and international law to the initiation and pursuance of small wars by the United States. Paper issues published since 1908, with slight differences in the name of this index.

40. *Social Sciences Full Text*. Bronx, NY: The H.W. Wilson Company. 1983– (abstracts); 1995– (some journals as full text).

   Focuses on social science journals that are widely held by academic and other libraries. Paper issues under different titles since 1907, at first as the *International Index*.

Sociology

41. *Sociological Abstracts*. Bethesda, MD: ProQuest LLC. 1953–.

   Broad scope, including articles, dissertations, papers, and book chapters.

Medical

42. *PILOTS: Published International Literature on Traumatic Stress*. Washington, DC: National Center for Post-Traumatic Stress, United States Department of Veterans Affairs. 1871–. Web address: www.ncpts.va.gov/ncmain/publications/pilots

   Mandatory for serious researchers of post-traumatic stress disorder (PTSD). Includes a wide variety of materials, including book chapters. Primarily an abstracting service, *PILOTS* does offer some full-text items.

43. *PubmedCentral*. Bethesda, MD: National Center for Biotechnology Information (NCBI). Web address: www.ncbi.nlm.nih.gov/pubmed

   Huge database that is highly significant for locating information about the medical aspects of United States small wars and other military operations. This is an index rather than a source of full-text materials.

Regional/country

44. *Asia-Studies Full-Text Online*. International Information Services. Coverage varies.

   Most extensive coverage began after 2000, but there are many citations to materials that appeared as early as 1979. Citations to reports were generally incorporated only from the 1990s.

45. *CountryWatch*. Houston, TX: CountryWatch, Inc. Includes information from up to 10 years ago.

Useful background information about countries where military operations are being undertaken or about those countries where United States military interventions may occur.

Dissertations and theses

46. *Dissertations and Theses*. Bethesda, MD: ProQuest Information and Learning Company. 1861–.

Coverage varies, depending upon when particular institutions submitted materials. Many items from European universities are included. Variable descriptions of these materials are provided. The usefulness of the source has been increased by the inclusion of "previews" for many items, in addition to the traditional abstracts. Print versions appear as *Dissertations Abstracts International* and *Masters Abstracts International*.

D. Bibliographies

47. Beisner, Robert L., ed. *American Foreign Relations Since 1600: A Guide to the Literature*. Santa Barbara, CA: ABC-CLIO, 2003. 2 volumes.

Superb bibliography sponsored by the Society for Historians of American Foreign Relations. Expertly arranged and annotated citations refer to archival collections and reference sources, as well as to substantive works. The coverage is broad, including institutions, individuals, and many other topics. A large number of citations refer to the small wars.

48. Bell, J. Bowyer. *Dragonwars: Armed Struggle and the Conventions of Modern War*. "Sources," pp. 431–45. New Brunswick, NJ: Transaction Publishers, 1999.

Highly incisive discussion of the nature of the literature concerned with unconventional warfare. This section should be mandatory reading for anyone interested in how this subject is discussed.

49. Burt, Richard, and Geoffrey Kemp, eds. *Congressional Hearings on American Defense Policy: 1947–1971*. Lawrence: University Press of Kansas, 1974.

Identifies sections of hearings that dealt with post–World War II guerrilla conflicts and the United States capability for waging such wars. It is a useful supplement to the *Congressional Information Service Index* that began publication in 1970. Unfortunately, there is no subject index.

50. Coletta, Paolo Enrico. *An Annotated Bibliography of U.S. Marine Corps History.* Lanham, MD: University Press of America, 1986.

Valuable bibliography with numerous citations, but few annotations, to materials related to small wars. There are a number of specific sections, such as the Boxer Uprising and Lebanon, but the Caribbean interventions are lumped into "Expeditionary and Police Duties, 1901–34," pp. 128–153.

51. Condit, D.M., Barbara Reason, Margaret Mughisuddin, Bum-Joon Lee Park, and Robert K. Geis. *A Counterinsurgency Bibliography.* Washington, DC: American University, Special Operations Research Office, 1963. Supplements issued through 1967.

Provides many references to small wars both before and after World War II. Its well-written annotations make it an impressive compilation. The vast majority of appropriate items are found in this bibliography.

52. Fletcher, Marvin. *The Peacetime Army, 1900–1941: A Research Guide.* New York: Greenwood Press, 1988.

Includes a brief segment titled "Conflicts," pages 26–39, with a number of useful citations to wars and incidents from the Philippine–American War through the Punitive Expedition into Mexico and tension along the Mexican–United States boundary.

53. Greenwood, John, comp. *American Defense Policy Since 1945: A Preliminary Bibliography.* Ed. Geoffrey Kemp, Clark Murdock, and Frank L. Simonie. Lawrence: University Press of Kansas, 1973.

Deals with general topics rather than particular wars. Sections I.B.2 and III.C examine bibliographies and substantive works on guerrilla warfare. The section on bibliographies is especially important. Not all are included in this work.

54. Higham, Robin. *A Guide to the Sources of United States Military History.* Hamden, CT: Archon Books, 1975.

Provides extensive bibliographies and impressive historiographical essays. The coverage is broad, including archival sources. Supplements, also issued by Archon Books, and co-edited by Higham and Donald J. Mrozek, appeared in 1986, 1993, and 1998.

55. Johnstone, John H. *An Annotated Bibliography of the United States Marines in Guerrilla-Type Actions.* Rev. ed. Washington, DC: Marine Corps, Headquarters, Historical Branch, G-3, 1962.

Includes items that are, for the most part, incorporated in this work. The title is quite misleading because a number of articles and books listed deal with foreign guerrillas and counterinsurgency operations.

56. Lane, Jack C. *America's Military Past: A Guide to Information Sources.* Detroit, MI: Gale Research Company, 1980.

Lists only a relatively few items pertaining to the small wars and to the general topic of counterinsurgency. Nevertheless, these are important, and this bibliography includes almost all of them.

57. Larson, Arthur D. *National Security Affairs: A Guide to Information Sources.* Detroit, MI: Gale Research Company, 1973.

Includes sections on intervention, guerrilla warfare, and limited warfare. The book focuses on what were then current materials. Relatively few historical treatments are listed. The Larson compilation is, nevertheless, an important bibliography.

58. Mickolus, Edward F. *The Literature of Terrorism: A Selectively Annotated Bibliography.* Westport, CT: Greenwood Press, 1980.

Provides a specific section on guerrilla warfare, but lists relevant items in many places in the bibliography. This is a very extensive listing of books, government publications, conference papers, special reports, and other items, many of them rather elusive.

59. Miller, Hope and William A. Lybrand, comp. *A Selected Bibliography on Unconventional Warfare.* Part I. Washington, DC: American University Special Warfare Research Division, Special Operations Research Office, 1961.

Emphasizes World War II, but contains many items relevant to small wars. Part II on counterinsurgency was apparently not issued.

60. Moran, John B. *Creating a Legend: The Complete Record of Writing about the United States Marine Corps.* Chicago: Moran/Andrews, 1973.

Includes a variety of materials, even novels, poetry, and films, but omits dissertations. Moran's bibliography should be consulted for many brief items on the small wars that have not been listed in this bibliography. An unorthodox classification scheme and the lack of a subject index make the bibliography somewhat difficult to use.

61. Pérez, Louis A., Jr. "Historiographical Essay, Intervention, Hegemony, and Dependency: The United States in the Circum-Caribbean, 1898–1980." *Pacific Historical Review* 51 (May 1982): 165–94.

Organizes a large number of dissertations, government documents, books, and periodical articles in a most impressive analysis of writing about United States intervention.

62. Robinson, Mary Ann, comp. *The American Military and the Far East.* Colorado Springs, CO: U.S. Air Force Academy Library, 1980.

Provides a modest number of books, articles, and government publications on small wars and counterinsurgency. Insofar as they could be identified, all pertinent items have been listed in this bibliography.

63. Shrader, Charles R. *U.S. Military Logistics, 1607–1991: A Research Guide.* New York: Greenwood Press, 1992.

    Identifies a number of articles and other materials related to the Boxer Uprising, the Philippine–American War and the Punitive Expedition. The subject index must be used to find those items, but this is a worthwhile source of information about a subject often overlooked by civilian researchers and readers.

64. Smith, Myron J., Jr. *American Naval Bibliography, Vol. V: The American Navy, 1918–1941.* Metuchen, NJ: Scarecrow Press, Inc., 1974.

65. ———. *The United States Navy and Coast Guard, 1946–1983: A Bibliography of English-Language Works and 16 mm Films.* Jefferson, NC: McFarland, 1984.

    These bibliographies contain many items related to various small wars, but those citations must be sought by using country names, such as "Haiti," in the extensive subject indexes.

66. Sutton, Ottie K., comp. *The United States in the Caribbean.* Colorado Springs, CO: U.S. Air Force Academy Library, 1970.

    Includes sections on Cuba and the Dominican Republic. The emphasis of the bibliography is upon popular magazines such as *Current History* and *Newsweek*, but some books and government publications appear.

67. Trask, David F., Michael C. Meyer, and Roger R. Trask. *A Bibliography of United States–Latin American Relations Since 1810: A Selected List of Eleven Thousand Published References.* Lincoln: University of Nebraska Press, 1968.

    Arranges material in chronological order and then by broad topics with generous cross-references. Some sections refer specifically to interventions, such as the Punitive Expedition.

68. U.S. Air Force Academy. Library. *Unconventional Warfare. Part I. Guerrilla Warfare.* Colorado Springs, CO: U.S. Air Force Academy Library, 1962.

    Focuses primarily on United States experiences. Most relevant items have been included in this bibliography. The list was a joint project of the Library and the History Department of the Air Force Academy.

69. U.S. Department of the Army. Adjutant General's Office. Army Library. *Military Power and National Objectives: A Selected List of Titles.* Washington, DC: Department of the Army, 1957.

Identifies some relevant articles and other materials, which have been examined, and, for the most part, have been incorporated into this bibliography.

70. U.S. Department of the Army. Adjutant General's Office. Army Library. *United States National Security.* Washington, DC: Department of the Army, 1956.

Contains approximately 1,000 books, articles, and other items dating from 1954 to 1956. Most entries are annotated. A number of articles and other materials were identified through this work and have been included in the present bibliography.

E. Libraries and research centers

Military and civilian federal libraries

Below is a section listing some of the most useful physical and web addresses for libraries with significant holdings in military affairs and United States foreign policy. Their websites provide highly useful resources. The Army War College Library, for example, offers an extensive bibliography titled "Insurgency/Counterinsurgency."

United States Air Force Academy Libraries. McDermott Library, 2354 Fairchild Drive, Ste 3A10, Colorado Springs, CO 80840–6214. Web address: www.usafa.af.mil/df/dflib/?catname+Dean%20ofFaculty

United States Air Force Muir S. Fairchild Research Information Center, 600 Chennault Circle, Maxwell Air Force Base, AL 36112–424. Web address: www.au.af.mil/au/aul/aul.htm

United States Army. Command and General Staff College. Combined Arms Research Library, Eisenhower Hall, 250 Gibbon Avenue, Fort Leavenworth, KS 66027–2314. Web address: http://cgsc.leavenworth.army.mil/CARL

United States Army. Training and Doctrine Command (TRADOC). Library Program Office, United States Army HQ TRADOC, Seven Bernard Road, Fort Monroe, VA 23651. Web address: www.tradoc.army.mil/library

United States Army. War College Library, 122 Forbes Avenue, Carlisle, PA 17013–5220. Web address: www.carlisle.army.mil/library/

United States Department of State, Ralph J. Bunche Library, 2201 C Street, NW, Room 3239, Washington, DC 20520.

United States Institute of Peace, Jeannette Rankin Library Program, 1200 17th Street, NW, Ste Washington, DC 20036–3011. Web address: www.usip.org/library.html

United States Marine Corps University Library, 2040 Broadway Street, Quantico, VA 22134. Web address: www.tecom.usmc.mil/mcu/MCRCweb/mculibrary/milres.htm

United States Military Academy Library, Building 757, West Point, NY 10996–1799. Web address: www.usmalibrary.usma.edu

United States. National Defense University Library. Fort Leslie J. McNair, 300 5th Avenue, Marshall Hall, Washington, DC 20319–5066. Web address: www.ndu.edu/library/index.cfm

United States. National Defense University. United States Joint Forces Staff College. Ike Skelton Library. Norfolk, VA 23511–1702. Web address: www.jfsc.ndu.edu/library/default.asp

United States Naval Academy. Nimitz Library, 589 McNair Road, Annapolis, MD 21402–6959. Web address: www.usna.edu/library

United States. Naval Postgraduate School, Dudley Knox Library, 411 Dyer Road, Monterey, CA 93943. Web address: www.nps.edu/library

United States. Naval War College Library, 686 Cushing Road, Newport, RI 02841–1207. Web address: www.nwc.navy.mil/library/

Military research centers

United States Air Force History Office. Air Force Department. Bolling Air Force Base, Building 5681, Washington, DC 20332–6089. Web address: www.airforcehistory.hq.af.mil/

United States Army and Marine Corps Counterinsurgency Center, 630 McClellen Avenue, Fort Leavenworth, KS 66027. Web address: http://coin.army.mil

United States Army Center of Military History. Fort Leslie J. McNair, Washington, DC 20319–5058. Web address: www.history.army.mil/about.html

United States. Department of the Navy. Naval Historical Center, 805 Kidder Breese SE, Washington Navy Yard, Washington, DC 20374–5060. Web address: www.history.navy.mil/nhc5.htm

United States Marine Corps. History Division, 3079 Moreell Avenue, Quantico, VA 22134. Web address: www.history.usmc.mil

United States. Marine Corps. Warfighting Laboratory. Wargaming Division. Small Wars Center of Excellence, 3255 Meyers Avenue, Quantico, VA 22134–5069. Web address: www.smallwars.quantico.usmc.mil/contact.asp

II

Histories and general studies

71. Axelrod, Alan. *America's Wars.* New York: John Wiley & Sons, Inc., 2002.

   Consists of brief, but thoroughly competent, surveys of numerous conflicts in which United States forces have been engaged. There is an extensive index, and a relatively short bibliography. Because of their brevity and the lack of notes or references, entries in this book have not been listed and annotated in *The Small Wars of the United States, 1899–2009*. This is a valuable reference work, nevertheless.

72. ———. *Political History of America's Wars.* Washington, DC: CQ Press, 2007.

   This is an innovative and much needed reference work. Given the purpose of the work, Axelrod explains the reasons for the war or other conflict and then discusses its various political facets and results while not ignoring the military elements of each story. There are brief biographies and excerpts from speeches and other documents, as well as concise but appropriate chronologies, that make this a highly informative source. The bibliographies are something of a weak point, however. Sometimes, the inclusion or exclusion of particular works is difficult to understand.

73. Barrett, Drew J., Jr. "Show of Force in Foreign Policy." *Military Review* 47, no. 8 (August 1967): 3–8.

   Examines a number of instances, including Lebanon in 1958 and Siberia in 1918–20. Barrett stresses the need for coordinating political and military planning and operations. Despite the passage of time, this is a fine analysis of the use of military power in support of diplomacy.

74. Beede, Benjamin R. *The War of 1898 and U.S. Interventions 1898–1934: An Encyclopedia*. New York: Garland Publishing, Inc. 1994.

In addition to dealing with the War of 1898, the volume provides detailed articles by recognized scholars about many of the episodes included in the present bibliography. There are a number of maps and several lists of military organizations and navy and coast guard ships that participated in various interventions and counterinsurgency campaigns.

75. Bell, J. Bowyer. *Dragonwars: Armed Struggle and the Conventions of Modern War*. New Brunswick, NJ: Transaction Publishers, 1999.

Sophisticated analysis of the roots of unconventional warfare waged against major powers, especially the United States, and of the United States' responses to such efforts. Only the sections dealing with Lebanon in the 1980s and "Sources," that is, the literature of unconventional warfare, are annotated separately, but this is a significant study.

76. Betts, Richard K. *Soldiers, Statesmen, and Cold War Crises*. Cambridge, MA: Harvard University Press, 1977.

Focuses on the role of the military elite in crises such as Lebanon and the Dominican Republic. Because Betts uses these episodes as examples and treats them only very briefly, these sections are not listed and annotated separately.

77. Blechman, Barry and Stephen S. Kaplan, eds., with David K. Hall, William B. Quandt, Jerome N. Slater, Robert M. Slusser, and Philip Windsor. *Force Without War: U.S. Armed Forces as a Political Instrument*. Washington, DC: Brookings Institution Press, 1978.

Analyzes troop movements to secure diplomatic objectives. Of the episodes included, only Lebanon (1958) and the Dominican Republic (1965–66) are listed and annotated in this bibliography. This is an extremely important book on the post–World War II small wars.

78. Boot, Max. *The Savage Wars of Peace: Small Wars and the Rise of American Power*. New York: Basic Books, 2003.

Popular account based on a thorough use of secondary sources and some primary materials. Boot discusses "another American way of war," thus contrasting his book with Russell Weigley's *The American Way of War*. This is a useful introduction, but it needs to be used with care owing to some highly debatable statements. Boot takes a more sanguine view of the United States' small wars than most authors.

79. Burgess, Susan Renee. "Challenging Judicial Supremacy: Departmental Constitutional Interpretation in the Abortion and War Powers Debates." Ph.D. diss., University of Notre Dame, 1989.

Burgess examines the long-standing view that each branch of the federal government can interpret the Constitution and, therefore, that the executive and legislative branches are not inferior to the judiciary, especially the Supreme Court, in constitutional construction. The dissertation focuses on the "departmental" or branch theory of constitutional interpretation as illustrated by the Hyde Amendment regarding abortion and the assertion of "war powers" authority by Congress.

80. Cable, James. *Gunboat Diplomacy: Political Applications of Limited Naval Force.* New York: Praeger Publishers, 1971.

Examines the concept of "gunboat diplomacy" and discusses a large number of examples, including many involving United States forces. Despite its British perspective, this is an important book for United States readers. A huge list of small wars is included.

81. Calhoun, Frederick S. *Power and Principle: Armed Intervention in Wilsonian Foreign Policy.* Ohio: Kent State University Press, 1986.

Portrays Wilson as an expert in applying limited military force to complex political situations abroad, although Calhoun admits that Wilson experienced a number of failures and learned from these episodes. Although Wilson listened to United States military leaders, he kept firm control of the armed forces, thereby avoiding war with Mexico among other possible disasters. A number of chapters from this book are annotated elsewhere.

82. Castle, William R., Jr. "Why Marines Are Landed." *Marine Corps Gazette* 19, no. 8 (August 1934): 16–18.

Takes sharp issue with the Good Neighbor Policy. Castle defends the use of marines to safeguard United States lives and property and warns that failure to continue such policies will further depress world economic conditions.

83. Colby, Elbridge. "American Interests in the West Indies." *United States Naval Institute Proceedings* 57, no. 4 (April 1931): 468–72.

Defends the United States role in policing the Caribbean. Colby begins with the increasing importance of United States influence in the area after the War of 1898 and gives very brief descriptions of the occupations of Haiti and the Dominican Republic. The article is an interesting example of a very late statement of the view that the United States has an absolute right to intervene.

Histories and general studies 27

84. Collins, John M., with the assistance of Frederick Hamerman and James P. Seevers. *America's Small Wars: Lessons for the Future.* Washington, DC: Brassey's (US) Inc., 1991.

Excellent reference work that retains much of its usefulness, nearly twenty years after its publication. Collins and associates examine sixty "cases" of low-intensity conflict in which the United States was involved from 1899 through the early 1990s. The brief discussions provide basic facts and cite important sources for further reading. In another segment of the book, the authors propose improvements in United States low-intensity capabilities. Seevers' description of "Congressional Actions" in the sixty "cases" is an important contribution. There is also a fine glossary.

85. Downie, Richard Duncan. "Military Doctrine and the 'Learning Institution': Case Studies in Low-Intensity Conflict." Ph.D. diss., University of Southern California, 1995.

This is a significant attempt to determine how the army assesses and recasts its doctrine relating to low-intensity conflict. The impact of field experiences with such conflict and initiatives by civilian decision-makers in defense policy are among the factors that can induce doctrinal modifications. Downie finds that the process of doctrinal transformation is complex. It cannot be explained by the influence of a single factor. He goes on to propose changes in doctrine and to suggest ways in which the army can improve its learning skills.

86. Dupuy, R. Ernest, and William H. Baumer. *The Little Wars of the United States.* New York: Hawthorn Books, Inc., 1968.

Examines a number of conflicts, beginning with the quasi-war with France. Many small wars of the twentieth century are included, such as the interventions in Russia and in the Caribbean countries. Nothing is said about the post–World War II era. Many individual sections are separately listed and annotated.

87. Eagleton, Thomas F. *War and Presidential Power: A Chronicle of Congressional Surrender.* New York: Liveright, 1974.

Attacks the War Powers Resolution for giving the president too much power. Chapter 4 discusses many of the small wars between the Boxer Rebellion and the interventions in Russia. The long bibliography includes many legal items.

88. Ellsworth, Harry Allanson. *One Hundred Eighty Landings of United States Marines 1800–1934.* Washington, DC: U.S. Marine Corps, Historical Section, 1934.

Constitutes a classic compilation with basic information on many of the episodes included in this bibliography. Ellsworth utilized archival materials, including ships' logs, extensively. The book has been reprinted several times since World War II.

89. Joes, Anthony James. *Resisting Rebellion: The History and Politics of Counterinsurgency.* Lexington: University Press of Kentucky, 2004.

    Acute, wide-ranging comparative analysis that includes much material concerning the United States experience with counterinsurgency, but the discussions of United States topics are too scattered to annotate any of the chapters except one.

90. ——. *Resisting Rebellion: The History and Politics of Counterinsurgency.* Chap. 12, "Deploying U.S. Troops in a Counterinsurgency Role," pp. 229–31. Lexington: University Press of Kentucky, 2004.

    Offers cogent advice to United States leaders who must consider how to deal with an insurgency threat and, in this connection, discusses and praises the United States advisory effort in Vietnam.

91. Kane, William Everett. *Civil Strife in Latin America: A Legal History of U.S. Involvement.* Baltimore, MD: Johns Hopkins University Press, 1972.

    Explores the complexity of the concepts of "intervention" and "nonintervention," reviews United States occupations between 1900 and 1917, and examines the post–World War II interventions in Guatemala, Cuba, and the Dominican Republic.

92. Kolb, Richard K. "'Restoring Order' South of the Border." *United States Naval Institute Proceedings* 110, no. 7 (July 1984): 56–61.

    Presents an interesting overview of United States involvement in Latin America from the 1820s to Grenada and the Honduras maneuvers of 1984. Combat action is summarized very briefly, and a table shows United States casualties in these campaigns.

93. Langley, Lester D. *The Banana Wars: United States Intervention in the Caribbean, 1898–1934.* Wilmington, DE: SR Books, 2002.

    The author is a leading specialist on United States–Caribbean relations. He focuses here on the military and naval officers who directed the various interventions. Political topics are dealt with as much as military subjects, which is essential for a study like Langley's. This is an expanded and updated version of Langley's *The Banana Wars: An Inner History of American Empire, 1900–1934*, which was published by University Press of Kentucky in Lexington, Kentucky in 1983 and revised in 1985. Several chapters of the 2002 volume are listed and annotated in appropriate sections of this bibliography.

94. Link, Arthur S. "The Caribbean: Involvement and Intervention," Chap. 15, pp. 495–550, in *Wilson: The Struggle for Neutrality 1914–1915*. Princeton, NJ: Princeton University Press, 1960.

   Interprets Wilson's interventions as being motivated by a desire to bring democracy to the Caribbean. Link notes the relative lack of knowledge about the Caribbean on Wilson's part and even within the State Department.

95. Mellman, Harry George. "The American Policy of Intervention in the Caribbean." Ph.D. diss., University of Illinois–Urbana, 1940.

   Surveys the diplomacy of United States interventions in Cuba, Nicaragua, the Dominican Republic, Haiti, Mexico, Panama, and Costa Rica. Mellman is very critical of the State Department and commends Hoover and Stimson for moving away from intervention. This is an ambitious, but very able, study.

96. Metcalf, Clyde H. "The Marine Corps and the Changing Caribbean Policy." *Marine Corps Gazette* 21, no. 11 (November 1937): 27–32, 68–72.

   Gives eight reasons for interventions and suggests six possible causes for the decline of interventionism, several of them relating to the advent of the Great Depression. Metcalf also discusses United States military intervention in the Caribbean generally.

97. Millett, Richard. "The State Department's Navy: A History of the Special Service Squadron, 1920–40." *American Neptune* 35 (April 1975): 118–38.

   Centers on naval operations in Nicaragua and ship movements around Cuba in 1933–34. Relationships between the State and Navy departments were not always cordial. Millett concludes with a review of goodwill cruises of the late 1930s.

98. Munro, Dana G. *Intervention and Dollar Diplomacy in the Caribbean 1900–1921*. Princeton, NJ: Princeton University Press, 1964.

   Defends vigorously United States policies of the period. Munro uses State Department archives and presidential papers to show how United States decision-makers viewed particular situations. This is an extremely important study because of its emphasis on State Department perceptions.

99. ———. *The United States and the Caribbean Republics 1921–1933*. Princeton, NJ: Princeton University Press, 1974.

   Provides another revealing portrait of personalities and perceptions in the State Department. Munro asserts that there was a steady movement away from intervention during the 1920s and argues that the United States did not support dictatorships in the Caribbean.

100. Musicant, Ivan. *The Banana Wars: A History of United States Military Intervention in Latin America from the Spanish–American War to the Invasion of Panama.* New York: Macmillan Company, 1990.

    One of the major general studies of the subject. Musicant made extensive use of primary materials including unpublished papers and oral histories, making this an extremely valuable study. The bibliography is excellent. Most chapters of this book are listed and annotated separately in the appropriate sections of this bibliography. The only deficiency of the book is the lack of detailed maps.

101. Nearing, Scott, and Joseph Freeman. *Dollar Diplomacy: A Study in American Imperialism.* New York: B.W. Huebsch and the Viking Press, 1925.

    Represents the classic indictment of United States policies in the Caribbean during the early twentieth century. The authors focus on the economic aspects of the United States penetration of the Caribbean and Central American countries.

102. Note. "Congress, the President, and the Power to Commit Forces to Combat." *Harvard Law Review* 81 (June 1968): 1771–1805.

    Examines the problem of how to synchronize the role of the president as commander-in-chief with the limitations inherent in the fact that only Congress can declare war. Section III deals with many of the small wars.

103. Offutt, Milton. *The Protection of Citizens Abroad by the Armed Forces of the United States.* Johns Hopkins University Studies in Historical and Political Sciences, Series XLVI, no. 4. Baltimore, MD: Johns Hopkins University Press, 1928.

    Covers military operations from 1813 to China in the 1920s. Chapters 1 and 5 discuss general principles. The bulk of the book contains very brief descriptions of small wars and other episodes.

104. Perkins, Dexter. *A History of the Monroe Doctrine.* New ed. Boston, MA: Little, Brown and Company, 1963.

    Constitutes the best-known treatise on the subject. Perkins discusses the use of the Monroe Doctrine to justify intervention in Latin America. He generally defends United States policies and asserts that the people of Nicaragua, for example, were better off under United States rule than under domestic dictators.

105. Perkins, Whitney T. *Constraint of Empire: The United States and the Caribbean Interventions.* Westport, CT: Greenwood Press, 1981.

    Considers interventions from the early 1900s through the 1970s. Perkins sees the interventions occurring primarily to establish democratic

Histories and general studies

governments and generally defends the actions of the United States in the area.

106. Reveley, W. Taylor III. *War Powers of the President and Congress: Who Holds the Arrows and Olive Branch?* Charlottesville: University Press of Virginia, 1981.

   Stresses the intent of the framers of the Constitution, but also contains frequent references to small wars, such as the Boxer Uprising, the Caribbean occupations, and the *Mayaguez* incident. A chapter reviews the significance of the War Powers Resolution.

107. Schlesinger, Arthur M., Jr. *The Imperial Presidency.* Boston, MA: Houghton Mifflin Company, 1973.

   Comments at length on the development of presidential war powers, although other aspects of executive domination are discussed. This is an essential volume for understanding the debate over such powers. Various small wars and military crises are touched upon, but they are not dealt with in any detail.

108. Simmons, Edwin H. "The Marines and Crisis Control." *United States Naval Institute Proceedings* 91, no. 11 (November 1965): 26–35.

   Distinguishes between pre–World War II interventions and the newer counterinsurgency operations, but grants that some of the lessons from the early years of the twentieth century are still valid. The article briefly reviews marine landings in the Dominican Republic and other areas.

109. Sturdevant, E.W. "Central America and the Marine Corps." *Marine Corps Gazette* 11, no. 3 (March 1926): 24–31.

   Reviews the political systems and economies of the Central American countries other than Panama. Sturdevant also discusses the economic domination of several nations by United States interests and warns of increasing Communist influence from Mexico.

110. Thomason, John W. "With the Special Service Squadron." *Marine Corps Gazette* 12, no. 6 (June 1927): 76–81.

   Describes vividly navy life along the coasts of Central and South America. The author was the best-known marine writer of the 1920s and 1930s.

111. U.S. Congress. Senate. Committee on Foreign Relations. *War Powers Legislation, Hearings on S. 731, S.J. Res. 18 and S.J. Res. 59.* 92nd Cong., 1st sess., 1972.

   Discusses the significance of presidential actions in various small wars as precedents for the Vietnam involvement. A number of lists of United States wars and related materials are included.

112. Waghelstein, John David. "Preparing for the Wrong War: The United States Army and Low-Intensity Conflict, 1755–1890." Ph.D. diss., Temple University, 1990.

Wide-ranging examination of low-intensity concerns of the army, even including British influence from before the Revolutionary War, that is intended to compare numerous experiences with this form of warfare to the United States effort in Vietnam. Waghelstein considers the attitude of major military leaders toward low-intensity conflict and the development of military doctrine on the subject. Given its breadth and organization, this study seems to be an important contribution to the literature.

113. "Wards of the United States: Notes on What Our Country Is Doing for Santo Domingo, Nicaragua, and Haiti." *National Geographic Magazine* 30 (August 1916): 143–77.

Reviews the problems that induced United States intervention. American reforms are briefly summarized. The article consists primarily of photographs, many of them intended to demonstrate the backwardness of the three nations.

114. Weed, Sylvia Lee. "Low-Intensity Conflict: An American Dilemma." Ph.D. diss., University of Alabama, 1988.

After reviewing United States involvements in those conflicts that may be termed "low-intensity," Weed discusses interpretations of those episodes. She examines the challenge of precisely defining the concept of "low-intensity conflict" at some length, questions relating to decisions about when the United States should enter such contests, and then current views of low-intensity conflict both in the armed services and among academic figures who concern themselves with the matter.

115. Wormuth, Francis D. "The Vietnam War: The President versus the Constitution." *Occasional Paper, Center for the Study of Democratic Institutions* 1 (1968): 2–63.

Includes an extensive examination of the small wars as precedents for the Vietnam commitment. Wormuth asserts that most of them were minor affairs in which military action was taken to prevent immediate harm to United States citizens.

116. Young, Oran R. *The Politics of Force: Bargaining During International Crises.* Princeton, NJ: Princeton University Press, 1968.

Analyzes a number of post–World War II incidents, most of them involving direct confrontations between the United States and the Soviet Union. Young rejects the assumption that there will be "no more Vietnams." Instead, he investigates the reasons for interventions.

III

Pre-World War II doctrines and general studies, 1899–1940

During the first three decades of the twentieth century, American armed forces landed again and again in Latin America and the Far East. While some important works such as the Marine Corps' *Small Wars Manual* emerged from these experiences, there was nothing comparable to the flood of literature on counterinsurgency that flowed during the 1960s.

A. Doctrines

117. Bickel, Keith B. *Mars Learning: The Marine Corps Development of Small Wars Doctrine, 1915–1940*. Boulder, CO: Westview Press, 2001.

    The *Small Wars Manual, 1940* that was published by the marines remains one of the most important works resulting from the United States experience with counterinsurgency. It has been reprinted several times, and it remains in print. Bickel skillfully relates the origin and refinement of the book. In addition to analyzing the marines' study of counterinsurgency, Bickel discusses late nineteenth- and early twentieth-century army counterinsurgency efforts, including the Philippine–American War and the campaigns in the Moro areas of the Philippines.

118. Birtle, Andrew J. *U.S. Army Counterinsurgency and Contingency Operations Doctrine, 1860–1941*. Washington, DC: Center of Military History, U.S. Army, 1998, 2004.

    Initially published in microfiche, this study is now available as a printed book. This volume and a companion study that is cited elsewhere constitute a major breakthrough in the study of army counterinsurgency doctrine. During much of the period covered, the army had ample experience with irregular warfare, ranging from the old West through extended encounters in the Philippines, and it documented its role in such conflicts, although it did not issue any publication equivalent to the *Small Wars Manual, 1940*, compiled and issued by the marine corps.

119. Bullard, Robert L. "Military Pacification." *Journal of the Military Institution of the United States* 46 (January–February 1910): 1–24.

    Foreshadows in many respects the Kennedy counterinsurgency doctrine of the 1960s. This important study was based on Bullard's experiences in Cuba and the Philippines.

120. Flint, Roy K. "The United States Army on the Pacific Frontier, 1899–1939," in *The American Military and the Far East: Proceedings of the Ninth Military History Symposium United States Air Force Academy 1–3 October 1980*, ed. Joe E. Dixon. Colorado Springs, CO: U.S. Air Force Academy Library and Office of Air Force History, 1980.

    Notes the very limited development of counterinsurgency doctrine by the army.

121. Gill, Burgo D. "Guerrilla Warfare." *Quartermaster Review* 13 (September–October 1933): 27–29.

    Presents an analysis of guerrilla warfare and counterinsurgency doctrine that seems to have been far ahead of much of the literature prepared during the pre–World War II period. Gill draws on American experience with Villa and in Nicaragua to some extent and refers to T.E. Lawrence's exploits in Arabia.

122. Schaffer, Ronald. "The 1940 Small Wars Manual and the 'Lessons of History.'" *Military Affairs* 36 (April 1972): 46–51.

    Examines the sources on which the *Manual* was based, and asserts it was the only systematic effort to distill counterinsurgency doctrines before World War II. Schaffer contends that greater knowledge of the *Manual* would not have influenced the conduct of the Vietnam War materially.

123. U.S. Marine Corps. *Small Wars Manual, United States Marine Corps, 1940*. Washington, DC: Government Printing Office, 1940.

    Contains very detailed chapters on patrols, the role of aviation, river operations, military government, and many other topics. The sections on the strategy and psychology of small wars are especially interesting. When initially issued, the *Manual* was considered "Restricted."

B. Organizing for expeditionary service

124. Evans, Frank E. "Motor Transportation in the Marine Corps." *Marine Corps Gazette* 2, no. 3 (March 1917): 1–12.

Examines the use of trucks in Haiti and the Dominican Republic and at Veracruz. Evans notes the expedients resorted to by marines to provide themselves with armored vehicles.

125. Manney, Henry N. "The Expeditionary Work of the Quartermaster's Department." *Marine Corps Gazette* 1 (June 1916): 175–80.

    Lists hints for officers helping to prepare an expeditionary force. Manney's major concern is the need for systematic stowing of cargo. This is a very early example of a study about what later would be called "logistics."

126. McGee, Vernon E. "Motor Transportation for Expeditionary Units." *Marine Corps Gazette* 13, no. 12 (December 1928): 271–76.

    Focuses on the work of a board that investigated the need for better expeditionary unit vehicles. China duty in the 1920s reinforced the view that marine land mobility had to be improved. McGee also reviews the trucks used in Haiti and the Dominican Republic.

127. Roosevelt, E.L. "Permanent Company Organization for the United States Marine Corps." *United States Naval Institute Proceedings* 33, no. 9 (September 1907): 917–26.

    Asserts that a structure of companies in the Marine Corps would expedite formation of expeditionary units. Roosevelt also deals briefly with the logistics problems associated with the sending of marines to Cuba and to Panama in 1906.

128. Smith, Oliver P. "In the Wake of the Expeditions." *Marine Corps Gazette* 12, no. 9 (September 1927): 146–59.

    Describes the considerable impact of the mounting of expeditionary forces for China and Nicaragua in 1927 upon general operations of the Marine Corps. Despite problems, expeditionary service provided valuable training and useful publicity, according to Smith's assessment.

C. Counterinsurgency tactics

129. Del Valle, Pedro A. "The Use of Gas in Minor Warfare." *Marine Corps Gazette* 15, no. 2 (February 1931): 25, 49–50.

    Reminds marines that no treaty bars use of chemical weapons, and asserts that tear gas and other mild chemical agents could break up guerrilla forces without loss of life and, therefore, without creating martyrs.

130. Ellis, E.H. "Bush Brigades." *Marine Corps Gazette* 6, no. 3 (March 1921): 1–15.

Formulates a general plan for landing in a small country and taking its major seaports. Ellis attacks the American public for uninformed criticisms of marine interventions.

131. Hamilton, Louis McLane. "Jungle Tactics." *Journal of the Military Institution of the United States* 37 (July–August 1905): 23–28.

Illustrates the problems of jungle warfare with examples from the Philippine fighting. Traditional methods for avoiding ambushes, such as flankers and a point man, were difficult to employ in the jungle. Hamilton offers various suggestions for improving jungle tactics.

132. Harrington, Samuel N. "The Strategy and Tactics of Small Wars." Parts 1, 2. *Marine Corps Gazette* 6, no. 12; 7, no. 3 (December 1921; March 1922): 474–91; 84–93.

Explicates marine operations in the Caribbean, especially the Dominican Republic. This is one of the most important pre–World War II studies. Part 2 moves away from American experience to European colonial campaigns and such episodes as the *Freikorps* suppression of revolutionaries in Germany.

133. Hirsch, Ralph. "Pack Artillery in the Tropics." *Field Artillery Journal* 15 (November–December 1925): 599–605.

Describes American pack guns and the organization of pack batteries in the army. Artillery units were needed that could move into the jungle during small wars with irregular forces.

134. Johnson, G.A. "Junior Marines in Minor Irregular Warfare." *Marine Corps Gazette* 6, no. 6 (June 1921): 152–63.

Constitutes a guide for young marine officers based on marine files and British writings. The short bibliography on "bush warfare" refers primarily to British experience.

135. Keyser, Ralph S. "March Security in Bush Warfare." *Marine Corps Gazette* 14, no. 3 (March 1929): 38–42.

Warns that in case of ambush the whole force should not be committed immediately. Flanking movements, moreover, are inherently dangerous in such situations because firm intelligence is generally lacking to the group ambushed.

136. Linsert, E.E. "Should the Marine Corps Use Chemical Agents in Guerrilla Warfare?" *Marine Corps Gazette* 18, no. 8 (August 1933): 36–41.

Notes that while an agreement had been signed by the United States not to use chemical weapons, this prohibition probably did not extend to bandit suppression. Linsert discusses both offensive and defensive use of chemical weapons in guerrilla warfare by both ground forces and air units.

137. Planier, George S. "Woodcraft and Bush Warfare." *Leatherneck* 16, no. 3 (March 1933): 14, 38.

Illustrates the numbers of environmental signs that can be used in tracking enemy patrols. Bits of chewed sugar, the movement of wood ticks, noise, and the presence or absence of water on vegetation may be clues about the movement of guerrillas. The article deals with operations in both wet and dry seasons.

138. Utley, Harold H. "An Introduction to the Tactics and Technique of Small Wars." Parts 1, 2, 3. *Marine Corps Gazette* 16, no. 5; 18, no. 8, no. 11 (May 1931; August, November 1933): 50–53; 44–48; 43–46.

Draws heavily on marine operations in the Caribbean area, including Nicaragua. Utley's series is important because it grew into a three-volume manuscript that led to the writing of the *Small Wars Manual*.

139. ———. "The Landing and Occupation of Seaports." *Marine Corps Gazette* 19, no. 2 (February 1935): 15–20, 54–56.

Describes a number of landings during interventions in the early twentieth century, including Veracruz, Port au Prince, and Santo Domingo City. The treatment of the last landing is the most extensive. Prophetically, Utley warns that landings against other kinds of enemies may not be as easy as those operations in the Caribbean.

140. White, Thomas D. "Jungle Fighting." *Infantry Journal* 29 (July 1926): 65–66.

Reports lessons learned from recent army exercises in Panama. Other articles published about the same time do not refer as specifically to small war environments and, therefore, have been omitted.

D. The role of air power

141. Campbell, H. Denny. "Aviation in Guerrilla Warfare." *Marine Corps Gazette* 15, no. 4; 16, no. 1, no. 3 (February, May, November 1931): 37–39, 71–75; 35–42; 33–40.

Recommends aggressive use of aviation in surprise attacks and to demonstrate mobility in unfriendly territory. Campbell uses British and other foreign experience extensively, but he also refers to United States operations in Nicaragua.

E. Military government/civic action

142. Davis, Henry C. "Indoctrination of Latin-American Service." *Marine Corps Gazette* 5, no. 6 (June 1920): 154–61.

Warns that Latin American culture cannot be changed quickly, if at all, by an American occupation. Marines should not debate policies with local inhabitants.

F. Other topics

143. Brown, Campbell H. "Bush Warfare Transportation." *Marine Corps Gazette* 19, no. 11 (November 1934): 34–38.

    Examines the problems of moving supplies in the Caribbean countries and puts forth solutions to them. Brown notes that former army cavalrymen and field artillerymen rarely were successful in handling animals and pack trains in the marines.

144. Gray, John A. "An Expeditionary Medal." *Marine Corps Gazette* 13, no. 9 (September 1928): 164–65.

    Argues that the Marine Corps should issue a medal to accompany its expeditionary ribbon.

IV

The Philippine–American War and the Moro campaigns, 1899–c.1918

This chapter deals with two separate events, the Philippine–American War that lasted from 1899 until as late as 1918 and the Moro campaigns that lasted from 1902 to 1913. The Philippine–American War was a protracted conflict that continued long after the United States had proclaimed its termination. Increasingly, members of the Philippine Scouts and the Philippine Constabulary, consisting largely of Filipino troops, fought the remaining nationalists and groups of Moros (Filipino Muslims in the southern islands). For years, however, elements of the United States Army other than the scouts were drawn into the fighting.

A. General studies

145. Coats, George Yarrington. "The Philippine Constabulary: 1901–17." Ph.D. diss., Ohio State University, 1968.

    Chronicles the military campaigns of the constabulary in great detail. Fighting outside the Muslim areas is given extensive coverage. Coats' research rests on wide use of manuscripts, and a fine bibliography of them is included. Secondary works are less fully dealt with.

146. Elarth, Harold Hanne. *The Story of the Philippine Constabulary*. Los Angeles, CA: Globe Printing Company, 1949.

    Contains lucid accounts of many campaigns, but focuses almost exclusively on the constabulary. Army participation is praised but not discussed fully. A long biographical section supplements the narrative.

147. Feuer, A.B., ed. *America at War: The Philippines, 1898–1913*. Westport, CT: Praeger, 2002.

Valuable collection of contemporary accounts of many facets of the War of 1898, the Philippine–American War, and the campaigns in the Muslim areas of the southern Philippines. These materials from various sources are much enhanced by Feuer's explanations and comments. Another significant feature is the inclusion of easy-to-read maps with almost all the segments of the book.

148. Hurley, Vic. *Jungle Patrol: The Story of the Philippine Constabulary.* New York: E.P. Dutton and Company, Inc., 1938.

    Deals extensively with the fighting in the northern and central islands as well as in the Moro areas. Hurley corresponded with or interviewed more than 40 constabulary officers in compiling this fascinating book.

149. Roth, Russell. *Muddy Glory: America's "Indian Wars" in the Philippines 1899–1935.* West Hanover, MA: Christopher Publishing House, 1981.

    Collects a number of independent essays, many of which focus on the Philippine–American War. The book, as the author states, is highly "episodic." Roth's bibliography is extensive, probably the best available.

150. Smith, Richard W. "Philippine Constabulary." *Military Review* 48, no. 5. (May 1968): 73–80.

    Reviews briefly the origins and development of the constabulary, but stresses its tactics and use of civic action. Smith argues that many lessons can be learned from the constabulary and applied fruitfully in Vietnam.

151. Twichell, Heath, Jr. Chap. 6, "Always Outnumbered: Never Outfought," pp. 117–46, in *Allen: The Biography of an Army Officer 1859–1930.* New Brunswick, NJ: Rutgers University Press, 1974.

    Discusses friction between the army, constabulary, and the civil government and Allen's efforts to integrate the Philippine Scouts and the Philippine Constabulary, a goal he never achieved. Allen put his stamp firmly on the constabulary as its first commander.

152. Woolard, James Richard. "The Philippine Scouts: The Development of America's Colonial Army." Ph.D. diss., Ohio State University, 1975.

    Gives an organizational history of the scouts. It is particularly useful in the context of this bibliography for the early 1900s before the constabulary took over so much responsibility for maintaining law and order.

B. The Philippine–American War, 1899–c.1918

After Admiral George Dewey's destruction of the Spanish naval squadron in the Philippines in the Battle of Manila Bay on May 1, 1898, the McKinley administration decided to annex the islands, despite the fact that Filipinos had largely eliminated Spanish authority from their country and had established a government. Months of tension followed as additional United States land forces arrived in the Philippines. Finally, early in 1899, as the Treaty of Paris between Spain and the United States transferred the Philippines to the latter, fighting broke out in the Manila area and spread through much of the island group. After nearly a year of conventional military resistance, the Filipino nationalists changed tactics to fight a guerrilla war.

Chronology

1892
July 7   Katipuman, a revolutionary organization, established to bring freedom to the Philippines. The most prominent leader was José Rizal y Mercado. For some time the group expanded secretly.

1896
August 23   The Katipuman rebelled openly in the wake of its detection by a Spanish friar. Violent Spanish repression followed.
December 30   Execution of Rizal, a landmark event in Philippine history. Despite his death, the revolt continued through most of 1897. Owing to his military abilities, Emilio Aguinaldo y Famy became the revolutionary movement's new leader.

1897   After a truce late that year, Aguinaldo agreed to leave the Philippines for Hong Kong. Both the Filipinos and the Spanish realized that the truce and Aguinaldo's departure did not mean the end of the anti-colonial conflict.

1898
May 1   Admiral George Dewey's squadron completely destroyed the Spanish fleet assembled in Manila Bay. Dewey worked with Philippine revolutionaries toward the elimination of Spanish power in the Philippine Islands.
May 19   Aguinaldo returned to the Philippines on one of Dewey's warships.
June 12   Revolutionary leaders, whose men controlled most of the Philippines by then, proclaimed the island nation to be an independent republic. Tensions between the revolutionaries and the United States began increasing, a process that culminated in the outbreak of war early in 1899.
June 30   The first of a number of contingents of United States troops arrived in the Philippines.
August 13   In accordance with a secret agreement made between General Wesley Merritt and the Spanish commander to stage a largely

sham battle and to prevent Philippine troops from entering the city, Manila was seized by United States forces. Skirmishes occurred between United States soldiers and members of the revolutionary army, but they ended fairly quickly.

1899

January 23 — Formal establishment of the Philippine Republic at Malolos north of Manila, two days after a constitution had been proclaimed.

February 4 — After a skirmish between United States and Philippine troops, full-scale fighting broke out in and around Manila. Owing to the limited size of the United States army in the Philippines, advances were limited to the Manila area. By the end of February Manila and its approaches had largely been secured by United States forces.

February 6 — The Treaty of Paris between Spain and the United States gave the Philippines to the United States, thus ensuring that there would be a protracted war between the new colonial power and Philippine nationalists.

March 31 — The Philippine revolutionary capital of Malolos was attacked and occupied by United States troops.

June 5 — General António Narciso Luna de St. Pedro, one of the military leaders of the nationalist movement was killed by rebellious troops, giving the Filipino campaign a decided setback.

November 13 — Aguinaldo ordered a change from conventional tactics against the United States to guerrilla operations, an approach that continued the Philippine–American War for years. By December, the new strategy was fully in effect.

December 15 — United States Senate Committee on the Philippines instituted. The committee conducted a detailed inquiry into United States policies and military actions in the Philippines in 1902.

1900

November — The re-election of William McKinley as president of the United States ended Philippine revolutionary hopes that a Democratic Party victory might bring a radical change in United States policy toward the Philippines.

1901

February 2 — Establishment of the Philippine Scouts, Filipino soldiers recruited for the United States Army with United States officers. Despite the misleading term "scouts," these men were simply Filipino members of the army. Significant numbers of Filipinos had already assisted United States forces, but this step marked increased recruitment and improved status for the volunteers.

March 24	In a daring operation, General Frederick Funston and a small party seized Aguinaldo, who soon agreed to cooperate with his captors.
April 30	Manuel Tinio, one of Aguinaldo's most able subordinate leaders, surrendered after waging a vigorous campaign in the Ilocano provinces.
July 4	William Howard Taft, Governor General of the Philippines, declared the end of martial law in the islands, but, in reality, fighting continued for years.
July 18	The Philippine Commission, which governed the islands as a United States colony, established the Philippine Constabulary, a paramilitary force that not only acted as a kind of state police but also conducted significant military operations, often in conjunction with the United States Army, especially the Philippine Scouts.
September 28	Philippine revolutionaries attacked the United States Army garrison at Balangiga on Samar Island, killing almost all of the soldiers. The successful assault brought severe treatment to many on Samar and resulted in charges that marine operations on the island had resulted in atrocities. Several United States officers were tried, and Congress began an inquiry into United States policy in the Philippines.

1902

April 16	Miguél Malvar, one of the leading commanders of Philippine revolutionary forces on Luzon, surrendered, marking a major step in the pacification of the Philippines.
July 4	President Theodore Roosevelt declared fighting in the Philippines to have ended.

Significant fighting continued for years against those whom the United States termed "bandits" and against the Moros. Aside from the Moros, Philippine guerrillas were either secular nationalists or those who combined nationalism and millenarianism. Almost every area of the Philippines was subject to unrest. United States soldiers, including Philippine Scouts, and units of the Philippine Constabulary encountered resistance until about 1918, although as time passed, more responsibility for suppressing outbreaks was given to the constabulary. Warfare on Luzon lasted until about 1910, in the Cebu area until about 1907, and in and around Samar until 1911.

1. Bibliographies

153. Geary, James W. "Afro–American Soldiers and American Imperialism, 1899–1902: A Select Annotated Bibliography." *Bulletin of Bibliography* 48, no. 4 (1991): 189–93.

A relatively long and well-informed introduction precedes a careful selection of publications which are commented upon in detail.

154. Meixsel, Richard B. *Philippine–American Military History, 1902–1942: An Annotated Bibliography.* Jefferson, NC: McFarland & Co., 2003.

    Most of the content relates to World War II, but there is a useful introduction and an important segment devoted to "Bibliographies and Other Reference Works," pp. 21–30, which includes much relevant material. A number of sources relate to the Philippine Constabulary, which played an increasingly significant role as the Philippine–American War continued.

155. Venzon, Anne Cipriano. *America's War with Spain.* Lanham, MD: Scarecrow Press, Inc., 2003.

    Two sections, "Philippines," pp. 117–30; and "Anti-Imperialism," pp. 163–70, list numerous government documents, books, articles, and pamphlets. Almost all of the items are annotated.

2. Encyclopedia

156. Keenan, Jerry. *Encyclopedia of the Spanish–American and Philippine–American Wars.* Santa Barbara, CA: ABC-CLIO, 2001.

    Provides considerable coverage of the campaigns of the Philippine–American War and the people involved in that conflict. The book includes an extensive bibliography, ample cross-references between entries, suggested readings, a chronology, a fine index, and other features.

3. Congressional hearings and official reports

157. Philippines. Military Governor. 1898–1900. *Report of Maj. Gen. E.S. Otis, United States Army, Commanding Division of the Philippines. September 1, 1899, to May 5, 1900.* Washington, DC: Government Printing Office, 1900.

158. ——. 1900–01. *Report of Major General Arthur MacArthur, U.S. Army, Commanding Division of the Philippines. Military Governor in the Philippine Islands. May 5–Oct. 1, 1900, October 1, 1900–June 30, 1901.* Manila: 1900–01.

    Major sources for an important period in the Philippine–American War.

159. Taylor, John R.M. *The Philippine Insurrection Against the United States: A Compilation of Documents with Notes and Introduction.* Pasay City, Philippines: Eugenio Lopez Foundation, 1971.

For many years this detailed study and assemblage of documents existed only as an archival resource, after its publication had been stopped during Theodore Roosevelt's second administration. This is an essential source for the Philippine–American War.

160. U.S. Congress. Senate. Committee on the Philippines. *Affairs in the Philippines. Hearings Before the Committee on the Philippines of the United States Senate.* Washington, DC: Government Printing Office, 1902.

Key source for the study of the Philippine–American War. The hearing's three large volumes contain detailed testimony by participants in events in the Philippines, offering a variety of points of view.

161. U.S. Philippine Commission. *Report of the Philippine Commission to the Secretary of War ... 1900–1915.* Washington, DC: Government Printing Office, 1901–16.

Detailed reports regarding events in the Philippines.

162. U.S. War Department. Office of the Adjutant General. *Correspondence Relating to the War with Spain and Conditions Growing Out of the Same, Including the Insurrection in the Philippine Islands and the China Relief Expedition, between the Adjutant-General of the Army and Military Commanders in the United States, Cuba, Porto Rico, China and the Philippine Islands, from April 15, 1898 to July 30, 1902.* Washington, DC: Government Printing Office, 1902.

This two-volume set provides the texts of many important messages and reports relating to the Philippine–American War.

4. General studies

163. Altieri, Jayson A., John A. Cardillo, and William M. Stowe III. "Practical Lessons from the Philippine Insurrection." *Armor* 116, no. 1 (January–February 2007): 27–34.

Emphasizes the success of the United States counterinsurgency effort in the Philippines. The authors provide a number of detailed lessons from the Philippine–American War and the Moro campaigns. Although many of the references are to standard works, the article cites some lesser known studies, thus attesting to the authors' careful examination of the literature about fighting in the Philippines. Some interesting comparisons are made between the Philippine–American War and both the Vietnam and Iraq wars.

164. Boot, Max. *The Savage Wars of Peace: Small Wars and the Rise of American Power.* Chap. 5, "'Attraction' and 'Chastisement': The Philippine Wars, 1899–1902," pp. 99–128. New York: Basic Books, 2003.

Provides a relatively detailed account. Boot rightly gives attention to the generally neglected, but important, role of the United States Navy in this conflict, but he makes the common error of asserting that fighting, except in the Moro Province, largely ended in 1902.

165. Bundt, Thomas S. "An Unconventional War: The Philippine Insurrection, 1899." *Military Review* 84, no. 3 (May–June 2004): 9–10.

Succinct but insightful discussion of the changes the United States Army underwent in order to win the Philippine–American War. Bundt explains the new tactics, weapons, and policies that were introduced.

166. Burdett, Thomas F. "A New Evaluation of General Otis' Leadership in the Philippines." *Military Review* 55, no. 1 (January 1975): 79–87.

Argues that Major General Elwell S. Otis has been unfairly criticized for his performance as United States commander during the early part of the Philippine–American War. He has often been attacked for having been overly wary in his execution of military operations, beginning with comments by United States journalists in the Philippines. Burdett's article is an interesting analysis, although his interpretation of Otis remains a minority view.

167. Caterini, Dino J. "Repeating Ourselves: The Philippine Insurrection and the Vietnam War." *Foreign Service Journal* 54, no. 12 (December 1977): 11–17, 31–32.

Impressive analysis. An experienced United States Information Agency official, Caterini believes that the Philippine–American and Vietnam Wars were quite similar and asserts that useful lessons can be drawn from those wars that are "applicable to the future should similar circumstances develop." (p. 11). Caterini reviews the problems of counterinsurgency, including the challenge of identifying guerrillas. He also devotes considerable attention to the roles of the press and of public opinion. His conclusions seem relevant to Iraq and other episodes since the article was published.

168. Crawford, James Grant. "The Warriors of Civilization: U.S. Soldiers, American Culture, and the Conquest of the Philippines, 1898–1902." Ph.D. diss., University of North Carolina, Chapel Hill, 2003.

Focuses on the experiences and changing views of United States soldiers in the Philippines, many of them short-term volunteers rather than regulars. Crawford traces their evolution from being recruits to becoming seasoned troops. He emphasizes the role of United States culture, including its general racism, in shaping the perceptions and actions of these men. This study is based heavily on soldiers' diaries and letters at the United States Military History Research Institute at Carlisle, Pennsylvania.

169. Deady, Timothy K. "Lessons from a Successful Counterinsurgency: The Philippines, 1899–1902." *Parameters* 35, no. 1 (Spring 2005): 53–68.

Presents an interesting comparison between the Philippine–American War and the Iraq War and other recent or current conflicts. Deady usefully reviews the strategies of the several United States commanders in the Philippines and discusses more briefly the Filipino nationalists' approach to fighting their war. This a valuable, although controversial, survey of the Philippine–American War.

170. Donovan, Consorcia Lavadia. "The Philippine Revolution: A 'Decolonized' Version." Ph.D. dissertation, Claremont Graduate School, 1976.

Vigorous defense of the Philippine nationalist movement that answers what Donovan views as the general dismissal of the legitimacy of the revolution in Western historical treatments. The author thoroughly examines John R.M. Taylor's *The Philippine Insurrection Against the United States* (item 159), seeking to refute the arguments made in that study. In carrying out that research, Donovan provides a detailed description of the Filipino efforts to win independence.

171. Faust, Karl Irving. *Campaigning in the Philippines*. San Francisco, CA: Hicks-Judd Company, 1899. Reprinted. New York: Arno Press/ The New York Times, 1970.

Highly detailed early account of the first stages of the Philippine War. Faust coordinated the collection of materials from official sources and wrote most of the work. The writing team received the cooperation of the United States armed forces. Numerous photographs and maps add considerable value to this book. There are brief sections about the operations of various components of the army, such as engineer troops and the Signal Corps. One chapter describes the capture of Manila from the Spanish.

172. Gates, John Morgan. "The Official Historian and the Well-Placed Critic: James A. Leroy's Assessment of John R. Taylor's *'The Philippine Insurrection Against the United States'*." *Public Historian* 7, no. 3 (Summer 1985): 57–67.

Discusses the writing and eventual suppression of Major John Rogers Meigs Taylor's work. The study was finished at a time when William Howard Taft was Secretary of War, and Taft wished to prevent publication of Taylor's research. The secretary did not want to recall the war to the public just before a congressional election or to court controversy with the Democrats about the Philippines issue. James A. LeRoy reviewed Taylor's manuscript and argued against its publication.

173. ———. *Schoolbooks and Krags: the United States Army in the Philippines, 1898–1902*. Westport, CT: Greenwood Press, 1973.

A path-breaking study that stresses what would now be termed "civic action," that is, operations by a military force to provide services to the civilian population to reduce hostility toward a military government or to a civilian regime challenged by an insurgency. Gates discusses the military side of the war, but he believes that positive measures by the United States in the Philippines, that is, the policy of "attraction," brought large numbers of Filipinos to accept United States rule and thereby helped to end the conflict.

174. ———. "Two American Wars in Asia: Successful Colonial Warfare in the Philippines and Cold War Failure in Vietnam." *War in History* 8, no. 1 (2001): 47–71.

Gates begins with a thoughtful review of the historiography of the Philippine–American War, including an assessment of his own work, especially *Schoolbooks and Krags*. Although he agrees that atrocities and abuses occurred during the Philippine–American War, he criticizes academics and journalists he believes have exaggerated those events. Gates describes similarities and differences between the Philippines and Vietnam, emphasizing the differences, and expresses the view that studying events in the Philippines would not have helped much in conducting the Vietnam War.

175. ———. "War-Related Deaths in the Philippines, 1898–1902." *Pacific Historical Review* 53, no. 3 (August 1984): 367–78.

Reviews the controversies about the number of Filipinos who died in the Philippine–American War. Both at the time of the war and later on, the figures cited in political and scholarly sources varied considerably. Numbers from the 1903 census provide useful results, although they may be misleading, at least for some parts of the Philippines. Gates analyzes the problems in arriving at reasonable estimates; he does not offer figures himself.

176. Ginsburgh, Robert N. "Damn the Insurrectors!" *Military Review* 44, no. 1 (January 1964): 58–70.

Skillfully compares the Philippine–American War with later counterguerrilla campaigns and asserts that counterinsurgency planners in the United States can learn much from the conflict. Ginsburgh gives Major General Arthur MacArthur credit for bringing the guerrilla effort under control. He quotes liberally from Filipino documents about the conduct of the war, thus giving much information from the Filipino perspective. Although a good deal has been written about the Philippine–American War since the 1960s, this is a still valuable analysis.

177. Kramer, Paul A. *The Blood of Government: Race, Empire, the United States and the Philippines.* Chap. 2, "'From Hide to Heart': The

Philippine–American War as Race War," pp. 87–158. Chapel Hill: University of North Carolina Press, 2006.

The whole book is an important contribution to the subject of Philippine–United States relations, but the chapter cited is of special significance for this bibliography. Kramer stresses the racism of United States participants in the war. This is a sophisticated analysis of racial justifications for annexation of the Philippines, moreover. Kramer notes usefully that many Filipinos were well aware of racism in the United States, as exemplified by the treatment of Native and African Americans.

178. Linn, Brian McAllister. *Guardians of Empire: The U.S. Army and the Pacific, 1902–1940*. Chap. 1, "The Guardians Arrive," pp. 5–22; and Chap. 2, "The Savage Wars of Peace," pp. 23–49. Chapel Hill: University of North Carolina Press, 1997.

Contrary to the implication of the title of this book, the chapters cited provide a relatively brief, but first-rate discussion of the Philippine–American War and the campaigns against the Moros in the southern Philippines. Linn summarizes the operations of the army and describes the increasing importance of the Philippine Constabulary, including the years well beyond Theodore Roosevelt's 1902 proclamation that the war was finished.

179. ———. "Intelligence and Low-Intensity Conflict in the Philippine War, 1899–1902." *Intelligence and National Security* 6, no. 1 (1991): 90–114.

This is an important contribution to the literature about the war. It is a thoroughly researched study that explores the various methods for gathering and processing intelligence, such as informants and scouting. Linn observes that "for much of the Philippine War, American intelligence was as diffuse, unconnected and disorganized as the resistance the [United States] soldiers encountered" (p. 108).

180. ———. *The Philippine War, 1899–1902*. Lawrence: University Press of Kansas, 2000.

This study provides full coverage of the war. Chapter 10, moreover, deals with the very early years of United States involvement with the Moros. There is a brief, but important, essay about the historiography of the Philippine–American War. This book is based on massive archival research, and it is likely to remain a leading study of the war for some time.

181. ———. "The Philippines: Nationbuilding and Pacification." *Military Review* 85, no. 2 (March–April 2005): 46–54.

Asserts that much can be learned from the Philippine–American War, but, owing to the complexity of that conflict, those who wish to draw on that experience "must truly think outside the box and be willing to engage in intensive study and self-reflection to learn the lessons from the conflict" (p. 53). Furthermore, Linn questions the appropriateness of applying current concepts, such as "nationbuilding," to United States policy in the Philippines during the late nineteenth and early twentieth centuries. President William McKinley and many other civil and military leaders regarded the Philippines simply as a colony.

182. McEnroe, Sean. "Painting the Philippines with an American Brush: Visions of Race and National Mission Among the Oregon Volunteers in the Philippine Wars of 1898 and 1899." *Oregon Historical Quarterly* 104, no. 1 (Spring 2003): 24–61.

Searing depiction of the role of racism in the Philippine–American War and the commission of atrocities by United States personnel. McEnroe discusses the dehumanization of the enemy that occurs during combat and asserts that Oregon (and other) troops "treated their enemies with extreme cruelty and were generally untroubled by it" (p. 27). United States soldiers identified the Filipinos with Native Americans and African–Americans and thus applied the racial code of the United States to a new setting. This is a sophisticated and important analysis.

183. May, Glenn Anthony. "Why the Filipinos Fired High: Popular Participation in the Philippine Revolution and the Philippine–American War." *Biblion* 7, no. 2 (1999): 87–106.

Argues that most authors in the United States who have written about the Philippine–American War have been too focused on United States actions when trying to explain why the Filipinos lost. This perceptive study emphasizes Filipino military tactics and the attitudes of the rank and file Filipino soldiers. Many in the Philippines have argued that during the period of the war the upper classes largely abandoned the struggle to save their property and status. May believes that the nationalist movement was supported by higher income people and that many ordinary Filipinos were never strongly attached to the goal of independence with a continuance of elite rule. The title of the article comes from May's belief that many Filipino soldiers simply fired in the direction of United States forces instead of aiming carefully.

184. ———. "Why the United States Won the Philippine–American War, 1899–1902." *Pacific Historical Review* 52, no. 4 (November 1983): 353–77.

Compares the Philippine–American and Vietnam Wars, although he does not regard them as identical conflicts. May reviews the relative strengths of opposing military forces in the two wars and discusses the awesome arsenal of weapons the United States employed in Vietnam. He points to Aguinaldo's stubborn adherence to conventional war tactics and his political ineptness, especially in regard to the peasantry, as major factors that helped defeat the Filipinos. The ethnic diversity of the Philippines was another barrier to nationalist victory.

185. Miller, Stuart Creighton. Chap. 1, "The American Soldier and the Conquest of the Philippines," pp. 13–24, in *Reappraising an Empire: New Perspectives on Philippine–American History*, ed. Peter W. Stanley. Cambridge, MA: Committee on American–East Asian Relations of the Dept. of History in Collaboration with the Council on East Asian Studies, Harvard University, 1984.

Stresses the fact that the Philippine–American War occurred during a period of vigorous nationalism worldwide. Despite numerous complaints about army life and about the Philippines, United States soldiers fighting the Filipino nationalists had a more positive attitude toward their role in the conflict than a good many men and women who were in the Vietnam War. Miller asserts that given their general enthusiasm for the war, lower-ranking officers and men had no difficulty applying tactics from the western campaigns against Native Americans while serving under the command of senior officers who had either led or at least had experience in those campaigns.

186. ———. *"Benevolent Assimilation": The American Conquest of the Philippines, 1899–1903*. New Haven, CT: Yale University Press, 1982.

Influential work that emphasises abuses and worse by members of the United States armed forces during the Philippine–American War. Miller includes many quotations from contemporary diaries, letters, newspaper stories, and reports. He also discusses the beginnings of the controversy about the behavior of United States service members in the Philippines.

187. Ninkovich, Frank. *The United States and Imperialism*. Chap. 2, "Failed Expectations: The Civilizing Mission in the Philippines," pp. 48–90. Malden, MA: Blackwell Publishers Inc., 2001.

Fine overview of Philippine–United States relations. Although only a few pages deal with the Philippine–American War, this is an important analysis that puts that war into the larger perspective of United States foreign policy and offers a balanced judgment about the whole range of relations between the Philippines and the United States.

188. Paulet, Anne. "The Only Good Indian is a Dead Indian: The Use of United States Indian Policy as a Guide for the Conquest and

Occupation of the Philippines, 1898–1905." Ph.D. diss., Rutgers–The State University of New Jersey, 1995.

Innovative study of the well-known but previously unanalyzed phenomenon that the United States actively tried to apply what seemed to be the lessons of dealing with Native Americans to the Philippines. Given the pervasive comparisons between Native Americans and Filipinos made at the time, this is an exceptionally valuable study that reflects extensive knowledge of the two strains of policy-making.

189. Pomeroy, William J. "American 'Pacification' in the Philippines, 1898–1913." *France–Asie–Asia* 21, nos. 189–90 (1967): 427–46.

Documents the determination of the McKinley administration to subjugate the Filipinos and examines the atrocities issue in some detail. Pomeroy contends that the war could have been avoided by a settlement that provided the Philippines with self-government and that gave the United States a protectorate over the islands.

190. Schirmer, Daniel. "How the Philippine–American War Began." *Monthly Review* 51, no. 4 (September 1999): 45–48.

Briefly examines the steps that the United States took to acquire the Philippines and to fight the Filipino nationalists. Schirmer asserts that McKinley intentionally harnessed United States nationalism to support the treaty between Spain and the United States that gave the Philippines to that country by engineering a clash between the United States and Filipino armies.

191. Sexton, William Thaddeus. *Soldiers in the Sun: An Adventure in Imperialism*. Harrisburg, PA: Military Service Pub. Co., 1939. Reissued as *Soldiers in the Philippines: A History of the Insurrection*. Washington, DC: Infantry Journal, 1944.

This was the first serious military study of the Philippine–American War. It is a clear narrative, based primarily on official records, which Sexton calls "the best evidence available" (second page of the Introduction). The book retains its value as a purely military treatment. Sexton is fiercely critical of the anti-imperialist movement, which he accuses of prolonging the war, partly through fostering Philippine nationalist expectations about a victory for their side in the 1900 presidential election in the United States.

192. Shaw, Angela Velasco and Luis H. Francia, eds. *Vestiges of War: the Philippine–American War and the Aftermath of an Imperial Dream, 1899–1999*. New York: New York University Press, 2002.

Extensive collection of material, including photographs and poems. Many essays are by specialists in Philippine studies. This is eloquent testimony about the continuing effects in the Philippines of the years of overt colonialism.

193. Silbey, David J. *A War of Frontier and Empire: The Philippine–American War, 1899–1902*. New York: Hill and Wang, 2007.

Informative survey of the conflict that provides the reader with the feel of the battles and skirmishes in the Philippines. It is a good choice for introducing undergraduates and others to the subject.

194. Smith, D.H. "American Atrocities in the Philippines: Some New Evidence." *Pacific Historical Review* 55, no. 2 (May 1986): 281–83.

After referring to a controversy between John M. Gates and Gore Vidal, Smith discusses an assertion by Richard E. Welch that use of the "water cure" (a form of torture) seems to have begun only in the fall of 1900 by Filipino auxiliary troops. Smith notes an instance in which the method was used in the spring of 1900, but he adds that further research is needed to determine the incidence of the practice.

195. Sweezy, Paul. "Kipling, the 'White Man's Burden,' and U.S. Imperialism." *Monthly Review* 55, no. 6 (June 2003): 1–11.

Draws parallels between the United States acquisition of the Philippines and the invasion of Iraq, and, more broadly, between traditional imperialism and "neoconservatism." Sweezy calls the Philippine–American War "one of history's most barbaric wars of imperial conquest" (p. 3). He underestimates how long the Filipinos attempted to utilize conventional military tactics and makes some other errors. Nevertheless, this is an interesting effort to compare contemporary United States foreign policy with that of a century ago.

196. Williams, John Hoyt. "National Guard Quells 1899 Philippine Insurrection." *National Guard* 42 (July 1988): 32–34, 36, 47.

Notes that Filipino nationalists took control of the island group from Spain before the arrival of United States land forces. Williams views the fighting value of the National Guard as high. He believes that the experience of numerous officers in campaigns against Native Americans was a significant factor in the United States victory. Another element was the United States tactics of using ethnic conflicts to keep certain groups out of the war and to recruit some men from the minorities to fight along with United States troops.

197. Wolff, Leon. *Little Brown Brother: How The United States Purchased and Pacified the Philippine Islands at the Century's Turn*. Garden City, NY: Doubleday, 1961.

This readable narrative remains a useful introduction to the Philippine–American War, but given the large number of works about the conflict that have been published since 1961, it has lost some of its value. Wolff's book is undocumented, but it does include an extensive bibliography.

198. Young, Kenneth Ray. *The General's General: The Life and Times of Arthur MacArthur*. Boulder, CO: Westview Press, 1994.

A significant segment of this biography deals with MacArthur's largely successful rule in the Philippines. Of special importance is Young's discussion of relations between MacArthur and William Howard Taft and the Philippine Commission that Taft headed. Young brought to his study of MacArthur a solid record of scholarship related to the Philippine–American War, which makes this study particularly valuable.

5. Campaigns

199. Bain, David Haward. *Sitting in Darkness: Americans in the Philippines*. Boston, MA: Houghton Mifflin, 1984.

Contrary to the title, Bain's focus is the capture of Filipino nationalist leader Emilio Aguinaldo y Famy by United States General Frederick Funston in a clever and daring operation. In addition to discussing Aguinaldo's capture, Bain has a good deal to say about the Philippines during the Marcos dictatorship decades after the Philippine–American War.

200. Birtle, Andrew J. "The U.S. Army's Pacification of Marinduque, Philippine Islands, April 1900–April 2001." *Journal of Military History* 61, no. 2 (April 1997): 255–82.

Marinduque, an island near Luzon, was the scene of a harsh United States Army campaign that involved forcible population movements. Birtle describes the people, geography, and economy of Marinduque. After United States troops arrived, a shadow nationalist government developed. Fighting was limited for a time, but in September 1900 guerrillas staged a highly successful ambush. Soon, the army attempted to capture and hold adult males. Later, efforts were made to destroy the economy of the interior, a tactic that persuaded many nationalist leaders to cease their resistance or support for the guerrillas. This is a model case study of a guerrilla movement and its suppression. Birtle observes that "only by realizing the uneasy, yet symbiotic relationship between attraction and compulsion can one shape a viable counterinsurgency policy" (p. 282).

201. Brown, John Clifford. *Gentleman Soldier: John Clifford Brown and the Philippine–American War*, ed. Joseph P. McCallus. College Station: Texas A&M University Press, 2004.

Well-edited and highly interesting diary and letters from an unusual soldier offer a different and important perspective on the Philippine–American War. Brown, a former officer, served as an enlisted man on Luzon for some time. His military cartography assignment allowed him to rove "as a sort of lone wolf mapmaker, able to jump from outfit to outfit" (p. 45). McCallus' expert introductions to each chapter and the excellent photographs and maps provided make this a significant contribution to the literature of the war.

202. Brown, Meredith. "A Killing in the Philippines, 1900: A Kentuckian Faces Insurgency and Military Justice." *Register of the Kentucky Historical Society* 104, no. 1 (Winter 2006): 43–76.

Although the author is a member of the Brown family, this attorney used his skills to produce an even-handed account of the killing of a Filipino by Captain Preston Brown and Brown's subsequent trial. Witnesses at the trial contradicted each other to such an extent that the nature of the initial event could not and still cannot be fully clarified. General Nelson A. Miles, the highest ranking officer of the army at the time, wanted Brown to be punished severely, but the Brown family's political connections helped win Preston Brown his freedom and his reinstatement in the army.

203. Chapman, Gregory. "Taking Up the White Man's Burden: Tennesseans in the Philippine Insurrection, 1899." *Tennessee Historical Quarterly* 47, no. 1 (Spring 1988): 27–40.

Heavily illustrated article, based largely on personal letters and newspaper accounts. The Tennessee troops fought in Cebu and Panay, occupying the major city of Iloilo, after participating in the initial battles near Manila. Chapman notes that the troops generally favored the war.

204. Chaput, Donald. "Private William W. Grayson's War in the Philippines, 1899." *Nebraska History* 61, no. 3 (Fall 1980): 355–66.

Detailed account of the beginning of fighting between the Filipinos and the United States on February 4, 1899. Grayson was the first United States soldier to fire on Filipinos on that date.

205. Filiberti, Edward J. "The Roots of US Counterinsurgency Doctrine." *Military Review* 68, no. 1 (January 1988): 50–61.

Award-winning contribution that attributes much of the success of the United States commitment in the Philippine–American War to General J. Franklin Bell in his Batangas campaign. The author describes the ruthless nature of United States operations, but he believes that the overall approach Bell took was necessary for victory. Filiberti also usefully describes the masterful blending of political and military tactics by the Filipino nationalists.

206. Fritz, David L. "Before the 'Howling Wilderness': The Military Career of Jacob Hurd Smith, 1862–1902." *Military Affairs* 43, no. 4 (Winter 1979): 186–90.

Argues that Smith's men killed relatively few people on Samar, despite Smith's order to make the island a "howling wilderness." Only a limited portion of the island was the scene of military operations, moreover. Smith was clearly a courageous officer, but his record included a number of incidents that suggested he often used poor judgment.

207. Ganley, Eugene E. "Mountain Chase." *Military Affairs* 24, no. 4 (Winter 1960–61): 203–10.

Discusses the successful United States effort to capture General Vicente Lukban, one of the major Filipino nationalist military leaders, who fought on and who was finally captured on Samar.

208. Hines, Richard K. "'First to Respond to Their Country's Call': The First Montana Infantry and the Spanish–American War and the Philippine Insurrection." *Montana: The Magazine of Western History* 52, no. 3 (Autumn 2002): 44–57.

Examines the infantry regiment's role in 1898 and 1899 during the tense months between the surrender of the Spanish in the Philippines and the outbreak of the Philippine–American War and the first period of fighting on Luzon, including the capture of Malolos, the Filipino nationalist capital. Hines asserts that the achievements of the militia troops in the Philippines gave the National Guard more significance within the armed forces.

209. Holli, Melvin G. "A View of the American Campaign Against 'Filipino Insurgents': 1900." *Philippine Studies* 17, no. 1 (1969): 97–111.

Holli presents the text of a long letter from Colonel Cornelius Gardner to Hazen S. Pingree, the governor of Michigan, describing his campaign in Luzon in 1900 and Gardner's views about United States policy toward the Philippines. Gardner, who commanded the 30th United States Infantry Regiment, was highly critical of the former Spanish colonial government, United States policies, and the conduct of his own men. Although he admitted that a Philippine government might not behave perfectly, at least at first, he considered the Filipinos capable of self-government and favored early independence for the Philippines. United States troops frequently disgraced themselves by looting. Holli's brief introduction to the Gardner letter makes useful comparisons between the Philippine–American and Vietnam wars.

210. Hunt, Geoffrey R. *Colorado's Volunteer Infantry in the Philippine Wars, 1898–1899*. Albuquerque: University of New Mexico Press, 2006.

A descriptive rather than an analytical account of the experiences of members of this regiment. Nevertheless, this is a detailed and well-written work. Despite the title, quite a bit of the book deals with the War of 1898 and the development of the Colorado militia before the establishment of a federal volunteer force, in good part from the ranks of the militia.

211. Johnson, John Reuben. "The Saga of the First Nebraska in the Philippines." *Nebraska History* 30 (June 1949): 139–62.

This useful study describes the organization of the regiment, which was a National Guard formation that had to be brought to full strength by recruiting additional men. There is much about the voyage to the Philippines. Johnson notes that the regiment's men wanted to be released after the end of fighting in the War of 1898, but the force was kept in the islands until June 1899. It took an active part in the capture of the Filipino capital of Malolos and subsequent fighting. The regiment had the highest casualties of any regiment in either the War of 1898 or the Philippine–American War.

212. Johnson, William R. "Three Years in the Orient: The Diary of William R. Johnson, 1898–1902." ed. Donald F. Carmony, Karen Tannenbaum, and Bess Sellers Johnson. *Indiana Magazine of History* 63, no. 4 (December 1963): 263–98.

Johnson, a member of the United States Volunteers with prior service during the War of 1898, was an enthusiastic supporter of United States policies in the Philippines. The editors of his diary provide an informative introduction to the document that records Johnson's service on Luzon. The diary entries are relatively short but do provide much detail about the fighting and about the life of the United States soldiers.

213. Lewis, Peter. *Foot Soldier in an Occupation Force: The Letters of Peter Lewis, 1898–1902*, compiled and arranged by H.R. Kells. Manila: Linguistics Office, de La Salle University, 1999.

After having been stationed in the United States and Cuba, Lewis was transferred to the Philippines where he served in Batangas and Laguna Provinces between 1900 and 1902. His letters provide a vivid picture of the fighting in the Philippine–American War, making the book an important source.

214. Linn, Brian McAllister. "The Pulahan Campaign: A Study in U.S. Pacification." *War in History* 6, no. 2 (April 1999): 45–71.

Groundbreaking research about a major revolt on Samar and Leyte between 1904 and 1907. Given the virtual absence of records from the Filipino side, it is difficult to determine the precise origins of the

Pulahan movement, but there were ties to earlier nationalist efforts. There is no doubt about the fighting prowess of the Pulahans, who were highly skilled guerrillas. The United States Army, now in a support role to work with the civil authorities, and the paramilitary Philippine Constabulary had learned from previous fighting on Samar. Restraint in dealing with Filipino adversaries and emphasis on civic action projects replaced severe repression. Linn underlines the importance of logistics in this counterinsurgency effort.

215. ———. *The U.S. Army and Counterinsurgency in the Philippine War, 1899–1902*. Chapel Hill: University of North Carolina Press, 1989.

Contrary to the title, this is not a general study of the Philippine–American War. Its scope is wide, though, describing the conflict in four large areas of Luzon. Linn argues that the war assumed a regional character on both sides, and, consequently, research needs to focus on particular parts of the large island group. Linn gives a good deal of attention to the development of the army's intelligence capability and effort. He has much to say, too, about another important development, the increasing use of Filipino troops and police to help suppress the nationalist struggle. This is one of the key books about the war. There is an extensive and important bibliography.

216. May, Glenn Anthony. *Battle for Batangas: A Philippine Province at War*. New Haven, CT: Yale University Press, 1991.

The Batangas campaign remains one of the most controversial aspects of a highly controversial war. On the one hand, United States military conduct has been praised for its effectiveness; on the other hand, the United States Army commanders and their subordinates have been accused of perpetuating widespread atrocities. May's book is a sophisticated analysis that discusses the issues in measured terms. May makes a good case for undertaking regional and local studies rather than grand surveys of Southeast Asian history. There is a brief but useful "Note about sources."

217. ———. "Private Presher and Sergeant Vergara: The Underside of the Philippine–American War," in *Reappraising an Empire: New Perspectives on Philippine–American History*, chap. 2, pp. 35–57, ed. Peter W. Stanley. Cambridge, MA: Committee on American–East Asian Relations of the Dept. of History in Collaboration with the Council on East Asian Studies, Harvard University, 1984.

Interesting comparison of the experiences of two enlisted men, one a United States soldier and the other a Filipino serviceman, in Batangas Province during the Philippine–American War. The principal sources for the study are a diary kept by Presher at the time of the war and an interview of Vergara conducted by May many years after the end of the conflict. As in his other writings, May stresses that the level of commitment to revolution and independence among Filipino military personnel varied considerably. Vergara, for example, was pressured into service by his patron who owned the land that Vergara worked.

218. Mojares, Resil B. *The War Against the Americans: Resistance and Collaboration in Cebu, 1899–1906*. Quezon City, Philippines: Ateneo de Manila University Press, 1999.

Thoroughly researched study of the Philippine–American War in a major area of the Philippines. Mojares consulted numerous sources in both the Philippines and the United States. This is an engrossing account of the coming of revolutionary nationalism to Cebu. The author does not neglect controversial subjects, such as divisions among the Cebuanos. An important feature of the book is its treatment of the fighting that continued after the supposed end of the war.

219. Ochosa, Orlino A. *The Tinio Brigade: Anti-American Resistance in the Illocos Provinces 1899–1901*. Quezon City, Philippines: New Day Publishers, 1989.

Absorbing account of Manuel Tinio, the youngest general among the Filipino nationalists, and the Philippine–American War in northwestern Luzon. The text of this fine book is supplemented by numerous maps and a large group of illustrations.

220. Owen, Norman G. "Winding Down the War in Albay, 1900–03." *Pacific Historical Review* 48, no. 4 (November 1979): 557–89.

For some time the nationalist movement in Alban, in southeast Luzon, was relatively weak. By 1900, significant resistance developed, however. By mid-1901 fighting ceased for a while, but conflict was reignited, and there was substantial violence between mid-1902 and the end of 1903. Both United States and nationalist troops behaved with restraint in 1900–01, but later campaigns were more brutal. Owen gives considerable attention to the economic situation in Alban and civic action programs of the United States Army, making this a thorough study of the war in one area.

221. Payne, Robert Bruce. "The Philippine War: The Diary of Robert Bruce Payne, 1899," ed. John Hall. *Nebraska History* 69, no. 4 (Winter 1988): 193–98.

Payne, a University of Nebraska student who enlisted in the army, kept a diary from February 4, through February 11, 1899, just at the beginning of the Philippine–American War. He was present at the initial skirmish. Payne was negative toward the United States effort to conquer the Philippines.

222. Ramsey, Robert D. III. *A Masterpiece of Counterguerrilla Warfare: BG J. Franklin Bell in the Philippines, 1901–1902*. Fort Leavenworth, KS: combat Studies Institute Press, 2007. http://purl.access.gpo.gov/GPO/LPS90965

Praises Bell highly, especially for his conduct of the campaign in Batangas Province, thereby echoing judgments by contemporary United States military and political leaders in the Philippines and by present-day historians. A detailed essay about operations in Batangas precedes a collection of documents related to the campaign.

223. ———. *Savage Wars of Peace: Case Studies of Pacification in the Philippines, 1900–1902*. Fort Leavenworth, KS: Combat Studies Institute Press, 2007. http://purl.access.gpo.gov/GPO/LPS104417

Focuses on the pacification of two areas of northern Luzon. The text is supplemented by excellent maps, an extensive chronology, a large bibliography, and copies of several key documents.

224. Reed, John Scott. "Burden and Honor: The United States Volunteers in the Southern Philippines, 1899–1901." Ph.D. dissertation, University of Southern California, 1994.

Examines the experiences and the achievements of the United States Volunteers on Leyte, northern Mindanao, Panay, and Samar. Scott praises the troops and their commanders for becoming effective soldiers who defeated the Filipino nationalists without extensive use of the ruthless tactics that were applied in some areas of the Philippines.

225. Reilly, Margaret Inglehart. "Andrew Wadsworth, Nebraska Soldier in the Philippines, 1898–99." *Nebraska History* 68, no. 4 (Winter 1987): 183–99.

Informative article based on Wadsworth's letters. Like many other United States soldiers, Wadsworth developed an increasingly antagonistic and racist view of the Filipinos. This is a detailed, graphic description of his experiences. There are fine photographs and a good map of the movements of the 1st Nebraska Regiment from December 5, 1898, to April 23, 1899.

226. Scott, Joseph L. *The Ordeal of Samar*. Indianapolis, IN: Bobbs-Merrill, 1965.

Popular, but detailed, account of the bloody attempt to suppress the Filipino nationalist movement on the island of Samar and to avenge a severe defeat that the nationalists had inflicted on a small United States Army garrison at Balangiga. This is an often cited and important account, but it has significant shortcomings. Scott takes a highly condescending view of the Filipinos at the time of the incidents related in his book. There is no real effort made to put the story he presents into its context, moreover. A reader of Scott's book would not learn that fighting continued for years on Samar after the events described, for example.

227. Strobridge, William F. "To San Francisco and Beyond: 51st Iowa Volunteers in 1898." *Annals of Iowa* 41, no. 5 (Summer 1972): 1039–47.

    Details the movement of the 51st Iowa Regiment of United States Volunteers to San Francisco. Most of the article deals with the Iowans' experiences at Camp Merritt.

228. Thomas, Lowell. *Woodfill of the Regulars: A True Story of Adventure from the Arctic to the Argonne.* Chap. IV, "The Battle of Balangiga," pp. 48–65. City, NY: Doubleday, Doran & Company, Inc., 1929.

    Describes the massacre of an infantry company on Samar and the population-concentration methods used by the United States forces to hamper the rebels.

229. Thompson, Henry O. *Inside the Fighting First: Papers of a Nebraska Private in the Philippine War,* ed. Thomas Solevad Nielsen. Blair, NE: Lur, 2001.

    Thompson served in the 1st Regiment of the Nebraska Volunteer Infantry. His letters and diaries from 1898 to 1899 are arranged in several chapters with informative introductions to each chapter. Thompson soon became extremely hostile toward the Filipinos.

230. Wise, H.D. "Notes on Field Service in Samar." *Infantry Journal* 4 (July 1907): 3–58.

    Gives a lucid picture of the fighting on Samar between 1904 and 1906 and of the bands fighting the Americans, including their organization and tactics. An appendix describes jungle traps used by the enemy and United States medical services.

6. Medical aspects

231. Gillett, Mary C. "Medical Care and Evacuation During the Philippine Insurrection, 1899–1901." *Journal of the History of Medicine and Allied Sciences* 42, no. 2 (April 1987): 169–85.

    Discusses the challenges that faced army medical officers during the Philippine–American War, partly because of the presence of diseases for which there were either no remedies or only limited remedies and partly because the war developed into a counterinsurgency that involved small units, which were difficult to provide with suitable medical support. Several hospitals were established, and much of the article deals with the hospital system. As the war continued, additional medical personnel were dispatched to the Philippines.

232. Smallman-Raynor, Matthew and Andrew D. Cliff. "'The Philippines Insurrection and the 1902–4 Cholera Epidemic,' Part 1: Epidemiological Diffusion Processes in War." *Journal of Historical Geography* 24, no. 1 (January 1998): 69–89; "'The Philippines Insurrection and the 1902–4 Cholera Epidemic,' Part 2: Diffusion Patterns in War and Peace." *Journal of Historical Geography* 24, no. 2 (April 1998): 188–210.

Highly detailed investigation of the diffusion of cholera. There were two waves of diffusion (1902–03 and 1903–04), and the authors discuss the effects of the Philippine–American War on patterns within the two waves.

7. Naval operations

233. Tomblin, Barbara B. "The United States Navy and the Philippine Insurrection." *American Neptune* 35, no. 3 (July 1975): 183–96.

Brief discussion based on secondary materials. Nevertheless, it is a useful introduction to this neglected, but important topic, because it draws together information that was scattered through the standard sources used.

234. Williams, Vernon Leon. "The U.S. Navy in the Philippine Insurrection and Subsequent Native Unrest, 1898–1906." Ph.D. diss., Texas A&M University, 1985.

Essential study for understanding the Philippine–American War. The United States Navy played a major role in the conflict by preventing nationalist troop movements and, equally, important, naval supremacy allowed the United States to control much of the economy of the Philippines by imposing blockades. Moreover, Williams shows that naval forces, including marines, fought on land and involved themselves extensively in civic action efforts.

8. Political aspects

235. Beisner, Robert L. *Twelve Against Empire: The Anti-Imperialists, 1898–1900*. Chicago: University of Chicago Press, 1968, 1985.

Focuses on a number of elite figures, most of whom were not political leaders in the sense of holding office. These men were largely dissident Republicans, such as Carl Schurz. The 1985 edition is especially worthwhile because of Beisner's brief but insightful review of his own work, which also refers to relevant studies from the 1970s and early 1980s.

236. Bresnahan, Roger G., comp. *In Time of Hesitation: American Anti-Imperialists and the Philippine–American War.* Quezon City, Philippines: New Day Publishers, 1981.

Much of the book is an anthology of literary works by anti-imperialists, including Mark Twain, Ambrose Bierce, and W.E.B. DuBois. This is also a useful introduction that briefly analyzes the Anti-Imperialist League among other topics.

237. Diokno, Maria Serena I. Chap. 4, "'Benevolent Assimilation' and Filipino Responses," pp. 75–88, in *Mixed Blessing: The Impact of the American Colonial Experience on Politics and Society in the Philippines,* ed. Hazel M. McFerson. Westport, CT: Greenwood Press, 2002.

Despite its brevity, this is an important analysis of one major facet of the Philippine–American War, the differing perspectives of the Filipinos who accepted United States rule without offering much or any resistance and those Filipinos who fought strenuously against the imposition of United States sovereignty. Those who adapted readily to continued foreign rule placed more emphasis on economic development than self-government, whereas the outright nationalists wanted both independence and economic growth. The whole book is required reading for anyone studying the Philippines, but most chapters are too removed from the fighting to be given separate entries in this bibliography. *Mixed Blessing* also has a long and highly useful chronology.

238. Escalante, Rene R. *The Bearer of Pax Americana: The Philippine Career of William H. Taft, 1900–1903.* Quezon City, Philippines: New Day Publishers, 2007.

Remarkably even-handed treatment of Taft and his policies toward the Philippines. Escalante portrays Taft as a man bent on pursuing United States interests yet making decisions that often benefitted the Philippines. The author carefully describes the positive effects of Taft's role, but he notes that "in the final analysis his approach to Philippine society was still elitist" (p. 263), thereby setting a pattern for future relations between the United States and the Philippines, even after the Philippines achieved independence in 1946.

239. Friedel, Frank. "Dissent in the Spanish–American War and the Philippine Insurrection." Chap. 3, pp. 65–97, in Samuel Eliot Morison, Frederick Merk, and Frank Friedel, *Dissent in Three American Wars.* Cambridge, MA: Harvard University Press, 1970.

Although several major books on the topic of opposition in the United States to both these wars have appeared since 1970, Friedel's study remains a useful introduction.

240. Fritz, David Lawrence. "The Philippine Question: American Civil/Military Policy in the Philippines, 1898–1905." Ph.D. diss., University of Texas at Austin, 1977.

Attempts to determine whether United States policies in the Philippines were intended primarily to serve United States or Filipino interests. Fritz stresses relationships between United States civilians and authorities in the Philippines. He asserts that Taft hoped for a permanent association of the Philippines with the United States but eventually abandoned his expectations. The dissertation includes an extended discussion of the atrocity issue.

241. Gates, John M. "Philippine Guerrillas, American Anti-Imperialists, and the Election of 1900." *Pacific Historical Review* 46, no. 1 (February 1977): 51–64.

Focuses on the impact that the campaign waged in the United States against imperialism had on Filipino nationalists. The nationalists were well aware of the activities of the anti-imperialists, and there was some cooperation between the anti-imperialists and the nationalists. As a result, the nationalists became overly optimistic about the possibility of William Jennings Bryan replacing William McKinley as president.

242. Mulrooney, Virginia Frances. "No Victor, No Vanquished: United States Military Government in the Philippines." Ph.D. diss., University of California–Los Angeles, 1977.

Although this dissertation focuses on the military government of the Philippines, it extends to the transition to rule by the Philippine Commission. Mulrooney skillfully discusses the complex implementation of military government and the role of Filipino institutions and personnel. She demonstrates a thorough grasp of the then existing literature about the early years of United States control of the Philippines.

243. Rystad, Goeran. "The Philippine Struggle for Independence and Its Effects on American Expansionism at the Turn of the Century." *Revue Internationale d'Histoire Militaire*, no. 70 (1988): 107–29.

Able summary of the arguments advanced by imperialists and anti-imperialists regarding United States policy toward the Philippines. The Philippine–American War brought home the costs of imperialism, importantly including the financial costs, and thus dampened enthusiasm for further ventures in imperialism.

244. Schirmer, Daniel B. *Republic or Empire: American Resistance to the Philippine War.* Cambridge, MA: Schenkman Publishing Company, 1972.

Wide-ranging examination of the anti-imperialist movement. Schirmer notes that he gave special attention to two newspapers: the *Boston Evening Transcript* and the *Springfield Republican.* He used a number of other sources, however, including manuscripts. Schirmer also provides a brief, but valuable, historiography of anti-imperialism at the time of the War of 1898 and the Philippine–American War.

245. Welch, Richard E., Jr. *Response to Imperialism: the United States and the Philippine–American War, 1899–1902*. Chapel Hill: University of North Carolina Press, 1979.

Highly important analysis of the reaction of major elements of society in the United States to the seizure of the Philippines and the subsequent protracted war. Although Welch provides a good deal of information about the anti-imperialists, his study is not limited to those activists. He looks at business, religious denominations, labor organizations, and newspapers generally. Perhaps of special interest is the diversity of views that existed among Christians in the United States. Welch gives due attention to the atrocities issue and ends with a provocative chapter about "The Problem of [the] Significance" of the war.

9. Racial and gender issues

246. Gatewood, Willard B. *Black Americans and the White Man's Burden, 1898–1903*. Urbana: University of Illinois Press, 1975.

Indispensable study that expands the ways in which the United States intervention in the Philippines has been understood. Gatewood has a good deal to say about the Philippines throughout his outstanding book, but for the purposes of this bibliography, chapter 10, "The Philippines: Black Soldiers and the White Man's Burden," pp. 261–92, is of special importance.

247. Hoganson, Kristin L. *Fighting for American Manhood: How Gender Politics Provoked the Spanish–American and Philippine–American Wars*. New Haven, CT: Yale University Press, 1998.

Taking a wide view of the nature of the conduct of foreign relations, Hoganson asserts that war and expansion came when they did, in part because of the need perceived by many male decision-makers and ordinary men to reassert masculinity in United States society. This view was partially a reaction to the increasing role of women in politics and related discussions of social issues. According to Hoganson, the stress on masculinity was rather short-lived.

248. Robinson, Michael C. "David Fagen: An Afro–American Rebel in the Philippines, 1899–1901." *Pacific Historical Review* 44, no. 1 (January 1975): 68–83.

Uses Fagen's defection from the United States Army to the Filipino nationalists as an example of African–American armed resistance to the racist policies and attitudes prevalent in the United States. Fagen was not the only African–American to join the Filipinos, but he seems to have been the most prominent. He achieved the rank of captain in the nationalist army. The United States Army was concerned about Fagen and searched the Philippines for him. Whether he was eventually killed by Filipinos or survived remains in doubt.

C. The Moro campaigns, 1902–c.1913

Fighting between the Moros, on the one hand, and the United States Army and the Philippine Constabulary, on the other hand, was entirely separate from the Philippine–American War. The Muslims (or Moros) of the southern Philippines were the most determined opponents of United States rule. The area included in the conflict became a testing ground for many well-known United States Army commanders, including John J. Pershing, Tasker Bliss, and Leonard Wood.

Chronology

1899
August 20   The "Bates Treaty" was signed between Brigadier General John C. Bates, United States Army, and a number of Muslim sultans. The Moros accepted United States rule over their area, but they retained a number of rights.

1902
May 2   Battle of Bayang, which followed a period of clashes between United States forces and the Moros. The insistence of some Moros on the retention of slavery was a major source of contention. Substantial numbers of United States troops deployed. Further fighting occurred, and, finally, a large-scale battle occurred. A number of Moros were killed, including many of their leaders. Jacob G. Shurman, who headed the Philippine Commission, thought that United States forces may have provoked the fighting.

1903
September 3   Moro Province, the Philippines, formally established. The United States military governors who ruled the province until 1913 exercised considerable, largely autonomous, authority.

1904
March 7   As the result of increasing tensions in the Moro area and some fighting, the United States unilaterally abrogated the Bates Treaty and thereby strengthened its legal hold over the

southern Philippine Islands. The sultans did not learn of the abrogation of the Bates Treaty until March 21.

1905
May 1–5 Datu Pala and his followers were killed in a series of fights on Jolo Island.

1906
March 6–8 Battle of Bud Daju, Jolo Island. Heavy fighting occurred when United States troops under Colonel Joseph W. Duncan attacked a Moro fort. There were some United States casualties, but the Moros, numbering about 600, were all killed. The battle aroused considerable criticism in the Philippines and the United States. Military governor Leonard Wood believed that the sultan of Sulu and other Moro leaders should have assisted the United States negotiators during talks prior to the attack.

October 22 After years of struggle against the United States military government, Datu Ali was killed at Simpetan, along with some of his followers. Others of his band were killed on October 31.

1909
July 2–4 In a notable battle, Jikiri, a Jolo Island Moro leader, was killed on Patián Island. The United States awarded Medals of Honor to several of its participants.

1911
September 8 Executive Order no. 24 of the Moro Province, by strengthening an earlier directive of the province's Legislative Council, directed the virtual disarmament of the Moros. Most Moros complied with the order, but some threatened resistance. Thousands of firearms were surrendered by late 1913.

1913
June 11–15 Battle of Bud Bagsak, Jolo Island. After several years of continued efforts by United States authorities to disarm the Moros and to destroy their forts, a number of Moros gathered at Bud Bagsak to contest further United States moves. Moro women and children were persuaded to leave the encampment, but the warriors held firm. Like most battles in the Moro areas, Bud Bagsak was a contest in which few United States troops and many Moros perished. The battle fueled further controversy about the U.S.A.'s handling of its responsibilities in the southern Philippines.

December 20   United States military government of Moro Province ended, and Moro Province became the Department of Mindanao and Sulu, thus signaling the greater integration of the area into the Philippines.

1. The American proconsuls

249. Bacevich, Andrew J. "Disagreeable Work: Pacifying the Moros, 1903–6." *Military Review* 62, no. 6 (June 1982): 49–61.

    Award-winning article written at the U.S. Army Command and General Staff College in 1981. The article begins with a few paragraphs about the Philippine–American War and then examines the early years of fighting in the Moro area of the Philippines. Bacevich focuses primarily on military governor Major General Leonard Wood and his attitude toward the Moros and his conduct while in Moroland. Wood disdained the Moros and intended to change their way of life quickly and radically. Much of the article discusses the battle of Bud Daju in 1906, during which many Moros of both genders and all ages were killed.

250. Byler, Charles. "Pacifying the Moros: American Military Government in the Southern Philippines, 1899–1913." *Military Review* 85, no. 3 (May–June 2005): 41–45.

    An incisive analysis of United States Army activities that seeks to draw lessons from the Philippines during the early twentieth century for current encounters with Muslim societies. Initially, under United States rule, the Moros preserved considerable autonomy. Once the non-Muslim areas of the Philippines had been, in part, pacified, the army sought greater control over the Moros. Military governor Leonard Wood's stern approach resulted in numerous Moro deaths. His successors, Tasker H. Bliss and John J. Pershing, attempted to reduce tension, although fighting continued.

251. Hagedorn, Hermann. *Leonard Wood: A Biography*. Vol. II. New York: Harper & Brothers, 1931.

    Argues that Wood began his governorship of Moro Province, 1903 to 1906, by greatly underestimating the problems of dealing with the Moros. Much of the text examines his extensive negotiations with Moro chieftains. Hagedorn includes a fairly detailed study of the battle of Bud Daju.

252. Jornacion, George William. "The Time of the Eagles: United States Army Officers and the Pacification of the Philippine Moros, 1899–1913." Ph.D. diss., University of Maine, 1973.

Asserts that the Americans used their frontier experience to formulate patterns of control over the Moros. Jornacion emphasizes army successes in keeping the peace.

253. Lane, Jack C. *Armed Progressive: General Leonard Wood.* San Rafael, CA: Presidio Press, 1978.

    Argues that Wood placed too much emphasis on military force and engaged in conflicts with the Moros when diplomacy might have achieved more than marches and attacks.

254. Murray, Robert Hammond. "The Pacifier of the Philippines." *World's Work* 16 (October 1908): 10773–78.

    Lauds Wood, especially for his ability to negotiate with the Moros in his capacity as governor of Moro Province. Murray also applauds what he regarded as Wood's strictness in dealing with the Moro chiefs. A good deal of the information in the article pertains to Wood's career generally.

255. O'Connor, Richard. *Black Jack Pershing.* Garden City, NY: Doubleday & Company, Inc., 1961.

    Emphasizes Pershing's devotion to peaceful methods of dealing with the Moros whenever possible. Pershing initially served in the Moro area between 1901 and 1903. Assaults on Moro strongholds at that time are described in detail. Pershing later served as governor of Moro Province between 1911 and 1913.

256. Palmer, Frederick. *Bliss, Peacemaker: The Life and Letters of General Tasker Howard Bliss.* New York: Dodd, Mead, 1934.

    Includes much material on Bliss' period as governor of Moro Province.

257. Smythe, Donald. *Guerrilla Warrior: The Early Life of John J. Pershing.* New York: Charles Scribner's Sons, 1973.

    Examines in detail Pershing's interaction with the Moros between 1901 and 1913. This excellent book, one of the best introductions to the Moro Conflict, rests on extensive use of manuscripts and interviews, as well as published materials. The bibliography covers the Moros in some detail, although they are not given a separate section.

258. Thompson, Wayne Wray. "Governors of the Moro Province: Wood, Bliss, and Pershing in the Southern Philippines, 1903–13." Ph.D. diss., University of California–San Diego, 1975.

    Presents probably the most critical interpretation of the governorships of these generals. They tended to use their positions to enhance their reputations and failed to recognize the differences between United States values and those of the Moros.

259. Vandiver, Frank E. *Black Jack: The Life and Times of John J. Pershing*. Vol. 1. College Station: Texas A&M University Press, 1977.

Focuses more specifically on Pershing the man than Smythe's study (item 257), which is more concerned with Pershing and the Moro Conflict. Nevertheless, Vandiver describes the Moro conflicts in detail, and his book should be used by any serious student.

260. Wood, Eric Fisher. *Leonard Wood: Conservator of Americanism: A Biography*. New York: George H. Doran Company, 1920.

Praises Wood for his extensive preparations for assuming the governorship of the Moro Province. This highly laudatory campaign biography asserts that Wood was a peacemaker in the Philippines. Lane's book above (item 253) should be consulted to balance Wood's account.

2. The campaigns

261. Barry, Richard. "The End of Datto Ali." *Collier's* 37 (June 9, 1906): 17–19.

Profiles Datto Ali's life, and examines Moro military methods. Frank McCoy conducted a successful campaign against Datto Ali, which ended in the latter's death in 1905, thereby ending several years of indecisive fighting.

262. Bateman, C.C. "Military Road Building in Mindanao." *Journal of the Military Service Institution of the United States* 33 (September–October 1903): 190–99.

Includes views of both the military and technical problems. Several attacks on camps and individual soldiers are mentioned. Moros were eventually hired to work on the roads as they grew used to the United States presence.

263. ———. "Military Taming of the Moros." *Journal of the Military Service Institution of the United States* 34 (March–April 1904): 258–66.

Analyzes Moro behavior. Bateman, an intelligence officer, warns against depending too much on Moro disunity. He points out that correspondence plays a major role in Moro military affairs and quotes a number of messages from Moro chieftains.

264. ———. "Progress Among the Moros." *Review of Reviews* 28 (December 1903): 689–93.

Emphasizes road building and the gradual accommodation of some Moro leaders to the military government. Bateman asserts that the governor of Moro Province should continue to be a soldier because civilian leaders will not be respected by the Moros.

265. Brownell, Atherton. "A Moro Experiment." *Outlook* 81 (December 23, 1905): 975–85.

Praises efforts of John P. Finley, an army officer who, as governor of Zamboanga District, was trying to move the Moros from fighting to the peaceful pursuits of agriculture and trade. Much of the article deals with a Moro exchange for trading produce. Brownell visited the area with Taft in the summer of 1905.

266. ———. "Turning Savages into Citizens." *Outlook* 86 (December 24, 1910): 921–31.

Praises Leonard Wood's efforts to pacify the Moros. The article reviews Finley's experiments with Moro exchanges (see item 265) and the great expansion of the program. Brownell also describes an expedition that Finley led against hostile Moros in 1909–10.

267. Bullard, Robert L. "Preparing Our Moros for Government." *Atlantic Monthly* 97 (March 1906): 385–94.

Depicts the chaos existing in Moro areas brought about by the divisions between petty rulers and the anarchism of individuals. Bullard attributes the early attacks on army road-building camps to the Moro belief that the Americans were spreading cholera. Most of the article deals with early contacts between the army and the Moros on Mindanao.

268. ———. "Road Building Among the Moros." *Atlantic Monthly* 92 (December 1903): 818–26.

Predicts early acceptance of United States rule by the Moros. Because of American influence, the role of the chieftains was already declining. Despite the title, Bullard says little about the construction process, beyond noting various methods for recruiting Moro laborers.

269. Chaput, Donald, and Konrad F. Schreier, Jr. "Marines in the Philippines." *Marine Corps Gazette* 59, no. 1 (January 1975): 49–51.

Describes a rebellion that occurred on an island near Mindanao during the early 1920s. Marines were landed from a gunboat to support Philippine Constabulary troops with automatic weapons. The authors regard this as an excellent example of cooperation between the navy, the marines, and the Philippine Constabulary.

270. Collins, James L. "The Battle of Bud Bagsak and the Part Played by the Mountain Guns Therein." *Field Artillery Journal* 15 (November–December 1925): 559–70.

Provides an excellent overview of the battle, with detailed charts of the mountain area that the American forces attacked. Collins emphasizes the great secrecy attending the attack.

271. Crockett, Cary I. "Jungle Warfare," in *The Infantry Journal Reader*, ed. Joseph I. Greene, 393–401. Garden City, NY: Doubleday, Doran & Company, Inc., 1943.

Describes a series of encounters between a detachment of the Philippine Constabulary and a large number of Moros during 1904. Crockett, one of the best-known officers in the constabulary, tells the story in semifictional form, but it is based on one of his own exploits.

272. Davis, Oscar K. "The Moros in Peace and War." *Munsey's Magazine* 27 (August 1902): 787–92.

Warns that conflict with the Moros will be serious. He describes the early fighting on Mindanao, notably Col. Frank D. Baldwin's campaign against Sultan Bayan, and cites General Chaffee's optimistic assessment of security on Mindanao in the wake of Baldwin's expedition.

273. Fleming, Thomas J. "Pershing's Island War." *American Heritage* 19 (August 1968): 32–35, 101–4.

Emphasizes Pershing's initiative and relative lack of racial prejudice as facilitating his peacemaking efforts. The article is primarily concerned with Pershing's early encounters with the Moros in 1902 and 1903.

274. Forbes, W. Cameron. *The Philippine Islands*. 2 volumes. Boston, MA: Houghton Mifflin Company, 1928.

Utilizes, in part, Forbes' experience as governor-general of the Philippines. In a somewhat disconnected way, the books relate much about the work of the United States Army and the Philippine Constabulary in pacifying the islands, especially the Moro areas.

275. Gowing, Peter Gordon. "Mandate in Moroland: The American Government of Muslim Filipinos, 1899–1920." Ph.D. diss., Syracuse University, 1968.

Examines the military governorships of Leonard Wood, Tasker Bliss, and John J. Pershing. Several chapters deal specifically with the United States campaigns against Muslims who revolted against foreign rule.

276. Hitt, Parker. "Amphibious Infantry: A Fleet on Lake Lanao." *United States Naval Institute Proceedings* 64, no. 2 (February 1938): 234–50.

Describes operations by the army against the Moros on Lake Lanao between 1904 and 1907. Small United States naval vessels were used until the army raised several sunken Spanish gunboats and reconditioned them. Several pages recount Spanish military experiences on the lake from the seventeenth to the late nineteenth centuries.

277. Hobbs, Horace P. *Kris and Krag: Adventures Among the Moros of the Southern Philippine Islands*, ed. Horace P. Hobbs, Jr., n.p., 1962.

Recounts conflicts with the Moros on Jolo and Mindanao between 1901 and 1905, based on Hobbs' memories and such sources as the Pershing and Hugh Scott papers. This is a readable and attractively illustrated book, which was issued in only a small printing.

278. "How Jikiri and His Band were Exterminated." *Literary Digest* 45 (December 7 1912): 1095–99.

Quotes at length from an anonymous letter that describes the actions for which four United States soldiers received belated Medals of Honor. Much of the account deals with manhandling guns on to vantage points and the final assault by the Moros when they realized that their situation was hopeless.

279. Howland, H.S. "Field Service in Mindanao." *Infantry Journal* 2 (October 1905): 36–80.

Relates the operations of an experimental company formed to study the problems of patrolling Moro country. Rations and transportation, including the use of Philippine sailing vessels, were the major concerns. The terrain and environment of Mindanao are also examined in some detail.

280. Hurley, Vic. *Swish of the Kris: The Story of the Moros.* New York: E.P. Dutton & Co., 1936.

Illustrates very ably the character of fighting in the Moro areas. Hurley reviews United States campaigns against the Moros between 1903 and 1917. This was probably the most useful general account until the preparation of the dissertations listed in this section of the bibliography.

281. Jenks, Jeremiah W. "The Philippine Constabulary and Its Chief." *Review of Reviews* 26 (October 1902): 436–38.

Provides a useful biographical sketch of Henry T. Allen, who led the constabulary in its early days. The article is based in part on Jenks' study of colonial administration in Southeast Asia.

282. Kolb, Richard K. "Campaign in Moroland: A War the World Forgot." *Army* 33 (September 1983): 50–59.

Provides an excellent introduction to the Moro Conflict. In contrast to most studies, Kolb's article gives due credit to the army, navy, and the constabulary. There is much on living conditions for troops in the wars with the Moros. The text is supplemented by excellent photographs.

283. Millett, Allan R. *The General: Robert L. Bullard and Officership in the United States Army 1881–1925.* Chap. 10, "Mindanao: The Road, the Lake, and the Moros, 1902–1904," pp. 165–186. Westport, CT: Greenwood Press, 1975.

Contends that Bullard sought to negotiate with the Moros rather than fight them as General Leonard Wood did. Bullard, nevertheless, participated in a number of campaigns in Moro areas, and these are described in some detail.

284. Pease, Frank Chester. "The Sentry: A Philippine Incident." *Harper's Weekly* 50 (May 19, 1906): 708–9.

Relates a typical episode in which an inattentive young soldier is killed by a Moro infiltrator. The story is told in considerable detail. The author's point of view is unclear, but he may be sympathetic to the officers he portrays as being enraged with the restrictions placed upon them in dealing with the Moros.

285. Reade, Philip. "The Cargador in Mindanao." *Journal of the Military Service Institution of the United States* 37 (July–August 1905): 114–20.

Describes the Moro carriers or cargadors and their carrying capacity, customs for packing loads, and rations. Reade warns that cargadors must never be armed. His observations derive from his service with a provisional company that carried out experimental field operations on Mindanao in 1904.

286. Rudolph, Jack W. "Capt. Cary Crockett: Jungle Fighter of Samar." *Army* 25 (November 1975): 35–38.

Pays tribute to Crockett for his jungle fighting ability. Much of the article relates to an attack on a fort held by Crockett and a company of Philippine Constabulary. Eventually, Crockett went on a harrowing march to seek relief for his men.

287. Smith, C.C. "The Mindanao Moro." *Cavalry Journal* 17 (October 1906): 287–308.

Warns that it is very difficult to avoid ambushes in the Mindanao jungle. Smith also examines the history of the Moros, emphasizing their fanaticism, and outlines Spanish attempts to pacify Mindanao in the 1880s and 1890s.

288. Smythe, Donald. "Pershing and Counterinsurgency." *Military Review* 46, no. 9 (September 1966): 85–92.

Uses the battle of Bud Bagsak to demonstrate United States capability for counterinsurgency operations. Much of the article deals with Pershing's efforts to reduce casualties among the women and children at Bud Bagsak. Smythe emphasizes the ferocity of the Moro attacks on American lines.

289. ———. "Pershing and the Disarmament of the Moros." *Pacific Historical Review* 31 (August 1962): 241–56.

Examines Moro behavior patterns that made disarmament an attractive policy for Pershing. Leonard Wood opposed disarmament, believing it would never work. The account stops at 1911, when many, but not all, Moros turned in their weapons.

290. "Soldiering with Pershing in the Philippines." *Literary Digest* 99 (November 17, 1928): 52, 54.

Quotes from Sergeant Lanckton's account of his service in the Philippines as General Pershing's orderly. He praises Pershing for his ability to win over Moro chieftains and his courage in the face of hostile Moros.

291. Tang, Samuel Kong. "The Muslim Armed Struggle in the Philippines, 1900–41." Ph.D. diss., Syracuse University, 1973.

Analyzes the sociological, psychological, and religious roots of conflict between the Muslims of the Philippines and other groups, not simply the American rulers of the islands. Tang limits his discussion of military operations to Mindanao.

292. Townsend, Henry Schuler. "Civil Government in Moro Province." *Forum* 36 (July 1904): 138–49.

Emphasizes the diversity of people and languages among the Moros and discusses several Moro leaders. He concludes with a warning that the authorities should not try to "Americanize" the Moros too quickly.

293. White, John Roberts. *Bullets and Bolos: Fifteen Years in the Philippine Islands.* New York: Century Co., 1928.

Focuses on White's service in the Philippine Constabulary from its beginning up to 1906, when elements of the force fought at the battle of Bud Daju. Much on the law enforcement functions of the constabulary is included in White's account.

V
Boxer Uprising, 1898–1901

When Chinese elements who were anti-foreign (the Boxers) began attacking the legations in Beijing, United States forces helped defend them. Marines had been posted in Beijing as early as 1898. Other Americans participated in efforts to relieve Beijing. The first "allied" forces (the Seymour column) failed to reach the city. When reinforcements arrived, a successful advance on Beijing took place. United States troops in the China Relief Expedition were commanded by General Adna R. Chaffee.

United States participation in the allied force was one of the least controversial of the small wars of the United States. The relief expedition was also one of the very few military efforts in which United States forces cooperated with foreign troops before the era of the World Wars.

Chronology

1899	After the defeat by Japan in the Sino–Japanese War of 1894–95 and continued interference in China by various European powers, an anti-foreign movement began in North China.
1900	
April–May	Boxer groups appeared near Beijing and Tianjin. When tension increased foreign powers planned to expand their military forces at Beijing to protect their legations.
June 10	An international force that was commanded by British admiral Edward H. Seymour began advancing from Tianjin to Beijing, but it encountered enough resistance to make further progress impossible. By June 23 the "Seymour column" had returned to Tianjin.

June 17	Foreign ships shelled and then captured the Dagu forts, and foreigners at Tianjin were besieged.
June 20	European, Japanese, and United States legations were besieged.
June 21	China issued what amounted to a declaration of war on the foreign powers, although the language used was ambiguous, thereby permitting many Chinese officials to work against the Boxers.
August	A strengthened foreign military force relieved the legations in Beijing.
1901 September	The Boxer Protocols between China and the foreign powers severely punished the former. Among other provisions were heavy indemnities and the establishment of foreign military forces in China for many years.

A. Published documentary collections and official reports

294. "Correspondence Relating to China Relief Expedition May 30, 1900 to June 1, 1901." In U.S. Army. Adjutant-General's Office. *Correspondence Relating to the War with Spain and Conditions Growing out of Same ... Between Adjutant-General and Military Commanders ... Apr. 15, 1898–July 30, 1902.* Vol. 1. Washington, DC: Government Printing Office, 1902.

    Contains full texts of messages in chronological order. Some items from the State and Navy departments are included, especially for the period before the China Relief Expedition arrived on the China coast.

295. U.S. War Department. *Annual Reports of the War Department for the Fiscal Year Ended June 30, 1901: Report of the Lieutenant-General Commanding the Army*, Part 4. Washington, DC: Government Printing Office, 1901.

    Includes reports from General Chaffee and other officers of the China Relief Expedition. General orders and circulars of the expedition are reproduced in full.

296. U.S. War Department. Adjutant General's Office. *Notes on China August, 1900.* Doc. no. 124 M.I.D. Washington, DC: Government Printing Office, 1900.

    Discusses the government, climate, and other facets of China, including the Chinese and foreign armed forces in China. Several pages describe the route taken by the allies between the coast and Beijing for the relief of the legations.

297. ———. *Reports on Military Operations in South Africa and China.* Document no. 143. M.I.D. Washington, DC: Government Printing Office, 1901.

Section on China is primarily excerpts from Chaffee's report. Most of the report describes the allied army. Of special interest is a report on the siege of the legations and a chronology of events, August–November, 1900.

B. General surveys

298. Agnew, James B. "Coalition Warfare—Relieving the Peking Legations, 1900." *Military Review* 56, no. 10 (October 1976): 58–70.

Focuses on decision making within the allied army. Agnew deals extensively with Chaffee's role and praises the general unstintingly for his handling of United States forces in China.

299. Boot, Max. *The Savage Wars of Peace: Small Wars and the Rise of American Power.* Chap. 4, "Red Summer: Boxer Rebellion, 1900," pp. 69–98. New York, Basic Books, 2003.

Competent summary that emphasizes the disunity of the allied forces. Boot probably devotes too much attention to the experiences and exploits of Smedley D. Butler than is warranted, thereby somewhat shortchanging other aspects of the story.

300. Braisted, William Reynolds. "The Navy and the Boxers." Chap. 2, pp. 75–114, in *The United States Navy in the Pacific, 1897–1909.* Austin: University of Texas Press, 1958.

Emphasizes the diplomatic and political aspects of naval involvement in the Boxer Uprising. This is an extremely important study of United States military history and foreign policy development.

301. Carter, William Harding. *The Life of Lieutenant General Chaffee.* Chicago: University of Chicago Press, 1917.

Includes the texts of appropriate orders and messages to and from Chaffee, who commanded the China Relief Expedition, chapters XXII–XXVI. Most of the account deals with occupation duty in Beijing and the difficulties of inter-allied cooperation. There is a detailed chapter on the taking of Beijing, but there is little on the skirmishing after its fall.

302. Daggett, Aaron S. *America in the China Relief Expedition.* Kansas City: Hudson-Kimberley Publishing Company, 1903.

Remains the most authoritative work on the United States role in suppressing the Boxer Uprising. Very little is said about the defense of the legations. The attack on Beijing is described in great detail. About a third of the text consists of appendices, many of them reports on fire fights with the Boxers.

303. Davidson, W.C. "Operations in North China." *United States Naval Institute Proceedings* 26 (December 1900): 637–46.

    Describes the attack on the Dagu forts and various other operations. It does not stress United States actions, but it is an important contemporary account. The article includes a number of detailed maps of North China.

304. Dupuy, R. Ernest, and William H. Baumer. Chap. 4, "The China Relief Expedition, 1900," pp. 100–122 in *The Little Wars of the United States*. New York: Hawthorn Books, Inc., 1968.

    Sketches the background of the revolt and reviews the events of the siege in Beijing and the various relief efforts. This account is based mainly on secondary materials.

305. Esherick, Joseph. *The Origins of the Boxer Rebellion*. Berkeley: University of California Press, 1987.

    Essential for understanding the complex movement. Esherick examines the political, social, and economic environment in Shandong, where the Boxers originated. After discussing the various interpretations of the Boxers, the author identifies "the popular culture of the west Shandong peasants" (p. 321) as the key factor in the origin and evolution of the Boxers and their ideology.

306. Hirschfeld, Burt. *Fifty-Five Days of Terror: The Story of the Boxer Rebellion*. New York: Julian Messner, Inc., 1964.

    Provides a useful overview of the Boxer Uprising, although the primary focus is on the siege of the legations. This is a highly readable account, with a brief bibliography.

307. Huidekoper, Frederic Louis. *The Military Unpreparedness of the United States: A History of American Land Forces from Colonial Times until June 1, 1915*. New York: Macmillan Company, 1915.

    Examines the major events of the Boxer Uprising and the China Relief Expedition, pp. 242–52. Huidekoper stresses the difficulty of supplying the United States troops in China.

308. Keown-Boyd, Henry. *The Fists of Righteous Harmony: A History of the Boxer Uprising in China in 1900*. London: Leo Cooper, 1991.

    Popular but detailed account that emphasizes the experiences of the people besieged in the legation area of Peiping. There is a brief chapter that discusses the development of the Boxer movement, but the book is useful largely for the legation story. Although a number of archives are mentioned it is difficult to determine how much of the book represents original research, owing to the lack of documentation. The maps and illustrations are outstanding.

309. Preston, Diana. *The Boxer Rebellion: The Dramatic Story of China's War on Foreigners That Shook the World in the Summer of 1900.* New York: Walker, 2000.

Solidly researched and documented account, based in part on archival materials from Great Britain, the United States, and Australia. This is a well-rounded and interesting book with three sections of important illustrations, and there is an extensive bibliography. Published in Great Britain by Constable in 1999 as *Besieged in Peking: The Story of the 1900 Boxer Rising.*

310. Sharf, Frederic Alan. *China, 1900: The Eyewitnesses Speak: The Boxer Rebellion as Described by Participants in Letters, Diaries and Photographs.* Mechanicsburg, PA: Stackpole Books, 2000.

Includes several accounts, including diaries, by United States military participants of their experiences in various theaters of the Boxer conflict.

311. Waite, Carleton Frederick. *Some Elements of International Military Co-Operation in the Suppression of the 1900 Antiforeign Rising in China with Special Reference to the Forces of the United States.* Los Angeles: University of Southern California Press, 1935.

Uses the documentary record skillfully to demonstrate the degree of cooperation existing at given times in the campaign. The United States forces worked rather closely with other foreign troops.

312. Wilson, James Henry. Chap. 18, "The Boxer War," pp. 517–43, in *Under the Old Flag: Recollections of Military Operations in the War for the Union, the Spanish War, the Boxer Rebellion, etc.* 2 volumes. New York: D. Appleton and Company, 1912. Reprinted. Westport, CT: Greenwood Publishing, 1971.

Narrates Wilson's experiences as second in command of the China Relief Expedition. His major combat duty was with a force that went to the Eight Temples in September, 1900.

313. Xiang, Lanxin. *The Origins of the Boxer War: A Multinational Study.* London: RoutledgeCurzon, 2003.

Remarkable contribution to the literature about the Boxer Uprising. This is a truly "multinational study" owing to Xiang's mastery of most of the many languages in which materials related to the episode exist and to his skill in applying his seemingly comprehensive knowledge. The bibliography is highly impressive and is of considerable value on its own. Xiang's preface presents a brief, but important, historiography of the Boxer Uprising.

C. The defense of the legations

314. Bainbridge, Mrs. William E. "Diary of an Iowan under Fire in Peking." *Annals of Iowa* 36 (1963): 613–40.

    Chronicles the events of the siege between May 6 and August 17, 1900, including defense preparations and combat contacts. Mrs. Bainbridge was the wife of the second secretary of the United States legation.

315. Hoe, Susanna. *Women at the Siege, Peking 1900*. Oxford: The Women's History Press–HOLO Books, 2000.

    Pioneering examination of the experiences of the women from China, Europe, Japan, and the United States who endured the siege. Until the publication of this book, the important role of the women had been largely ignored. Fortunately, they did leave a substantial written record that Hoe successfully utilized.

316. Lindsey, J.R. "Report on the Siege of Peking." In *Reports on Military Operations in South Africa and China*. Document no. 143. M.I.D. Washington, DC: Government Printing Office, 1901.

    Examines the origins of the revolt and describes the defense efforts by foreigners in the legations. The report includes two very precise charts of the legation area and its defenses.

317. Lowry, Mrs. E.K. "A Woman's Diary of the Siege of Peking." *McClure's Magazine* 16 (November 1900): 65–76.

    Consists of a diary from May 28 to August 14, 1900. It discusses the fortification of the Methodist mission and the defense of the British legation where foreigners gathered. The narrative is supplemented by several detailed plans.

318. Myers, John T. "Military Operations and Defenses of the Siege of Peking." *United States Naval Institute Proceedings* 28, no. 9 (September 1902): 541–51.

    Emphasizes that there was little formal fighting during the siege, thereby providing a useful corrective to other accounts. Myers includes a list of casualties by nationality and a map of the legation area.

319. Payen, Cecile E. "Besieged in Peking: The Diary of a Visitor at the United States Legation." *Century* 61 (January 1901): 453–68.

Constitutes probably the most interesting of the diaries kept during the siege. Payen discusses the role of the women defenders, who made and filled sandbags. There is quite a bit of information about the activities of the marines in the garrison, including a brief account of the attack on the Boxers, July 3, 1900.

320. Pethick, W.N. "The Struggle on the Peking Wall: An Episode of the Siege of the Legations." *Century* 61 (December 1900): 308–13.

Provides one of the most analytical discussions of the defense. Pethick assigns much blame to the defenders for poor planning at times and for panic at other times. This criticism is directed in part at the marines and especially at Captain Hall, although he is not named.

321. Woodward, Mrs. M.S. "The Personal Side of the Siege of Peking." *Independent* 52 (November 22, 1900): 2782–91.

Records Mrs. Woodward's experiences as a nurse when her visit to China was interrupted by the uprising. She describes the fighting in Beijing and the evacuation of the civilians after the siege was lifted. The article includes a number of photographs of United States Marine Corps officers and enlisted men.

D. The Seymour Expedition

322. Taussig, J.K. "Experiences During the Boxer Rebellion." *United States Naval Institute Proceedings* 53, no. 4 (April 1927): 403–20.

Focuses on the initial advance and subsequent retreat of Seymour's force. The article is based on a journal Taussig prepared while recuperating from wounds received during the campaign. Two maps and illustrations add to the value of this often cited article.

323. Wurtsbaugh, Daniel W. "The Seymour Relief Expedition." *United States Naval Institute Proceedings* 28, no. 6 (June 1902): 207–19.

Furnishes a very detailed, readable account of the United States naval contingent with the Seymour column between May 27 and June 28, 1900.

E. The attack on Tianjin

324. Butler, Smedley D. *General Smedley Darlington Butler: The Letters of a Leatherneck, 1898–1931*, Chap. 2, "The Happiest Man Alive," pp. 13–29, ed. Anne Cipriano Venzon. New York: Praeger, 1992.

Venzon provides a concise survey of the initial stages of the Boxer Uprising and then presents several of Butler's letters. Butler was delighted to be commanding a company going into combat in China.

His subsequent letters offer a description of the fighting at Tianjin. Venzon has thoroughly annotated each letter, helping to make the brief group of letters an important primary source.

325. Leonard, Henry. "The Visit of the Allies to China in 1900." *Journal of the Military Service Institution of the United States* 29 (July 1901): 40–55.

    Provides a graphic description of the fighting at Tianjin and the advance of the Seymour column. The Tianjin fortifications are discussed in detail. Leonard was a marine officer.

F. Subsequent fighting in the Tianjin area

326. Forsyth, William W. "American Cavalry in China." *Cavalry Journal* 14 (July 1903): 5–20.

    Recounts experiences of dismounted cavalry left at Tianjin to await arrival of their horses. Fighting developed when Chinese troops moved into the Tianjin area in the absence of the main body of allied troops. Forsyth also touches upon patrol operations in the Beijing area and the cavalry equipment used in China.

G. The advance to Beijing

327. Bevan, James A. "With the Marines on the March to Peking, China—1900." Parts 1, 2. *Leatherneck* 18, no. 6, no. 7 (June, July, 1935): 5–7, 55–56; 14–15, 50.

    Provides a vivid eyewitness account of the suffering endured by the troops along the line of march. Although the advance is discussed thoroughly, there is little on the attack. There is some material on life in Beijing during 1900 and 1901, after the city's capture.

328. Cabell, De R.C. "Troop 'M' Sixth Cavalry in the Chinese Relief Expedition of 1900." *Cavalry Journal* 15 (July 1904): 48–71.

    Includes much background information on the advance, along with data on the transport capabilities of the allies. Cabell briefly describes the battle of Beijing.

329. Davis, Oscar King. "Reporting a Cosmopolitan War." Parts 1, 2, 3. *Harper's Weekly* (July 27, August 3, 10, 1901): 748–49; 772; 796.

    Stresses the problems of American war correspondents, especially the lack of any recognized status with the United States armed forces. The third installment is particularly useful for its account of the march to Beijing.

330. ——. "The Taking of Peking." *Harper's Weekly* 44 (November 3, 1900): 1036–37.

Emphasizes the hardships of the march, especially for the United States troops who were last. Several paragraphs give a full account of the attack. The final section of the article deals with the situation of the legation garrison in terms of food and manpower on the eve of the relief. Food supplies were reasonably good, but there was a dangerous lack of fighting men.

331. Dickman, Joseph T. "Experiences in China." *Cavalry Journal* 13 (July 1902): 5–40.

Describes the legation area after the capture of Beijing, the allies march into the Imperial City to symbolize their victory, and skirmishes in the Beijing area. Dickman did not reach China until mid-August, 1900, when most of the fighting was over, but his account, nevertheless, is worthwhile.

332. Palmer, Frederick. "With the Peking Relief Column: A Correspondent's Notes of the Fighting in China." *Century* 61 (December 1900): 302–7.

Consists of a diary, July 13–14, August 10–11, 15, 20, 1900, recording the Tianjin and Beijing assaults. Palmer is contemptuous of the Boxers and respectful of the Japanese. Interestingly, he notes the existence of disharmony between the relieving forces and the legation garrison.

333. Rhodes, Charles D. "China Notes (1900)." *Cavalry Journal* 13 (April 1903): 580–615.

Consists of diary entries from July to November, 1900. The journal began with the voyage of a ship transporting cavalry horses. The heavily illustrated article discusses a number of combat patrols near Tianjin and Beijing between August and October, 1900.

H. The role of the marines

334. Bevan, James. "From Filipinos to Boxers in 1900." *Leatherneck* 18 (April 1935): 5–7, 65–66.

Tells the story of the marines dispatched from the battlefields of the Philippines to China. Bevan describes his service at the battle of Tianjin in which he helped manhandle a gun into place. The heavy fighting and United States losses are noted. The article concludes with a discussion of preparations for the march to Beijing.

335. Biggs, Chester M., Jr. *The United States Marines in North China, 1894–1942.* Jefferson, NC: McFarland & Company, Inc., 2003.

Despite the title, the book is largely devoted to the story of the marines during the Boxer Uprising, both in the defense of the legations in Peiping and in the relief forces. Biggs follows the activities of the marines in considerable detail and supplements the text with many unusual photographs and a number of maps. The maps are a bit rough by academic standards, but they are clear and they are essential. There is a long chronology.

336. "Boxer Rebellion Recalled ... " *Marine Corps Gazette* 84, no. 6 (June 2000): 56–57.

    Brief but interesting and well-illustrated account.

337. Butler, Smedley D. *Old Gimlet Eye: The Adventures of Smedley D. Butler As Told to Lowell Thomas.* New York: Farrar & Rinehart, Inc., Publishers, 1933.

    Recounts the attack on Tianjin and the advance to Beijing in terms of Butler's personal experiences, chapters V–VII.

338. Clark, George B. *Treading Softly: U.S. Marines in China, 1819–1949.* Chap. 2, "The Boxer Rebellion," pp. 19–45. Westport, CT: Praeger, 2001.

    Fine treatment of marine corps operations. Clark offers careful characterizations of the principal marine officers on the scene. In addition to giving due attention to the achievements of the marines, mentioning many individuals by name, the author discusses the fighting prowess of the Chinese soldiers and irregulars (the "Boxers").

339. Fenn, Courtenay Hughes. "The American Marines in the Siege of Peking." *Independent* 52 (November 29, 1900): 2845–49.

    Praises the marines as the most important element of the legation defense force. They began by bringing many Christian Chinese behind the walls and ended with a vigorous attack on the Boxers to support the relief column as it neared the legations.

340. Hargreaves, Reginald. "Comrades in Arms." *Marine Corps Gazette* 48, no. 10 (October 1964): 50–55.

    Reviews the close cooperation between the United States and Royal Marines during the siege. Much of the article concerns an Anglo–American–Russian attack on a barricade that had been temporarily lost to the Boxers. A short section deals with the Seymour Expedition.

341. Heinl, Robert Debs, Jr. "Hell in China." *Marine Corps Gazette* 39 (November 1959): 55–68.

    Constitutes an expanded version of the section in Heinl's book, *Soldiers of the Sea* (item 342), on the Boxer uprising. The article deals briefly with the defense of the legations and emphasizes the relief of Beijing. There is relatively little on the attack on Tianjin.

342. ———. *Soldiers of the Sea: The United States Marine Corps, 1775–1962.* Annapolis, MD: United States Naval Institute, 1962.

Gives a lucid discussion of marine operations during the campaign, pages 127–46. Of special value is the examination of marine Captain Hall's alleged lack of fighting spirit at Beijing. A number of quotations are taken from young Smedley Butler's letters.

343. Holbrook, Francis X. "Translated: Brave Hearts and Bright Weapons." *Marine Corps Gazette* 57, no. 11 (November 1973): 56–65.

Deals with initial marine landings at Tianjin to prepare the way for relieving the legations. The article includes material on the Seymour column and a number of interesting photographs.

344. Jaunal, Jack W. "Bill Horton and the Boxer Rebellion." *Marine Corps Gazette* 57, no. 11 (November 1973): 29–32.

Tells the story of Sergeant Horton, who won the Medal of Honor during the siege of the legations. The article is based on Horton's own account in 1964. He replaced the United States flag three times when it was shot away over the legation.

345. Metcalf, Clyde H. *A History of the United States Marine Corps.* New York: G.P. Putnam's Sons, 1939.

Reviews very briefly the Chinese situation in 1900, and then describes the fighting at Beijing and the relief of the legations, pages 279–85. Another brief section, pages 307–9, discusses marine operations in China between 1901 and 1912.

346. Miller, J. Michael. "Battle Atop Tartar Wall: Boxer Rebellion," *Leatherneck* 83, no. 8 (August 2000): 28–36.

Relates the heroism that won Daniel Joseph Daly the first of his two awards of the Congressional Medal of Honor. Daly served in the marines through World War I and displayed his courage repeatedly.

347. ———. "Marines in the Boxer Rebellion," *Leatherneck* 83, no. 6 (June 2000): 48–54.

Discusses the accomplishments of the marines who formed an important segment of the troops who successfully defended the foreign legations for nearly two months before allied forces reached them.

348. ———. "Rescue the Legations—Boxer Rebellion," *Leatherneck* 83, no. 7 (July 2000): 40–47.

Reviews the various efforts made to relieve the legations and the significant part the marines played in them.

349. Neville, Edwin L., Jr., comp. "The Diary of Private Sullivan." *Marine Corps Gazette* 52, no. 11 (November 1968): 68–74.

Testifies vigorously to the high morale of marines in the Boxer Uprising. Much of the diary refers to the attack on Tianjin, in which Sullivan participated. Sullivan's death terminated the diary.

350. Plante, Trevor K. "U.S. Marines in the Boxer Rebellion," *Prologue* 31, no. 4 (Winter 1999): 284–89.

Succinct, but valuable, examination of the marines' service in China, both at the legations and with the relief forces. The missionaries who experienced the legation siege, for example, formulated a resolution praising the marines' courage and military skill. No fewer than 33 marines received the Congressional Medal of Honor.

351. Rentflow, Frank Hunt. "'In Many a Strife We've Fought for Life.'" *Leatherneck* 14 (July 1931): 10–11, 47–49.

Discusses the experiences of the marines who were rushed to Beijing just before the siege began and the situation that led to this reinforcement of the legation guard forces. Rentflow credits the United States Marines with primary responsibility for holding the legations.

352. Schmidt, Hans. *Maverick Marine: General Smedley D. Butler and the Contradictions of American Military History.* Chap. 3, "The Teenage Hero: China," pp. 14–26. Lexington: University of Kentucky, 1987.

In the course of narrating the achievements of Butler and the marines, Schmidt provides a solid overview of the Boxer Uprising. The chapter is firmly based on the extensive use of archival materials.

I. Medical aspects of the China Relief Expedition

353. Banister, W.B. "Surgical Notes on the China Relief Expedition." *Cavalry Journal* 13 (April 1903): 616–35.

Describes the medical staff and medical supply procedures. Banister compares patterns of wounds in China, the War of 1898, Philippine–American War, Civil War, and South African War.

354. Seaman, Louis Livingston. "Observations in China and the Tropics of the Army Ration and the Post Exchange or Canteen." *Journal of the Military Institution of the United States* 29 (September 1901): 289–93.

Argues that a tropic ration is needed and that unless liquor is provided by the canteen, soldiers will procure dangerous, even poisonous, drinks.

J. Military government

355. Hunt, Michael H. "The Forgotten Occupation: Peking, 1900–1901." *Pacific Historical Review* 48, no. 4 (November 1979): 501–29.

    Depicts the occupation as an effective operation, based on Chinese records and the archives of the China Relief Expedition. Much of the article deals with Chinese who worked with the Americans. Hunt notes that General Chaffee generally rejected the harsh posture toward the Chinese demanded by missionaries and businessmen.

VI
China, 1901–41

A United States military and naval presence lasted for decades after the Boxer Rebellion. Marines served as legation guards at Beijing as well as on the coast. The army participated in China duty from 1901 to 1905 and again from 1912 to 1938. The navy maintained gunboats on Chinese rivers and a large squadron along the coast.

The gunboat crews and detachments of naval personnel serving as armed guards on United States merchant vessels saw most of the action, but the marines and army, nevertheless, were always aware of the potential for violence.

Bandits were a danger for much of the period. At times, the Nationalists or the Communists were very menacing. Finally, in the 1930s clashes with the Japanese became more and more of a possibility. China service was always a challenge for the armed forces.

Chronology

1901–5   A company of army infantry provided security for the United States legation in Beijing in the wake of the Boxer Uprising.

1905–41 Marines replaced the soldiers as the legation guard. Until 1911, the only other United States forces in China consisted of marines; so logistical and command reasons dictated the placement of marines at the legation.

1911
November Because of the unrest associated with the Chinese Revolution, the United States dispatched elements of the army's 15th Infantry Regiment to Tianjin. The rest of the regiment arrived

	early in 1912. The regiment had already been at Tianjin for a time during the Boxer Uprising.
1919 December	Although United States warships had been in Chinese waters for many years, the Yangzi Patrol was only established at this time. It consisted of a number of gunboats that cruised the larger rivers of China to protect United States interests.
1927–28	Because the older vessels were wearing out, six new gunboats were constructed in Shanghai and put into service.
1927 March 24–25	United States warships cooperated with British and Japanese naval vessels in briefly bombarding Chinese Nationalist forces at Nanjing. The "Nanjing Incident" remains controversial.
March 31	The first elements of the marines' West Coast Expeditionary Force (3rd Marine Brigade) disembarked at Shanghai to be followed by other components. The purpose of the build-up was to be ready for possible unrest in China in the face of the increasing strength of the Chinese Nationalists.
1928	During the second half of 1928 the 3rd Marine Brigade was drawn down substantially, and early in 1929 it was abolished. Marines remained at Tianjin, however.
1932 February 4	The United States Army's 31st Infantry Regiment arrived at Shanghai to reinforce marines during a period of sharp fighting in the area between Chinese and Japanese forces.
July 5	After the fighting between the Chinese and Japanese had ceased, the 31st Infantry Regiment was sent back to the Philippines.
1937 December 12	The USS *Panay* was sunk by Japanese aircraft as the gunboat was escorting some Standard Oil Company vessels during the Sino–Japanese War. Japan apologized for the error and made restitution.
1938 March 2	The continuing Sino–Japanese War ended the usefulness of the 15th Infantry Regiment as a deterrent, and so it was transferred to Fort Lewis, Washington. Marines were still stationed at Beijing and Tianjin for a time.

1941

November — Because of the increasing tension between the United States and Japan, the marines at Tianjin were transferred to the Philippines. The legation guard continued its service, however.

December 8 — The gunboat USS *Wake* was captured at Shanghai by Japanese forces. Three of the remaining gunboats had sailed to the Philippines shortly before. They were sunk in May 1942.

December 8 — In the face of the overwhelming strength of the Japanese in the area, the legation guard at Beijing surrendered without fighting. Attempts to repatriate the marines on the ground that they were diplomatic personnel failed, and they remained prisoners of the Japanese until the end of World War II.

A. General studies

356. Boot, Max. *The Savage Wars of Peace: Small Wars and the Rise of American Power.* Chap. 11, "By Bluff Alone: China, 1901–41," pp. 253–78. New York: Basic Books, 2002.

    Provides background information that puts the varied United States military involvements in China into perspective. Boot discusses the naval (the Yangzi Patrol), marine, and army presences in some detail. A lengthy section dealing with Smedley D. Butler is probably misplaced in this chapter.

357. Carlson, Evans F. "Legal Bases for Use of Foreign Armed Forces in China." *United States Naval Institute Proceedings* 62, no. 11 (November 1936): 1544–56.

    Reviews various pertinent treaties and cites a number of instances in which foreign troops intervened in China. Several pages focus specifically on the Yangzi Patrol and the legation guards of the United States.

358. Cole, Bernard D. *Gunboats and Marines: The United States Navy in China, 1925–1928.* Newark: University of Delaware Press, 1983.

    Examines the operations of the Yangzi Patrol in detail during its greatest crisis and the landing of marines under Smedley Butler. This is the most important study of United States naval involvement in China during the 1920s.

359. Morton, Louis. "Army and Marines on the China Station: A Study in Military and Political Rivalry." *Pacific Historical Review* 29, no. 1 (February 1960): 51–73.

    Traces the conflicts that arose because the army and marines had independent missions, but were both subject to State Department control. Morton gives a brief description of the kinds of quasi-combat situations the army and marines faced.

360. Noble, Dennis L. *The Eagle and the Dragon: The United States Military in China, 1901–1937.* New York: Greenwood Press, 1990.

    Deals with army, navy, and marine forces stationed in China until the outbreak of the Sino–Japanese War, when conditions changed dramatically and the army contingent was ordered home. This book is largely a study in social history. Noble demonstrates that the combat experiences of the army, navy, and marine personnel were quite limited during their China duty.

361. Smith, Roy C. "The Protection of American Nationals in China." *United States Naval Institute Proceedings* 56, no. 12 (December 1930): 1097–1104.

    Defends use of naval power on the ground that China cannot afford the requisite security. Various incidents are used to illustrate facets of this function. Smith describes the kinds of property to be protected and discusses the plans for protecting foreigners at Nanjing in 1927.

B. Army service in China

362. Boehnke, E.C. "Mascot of the 'Dragon' Regiment." *Infantry Journal* 27 (September 1925): 324–26.

    Tells the story of "Bill Howitzer," a Chinese boy who was adopted by the 15th Infantry at Tongshan. At the time this article was published, he was about 11 years old and assisted the troops by running errands and interpreting.

363. Cahill, Howard F.K. "The Thirty-First Infantry in Shanghai." *Infantry Journal* 39 (May–June 1932): 165–75.

    Describes the temporary service of this regiment during the fighting between China and Japan in 1932. Its role was limited to security duties in the International Settlement. Three charts show its position on the defense perimeter, and a number of photographs are included.

364. Coffman, Edward M. "The American 15th Infantry Regiment in China, 1912–38." *Journal of Military History* 58, no. 1 (January 1994): 57–74.

    Thorough overview of the regiment's service in China. Coffman's research included extensive use of interviews with former members of the regiment. This was an elite force within the United States Army with a unique history.

365. "Conditions of Service in China." *Infantry Journal* 29 (August 1926): 167–74.

Emphasizes that the 15th Infantry is sometimes exposed to combat situations, especially along the rail line from Tianjin to Beijing. Much of the article covers living conditions. The author is less confident about the cheapness of service in China than most other observers.

366. Cornebise, Alfred E. *The United States 15th Infantry Regiment in China, 1912–1938*. Jefferson, NC: McFarland & Co., 2004.

Thoroughly researched study of the experiences of members of this force. The regiment's story is less well known than the adventures of the marines in China during approximately the same period, but its stationing was a major commitment of the United States to having a voice in the Far East. A number of officers who became generals later helped lead the regiment at various times, making the story of this service all the more important. The force even had its own newspaper, and Cornebise devotes a chapter to that enterprise.

367. Finney, Charles G. *The Old China Hands*. Garden City, N.Y.: Doubleday & Company, Inc., 1961.

Takes the form of thinly disguised fact in short stories. They are based on the author's service with the 15th Infantry in 1927–29, as well as on correspondence with former officers and some source materials. Chapter 10 describes relations between the soldiers and the 4,000 marines who landed at Tianjin in 1928.

368. "An Officer of the China Expeditionary Forces." "On the China Coast." *Infantry Journal* 19 (November 1921): 502–7.

Reviews army involvement in China after 1912. Living conditions for servicemen are discussed in some detail. Most important is the section on the impact of fighting in 1920 between rival Chinese factions on United States troops.

369. Tuchman, Barbara W. *Stillwell and the American Experience in China, 1911–45*. Chap. 5, "The 'Can Do' Regiment and the Rise of Chiang Kai-shek, 1926–29." New York: Macmillan Company, 1970 (reissued Bantam, 1984).

Relates Stillwell's experience as a battalion commander in the 15th Infantry. Much of the chapter deals with the response of United States forces to the Nationalist march to the north to destroy the warlords.

370. Wheeler, W.R. "China Service." *Journal of the Military Service Institution of the United States* 59 (July–August, September–October, and November–December, 1916): 80–93, 234–50, 378–93.

Includes a historical treatment of European military commitments in China from the First Opium War onward and describes other foreign units in China. Wheeler examines closely the assignment of United States troops to China in 1912. He includes a short section on living conditions for United States servicemen.

371. Williams, Laurin L. "East Meets West." *Infantry Journal* 27 (August 1925): 157–61.

Reviews the defense of Tianjin by the 15th Infantry in the fall of 1924, when raids by former Chinese soldiers became a concern. The inhabitants in the United States sector erected a monument in Tianjin to the American defense.

C. The Yangzi Patrol

372. Borg, Dorothy. *American Policy and the Chinese Revolution 1925–1928*. New York: American Institute of Pacific Relations, 1947.

Discusses the anti-foreign movement and the measures Western powers and Japan took to protect their nationals. Several chapters deal with the "Nanking Incident," in which United States gunboats shelled Chinese Nationalist forces.

373. Braisted, William Reynolds. *The United States Navy in the Pacific, 1909–1922*. Austin: University of Texas Press, 1971.

Reviews the internment of American gunboats by Chinese authorities between April and August, 1917, when China entered World War I, pages 313–16.

374. Ellis, E.M. "New Gunboat Construction at Shanghai." *United States Naval Institute Proceedings* 55, no. 9 (September 1929): 783–85.

Begins with an analysis of the shortcomings of the old vessels and the attractive features of the new ships. The armament and composition of the crews of the gunboats under construction are also discussed.

375. Gale, Esson M. "The Yangtze Patrol." *United States Naval Institute Proceedings* 81, no. 3 (March 1955): 307–15.

Provides an overview of the history of the Yangzi Patrol. The first section deals extensively with living conditions and social life in pre–World War II China. Gale describes several incidents, primarily from the 1920s, in which gunboats gave protection to foreign nationals. He also briefly mentions the Boxer Uprising and the *Panay* attack.

376. Gardner, K.N. "The Beginning of the Yangtze River Campaign of 1926–27." *United States Naval Institute Proceedings* 58, no. 1 (January 1932): 40–44.

Deals with operations of the destroyer *Stewart* while convoying merchant ships. The heart of the article is a description of the vigorous exchange of fire with Chinese guns on September 10, 1926. The *Stewart* was also fired upon near Hankou.

377. Howell, Glenn. "Ascent of the Min." *United States Naval Institute Proceedings* 65, no. 5 (May 1939): 709–13.

Chronicles the efforts of the *Palos* to ascend the Min, which had not been visited by a United States gunboat for years. Relatively little was known about the Min in 1920, but the navy persisted. The result was a tense voyage against the advice of the Chinese pilot. The vessel went 1,635 miles north and west of Shanghai.

378. ———. "Chungking to Ichang." *United States Naval Institute Proceedings* 64 (September 1938): 1312–16.

Discusses the persistent engineering problems of the old river gunboat *Palos* in 1920. At one time, the *Palos* had to travel with a Standard Oil vessel for possible assistance in case its engines broke down.

379. ———. *Gunboat on the Yangtze: The Diary of Captain Glenn F. Howell of the USS* Palos, *1920–1921*, ed. Dennis L. Noble. Jefferson, NC: McFarland & Co., 2002.

This diary is part of the monumental collection of records Howell left of his naval service. Noble provides a biography of Howell and briefly discusses developments in China up to the time the diary was written and foreign military involvements in that country. *Palos* did not see combat in 1920–21, but this is an important record of gunboat life.

380. ———. "Hwang Tsao." *United States Naval Institute Proceedings* 64 (August 1938): 1151–55.

Describes the limitations of the old river gunboats *Palos* and *Monocacy* in passing gorges. Draft problems prevented the engines from developing enough power. The article deals with such an episode in 1921. The skill of the Chinese pilot finally brought the vessel safely through the gorge.

381. ———. "Operations of the United States Navy on the Yangtze River—September, 1926, to June, 1927." *United States Naval Institute Proceedings* 54, no. 4 (April 1928): 273–86.

Describes the growth in anti-foreign sentiment during the mid-1920s. Howell provides a day-by-day account of fighting between United States ships and Nationalist forces and mentions the various landing parties that were used. There is a long section on the battle at Nanjing in March, 1927.

382. Johnson, Felix L. "Naval Activities on the Yangtse." *United States Naval Institute Proceedings* 53 (April 1927): 506–14.

Includes descriptions of the major ports on the Yangzi. Three maps illustrate the areas discussed. The article also describes the older river gunboats. There is an extensive, unpaged section of photographs of typical river scenes in the same issue.

383. Pfaff, Roy. "Sea Duty on the Yangtze." *United States Naval Institute Proceedings* 59, no. 11 (November 1933): 1612–24.

Emphasizes navigational difficulties on the river and the special kind of seamanship required for river sailing. Pfaff also discusses patrol activities, defensive capabilities of the gunboats, and the organization of landing parties.

384. Pineau, Roger. "U.S.S. *Noa* and the Fall of Nanking." *United States Naval Institute Proceedings* 81, no. 11 (November 1955): 1221–28.

Praises Lieutenant Commander Roy C. Smith, captain of the destroyer *Noa*, for his resolute action in protecting foreign residents from the Nationalists. Pineau stresses Communist influence on the Nationalists. Much of the article deals with the background of the crisis at Nanjing.

385. Smith, Roy C., Jr. "Nanking, March 24, 1927." *United States Naval Institute Proceedings* 54, no.1 (January 1928): 1–21.

Relates the evacuation of refugees by the destroyer *Noa* and the shelling of Socony Hill to protect foreigners. The article includes the texts of many messages concerning the evacuation and the shelling. Smith emphasizes the close cooperation between the Americans and British.

386. Sutliff, R.C. "Duty in a Yangtze Gunboat." *United States Naval Institute Proceedings* 61, no. 7 (July 1935): 981–84.

Concentrates on the duties of the officers of a gunboat in China and the peculiar challenges they face. A long section deals with navigational problems, and a shorter one discusses armed guard duty on United States merchant ships. Pages 973–80 contain photographs of many American gunboats.

387. Tolley, Kemp. "YangPat–Shanghai to Chungking." *United States Naval Institute Proceedings* 89, no. 6 (June 1963): 80–94.

Summarizes the history of the Yangzi Patrol from the quiet of the late nineteenth century to the constant pressures of the 1920s and 1930s. Much of the article focuses on the *Panay* attack and the end of the gunboat era in 1937–41.

388. ———. *Yangtze Patrol: The U.S. Navy in China.* Annapolis, MD: Naval Institute Press, 1971.

Constitutes the most complete history of the force. Tolley's book is written in an anecdotal style and is only partially documented. The book goes back to the earliest United States presence on the China coast and follows the last gunboats to their capture or destruction in China and in the Philippines. There is an extensive bibliography.

389. Villere, Paul, Jr. "My Yangtze Armed Guard Detail." *Leatherneck* 17, nos. 8, 9 (August, September 1934): 3–5, 47–49; 5–7, 47–48.

Describes the pirates and trigger-happy Chinese soldiers that made the armed guard details for United States merchant ships necessary. Villere was one of four marines attached to a United States lighter going between Shanghai and Yizhang. The author describes the scenery on the voyage and the appearance of Yizhang. For a time he was billeted on the gunboat *Panay*.

390. Wharton, Wallace S. "Our Chinese Navy." *United States Naval Institute Proceedings* 51, no. 1 (January 1925): 68–77.

Cites a number of instances in which United States merchant vessels and gunboats were fired upon by Chinese troops. Wharton includes a brief description of the older gunboats and their limitations, quoting from a lengthy letter from Secretary of State Hughes on the need for new gunboats.

391. Winslow, Cameron McR., Jr. "Action on the Yangtze." *United States Naval Institute Proceedings* 63, no. 4 (April 1937): 491–98.

Narrates Winslow's experiences with a navy armed guard detachment on a United States steamer sailing from Chongqing to Yizhang. Chinese troops fired on the ship several times, and the author was wounded.

D. The *Panay* incident

392. Grover, David H. "The 'Panay' Revisited: A Maritime Perspective." *American Neptune* 50, no. 4 (Fall 1990): 260–69.

Discusses the circumstances surrounding the attack, noting that the Japanese were attacking what they perceived as a group of ships evacuating Chinese Nationalist troops, not a convoy of United States merchant vessels protected by a United States warship.

393. Koginos, Manny T. *The Panay Incident: Prelude to War*. Lafayette, IN: Purdue University Press, 1967.

Deals primarily with the background of the attack and its political effects. One chapter discusses the Ludlow Amendment to the Constitution that was intended to make United States entrance into another war more difficult. There is an extensive bibliography.

394. Perry, Hamilton Darby. *The Panay Incident: Prelude to Pearl Harbor.* New York: Macmillan Company, 1969.

Focuses almost exclusively on the sinking of the *Panay*, in contrast to Koginos' study (item 393). It is a popularly written, undocumented but, nevertheless, thorough treatment based on material in the Naval History Division, contacts with *Panay* survivors, and research in Japan.

395. Swanson, H.J. "The *Panay* Incident: Prelude to Pearl Harbor." *United States Naval Institute Proceedings* 93, no. 12 (December 1967): 26–37.

Begins with the failure of the *Panay* to complete a radio transmission, the first sign to the outside world that the gunboat had encountered difficulties. The author emphasizes, quite rightly, that the incident was quickly forgotten.

E. The end of the gunboat era

396. Clay, James P. "Pearl River Log: A Different Navy, a Different World." *United States Naval Institute Proceedings* 96, no. 9 (September 1970): 58–67.

Discusses the gunboat *Mindanao*'s rescue mission to Canton during the Japanese attack in 1938. Several pages deal with the occupation of Canton. There is also much about life in the old China squadron.

397. Gulliver, Louis J. "The Yangtze U.S. Gunboats." *United States Naval Institute Proceedings* 68 (September 1942): 1285–87.

Reviews the loss of three gunboats from the Yangzi Patrol in the Philippines, the capture of the *Wake* at Shanghai after Pearl Harbor, and the *Tutuila*'s service at Chongqing before being turned over to the Chinese in March 1942. About half the article deals with the *Panay* affair.

398. Lederer, W.J. "The American Navy is in the Middle of China." *United States Naval Institute Proceedings* 68 (August 1942): 1142–54.

Describes the mission of the gunboat *Tutuila* in providing radio service for the United States embassy at Chongqing from 1938 on. Lederer reminds readers that it was the last Western gunboat in Chinese waters. The article is essentially a human interest story, dealing in part with an air raid on Chongqing.

399. McCracken, Alan R. "Canton Flower Boat." *United States Naval Institute Proceedings* 78, no. 3 (March 1952): 369–77.

Presents a nostalgic picture of life in the Yangzi Patrol. The final section of the article tells of the preparations that McCracken, captain of the gunboat *Mindanao*, took when the vessel was ordered to leave China for the Philippines.

400. Sheehan, J.M. "From the Side Lines." *United States Naval Institute Proceedings* 65, no. 1 (January 1939): 33–37.

   Depicts the challenges of river navigation as United States gunboats stood watch over Americans potentially threatened by disorder and military operations. Several pages discuss the plight of Chinese refugees seeking to escape the Japanese. Sheehan sees the fighting and suffering in China, a result of the China Incident, as simply a continuation of the woes of Chinese history.

401. ———. "Nanking." *United States Naval Institute Proceedings* 69, no. 9 (September 1942): 1189–95.

   Describes life on the eve of the China Incident and provides a detailed description of the devastation in Nanjing when the Japanese attacked. Sheehan was on the gunboat *Oahu* when it gathered United States refugees during the Nanjing fighting.

F. Other naval operations in China

402. Roberts, Stephen S. "The Decline of the Overseas Station Fleets: The United States Asiatic Fleet and the Shanghai Crisis, 1932." *American Neptune* 37 (July 1977): 185–202.

   Analyzes movements of the Asiatic fleet in reaction to the Japanese attack on Shanghai. The naval command was not alarmed initially by Japanese moves. In any case, the fleet could not protect United States interests against the powerful Japanese, although it could control Chinese rioters.

G. Marine operations in China

403. Biggs, Chester M., Jr. *The United States Marines in North China, 1894–1942*. Chap. 10, "Marines Return to Peking," pp. 137–67; chap. 11, "The End of an Era—The Old China Marine," pp. 168–209. Jefferson, NC: McFarland & Company, Inc., 2003.

   Although the emphasis of the book is on the marines' experiences in the Boxer Uprising, these two long chapters provide much information about the later years, from 1901 through the removal of most of the marines in 1941 and the surrender of the others on December 8, 1941. Biggs was one of the marines who was forced to surrender to the Japanese. His account benefits from his personal knowledge of the last period of pre–World War II marine involvement in China.

The inclusion of many photographs, most of them from Biggs' collection, greatly enhances the value of the book.

404. Buckner, David N. *A Brief History of the 10th Marines.* Washington, DC: Headquarters, U.S. Marine Corps, History and Museums Division, 1981.

Focuses on human interest aspects of the marine artillery in China, 1927–28, pages 34–39.

405. Burton, F. "News from the Shanghai Front." *Leatherneck* 15, no. 4 (April 1932): 13–14.

Explains defense procedures of the marines and discusses the danger of their situation in Shanghai.

406. Butler, Smedley D. "American Marines in China." *Annals* 144 (July 1929): 128–34.

Emphasizes the neutrality of United States forces in the Chinese civil war and the lack of political guidance given the marines by American political leaders. Butler commanded the marines in China during the crises of the late 1920s.

407. ———. *General Smedley Darlington Butler: The Letters of a Leatherneck, 1898–1931.* Chap. 13, "Diplomat in China," pp. 259–94, ed. Anne Cipriano Venzon. New York: Praeger, 1992.

Venzon provides a fine summary of the events that led to the strengthening of the marine force in China in the late 1920s and then offers a number of Butler's letters. Butler had to deal with a tense and highly complex situation, but he accomplished his mission successfully. Both the Japanese and Chinese gave him unusual honors for his services. Venzon includes detailed notes that identify people mentioned in the letters and that provide other needed information.

408. ———. "Notes on Staff Activities of Third Brigade, U.S. Marine Corps, Serving in China." *Marine Corps Gazette* 14, no. 6 (June 1929): 95–108.

Gives a readable, level-headed critique of the organizational problems of expeditionary forces. The article consists of excerpts from Butler's final report for the Third Brigade.

409. ———. *Old Gimlet Eye: The Adventures of Smedley D. Butler as Told to Lowell Thomas.* Chap. 23, "Treading Softly in China," pp. 287–98. New York: Farrar & Rinehart, Inc., 1933.

Emphasizes the successful efforts of the marines to remain neutral in the Chinese civil war. While the marines had to be ready for every eventuality, their activity was limited to road-building, parades, and fire-fighting. Butler makes some comparisons between China in the 1920s and at the time of the Boxer Uprising.

410. "The Butler Bridge at Peichang." *Leatherneck* 12 (January 1929): 6–7, 43.

Describes the ceremony that took place when the marines turned over a bridge that they had built to the Chinese.

411. Carlson, Evans Fordyce. "The Fourth Regiment's 'Little Colonel.'" *Leatherneck* 11, no. 12 (December 1928): 7, 55.

Pays tribute to Mrs. James L. Underhill, an officer's wife who took a special interest in the welfare of enlisted men.

412. ———. "Marines as an Aid to Diplomacy." *Marine Corps Gazette* 20, no. 2 (February 1936): 27–30, 47–53.

Covers marine landings in a cursory way. Much of the article deals with Chinese history. Carlson asserts that the marines are used to aid diplomatic moves because of their availability as part of the fleet.

413. "China Side 1927." *Marine Corps Gazette* 33, no. 4 (April 1949): 46–52.

Includes a lengthy description of the marine landing at Tianjin in 1927.

414. Clark, George B. *Treading Softly: U.S. Marines in China, 1819–1949.* Chap. 3, "Early Twentieth Century, 1905–29," pp. 47–82; chap. 4, "The China Marines, 1930–41," pp. 83–130. Westport, CT: Praeger, 2001.

Excellent survey of the unusual services and experiences of marines stationed in China. Although he does not neglect the "human interest" facets of the marine story, Clark provides full descriptions of the numerous serious incidents that faced the marines. The book includes precise maps that are essential to an understanding of what occurred in North China. There is also a fascinating section of contemporary photographs, many of them reproduced from Clark's personal collection.

415. Greene, Wallace M. "The Employment of the Marine Rifle Company in Street Riot Operations." *Marine Corps Gazette* 24 (March 1940): 49–53, 62.

Examines the organization and training of China marines for riot duty in 1937–38.

416. "History of the Fourth Marines in China." *Leatherneck* 15, no. 4 (April 1932): 20–21.

Recalls the role of the Fourth Marines in defending Shanghai in the late 1920s. Of particular interest is the brief story of the capture of a Standard Oil tug by marines when it was seized by Chinese soldiers.

417. Johnston, Lucius F. "Shanghai Revisited." *Leatherneck* 29, no. 8 (August 1946): 4–7.

Discusses customs, life, and liberties during the years just before World War II as experienced by marine enlisted men.

418. Kengla, W.A. "Mission in Tientsin." *Marine Corps Gazette* 28, no. 10 (October 1944): 57–61.

Describes marine relief efforts during a flood near Tianjin in 1939. Kengla reviews his relations with the Japanese during these operations.

419. Lasswell, A. Bryan. "A Shanghai Incident." *Marine Corps Gazette* 64, no. 9 (September 1980): 54–57.

Relates a remarkable series of events in which Lasswell and another marine arrested 16 Japanese military policemen in 1940 because of their intrusion into the United States sector of Shanghai.

420. Leventhal, Robert M. "China Marine." *Marine Corps Gazette* 56, no. 11 (November 1972): 36–43.

Tells the story of marine service in China between the World Wars through a biographical sketch of Albert C. Marts, a senior marine noncommissioned officer. The text is supplemented by a number of contemporary photographs.

421. ——. "Shanghai Duty 1937–38: How Bittersweet It Was." *United States Naval Institute Proceedings* 100, no. 11 (November 1974): 79–91.

Consists primarily of photographs taken by Albert C. Marts, a marine noncommissioned officer, during the late 1930s. The pictures show marine activities, including military maneuvers, guard duty, and ceremonial functions. A few photographs relate to the emergency in Shanghai in 1932.

422. "Living Conditions in China: Circular of Information Relative to Tientsin, China." *Marine Corps Gazette* 13, no. 3 (March 1928): 67–71.

Emphasizes the cost of food, fuel, water, electricity, and other daily needs. There is also a section on living costs at Shanghai.

423. "Living with the Marines in Peking, China." *Leatherneck* 7 (January 1, 1924): 2, 5.

Describes briefly training and the physical surroundings of the legation guards.

424. McClellan, Edwin N. "That Mysterious 'Peking Mutiny.'" *Marine Corps Gazette* 20, no. 2 (February 1936): 41–42.

Suggests that Yuan-Shih-Kai may have instigated mutinies because he wanted to be inaugurated president of China in Peiping instead of in Nanjing. Marines and soldiers were alerted in February–March 1912 because of the Chinese mutinies.

425. ———. "With the Marines in China from 1841 to 1924." *Leatherneck* 7 (January 30, 1924): 1–2.

   Focuses on marine operations between 1905, when the marines began guarding the legation in Peiping, and 1911–12, when additional units went to China during the revolution.

426. McHugh, James M. "Living Conditions in the Marine Corps: The Marine Detachment, American Legation, Peking, China." *Marine Corps Gazette* 12, no. 6 (June 1927): 110–14.

   Describes the detachment's voyage to China and the shopping opportunities and housing the marines found there.

427. Millett, Allan R. *Semper Fidelis: The History of the United States Marine Corps.* Rev. ed. Chap. 8, "The Marines in China, 1905–41," pp. 212–35. New York: Free Press, 1991.

   Stresses the crises of the late 1920s when the marines reached peak strength in China. Millett provides useful background data that enable the reader to see marine activities in their full political and military perspective. His work is based in good part on manuscript sources.

428. Neufeld, Gabrielle M., and James S. Santelli. "Smedley Butler's Air Corps: The First Marine Aviators in China." *United States Naval Institute Proceedings* 103, no. 4 (April 1977): 48–59.

   Reviews the use of marine planes to watch the contending Chinese warlord armies. The primary thrust of the article is that Butler fought successfully to have marine flyers used effectively, despite the opposition of local navy commanders.

429. "Peiping's Mounted Marines Disband." *Leatherneck* 21 (April 1938): 7, 64.

   Tells the story of the detachment from 1909 to 1938. It was abolished when some marines went to Tianjin to replace the 15th Infantry Regiment that had been withdrawn by the army. The force saw little combat action. The article concludes with a description of the horses used. The story is taken from the *Baltimore Evening Sun*.

430. Rothwell, Richard D. "Shanghai Emergency." *Marine Corps Gazette* 56, no. 11 (November 1972): 44–53.

   Analyzes marine operations at Shanghai in 1937 as a model of the employment of military power without actually fighting. Troop movements and peacekeeping activities are discussed at length. In addition, Rothwell provides a lucid description of the developing crisis in Shanghai.

431. Schmidt, Hans. *Maverick Marine: General Smedley D. Butler and the Contradictions of American Military History.* Chap. 13, "The Marines

Who Wouldn't Fight: China," pp. 173–201. Lexington: University Press of Kentucky, 1987.

The chapter title is misleading, because the marines in China during the late 1920s were under strict orders not to become involved in the Chinese civil war. Nevertheless, this chapter is a fine discussion of the highly complex situation that challenged the marines. Schmidt includes much information about the major United States and British personalities at the time. He also discusses Japanese activity, which concerned Butler and other United States leaders.

432. Shoup, David M. *The Marines in China, 1927–1928: The China Expedition Which Turned Out to be the China Exhibition: A Contemporaneous Journal*, ed. by Howard Jablon. Hamden, CT: Archon Books, 1987.

Recently commissioned David M. Shoup recorded his impressions of China. Shoup eventually became commandant of the United States Marine Corps.

433. Tallent, Robert W. "The Fourth Marines." *Leatherneck* 36, no. 2 (February 1953): 16–21, 77.

Details the earlier history of the regiment, but concentrates on its service in China from 1927 to World War II. A number of illustrations depict marine activities during the Sino–Japanese War, 1937–41.

434. "The 10,000 Men Umbrella." *Leatherneck* 11, no. 12 (December 1928): 8–9.

Explains the changed view of the Chinese people toward the presence of the marines in their country. Initially, there was suspicion, based in part on their experiences during the Boxer conflict. Eventually, United States discipline and concern for the Chinese brought about a formal declaration by some Chinese leaders that they appreciated the peaceful behavior of the marines. This was symbolized by the presentation of a ceremonial umbrella to General Smedley Butler.

435. U.S. Marine Corps Headquarters. Division of Operations and Training. "Protection of American Interests." *Marine Corps Gazette* 12, no. 9 (September 1927): 175–83.

Provides a detailed listing of marine units sent to China.

436. Waters, Fleming A. "With the Marines During the Recent Outbreak in China." *Leatherneck* 8 (September 25, 1925): 6–7, 20.

Describes the very tense atmosphere encountered by the marines in China during the spring of 1925. The article includes quotations from Waters' diary.

437. Williams, Robert Hugh. *The Old Corps: A Portrait of the U.S. Marine Corps Between the Wars.* Annapolis, MD: Naval Institute Press, 1982.

Details everyday life among the marines on China service, chapters IV–V. Williams' account is heavily illustrated. He emphasizes duty at Shanghai between 1932 and 1935, but also includes a description of his duties as commander of an armed guard detachment aboard a United States merchant ship on the Yangzi.

VII

Cuba, 1906–09

When the United States military government ended in 1902, Cuba was required to agree that the United States could reoccupy the country whenever there was a breakdown in public order. This was the famous Platt Amendment that continued in force until 1934.

A civil war led to the "Second Intervention" in 1906. A provisional government was established, and steps were taken to move toward the restoration of a democratic political system.

After a new election in 1908, the provisional government was dissolved, and the last American troops departed in 1909.

While there was no fighting, the United States Army was much occupied with reorganizing the Rural Guard and the Cuban Army.

Chronology

1906

August	Liberals revolted against the Moderate government that had manipulated the 1905 elections, hoping to bring about a United States intervention that would satisfy their demands. Moderates also hoped for such an intervention.
September 8	Cuban president Estrada Palma made a formal request for United States intervention. President Theodore Roosevelt hesitated to send United States forces to Cuba, other than some warships which did not land troops.
September 28	After the failure of mediation efforts by the United States, Estrada Palma resigned the Cuban presidency. Secretary of War William Howard Taft was then appointed provisional governor.

October 6	The Army of Cuban Intervention, consisting of United States troops, later termed the Army of Cuban Pacification, arrived in Cuba.
October 13	Charles E. Magoon, who had been governor of the Panama Canal Zone, was appointed governor. Efforts were undertaken to restore and improve the economic situation in Cuba.

1908

April 4	The Armed Forces of Cuba were formed, constituting an army separate from Guardia Rural, a paramilitary police organization, which continued in existence. In 1906, the Guardia had proved itself unable to cope with widespread unrest.
May 25	Regional elections were held.
November 14	National elections brought a sweeping victory to the Liberals.

1909

January 28	Governor Magoon surrendered his powers to the duly elected Cuban government and left Cuba immediately.
April 1	The last elements of the Army of Cuban Pacification left the country.

A. The Platt Amendment

438. Foner, Philip S. *The Spanish–Cuban–American War and the Birth of American Imperialism 1895–1902*. Vol. II: 1898–1902. New York: Monthly Review Press, 1972.

    Chronicles the origins of the Platt Amendment, chapters XXV–XXVIII. Foner interprets it as a device to protect United States investments in Cuba. The Cuban Constitutional Convention opposed the principle until further pressure was exerted by the United States.

439. Munro, Dana G. *Intervention and Dollar Diplomacy in the Caribbean 1900–1921*. Princeton, NJ: Princeton University Press, 1964.

    Provides another view of the background of the Platt Amendment, which was the basis for several United States military interventions in Cuba.

440. Pérez, Louis A., Jr. *Cuba Under the Platt Amendment, 1902–1934*. Pittsburgh, PA: University of Pittsburgh Press, 1986.

    The book does not say much about the intervention between 1906 and 1909, instead focusing on events after 1909. Nevertheless, this is the definitive work about the Platt Amendment by the leading scholar for Cuba–United States relations.

B. General studies

441. Callcott, Wilfrid Hardy. *The Caribbean Policy of the United States, 1890–1920*, pages 230–40. Baltimore, MD: Johns Hopkins University Press, 1942.

    Summarizes the work of the Second Intervention. Callcott bases his study in part on primary materials.

442. Chapman, Charles E. *A History of the Cuban Republic: A Study in Hispanic American Politics*. New York: Macmillan Company, 1927.

    Supplies a relatively detailed account of the Second Intervention, chapters IX–XI. The author says little about the military aspects. His research included a number of interviews, as well as more traditional sources.

443. Fitzgibbon, Russell H. "Intervention by the United States." Chap. 5 in *Cuba and the United States*. New York: Russell & Russell, 1935, 1964.

    Reviews, at some length, the events that led to intervention. Fitzgibbon strongly defends Taft and Roosevelt against criticism. He briefly describes the disarming of insurgent troops and the retraining of the Cuban Army by United States forces.

444. Hill, Howard C. "Cuba and Intervention." Chap. 4, pp. 69–106, in *Roosevelt and the Caribbean*. Chicago: University of Chicago Press, 1927.

    Focuses on the negotiations that led to the provisional government in 1906. Hill takes an extremely favorable view of the intervention, believing that the Cubans made great strides toward democracy between 1906 and 1909. A long preliminary section of the chapter discusses the Platt Amendment.

445. Howland, Harold J. "Saving a People from Themselves: Impressions of Cuba under American Intervention." *Outlook* 84 (October 27, 1906): 454–64.

    Praises the peaceful nature of the occupation and the diplomatic stance of the United States authorities in Cuba. Howland provides a very early look at the intervention regime. He warns that disarmament is far short of what it should be.

446. Jenks, Leland Hamilton. *Our Cuban Colony: A Study in Sugar*. Chap. 6, "Second Intervention," pp. 85–103. New York: Vanguard Press, 1928.

    Asserts that annexation was a possibility in 1906, but Roosevelt and Taft averted it by the intervention. Jenks discusses, in some detail, the ultimately unsuccessful efforts of the provisional government to buy loyalty through the generous use of patronage.

447. Johnston, Harry. "An Englishman's Impressions of American Rule in Cuba." *McClure's Magazine* 33 (September 1909): 496–504.

Expresses warm approval for the achievements of the Americans in Cuba. Johnston discusses civic action operations such as the digging of wells and the construction of highways. A large part of the article deals with the place of blacks in Cuban society.

448. Langley, Lester D. *The Banana Wars: United States Intervention in the Caribbean, 1898–1934*. Chap. 3, "The Second Cuban Intervention, 1906," pp. 27–35. Wilmington, DE: SR Books, 2002.

Emphasizes the reactions of Generals Robert L. Bullard and Frederick Funston to the Cubans and the problem of promoting good government in Latin America, chapters 3–4. Langley takes a dim view of the Liberals and, generally, of the competence of Cuban political parties.

449. Lockmiller, David A. *Magoon in Cuba: A History of the Second Intervention, 1906–1909*. Chapel Hill: University of North Carolina Press, 1938.

Defends Charles E. Magoon, the United States governor, against Cuban attacks. This is a well-documented political and economic study. It includes a brief discussion of the Army of Cuban Pacification, which constituted the American occupation force.

450. Millett, Allan Reed. *The Politics of Intervention: The Military Occupation of Cuba, 1906–1909*. Columbus: Ohio State University Press, 1968.

Provides the best account of the intervention for students of military affairs. Chapters IV and VIII deal with the role of the Army of Cuban Pacification and the reorganization of the Cuban Army, respectively.

C. Military studies

451. Bullard, Robert L. "The Army in Cuba." *Journal of the Military Service Institution of the United States* 41 (September–October 1907): 152–57.

Emphasizes that as a force in being the American Army of Cuban Pacification played an important role in the intervention. He also notes the extensive efforts made by the army to map Cuba, to be ready in case of trouble.

452. Hanna, Matthew Elting. "The Necessity of Increasing the Efficiency of the Cuban Army." *Journal of the Military Service Institution of the United States* 35 (July–August 1904): 28–36.

Describes the Cuban Army and the Rural Guard and their shortcomings. As the United States military attaché in Cuba, Hanna had direct knowledge of these forces.

453. Inglis, William. "How the 'Warlike' Cubans Gave Up Their Arms." *Harper's Weekly* 50 (November 3, 1906): 1564–66.

Describes the disarmament of warring factions by United States Marines. Inglis stresses the poor armament of the Cubans. He notes that the army is on the way, but he gives full credit for peacemaking in Cuba to the marines.

454. Metcalf, Clyde H. "Interventions in Cuba under the Platt Amendment." Chap. 12, pp. 312–38, in *A History of the United States Marine Corps*. New York: G.P. Putnam's Sons, 1939.

Contains detailed, but undocumented, accounts of marine participation in the intervention of 1906–09, marine landings in 1912, and marine security duty between 1917 and 1922. Most of the chapter deals with the years 1906–09.

455. Millett, Allan R. "The General Staff and the Cuban Intervention of 1906." *Military Affairs* 31 (Fall 1967): 113–19.

Analyzes planning for the intervention, which began as early as 1905, and the continuing involvement of the General Staff in operations of the Army of Cuban Pacification. Not only did it plan, but it also carefully supervised the execution of its plans.

456. Pérez, Louis A., Jr. Chap. 2, "Genesis of the Cuban Army (II), 1906–9," pp. 21–28, in *Army Politics in Cuba, 1898–1958*. Pittsburgh, PA: University of Pittsburgh, 1976.

Considers the reorganization of the Rural Guard and the constitution of an army. The provisional government initially opposed formation of an army but gave way when it was realized that the Rural Guard, while useful for protecting property, could not discourage revolutionary movements.

457. U.S. War Department. Office of the Chief of Staff. Second Section, General Staff, no. 16. *Military Notes on Cuba 1909*. War Department Document 345. Washington, DC: Government Printing Office, 1909.

Covers all aspects of Cuba, including geography, population, and economic and governmental information. The individual provinces are described in detail. The information was collected by the Military Information Division, Army of Cuban Pacification, between 1906 and 1908.

458. ———. *Road Notes Cuba—1909*. War Department Document 349. Washington, DC: Government Printing Office, 1909.

Provides an extremely detailed description of the roads, including mileages. The survey is based on notes prepared by the Army of Cuban Pacification in 1906–07.

459. Wieczorek, George A. "Field Wireless Operations in Cuba." *Cavalry Journal* 17 (January 1907): 526–29.

Reviews the organization of the first wireless unit of the United States Army as part of the Army of Cuban Pacification.

D. Other activities

460. Lockmiller, David A. "The Advisory Law Commission of Cuba." *Hispanic–American Historical Review* 17 (February 1937): 2–29.

Reviews the work of the commission, which included both Cuban and United States members, between 1906 and 1909. The critical task for the commission was creation of an election law to prevent disputes of the sort that led to the intervention of 1906. After completing this assignment, the commission turned to other legal matters. Lockmiller believes it made a long-lasting contribution.

VIII

Nicaragua, 1909–25

Marines landed at various Nicaraguan ports during the nineteenth century. Other landings occurred in 1909 and 1910 as the United States became more interested in the possibility of another canal being built through Nicaragua.

In 1912, naval forces, on what was then a large scale, became directly involved with Nicaraguan affairs as they intervened in the current civil war. The Liberal rebels were decisively defeated by the marines.

To maintain stability, the United States stationed approximately 100 marines in the Nicaraguan capital of Managua to symbolize continuing American concern with the country. While unrest continued, the presence of the legation guard until 1925 was, and still is, widely regarded as having prevented social upheaval in Nicaragua for more than a decade.

Chronology

1909

October — Revolt led by Juan J. Estrada, governor of the Department of Zelaya, began against President José Santos Zelaya, who had incurred the ill will of the United States.

December 1 — The United States ended diplomatic relations with Nicaragua after two of its citizens who had been officers in Estrada's army were killed by Zelaya's forces.

1910

May 31 — Marines landed at Bluefields, Nicaragua, to protect United States citizens and their property, but the real reason for the landing was to establish a safe haven for Estrada's army, which was losing the civil war.

August 20	José Madriz, a follower of Zelaya, who had assumed the presidency, resigned. Various political maneuvers ensued for some time.
1911	
January 1	The United States recognized the new Estrada government, in which Adolfo Díaz served as vice-president. Estrada did not retain the presidency for long. In a few months, he was forced from office by Luis Mena, who headed the army, and Díaz became president.
1912	New revolts broke out and United States forces landed to support President Adolfo Díaz in a highly complicated political situation.
July	Mena brought troops to Managua to bolster his bid for the presidency. Skirmishes between his men and government forces ceased temporarily on July 29 as the result of a United States-mediated agreement. Mena continued his quest for the presidency, however.
August 4	A detachment of marines that had been dispatched to support Díaz, arrived at Managua.
August 11–14	Despite the presence of the marines, Díaz's foes shelled Managua, and the civil war intensified. Additional United States forces moved inland from Bluefields and skirmished from time to time with rebel troops. Insurgents blocked several railway lines. Finally, a showdown occurred when one insurgent army severed the railway between Granada and Managua. Several confrontations occurred before the marines reached Granada. Even then, the rebel army continued to hold strong positions near the city.
October 4	After a long bombardment of the insurgents, the marines assaulted the insurgent positions at El Coyotepe.
October 6	León, the last center of resistance, fell to United States forces.
November 23	A national election was held in Nicaragua. Conditions were reasonably calm by then, and most United States forces left the country, leaving behind a legation guard in Managua to remind the Nicaraguans of United States interest in their country and to discourage further revolutions and coup attempts.
1921	
February 9	A number of the legation guards wrecked the offices of *La Tribuna*, a newspaper which had printed stories that offended many of the marines.
December	Clashes between a few legation guards and the Nicaraguan police left one marine and several police officers dead.

1922

October   Opponents of the Nicaraguan government, who had seized the high ground commanding Managua, left their positions at the insistence of the United States authorities.

1925

August 1   The legation guard was withdrawn from Managua on the assumption that a paramilitary constabulary intended to maintain order would be established with United States assistance.

A. Interventions of 1909, 1910, and 1912

461. "American Blood Spilt in Nicaragua." *Literary Digest* 45 (October 19, 1912): 657–58.

    Describes the battle of Coyotepe at some length and discusses several other armed clashes between United States naval forces and the insurgents more briefly. Press comment in the United States is summarized.

462. Bookout, Jack Eugene. "Outpost to Empire: United States–Nicaraguan Diplomatic Relations in the Age of Zelaya, 1893–1909," M.A. thesis, California State University, Fullerton, 1990.

    Discusses past relations in light of the then current hostility between the United States and Sandinista Nicaragua. Bookout asserts that the United States took an aggressive role throughout the period of Zelaya's presidency.

463. Butler, Smedley D. *General Smedley Darlington Butler: The Letters of a Leatherneck, 1898–1931*. Chap. 7, "Working for Brown Brothers and Uncle Sam," pp. 95–133, ed. Anne Cipriano Venzon. New York: Praeger, 1992.

    After providing a succinct but thorough summary of events leading to the intervention, Venzon presents a number of letters from Butler, which constitute excellent source material for the campaign in Nicaragua. Venzon supplements these important letters with full annotations that identify people mentioned in passing and supplies other relevant information.

464. ———. *Old Gimlet Eye: The Adventures of Smedley D. Butler As Told to Lowell Thomas*. New York: Farrar & Rinehart, Inc., 1933.

    Focuses on Butler's efforts to keep the American-owned railroad running, but also deals with his removal of United States mercenaries. Chapters XII–XIV discuss the interventions of 1909, 1910, and 1912. The last chapter cited describes the battle of Coyotepe in 1912.

465. Fox, Frank W. *J. Reuben Clark: The Public Years*. Provo, UT: Brigham Young University Press, 1980.

Describes the historical and legal background to the intervention of 1910 in some detail, pages 99–108, and the 1912 intervention, including naval operations, very briefly, pages 189–91. Fox notes that in 1910 occupation of the entire country was considered.

466. Gobat, Michel. *Confronting the American Dream: Nicaragua Under U.S. Imperial Rule.* Durham, NC: Duke University Press, 2005.

    Major study ranging from attacks on Nicaragua by William Walker in the mid-nineteenth century through the intervention between 1927 and 1933. Gobat skillfully follows the highly complex political and economic relationships between Nicaragua and the United States. Much of his emphasis is on the accommodations and adjustments Nicaraguans made to the long-term interference of the United States with their country.

467. Harrison, Benjamin. "The United States and the 1909 Nicaraguan Revolution," *Caribbean Quarterly* 41, nos. 3–4 (September–December 1995): 45–63.

    Argues that as early as Theodore Roosevelt's administration, the United States was hostile toward Zalaya and that the Taft administration and United States business interests encouraged the revolutionaries in 1909. Harrison takes a highly critical view of Dana G. Munro's discussion of United States policy in 1909 and 1910 (item 473).

468. Heinl, Robert Debs, Jr. *Soldiers of the Sea: The United States Marine Corps, 1775–1962.* Annapolis, MD: United States Naval Institute, 1962.

    Provides a succinct account of marine operations in Nicaragua in 1909–10, page 167, and 1912, pages 168–70. An excellent map of the country is included.

469. Langley, Lester D. *The Banana Wars: United States Intervention in the Caribbean, 1898–1934.* Wilmington, DE: SR Books, 2002.

    Concludes that the interventions and the continued presence of the legation guard did nothing to change Nicaraguan political life, chapters 5–6. Chapter 5 discusses Nicaraguan–United States relations from the late nineteenth century up to the intervention of 1909 and 1910. The campaign of 1912 is considered briefly.

470. McClellan, Edwin N. "American Marines in Nicaragua." Parts 1, 2. *Marine Corps Gazette* 6 (March, June, 1921): 48–65; 164–87.

    Constitutes a highly detailed account of the marine interventions of 1909, 1910 and 1912. The first part deals with the early landings and begins the story of the 1912 campaign, which the second part completes. McClellan relies heavily on official documents, including original cable messages regarding the Nicaraguan situation.

471. McDevitt, John Richard. "American–Nicaraguan Relations from 1909 to 1916." Ph.D. diss., Georgetown University, 1954.

Contends that United States involvement arose from security rather than financial considerations. He is highly critical of the Taft administration, asserting that the Díaz government was imposed on Nicaragua by armed force. McDevitt made extensive use of State and Navy department archives and United States and Latin American newspapers.

472. Meriwether, Walter Scott. "Admirals as Diplomats." *United States Naval Institute Proceedings* 67 (March 1941): 378–81.

Describes the dilemma of the commander of United States naval forces dispatched to Nicaragua in 1910 because of the ambiguity of his orders. Admiral Kimball bluffed his way through the crisis by planting rumors that a much larger force was on the way.

473. Munro, Dana G. "Dollar Diplomacy and Intervention in Nicaragua, 1909–13." Chap. 5, pp. 160–216, in *Intervention and Dollar Diplomacy in the Caribbean 1900–1921*. Princeton, NJ: Princeton University Press, 1964.

Sees Secretary of State Knox as an instigator of intervention. The last section of the chapter deals with the intervention of 1912 and the formation of a legation guard. Munro regards the intervention as "a turning point in American policy in the Caribbean" (p. 215) because for the first time a revolution was put down by United States forces.

474. ———. "The Interest of the United States in Nicaragua." Chap. 11, pp. 227–64, in *The Five Republics of Central America: Their Political and Economic Development and Their Relation with the United States*. New York: Carnegie Endowment for International Peace, 1918.

Discusses political events from 1909 to 1912. The problems of intervention are explored rather fully.

475. Musicant, Ivan. *The Banana Wars*, Chap. 4, "Nicaragua 1912," pp. 137–56. New York: Macmillan Company, 1990.

Places the fighting in 1912 into the context of United States–Nicaraguan relations, and more broadly, United States–Central American, relations. Musicant skillfully interweaves political and military events, using many materials from marine commander Joseph Pendleton's papers.

476. "The Nicaraguan Campaign." Parts 1, 2. *Leatherneck* 8 (February 7, 14, 1925): 1–3, 10; 1–3, 11, 16.

Provides a fairly detailed, but not definitive, account of the intervention of 1912.

477. Perkins, Whitney T. "Nicaragua." Chap. 2, pp. 100–16, in *Constraint of Empire: The United States and the Caribbean Interventions*. Westport, CT: Greenwood Press, 1981.

Gives a brief look at the interventions of 1909, 1910, and 1912 and the consideration given to making Nicaragua a United States protectorate during the Wilson administration. The emphasis is on the political, rather than the military, aspects of the Nicaraguan situation.

478. Rice, Michael David. "Nicaragua and the United States: Policy Confrontations and Cultural Interactions, 1893–1933." Ph.D. diss., University of Houston, 1995.

Offers a careful analysis of the many issues that caused conflict between Zelaya and the United States in the early twentieth century. After the ouster of Zelaya, United States influence increased in Nicaragua, but many Nicaraguans responded by developing a new form of nationalism that was embodied in Sandino's movement.

479. Schmidt, Hans. *Maverick Marine: General Smedley D. Butler and the Contradictions of American Military History.* Chap. 5, "The American Kitchener: Nicaragua," pp. 38–57. Lexington: University Press of Kentucky, 1987.

Given the highly complex political situation in Nicaragua and the continued United States involvements, Schmidt necessarily discusses these matters before describing Butler's significant role. This chapter is impressively documented with numerous citations to archival materials.

480. "Taking Nicaragua Under the Eagle's Wing." *Literary Digest* 47 (August 2, 1913): 155–57.

Summarizes press comment on proposals to make Nicaragua a United States protectorate.

481. U.S. Congress. Senate. Foreign Relations Committee. *Authorizing Committee to Investigate as to Alleged Invasion of Nicaragua by Armed Sailors and Marines of the United States, Hearing Pursuant to S.R. 385.* Washington, DC: Government Printing Office, 1913.

Reviews the intervention of 1912.

482. Valentine, Lincoln G. "Meddling with Our Neighbors." *Century* 90 (October 1915): 801–8.

Reviews United States military action in 1909, 1910, and 1912, and assails the interventions.

B. The legation guard as an American presence in Nicaragua, 1912–25

483. Birnbaumer, F.F. "Marines in Nicaragua, 1924." *Leatherneck* 11 (March 1928): 7–8.

Relates the events of a patrol made to observe one of Nicaragua's revolutions. Birnbaumer testifies to the very primitive conditions in the Nicaraguan jungle and the hardships encountered by the marines. The "armies" on both sides of the revolution are examined in terms of composition and arms.

484. Greer, Virginia L. "State Department Policy in Regard to the Nicaraguan Election of 1924." *Hispanic–American Historical Review* 34 (November 1954): 445–67.

Describes the Nicaraguan situation in 1925. Greer discusses the significance of the marines in Nicaragua between 1912 and 1923. This is an important study, with a scope much broader than the title suggests.

485. Hill, Roscoe R. "American Marines in Nicaragua, 1912–25." In *Hispanic American Essays: A Memorial to James Alexander Robertson*, edited by A. Curtis Wilgus, pp. 341–60. Chapel Hill: University of North Carolina Press, 1942.

Interprets United States involvement with Nicaragua as stemming almost entirely from a concern with security rather than preventing revolution. Hill points to the fact that many disturbances occurred while the legation guard was on duty.

486. Munro, Dana G. "Getting the Marines Out of Nicaragua." Chap. 6, pp. 157–86, in *The United States and the Caribbean Republics 1921–1933*. Princeton, NJ: Princeton University Press, 1974.

Reviews at length the role of the legation guard and the reasons the State Department had for wanting to withdraw it. There is some information on early plans for a Nicaraguan constabulary.

487. "Nicaragua and the United States Marines." *Leatherneck* 7 (March 15, 1924): 1–2, 8.

Describes the life of members of the legation guard in Managua and gives a very brief history of the various marine landings in Nicaragua.

488. "Nicaragua in 1925." *Leatherneck* 11 (July 1928): 6–7.

Emphasizes the role of the legation guard as a peacekeeping force that was appreciated by the Nicaraguan people. The article describes the camp of the small marine detachment at former dictator Zelaya's estate. Overall, this is a good picture of the guard on the eve of its departure from Nicaragua.

489. Weinhold, H.W. "Highlights on Tropical Service—Duty in Nicaragua." *Leatherneck* 9 (June 1926): 29.

Discusses the scenery, hunting, and travel opportunities of the legation guard in Managua.

IX

The Mexican border, 1911–19

The outbreak of the Mexican Revolution of 1910 brought a series of crises that occurred because of continuing United States involvement with Mexican politics and because of the existence of a number of towns in Mexico and the United States that were in extremely close proximity. President William Howard Taft increased military readiness on the border and considered direct intervention. President Woodrow Wilson took a much more active role in United States–Mexican relations. The United States played a major part in driving General Victoriano Huerta from the Mexican presidency through the seizure of Veracruz, a subject that is treated in the next chapter.

Scattered fighting occurred along the border for some time, and after Pancho Villa's raid into the United States in 1916, the Punitive Expedition, under General John J. Pershing's command, penetrated deep into Mexico. For a time, there was a real possibility of war between the United States and Mexico, but neither nation wanted an outright conflict. The evacuation of Pershing's troops early in 1917 did not end border troubles. Further raids occurred. Some United States political leaders advocated occupation of Mexico, but despite continuing friction between the two countries for years, there was little prospect of an occupation after 1920.

Chronology

1910

November 20   A proclamation by political leader Francisco I. Madero marked the beginning of the Mexican Revolution, which initiated a long period of tensions between Mexico and the United States.

1911
March — President William H. Taft ordered the United States Army to form a Maneuver Division to prepare for possible intervention in Mexico but, more importantly, to act as a force in being to deter Mexican forces from clashes on the border. The division was dissolved in July–August.

1913
March 13 — United States troops forced the surrender of some Constitutionalist soldiers when they fired into Nogales in the United States while fighting in Nogales, Mexico.

1914
October 11 — United States cavalry troopers skirmished with followers of Pancho Villa at Naco, Mexico, when artillery shells from Mexico began killing United States soldiers and civilians.

1915
January 24 — An investigation at McAllen, Texas, uncovered the "Plan of San Diego," a scheme by Mexican radicals to retake the areas of their country that had been ceded to the United States in 1848 as a result of the Mexican–American War.

July — Incursions into the United States began as supporters of the "Plan of San Diego" endeavored to put their strategy into effect.

October — Once the United States recognized the Carranza administration (Constitutionalists) as the government of Mexico, raiding quickly ended, at least for a time.

1916
March 9 — Angered by the fact that the United States had recognized the Carranza government, Pancho Villa led his forces in a serious attack on Columbus, New Mexcio. United States troops pursued the raiders, killing a number of them.

March 15 — United States forces under the command of General John J. Pershing, called the Punitive Expedition, entered Mexico with the purpose of dispersing Villa's army.

March 29 — Sharp fight at Guerrero, Mexico, between United States troops under Colonel George A. Dodd and followers of Villa. The battle was a distinct defeat for Villa's forces.

April 12 — Battle at Parral Mexico, between United States troops and Mexican forces and Mexican civilians. Although the Mexicans were not all Villa supporters, they strenuously objected to United States cavalry entering the town of Parral.

May 5 — United States forces under Major Robert L. Howze defeated a group of Villa's men at the Ojos Azules Ranch in Mexico.

May 5	Mexican attack on Glenn Springs, Texas, led to a second and smaller "Punitive Expedition" being sent into Mexico under the command of Major George T. Langhorne.
June 18	President Woodrow Wilson called up the National Guard for border duty from all the states that had not already mobilized their guards.
June 21	Increasing tension between Mexico and the United States because of the incursion of the Punitive Expedition erupted into significant fighting between United States and regular Mexican troops in the battle of Carrizal. Unwise decisions by the United States commander, Captain Charles T. Boyd, had brought a confrontation. The affair severely stressed Mexican–United States relations for a time, but both Mexico and the United States recognized the danger of a major war and chose to establish an American–Mexican Commission to settle differences.

1917

January 15	The American–Mexican Commission ended its work after having deliberated since September 1916. The negotiations were inconclusive, but they doubtless helped prevent a worse crisis in Mexican–United States relations.
February 5	Pershing's Punitive Expedition left Mexico as relations between the United States and Germany deteriorated.
February 15	The National Guard mobilization was ended, and the final members of the various guard contingents returned to their states in March.
December	Skirmish between United States forces and Mexican irregulars near Indio, Texas.

1918

March	Skirmish between United States forces and Mexican irregulars near Van Horn, Texas. The Indio and Van Horn incidents were only two of a number of military encounters that sometimes involved United States soldiers entering Mexico to the dismay of the Mexican government.
May	The Wilson administration sharply restricted the United States Army's authority to follow irregulars into Mexico.

1919

June 15–16	Troops under the command of Pancho Villa fired into El Paso, Texas, resulting in a large United States incursion into the area of Mexico close to El Paso. The Villa forces were severely defeated.

A. General treatments

490. Bruscino, Thomas A., Jr. "A Troubled Past: The Army and Security on the Mexican Border, 1915–17." *Military Review* 88, no. 4 (April 2008): 31–44.

    Valuable overview of low-intensity conflict on the border and the Punitive Expedition. Bruscino makes good use of secondary and some primary materials. The significance of the article is considerably enhanced by the author's effort to relate events of the time to the current deployment of elements of the United States military to the same border.

491. Clendenen, Clarence C. *Blood on the Border: The United States Army and the Mexican Irregulars.* New York: Macmillan Company, 1969.

    Examines troop movements on the border, skirmishes along the frontier, the Veracruz occupation, and the Punitive Expedition. The study is readable and well documented, including as it does much material from interviews with United States participants.

492. Coerver, Don M. and Linda B. Hall. *Texas and the Mexican Revolution: A Study in State and National Border Policy 1910–1920.* San Antonio, TX: Trinity University Press, 1984.

493. Hall, Linda B. and Don M. Coerver. *Revolution on the Border: The United States and Mexico, 1910–1920.* Albuquerque, NM: University of New Mexico Press, 1988.

    Despite some overlapping, these important volumes have significant differences in emphasis. Coerver and Hall provide a good deal of information about the Taft administration's stance toward Mexico, for example, while Hall and Coerver give much more attention to the Punitive Expedition. These studies are based on deep research in the archives of Mexico, Great Britain, and the United States. Hall and Coerver deal in detail with the economic facets of relations between the United States and Mexico.

494. Eisenhower, John S.D. *Intervention! The United States and the Mexican Revolution, 1913–1917.* New York: W.W. Norton, 1993.

    Based mainly on secondary sources and published primary materials, this is a detailed survey of the subject. The book might usefully have covered a longer time span to include initial United States military moves by the Taft administration and border incidents through 1919. The book focuses largely on military events, and it does well in treating those. There is an excellent chronology and an interesting appendix that describes some of the major places in Mexico and the United States that are mentioned in the book, such as Columbus,

New Mexico. The chapters dealing with the Veracruz occupation are listed in the appropriate section of this bibliography.

495. Johnson, Benjamin Heber. *Revolution in Texas: How a Forgotten Rebellion and Its Bloody Suppression Turned Mexicans into Americans.* New Haven, CT: Yale University Press, 2003.

    Important examination of the complex events that played out along the Mexican–United States border during the World War I era. This is largely a political rather than a military study of the Plan of San Diego and its ramifications, but it is vital for anyone wishing to understand the long-term effects of the savage repression of Mexican-heritage people in Texas.

496. Mulcahey, Mark. "The Overlooked Success: A Reconsideration of the U.S. Military Interventions in Mexico during the Wilson Administration," pp. 43–73 in *Defending the Homeland: Historical perspectives on Radicalism, Terrorism, and State Responses*, ed. Melinda M. Hicks and C. Belmont Keeney. Morgantown: West Virginia University Press, 2007.

    Praises what Mulcahey views as Wilson's skill in applying military force for limited goals. He does not say much about the purely military aspects of the Veracruz occupation and the Punitive Expedition, focusing rather on Wilson's objectives in Mexico. Mulcahey demonstrates an impressive use of archival materials.

497. Sandos, James A. *Rebellion in the Borderlands: Anarchism and the Plan of San Diego, 1904–1923*. Norman: University of Oklahoma Press, 1992.

    Pioneering study firmly based on deep archival research in both Mexico and the United States. Although stressing an anarchist movement, Sandos eloquently describes the situation of United States citizens of Mexican heritage that resulted in considerable unrest at the time of the Mexican Revolution. Although Sandos does not deal in detail with the military aspects of the border region, his work is essential reading for anyone committed to understanding the economic and political ramifications of the low-intensity conflict. Sandos skillfully places his story in the context of United States and Mexican politics.

B. Taft and the intervention question

498. Fox, Frank W. *J. Reuben Clark: The Public Years.* Provo, UT: Brigham Young University Press, 1980.

    Reviews efforts by Secretary of State Philander Knox and Solicitor General J. Reuben Clark to keep Taft from moving troops into Mexico in 1911 and 1912.

499. Haley, P. Edward. *Revolution and Intervention: The Diplomacy of Taft and Wilson with Mexico, 1910–1917*. Cambridge, MA: MIT Press, 1970.

Reviews Taft's mobilization of troops on the border and the deliberations on intervention that took place in 1911–12.

500. Ross, Stanley R. *Francisco I. Madero: Apostle of Mexican Democracy*. New York: Columbia University Press, 1955.

Describes the effects that fears of United States intervention had upon Mexican governmental leaders during the early stages of the revolution, 1911–13.

C. Border skirmishes and the broader problem of intervention

501. Chamberlain, Weston P. "Hints for Line Officers Regarding the Diseases Liable to be Met with in Mexico and the General Methods of Combating Them." *Journal of the Military Service Institution of the United States* 54 (May–June 1914): 341–59.

Makes some comparisons with British experience in the tropics.

502. Cumberland, Charles C. "Border Raids in the Lower Rio Grande Valley—1915." *Southwestern Historical Quarterly* 57 (January 1954): 285–311.

Discusses United States military deployment along the border and the little-known but extensive attacks by Mexicans in Texas. Fighting was sometimes heavy. Cumberland includes interesting material on the role of aircraft in this period of border conflict.

503. Estes, George H. "The Internment of Mexican Forces in 1914." Parts 1, 2, 3. *Infantry Journal* 11, 12 (May–June, July–August, September–October, 1915): 747–69; 38–57; 243–64.

Deals with the administrative problems created by a large body of Huerta's troops who were driven into the United States.

504. Harris, Charles H. III and Louis R. Sadler. *The Texas Rangers and the Mexican Revolution: The Bloodiest Decade, 1910–1920*. Albuquerque: University of New Mexico Press, 2004.

A shade more favorable to the rangers than Utley's book (item 517), this is a thoroughly academic study with extensive notes and a large bibliography. It is necessary for anyone seriously interested in border events of the period. The authors note the highly controversial nature of ranger operations during these years. There has been both supportive and critical writing about them. The authors suggest "that there is a wealth of material to support either version" (p. 506).

505. Hopper, James. "Twin Towns of the Border." *Collier's* 57 (August 19, 1916): 5–7, 27–28.

Describes a skirmish between United States troops and Mexicans at Nogales, November 1915, and an exchange of shots at Laredo, April 1914. The fighting in such twin cities often brought with it firing across the border.

506. Kestenbaum, Justin Louis. "The Question of Intervention in Mexico, 1913–17." Ph.D. diss., Northwestern University, 1963.

Considers "intervention" in political as well as military terms. Kestenbaum identifies groups in the United States that were concerned with intervention. There are chapters on the Veracruz occupation and the Punitive Expedition. Perhaps of particular interest is Kestenbaum's examination of the intervention issue and the election of 1916.

507. Lou, Dennis Wingsou. "Fall Committee: An Investigation of Mexican Affairs." Ph.D. diss., Indiana University, 1963.

Examines closely the debate over intervention in Mexico following World War I. The Fall Committee was the focus of it.

508. "Mexico." *Infantry Journal* 9 (March–April 1913): 711–13.

Discusses the grim prospects for an effective intervention in Mexico. Guerrilla warfare would create a demand for at least 300,000 troops. The United States armed forces were so weak that it would take a long time to raise the needed soldiers and to train them. Much of the army was committed to coast defense duties, and retraining this group of soldiers would be difficult and would require a good deal of time.

509. Millett, Allan R. *The General: Robert L. Bullard and Officership in the United States Army 1881–1925*. Westport, CT: Greenwood Press, 1975.

Examines Bullard's participation in the mobilization of 1911 and his later service on the border between 1912 and 1917. The Second Division, Bullard's command, would have been the spearhead of any United States incursion into Mexico.

510. Pickering, Abner. "The Battle of Agua Prieta." *Infantry Journal* 12 (January 1916): 707–10.

Describes United States troops going on the alert when Villa's men and Mexican federal troops fought on the border during October and November 1915. The Americans were given a unique view of the battle but also suffered from heavy fire.

511. Robertson, O. Zeller, Jr. "Mexico and Non-Intervention 1910–19: The Policy, the Practice and the Law." Ph.D. diss., University of California, Los Angeles, 1969.

Examines both United States and Mexican doctrines and attitudes regarding intervention and their relationships to such episodes as the Punitive Expedition.

512. Salisbury, Nathan F. "The Bandit Raid on Ojo de Agua." *Cavalry Journal* 47 (May–June 1938): 238–42.

Recounts the events of an attack on a signal corps station at Ojo de Agua on October 21, 1915. A participant tells the story of the uneven battle between 20 soldiers and from 50 to 100 Mexicans. There are several sketches showing the Mexican attacks.

513. Schwarzkopf, Olaf. "The Hygiene and Treatment of the Horses of the Maneuver Division, San Antonio." *Cavalry Journal* 22 (March 1912): 852–61.

Compares the division's experiences with those of the army during the War of 1898. The health of the men and animals was better in 1911 than in 1898.

514. Smith, Harry A. "The Texas Bandit Problem." *Infantry Journal* 12 (March 1916): 845–52.

Presents a series of tactical problems arising from a hypothetical incursion of Mexican irregulars into the United States. The border situation is discussed in some detail, and Smith makes some highly perceptive remarks about counterinsurgency warfare generally.

515. Smith, Robert Freeman. *The United States and Revolutionary Nationalism in Mexico, 1916–1932*. Chicago: University of Chicago Press, 1972.

Examines the Punitive Expedition and later tensions between Mexico and the United States in 1918 and in 1919–20. Smith views the Punitive Expedition as a device to prevent the Mexicans from nationalizing their oil industry as much as for border protection. The activities of Albert B. Fall, prominent interventionist, and others are examined in considerable detail.

516. U.S. War Department. Office of the Chief of Staff. War College Division, General Staff. *Monograph on Mexico*. Washington, DC: Government Printing Office, 1914.

Details the history, geography, resources, population, and industries at length. It includes in-depth information about each Mexican state. The Mexican army and navy are given two chapters.

517. Utley, Robert Marshall. *Lone Star Lawmen: The Second Century of the Texas Rangers*. Oxford: Oxford University Press, 2007.

The first four chapters discuss ranger activities during the years of the Mexican Revolution, which caused numerous border incidents. Despite its relative brevity, Utley's account is vital because of its objectivity. Utley does not hesitate to describe what he terms as "the atrocities of 1910–20" (p. x). There are detailed references and an important bibliography.

518. Welsh, William E. "The Military Geography of Mexico." *Journal of the Military Service Institution of the United States* 50 (January–February 1912): 44–56.

    Interprets "military geography" broadly as a description of a country that could serve as the basis for a campaign plan. Communications, trade, resources, and other topics are examined. The article was to have been continued, but apparently it never was.

D. The attack on Columbus, New Mexico, 1916

519. Castleman, James T. "The Columbus Raid." *Cavalry Journal* 27 (April 1917): 490–96.

    Discusses both the attack and the resulting pursuit into Mexico. Castleman was Officer of the Day at Columbus when the attack took place, but the material in his article comes from sources other than his own observations.

520. "Cavalry Fight at Columbus." *Cavalry Journal* 27 (November 1916): 183–85.

    Deals solely with the pursuit. The article emphasizes that a small number of cavalrymen succeeded in disorganizing Villa's forces and in inflicting very heavy casualties without much loss to themselves.

521. Harris, Larry A. *Pancho Villa and the Columbus Raid*. El Paso, TX: McMath Company, Inc., 1949.

    Includes an abundance of unusual photographs, which is perhaps the most useful feature of the book for researchers. Two chapters are concerned with the raid, but most of the text revolves around Villa's earlier career.

522. Hopper, James. "New Columbus and the Expedition." *Collier's* 57 (August 5, 1916): 10–11, 35.

    Describes Columbus in its new status as a boom town. Not only did it attract considerable attention, because of the attack, but it became a supply point for the Punitive Expedition.

523. Hurst, James W. *The Villista Prisoners of 1916–1917*. Las Cruces, NM: Yucca Tree Press, 2000.

    Explores the story of 19 of Villa's men who were captured during the Columbus, New Mexico, battle that prompted the Punitive Expedition.

The men were tried for murder. Several were executed, and others were given long prison terms. After World War I, the prisoners were released, re-arrested, and after a second trial were found not guilty, at which time they returned to Mexico.

524. Mahoney, Tom. "The Columbus Raid." *Southwest Review* 17 (January 1932): 161–71.

    Reconsiders the question of how much information the United States had about Villa's advance to the border before the attack. Mahoney also retells some of the most spectacular events of the battle, praising individuals on both sides for their courage.

525. Sandos, James A. "German Involvement in Northern Mexico, 1915–16: A New Look at the Columbus Raid." *Hispanic–American Historical Review* 50 (February 1970): 70–88.

    Important investigation of the reasons for Villa's attack.

E. The Punitive Expedition: general studies

526. Archer, Jules. *Mexico and the United States.* New York: Hawthorn Books, Inc., 1973.

    Discusses the expedition very briefly.

527. Atkin, Ronald. *Revolution: Mexico 1910–20.* Chap. 4, "The Pershing Expedition," pp. 274–92. New York: The John Day Company, 1970.

    Bases his study on published works and United States and Mexican newspapers. It is a useful British perspective.

528. Boot, Max. *The Savage Wars of Peace: Small Wars and the Rise of American Power.* Chap. 8, "The Dusty Trail: The Pancho Villa Punitive Expedition, 1916–17," pp. 182–204. New York: Basic Books, 2002.

    Boot's detailed and balanced narrative is somewhat disconnected owing to the author's fascination with picturesque aspects of the episode. Nevertheless, this is a useful introduction to the complex topic.

529. Braddy, Haldeen. *Pershing's Mission in Mexico.* El Paso: Texas Western Press, 1966.

    Continues to be a work of real importance, partly because Braddy utilized material from a number of interviews. It is worth reading, despite the eccentric writing style.

530. Calhoun, Frederick S. *Power and Principle: Armed Intervention in Wilsonian Foreign Policy.* Chap. 2, "The Power of Civilian Control: Mexico," pp. 51–67. Kent, OH: Kent State University Press, 1986.

Penetrating examination of Wilson's purposes in ordering and controlling the Punitive Expedition. Calhoun describes Wilson's almost surgical use of military measures to reach his political goals. Most United States military leaders favored a stronger hand against Mexico than Wilson would allow, but, eventually, they realized how Wilson meshed their efforts with his diplomatic offensive.

531. Clendenen, Clarence C. *The United States and Pancho Villa: A Study in Unconventional Diplomacy.* Ithaca, NY: Cornell University Press, 1961.

    Asserts that the expedition was successful because it prevented further attacks. This is an extremely important study. Many chapters deal with the Columbus raid and the Punitive Expedition.

532. Cramer, Stuart W. "The Punitive Expedition from Boquillas." *Cavalry Journal* 27 (November 1916): 200–227.

    Provides a detailed treatment of the Boquillas incursion into Mexico. Cramer served in the force that pursued Mexican raiders after an attack on Texas. The cavalry detachment was in Mexico for 10 days in May 1916.

533. Dupuy, R. Ernest, and William H. Baumer. "Mexican Interventions, 1914–17." Chap. 5, pp. 123–43, in *The Little Wars of the United States.* New York: Hawthorn Books, Inc., 1968.

    Discusses five battles in Mexico.

534. Edwards, Warrick Ridgely, III. "United States–Mexican Relations, 1913–16: Revolution, Oil, and Intervention." Ph.D. diss., Louisiana State University and Agricultural and Mechanical College, 1971.

    Provides indispensable background information from State Department archives on the tensions between the United States and Mexico. Several chapters deal with the prelude to the Punitive Expedition.

535. Elser, Frank B. "General Pershing's Mexican Campaign." *Century* 99 (February, 1920): 433–47.

    Reveals many stories Elser could not write as a correspondent with the Punitive Expedition because of the strict press censorship.

536. Gilderhus, Mark T. *Diplomacy and Revolution: U.S.–Mexican Relations under Wilson and Carranza.* Chap. 3, "The Punitive Expedition." Tucson: University of Arizona Press, 1977.

    Puts the expedition into the broader context of Mexican–American relations.

537. Haley, P. Edward. *Revolution and Intervention: The Diplomacy of Taft and Wilson with Mexico, 1910–1917.* Cambridge, MA: MIT Press, 1970.

Explores in great detail the negotiations that led to the withdrawal of United States forces, chapters 9, 10.

538. Howe, Jerome W. "Campaigning in Mexico, 1916," *Journal of Arizona History* 7, no. 3 (Autumn 1966): 123-38.

    Howe recalls his participation in the expedition. An officer in the 2nd Cavalry Brigade, he discusses the battle with Mexican federal troops at Carrizal.

539. Hurst, James W. *Pancho Villa and Black Jack Pershing: The Punitive Expedition in Mexico*. Westport, CT: Praeger, 2008.

    Although not seen by the compiler, this book appears to be an important contribution to the literature about the Punitive Expedition. Hurst gives a generally favorable interpretation of the expedition and its results. Villa's army was scattered fairly quickly, which prevented it from mounting further large-scale attacks in the United States. One of the most valuable features of the book is the attention to the intelligence operations during the incursion into Mexico.

540. Johnson, Robert Bruce. "The Punitive Expedition: A Military, Diplomatic, and Political History of Pershing's Chase after Pancho Villa, 1916-17." Ph.D. diss., University of Southern California, 1964.

    Constitutes one of the most important studies of the expedition. The bibliography is now somewhat dated, but it appears to be the most extensive on the subject.

541. Katz, Friedrich. *The Life and Times of Pancho Villa*. Stanford, CA: Stanford University Press, 1998.

    Thoroughly grounded in both Mexican, United States, and other sources, this massive study is likely to remain the authoritative biography of Villa for many years. "On the Archival Trail of Pancho Villa," pp. 821-34, reflects the long-term efforts Katz devoted to his subject. Chapters 14 and 15 are the most important segments of the book for the purposes of this bibliography, but the whole volume is required reading for anyone concerned with Villa's role in relations between Mexico and the United States during the revolutionary period.

542. Link, Arthur S. *Wilson: Confusions and Crisis 1915-1916*. Princeton, NJ: Princeton University Press, 1964.

    Emphasizes Wilson's moderation in the face of anti-American incidents, chapters 7, 10.

543. Lister, Florence C. and Robert H. Lister. *Chihuahua: Storehouse of Storms*, pp. 247-63. Albuquerque: University of New Mexico Press, 1966.

    Provides a useful introduction to the Punitive Expedition.

544. Mason, Herbert Molloy, Jr. *The Great Pursuit*. New York: Random House, 1970.

Presents a semi-scholarly treatment of the Punitive Expedition, based in part on interviews with unidentified persons. Appendices list the participating units and the depth of their penetration into Mexico.

545. O'Connor, Richard. *Black Jack Pershing*. Garden City, NY: Doubleday & Company, Inc., 1961.

Discusses Pershing's meetings with Villa and General Obregon in 1914–15. Villa's motivations for the Columbus raid are examined in detail.

546. Scott, Hugh Lenox. *Some Memories of a Soldier*. New York: Century Co., 1928.

Reviews the Columbus raid and describes Scott's negotiations with Obregon for the halting of the Punitive Expedition. Several pages deal with Scott's decision to supply the expedition by truck after Mexican President Carranza closed the railways to the United States forces.

547. Smythe, Donald. *Guerrilla Warrior: The Early Life of John J. Pershing*. New York: Charles Scribner's Sons, 1973.

Emphasizes the hardships of the campaign and Pershing's great determination to capture Villa, chapters 15, 16, and 17. Pershing would have preferred an invasion of Mexico on a large scale and regretted having to withdraw in 1917. This is probably the best account from Pershing's perspective.

548. Stout, Joseph A., Jr. *Border Conflict: Villistas, Carrancistas and the Punitive Expedition, 1915–1920*. Fort Worth: Texas Christian University Press, 1999.

A significant addition to the relatively large literature on the Punitive Expedition. Stout describes in detail the Mexican armies of Carranza and Villa and their operations. The book is based on extensive use of sources in Mexico, thereby importantly complementing other studies.

549. Thomas, Robert S., and Inez V. Allen. *The Mexican Punitive Expedition under Brigadier General John J. Pershing United States Army 1916–1917*. Washington, DC: Department of the Army, Office of the Chief of Military History, War Histories Division, 1954.

Constitutes one of the essential studies of the expedition. It is a fine piece of military history, based on the records of the expedition, the Wilson, Borah, and Daniels papers, interviews, and other materials.

550. Tompkins, Frank. *Chasing Villa: The Story Behind the Story of Pershing's Expedition into Mexico*. Harrisburg, PA: Military Service Publishing Company, 1934.

Forms a useful corrective to other works because it looks at United States units not directly under Pershing's control as well as the main body. Its appendices contain much information on aviation, cavalry, and motor transport units.

551. Toulmin, Harry Aubrey, Jr. *With Pershing in Mexico.* Harrisburg, PA: Military Service Publishing Company, 1935.

    Provides a fairly detailed account of the expedition, with separate chapters on the battles at Parral, Carrizal, and Ojos Azules. The political background and the Columbus raid are treated very briefly. Sketches help to convey the atmosphere of the expedition.

552. Twichell, Heath, Jr. *Allen: The Biography of an Army Officer 1859–1930.* Chap. VIII, "Alarums and Excursions," pp. 182–95. New Brunswick, NJ: Rutgers University Press, 1974.

    Describes General Allen's experiences with the 13th Cavalry on the border and summarizes very ably the operations of the Punitive Expedition.

553. Vandiver, Frank E. *Black Jack: The Life and Times of John J. Pershing.* Vol. 1. College Station: Texas A&M University Press, 1977.

    Provides a useful overview of the expedition in chapter 17. The index should also be consulted for additional material.

554. Welsome, Eileen. *The General and the Jaguar: Pershing's Hunt for Pancho Villa: A True Story of Revolution and Revenge.* New York: Little, Brown and Co., 2006.

    Impressive study that presents a wealth of new information based on intensive and innovative archival research. Among the useful sources that Welsome consulted were claims submitted by inhabitants of Columbus, New Mexico, because of losses suffered during Pancho Villa's attack on the town in 1916. The maps are easy to use, and the photographs are informative. This is a highly valuable work, especially in respect to military operations, but it is one-sided owing to the fact that Mexican resources were not utilized.

555. Wolff, Leon. "Black Jack's Mexican Goose Chase." *American Heritage* 13 (June, 1962): 22–27, 100–106.

    Gives a popular account of the expedition. The article says little of earlier border tensions or the broader problems of United States–Mexican relations.

F. The Punitive Expedition: major battles

556. "The Attack on Our Cavalry at Parral." *Cavalry Journal* 27 (November 1916): 249–53.

Accuses Carranzista forces of treachery. This is one of the first published military accounts of the battle between United States forces and Mexican regulars in April 1916.

557. Craig, Edward A. "The Fight at Carrizal." *Leatherneck* 22 (March 1939): 6, 54–55.

Compares the tense situation confronting the cavalry patrol with similar challenges to the marines during their small wars in the Caribbean and elsewhere. After giving a historical introduction, Craig presents a series of problems based on the episode for solution by the reader.

558. Morey, Lewis S. "The Cavalry Fight at Carrizal." *Cavalry Journal* 27 (January 1917): 449–56.

Covers the preliminary movements of the Americans and their tactics in battle in considerable detail. Morey, one of the officers present at Carrizal, contributed the article at the express invitation of the editors of the *Cavalry Journal*.

559. Williams, S.M. "The Cavalry Fight at Ojos Azules." *Cavalry Journal* 27 (January 1917): 405–8.

Consists in considerable part of charts relating to this battle between United States cavalry and a Villista band in May, 1916. Williams notes the lack of cooperation on the part of Carranzista troops in the area.

560. Young, Karl. "A Fight That Could Have Meant War." *American West* 3 (Spring 1966): 16–23, 90.

Depends primarily on an interview with L.H. Spilsbury, one of Pershing's scouts. Young also consulted the National Archives.

G. The Punitive Expedition: weapons and equipment

561. Evans, Ellwood W. "Cavalry Equipment in Mexico." *Cavalry Journal* 27 (November 1916): 230–32.

Emphasizes saddles.

562. "Field Notes from Mexico and Border." *Cavalry Journal* 27 (November 1916): 171–82.

Focuses on experiences with army-issue equipment. A few paragraphs deal with the battle of Ojos Azules.

563. Fleming, Lawrence J. "The Automatic Pistol in the Punitive Expedition." *Cavalry Journal* 27 (April 1917): 497–514.

Reproduces a manual written to govern use of the famous "forty-five." At Pershing's direction, a questionnaire was circulated to all cavalry officers of the expedition. The consensus was that the pistol had proved itself.

564. "Notes from Here and There in Mexico." *Infantry Journal* 13 (February 1917): 457–62.

Emphasizes transportation, equipment, and establishing camps. The notes were compiled from incoming correspondence. This article provides one of the few perspectives on the infantry experience in the Punitive Expedition.

565. "Notes on Campaigning in Mexico." *Cavalry Journal* 27 (July 1916): 63–66.

Consists of short notes from officers with Pershing. Most of them deal with equipment and food. Two paragraphs describe the terrain between the border and Parral.

566. Weissheimer, J.W. "Field Ovens in Mexico." *Infantry Journal* 13 (February 1917): 470–73.

Describes various types of ovens and the various field expedients employed for cooking.

H. The Punitive Expedition: logistics

567. Martin, C.H. "Rapid Transportation of Infantry." *Infantry Journal* 13 (November–December 1916): 333–37.

Describes movement of an infantry battalion by truck in September, 1916, on the border. The riflemen moved three times faster than cavalry.

568. Millard, George A. "US Army Logistics During the Mexican Punitive Expedition of 1916." *Military Review* 60 (October 1980): 58–68.

Chronicles the first extensive use of trucks in a United States military operation. Even with the innovative use of trucks, however, Pershing's movements were severely limited by logistics problems.

569. "The Motor Truck in Mexico." *Literary Digest* 52 (May 27, 1916): 1599.

Quotes a story from the *New York Evening Post* on the effective use of trucks in Mexico. They were more efficient than mules because of the weight of their own feed that the mules had to carry.

570. "Movement of Kansas Regiments by Motor Trucks."/"Movement of 1st Battalion, 23d Infantry by Motor Trucks." *Infantry Journal* 13 (November–December 1916): 338–39.

Describes experiments in moving troops by truck during August and September, 1916. The vehicles and the marches performed are treated in considerable detail. Despite minor problems, most of them mechanical, the movements were accomplished in good order, and the soldiers were ready for active duty even after being confined in the trucks.

571. Porter, John A. "The Punitive Expedition." *Quartermaster Review* 12 (January–February 1933): 18–30.

Surveys the story of the expedition and then concentrates on logistics. Tables indicate the daily supply requirements of Pershing's force. A detailed map illustrates supply depots along the route and the engagements that were fought with the Villistas. The various types of transportation utilized are described.

I. The Punitive Expedition: patrols and skirmishes

572. Adams, Cyrus C. "Northern Mexico, the Scene of Our Army's Hunt for Villa." *Review of Reviews* 53 (April 1916): 421–23.

Describes the geography of the border area, and speculates about Villa's possible courses of action.

573. "A Continuing Mexican Trouble." *World's Work* 37 (June 1916): 137–38.

Examines the political problems of both the Americans and Mexicans in fairly objective terms. Carranza, it was recognized, could cooperate with the United States to only a limited degree. The article also considers the obstacles United States troops faced in trying to capture Villa.

574. Dallam, Samuel F. "The Punitive Expedition of 1916: Some Problems and Experiences of a Troop Commander." *Cavalry Journal* 36 (July 1927): 382–98.

Describes in minute detail preparations for the incursion into Mexico. Dallam discusses the pursuit of Villa and the battle of Guerrero, one of the most successful skirmishes of the "war." The conclusion includes a number of reminders for a troop unit preparing for expeditionary duty.

575. Dunn, Robert. "With Pershing's Cavalry." *Collier's* 58 (September 23, 1916): 8–9, 25–26, 28.

Presents a graphic picture of living conditions on the march into Mexico, encounters with suspected bandits, and minor incidents before the battle of Parral.

576. Marvin, George. "Bandits and the Borderland." *World's Work* 32 (October 1916): 656–63.

Asserts that a few bandits have tied up thousands of United States troops and examines some reasons for Mexican banditry. Marvin, an especially acute observer, also discusses the physical characteristics of the border and describes the kind of people who are viewed as bandits.

577. Shaw, W.B. "Pershing on the Trail." *Review of Reviews* 53 (April 1916): 419–21.

Compares the pursuit of Geronimo in the 1880s with the Punitive Expedition. This useful article also notes Pershing's experiences in the Moro wars.

J. The Punitive Expedition: commanders and others

578. "Funston and Pershing, the Generals in Charge of the Chase after Villa." *Current Opinion* 60 (May 1916): 318–20.

Reviews the careers of both men, and emphasizes the value of both their experience and their characters for dealing with the complex border situation.

579. Patton, George E. *The Patton Papers I: 1885–1940*. Ed. Martin Blumenson. Boston, MA: Houghton Mifflin Company, 1972.

Skillfully weaves Patton's letters and diary entries together with his subsequent speeches and comments to form a connected narrative. Patton's service in Mexico was conspicuous, and his account of the expedition is, therefore, important.

K. The Punitive Expedition: the role of aviation

580. Hennessy, Juliette A. *The United States Army Air Arm, April 1861 to April 1917*. Montgomery, AL: Air University Research Studies Institute, 1958.

Chronicles the misadventures of the first United States combat aircraft. In addition to chapter IX, pages 145–50, 177, and 188–89 should be consulted for material.

581. Miller, Roger G. *A Preliminary to War: The First Aero Squadron and the Mexican Punitive Expedition of 1916*. Washington, DC: Air Force History and Museums Program, 2003.

Thorough examination of aviation support for the Punitive Expedition. Miller is not uncritical of Benjamin D. Foulois, the squadron commander, especially because of his decision to fly his aircraft to Pershing without waiting for daylight. In addition to discussing its aviation aspects, Miller provides a good survey of the Punitive Expedition. A fine map and numerous illustrations are other important features of this first-rate study.

L. The Punitive Expedition: medical aspects

582. Marble, William Sanders. "Medical Support for Pershing's Punitive Expedition in Mexico, 1916–17." *Military Medicine* 173, no. 3 (March 2008): 287–92.

Despite the ad hoc character of the Punitive Expedition and the rapidity with which it was assembled, the force had no major medical problems. Marble carefully describes army medical services of the period. The medical staff of the expedition was short-handed, but the health record of the troops assigned to this incursion into Mexico was excellent overall.

M. The Punitive Expedition: the horses

583. Case, Frank L. "Remounts in Southern Department." *Cavalry Journal* 27 (October 1917): 235–40.

Reviews the measures taken to provide remounts for the Punitive Expedition and other military needs in the border area. Case also discusses efforts made to keep the horses in a healthy state.

584. Lininger, Clarence. "The Type of Cavalry Horse for Campaign." *Cavalry Journal* 27 (April 1917): 581–86.

Examines experiences with the horses that went to Mexico with the Punitive Expedition. Lininger compares the best and the worst animals.

N. Mobilization of the National Guard for border service

585. Ainsworth, Troy. "Boredom, Fatigue, Illness, and Death: the United States National Guard and the Texas–Mexico border, 1916–17." *Journal of Big Bend Studies* 19 (2007): 81–96.

As the title suggests, this study examines the impact of sudden active service on the citizen soldiers of the National Guard. The climate and changes in living conditions took a toll on these men.

586. "Calling Out the Guard." *World's Work* 32 (August 1916): 437–52.

Presents a fascinating collection of photographs showing mobilization of the National Guard for border service. A few illustrations relate to the Punitive Expedition. Notes below the illustrations generally refer to the shortcomings of the Guard.

587. Case, Frank L. "Experiment with 'Reduced' or 'Minimum' Type of Horses." *Cavalry Journal* 27 (January 1917): 342–47.

Considers the problem of procuring mounts for the National Guard when it moved to the border. Reduced standards had to be applied because of the acute need for horses.

588. Colby, Elbridge. "Border Mobilization of 1916," Chap. 5 in *The National Guard of the United States: A Half Century of Progress*. Manhattan: Military Affairs/Aerospace Historian, Kansas State University, 1977.

Interprets the mobilization as an effort by Wilson to prepare for United States involvement in World War I. Admitting that there were problems, Colby praises the National Guard for moving as quickly as it did. He stresses the value of the long period of intensive training for the soldiers.

589. "Dry Run for Destiny." *National Guardsman* 20 (June 1966): 2–9.

Examines the mobilization process and the problems associated with it. The Regular Army was no more ready to receive the militiamen than most of them were ready to enter the army. Some units were prepared for immediate induction and movement.

590. Houston, Donald E. "Oklahoma National Guard on the Mexican Border, 1916." *Chronicles of Oklahoma* 53, no. 4 (Winter 1975–76): 447–62.

The Oklahoma guardsmen willingly accepted mobilization. They were stationed in Texas, receiving training and engaging in a limited amount of fighting with hostile forces.

591. "Life on the Border—A Guardsman's Letter." *Outlook* 113 (August 23, 1916): 946–47.

Describes the climate, food, and wildlife on the border, as well as the very routine duties the writer, a New York State militiaman, and his fellow soldiers performed.

592. Marvin, George. "Marking Time with Mexico." *World's Work* 32 (September 1916): 526–33.

Considers the mobilization of the National Guard and its deployment on the border as a successful bluff that overawed Mexico, despite the poor state of the Guard. Marvin presents a most unflattering portrait of the Guard, especially in comparison to European reserve forces.

593. Mason, Gregory. "Our Citizens in Arms." *Outlook* 113 (July 5, 1916): 546–49.

Stresses the spontaneous enthusiasm of the guardsmen during mobilization, despite shortages of equipment and other problems.

Mason includes the text of the oath the guardsmen took that would permit them to cross into Mexico, if the need arose.

594. "The National Guard Demonstrates Some Facts." *World's Work* 32 (September 1916): 485–86.

   Takes a more favorable view of the performance of the National Guard than did most other accounts, although it was admitted that there were some serious deficiencies. The article suggests a number of changes in structure and organization for the Guard.

595. "Our Unprepared Militia." *Literary Digest* 52 (June 3, 1916): 1617.

   Describes the equipment shortages, the refusals to report for duty, and the general disorganization that emerged when the National Guard in Texas, Arizona, and New Mexico was mobilized. Various newspaper articles are cited or quoted on aspects of this mobilization.

596. Parker, James. "Cavalry and Auto Trucks." *Cavalry Journal* 27 (January 1917): 348–60.

   Reviews an exercise held near Brownsville, Texas, on October 6, 1916. A truck-mounted infantry unit pursued a cavalry force. Parker concludes that truck trains require cavalry escorts in hostile areas and that trucks can increase the mobility of cavalry forces by carrying supplies for them.

597. "A Report on a Two Hundred Mile Experimental Cavalry March." *Cavalry Journal* 27 (January 1917): 361–64.

   Discusses an exercise near Brownsville, Texas, that involved a cavalry truck force on November 4–7, 1916. Soldiers alternated between leading horses and riding in trucks. The vehicles also carried supplies and equipment. There are several references to the use of motor vehicles by the Punitive Expedition.

598. Roberts, Richard C. "The Utah National Guard on the Mexican Border in 1916." *Utah Historical Quarterly* 46 (Summer 1978): 262–81.

   These guardsmen served in Arizona during their mobilization where they received much training in additional to helping protect the border.

599. "The Spirit of the Guardsman." *Outlook* 113 (August 9, 1916): 815.

   Consists for the most part of a letter from a guardsman who acknowledged some hardships and some disorganization but who wanted border problems ended.

600. Stenberg, Richard K. "Dakota Doughboys in the Desert: The Experiences of a North Dakota National Guard Company During the Mexican Border Campaign of 1916–17." *North Dakota History* 71 (Spring–Summer 2004): 50–64.

In addition to guarding the border in Texas, the men of Company E of the 1st North Dakota Infantry Regiment underwent extensive training during their long mobilization.

601. "The Swearing in of the Militia." *Literary Digest* 52 (June 24, 1916): 1866–68.

    Quotes heavily from the *Dallas News* for a description of the entrance of the Texas State Militia into federal service. Some men refused to take the oath, misunderstanding the terms of service.

602. Tackenberg, William D. "'Sore as a Boil But Solid as a Rock': The Arizona National Guard on the Mexican Border, 1916–17." *Journal of Arizona History* 42, no. 4 (Winter 2001): 429–44.

    Arizona troops spent many months on the border, replacing regular army men, who were employed on other duties, including the Punitive Expedition. Limited activity undermined morale, and some desertions occurred.

X
Veracruz, 1914

Ostensibly, the occupation of Veracruz, Mexico, by United States forces between April and November 1914, arose from the Mexican failure to make amends for the temporary arrest of some American sailors at Tampico and several lesser incidents.

In actuality, the occupation was designed to drive General Victoriano Huerta out of the Mexican presidency and was eventually successful. The landing was opposed, and there were casualties on both sides. For a time, another march to Mexico City was discussed, but Huerta's fall and the military weakness of the United States prevented any further moves.

A military government was established, which temporarily improved the daily life of the city's inhabitants and which gave some United States officers valuable experience. At length, after mediation by the major Latin American nations, the United States removed its forces from Veracruz.

Chronology

1913
February 19   A coup d'état by Mexican general Victoriano Huerta brought him to the presidency of Mexico. The Wilson administration in the United States bitterly opposed Huerta throughout his brief regime, making every effort short of complete military occupation to bring about his overthrow. For some time, Huerta held on to power, but eventually Wilson used military force to oust him.

1914
April 9   A few United States sailors were briefly detained by Mexican forces at Tampico, Mexico, as the result of a misunderstanding.

	This minor affair presented the Wilson administration with an opportunity to intervene in Mexico against Huerta.
April 21	After various diplomatic exchanges between the United States and Mexico about the "Tampico incident," the United States Atlantic Fleet, which had been concentrated at Veracruz, landed marines and sailors. Although the object of the incursion had only been to prevent an arms shipment from Germany, significant fighting developed, and all of Veracruz was occupied. In the event, the arms were landed at Puerto Mexico on May 27, but the delay in their delivery diminished their usefulness to Huerta's forces.
April 22	Fighting at Veracruz came to an end. Because Mexican authorities and other Mexicans refused to assist the United States in governing Veracruz during its occupation of the city and the immediate area, United States military officers administered the city. A good deal of what would now be termed "civic action" by military organizations was undertaken. Among many other programs was an extensive public health effort that went far to eliminate malaria.
May 20	Beginning of the "Niagara Falls Conference" in Canada between Mexico and the United States to resolve the Veracruz situation. In addition to the Mexican and United States delegates, there were mediators from Argentina, Brazil, and Chile.
July 1	End of the Niagara Falls Conference. The negotiations were complicated, due in large part to the Wilson administration's efforts to use them to push Huerta out of his presidency. Nevertheless, the conference may have helped to prevent an outright war between the United States and Mexico.
July 15	Huerta's presidency of Mexico ended when he left his country for exile.
November 9	The United States removed its military forces from Veracruz and transferred control to supporters of Venustiano Carranza. The city became Carranza's base of operations for a time.

A. General studies

603. Boot, Max. *The Savage Wars of Peace: Small Wars and the Rise of American Power*, pp. 148–55. New York: Basic Books, 2002.

    Views the Veracruz occupation as a success, because Wilson's goal of ousting Huerta was achieved.

604. Butler, Smedley D. *General Smedley Darlington Butler: The Letters of a Leatherneck, 1898–1931*. Chap. 8, "Watchful Waiting," pp. 135–48, ed. Anne Cipriano Venzon. New York: Praeger, 1992.

This chapter contains important letters from Butler, who eventually commanded a part of the marines deployed in the landing at Veracruz. Before the landing, Butler was assigned the task of developing a plan to advance to Mexico City, an assignment that involved Butler going to that city in civilian clothes to assess the political and military situation in the area between the coast and the capital. Despite his own military accomplishments, Butler believed Wilson overreacted to Mexican hostility.

605. Calhoun, Frederick S. *Power and Principle: Armed Intervention in Wilsonian Foreign Policy.* Chap. 2, "The Power of Civilian Control: Mexico," pp. 39–51. Kent, OH: Kent State University Press, 1986.

Interprets the occupation as Wilson's deft, initial use of military power while avoiding extremes, in this case, the outright invasion of large areas of Mexico. Calhoun reviews the steps Wilson took to dislodge the Mexican dictator Victoriano Huerta and the plans the army prepared for an all-out invasion. Wilson restrained the armed forces and kept the situation under his strict control.

606. Donnell, Guy Renfro. "United States Intervention in Mexico, 1914." Ph.D. diss., University of Texas, 1951.

Describes the attack on Veracruz and the formation of the military government in detail. Much of the dissertation deals with Wilson's relations with Huerta. The Tampico incident, United States and Mexican war plans for a wider conflict, and the debate over the congressional resolution on the Tampico affair are thoroughly discussed.

607. Eisenhower, John S.D. *Intervention! The United States and the Mexican Revolution 1913–1917.* New York: W.W. Norton & Company, 1993.

Although the bulk of this book deals with skirmishes along the Mexican–United States border and the Punitive Expedition, chapters 9 to 12 explore in some detail the Veracruz attack and occupation and the end of the Huerta regime. Eisenhower's account is based in part on archival investigations and is a useful summary of events. A fine map increases the value of the chapter dealing with the attack.

608. Forster, Merlin H. "US Intervention in Mexico: The 1914 Occupation of Veracruz." *Military Review* 57, no. 8 (August 1977): 88–96.

Asserts that the occupation ran smoothly but that worsened relations with Mexico and Latin America far offset any benefits. Forster uses the Quirk and Sweetman studies (items 610, 612) extensively. He examines the negotiations for United States withdrawal in detail.

609. Knopp, Anthony. "The Will of the People: The American Intervention and Mexico, 1914." Ph.D. diss., Texas Technical University, 1973.

Pursues a far-ranging survey of public opinion about the Veracruz occupation. Knopp examines American, Mexican, and European views. He points out that the Veracruz population was unenthusiastic about the occupation. Chapter VII includes several pages on the fighting in the city.

610. Quirk, Robert E. *An Affair of Honor: Woodrow Wilson and the Occupation of Veracruz.* Lexington: University of Kentucky Press, 1962.

Discusses all aspects of the occupation, administration, and evacuation of Veracruz, even sanitary measures and postal service. Quirk is also very thorough in giving physical descriptions of the city. There is a fine bibliographic essay.

611. Schmidt, Hans. *Maverick Marine: General Smedley D. Butler and the Contradictions of American Military History.* Chap. 6, "The Spy: Panama, Mexico," pp. 58–73. Lexington: University Press of Kentucky, 1987.

Reviews Butler's experiences at Veracruz and examines at some length his earlier plan for employing marines to seize Mexico City in a surprise operation. Butler went to Mexico City wearing civilian clothes to investigate the rail route from Veracruz to the Mexican capital, but his ambitious strategy was never used.

612. Sweetman, Jack. *The Landing at Veracruz, 1914: The First Complete Chronicle of a Strange Encounter in April, 1914, when the United States Navy Captured and Occupied the City of Veracruz, Mexico.* Annapolis, MD: United States Naval Institute, 1968.

Contains a detailed account of fighting at Veracruz. There is less on the occupation than in Quirk's study (item 610), however. The book is based on extensive archival research, as reflected in the extensive bibliography, and interviews with participants.

613. Worcester, Dean C. "The Mexican Question in the Light of Philippine Experience." *Outlook* 107 (July 11, 1914): 602–8.

Compares Veracruz's occupation with the beginning of the Philippine–American War. Worcester argues strongly for protecting American citizens abroad, even at the cost of fighting, and emphasizes the civilizing mission of the United States in the Philippines and, potentially, in Mexico.

B. Diplomatic aspects of the occupation

614. Coker, William S. "Naval Diplomacy During the Mexican Revolution: An Episode in the Career of Admiral Frank Friday Fletcher." *North Dakota Quarterly* 40 (Spring 1972): 51–64.

Analyzes conflicts between Great Britain and the United States over the policy to be followed toward Mexico. Coker focuses on relations between Fletcher and the British admiral Cradock during 1913–14. Little space is given to the Veracruz landing; nevertheless, the article is useful background.

615. Grieb, Kenneth J. *The United States and Huerta*. Lincoln: University of Nebraska Press, 1969.

Constitutes an essential source for understanding the background of the occupation, pages 142–77. Especially evident is Wilson's determination to use the Tampico incident to unseat Huerta.

616. Haley, P. Edward. "Comparative Intervention: Mexico in 1914 and Dominica in 1965." In *Intervention or Abstention: The Dilemma of American Foreign Policy*, ed. Robin Higham, pp. 40–59. Lexington: University Press of Kentucky, 1975.

Emphasizes the environment in which United States interventions take place and their objectives. In both countries the American goal was to bring to power a moderate, democratic regime. Haley postulates a sequence of demands that the United States makes upon participants in a civil war before intervening.

617. Hinckley, Ted C. "Wilson, Huerta, and the Twenty-One Gun Salute." *Historian* 22 (February, 1960): 197–206.

Reviews the temporary seizure of American sailors at Tampico that led to Admiral Mayo's insistence that the United States receive a special salute by Mexican forces as an act of atonement. The attack on Veracruz that followed the Mexican refusal to provide such an acknowledgment is discussed briefly.

618. Link, Arthur S. *Wilson: The New Freedom*. Princeton, NJ: Princeton University Press, 1956.

Develops the background of the Veracruz episode, notably relations between Wilson and Huerta, at considerable length in chapters 11 and 12. By the end of 1913, Wilson had resolved to take military and naval action against Huerta. Relatively little is said about the fighting at Veracruz or the occupation of that city.

619. Meyer, Michael C. "The Collapse of a Regime," chap. 10, pp. 100–209, in *Huerta: A Political Portrait*. Lincoln: University of Nebraska Press, 1972.

Supplements other treatments such as Quirk's (item 610), even though Meyer's analysis of the Veracruz crisis is very brief. There is a splendid bibliographic essay that identifies major United States and Mexican archival materials and other items.

620. Slayden, James L. "The A.B.C. Mediation." *American Journal of International Law* 9 (January, 1915): 147–52.

Praises the mediation by Argentina, Brazil, and Chile, and suggests that an effort be made to replace unilateral intervention by the United States with multinational action. Slayden points several times to the cooperation among nations during the Boxer uprising as a model of what should happen when problems emerge in the Western Hemisphere.

C. The attack on Veracruz

621. Aguirre, Cesar Reyes. "The Seizure of Vera Cruz: A Mexican Version." *Infantry Journal* 12 (November 1915): 411–23.

Argues quite correctly that the attack was intended to weaken Huerta's position. The fighting is described in some detail, about half the article being devoted to the battle at the naval school, where Mexican cadets bravely sacrificed themselves.

622. "The Tragic Days of Vera Cruz." Translated by R.W. McNelly. *United States Naval Institute Proceedings* 40 (May–June 1914): 741–58.

Constitutes the first Mexican account of the attack. It is taken from the Veracruz newspaper, *La Opinion*, April 23, 1914. The article contains a full account of the fighting, including the role of the naval cadets.

623. "Under Fire." *Outlook* 107 (June 20, 1914): 380–81.

Praises the courage of United States sailors and marines in the battle of Veracruz and the quality of Admiral Fletcher's report on the capture of the city. The article quotes extensively from the report as to the heroism of a naval officer and two enlisted men.

D. Military government/civic action

624. Cotten, Lyman A. "Handling Refugees." *United States Naval Institute Proceedings* 43, no. 3 (March 1917): 473–82.

Recounts the assistance given to Mexican refugees at Veracruz by the United States battleship *Nebraska*. Cotten also suggests procedures that officers faced with similar situations might adapt to their needs.

625. "How Invasion Helps Vera Cruz." *Literary Digest* 49 (July 25, 1914): 156–57.

Gives much credit to the United States Army for organizing and improving life in Veracruz. Among the areas noted are poor relief, education, and sanitation.

626. Langley, Lester D. *The Banana Wars: United States Intervention in the Caribbean, 1898–1934*. Chap. 8, "Veracruz," pp. 85–95; Chap. 9, "The Rulers of Veracruz," pp. 97–108. Wilmington, DE: SR Books, 2002.

Describes the landing and the resultant occupation, but puts most of its emphasis upon the military government of Veracruz. Langley also notes the trivial nature of the incidents that led to the occupation and credits President Carranza, Huerta's successor, with preventing a United States advance to Mexico City.

627. London, Jack. "The Mexican Conflict." In *Jack London Reports: War Correspondence, Sports Articles, and Miscellaneous Writings*, ed. King Hendricks and Irving Shepard, 126–210. Garden City, NY: Doubleday & Company, Inc., 1970.

Praises the military government of Veracruz. The articles, reprinted from *Collier's*, convey very effectively the atmosphere along the border and at Veracruz right after the attack and occupation, May–June, 1914.

628. McCormick, Medill. "The Army in Vera Cruz." *Outlook* 107 (May 30, 1914): 233–34.

Comments on the lack of preparedness of the army for operations such as the occupation of Veracruz. McCormick also notes the tendency for the Americans to underestimate the military abilities of the Mexicans. Finally, the article refers to the American talk of marching to Mexico City, as occurred in 1847 during the Mexican War.

629. ———. "Vera Cruz: A Crusade for Decency." *Outlook* 107 (July 4, 1914): 527–28.

Discusses some typical cases brought before the United States military courts of Veracruz. The article, like those of Jack London (item 627), does an excellent job of conveying the atmosphere in Veracruz during the temporary occupation. American civilians, Mexican refugees, and other groups are considered.

630. McMillen, Fred E. "San Juan de Ulva Under the American Flag." *United States Naval Institute Proceedings* 62, no. 8 (August 1936): 1147–66.

Describes the capture of the prison by marines from the *North Dakota*, April 27–28, 1914. It became the U.S. Naval Station, Veracruz, until it was turned over to the army on September 19, 1914. The article is in large part a description of the terrible conditions in the prison before the Americans assumed control.

631. Palmer, Frederick. "The American Spirit in Vera Cruz." *Everybody's Magazine* 30 (June 1914): 806–10.

Emphasizes that life is proceeding very normally at Veracruz under the United States occupation. Good treatment of the Mexicans by the United States servicemen ended sniping quickly. Palmer portrays the people of Veracruz as being enthusiastic about United States rule.

632. ———. "Army Housekeeping." *Everybody's Magazine* 31 (August 1914): 198–205.

Discusses the feelings of the army as it waits and governs Veracruz. The soldiers want to fight, but the army is short of vehicles and horses, making an advance to Mexico City impossible.

633. ———. "Watchful Perspiring at Vera Cruz." *Everybody's Magazine* 31 (July 1914): 65–79.

Discusses the daily life in Veracruz at length. The problem of recruiting Mexicans for government posts is mentioned along with a decline in the enthusiasm of the people for American rule. There is some material on the appearance and equipment of the United States troops.

E. Other topics

634. "Army Organization." *Infantry Journal* 11 (July–August 1914): 124–28.

Reflects on army organization in light of the Veracruz operation. Much of the personnel of the army is dedicated to coast defense. Relatively few soldiers are available to fight the lengthy guerrilla campaign that would occur if the United States and Mexico became involved in a war.

635. "Hydroplaning Over Vera Cruz." *Literary Digest* 48 (June 6, 1914): 1375, 1377, 1378–79.

Records the impressions of an American correspondent who viewed Veracruz from a navy seaplane. Such planes were fired upon by the Mexicans during their scouting missions.

636. "Tribute to Those Who Died at Vera Cruz." *Outlook* 107 (May 23, 1914): 139–40.

Describes the funeral in New York City for the 17 sailors and marines who were killed in the battle. Wilson's remarks at the funeral are summarized.

XI
Haiti, 1915–34

American intervention in Haiti was one of the most controversial of the small wars of the United States, and it may have been among the bloodiest.

A significant amount of fighting occurred in 1915–16 and 1918–20. Rioting occurred in 1929, and it helped bring about the United States withdrawal. Aircraft and motor vehicles were employed, but much of the marine service consisted of foot or mounted patrols in the jungle. As in many other countries where its forces were present, the United States formed a constabulary.

The State and Navy departments made a considerable investment in Haiti. While there was no military government, the Navy Department did operate government bureaus for some time. Despite efforts to improve life in Haiti and to introduce democracy, the United States impact ultimately was rather limited.

Chronology

1915

July 28 — United States marines begin landing in Haiti. Continued disorder in that country concerned the Wilson administration, owing to fears that third countries might intervene.

August — Some Haitians began a guerrilla campaign against occupying United States forces, a conflict that lasted until September 1916. Resistance of another sort was demonstrated by the *Union Patriotique* that was established the same month. That organization, representing the point-of-view of the Haitian elite, was an effective voice in Haiti and in the United States against the occupation, and it continued its existence until the United States withdrawal in 1934.

August 24	Haiti and the United States signed a Gendarmerie Agreement that provided for the establishment of a paramilitary police force under United States direction to replace the former Haitian army. The force, initially known as the *Garde d'Haiti*, eventually became the *Gendarmerie d'Haiti*.
November 29	A bilateral treaty effectively made Haiti a protectorate of the United States, and the United States Senate ratified it on May 3, 1916.
1918	
September	Armed resistance to the United States occupation was renewed under the leadership of Charlemagne Massena Péralte, and fighting continued until the middle of 1920. Numerous Haitian insurgents were killed, some of them through air attacks.
1919	
October 7	Haitian nationalist forces made an unsuccessful attack on Port-au-Prince, the capital.
1920	
January 15–16	Second and larger scale nationalist attack on Port-au-Prince. The attackers were repelled with significant losses.
1921	
July 19	Senate Select Committee on Haiti and Santo Domingo created as a subcommittee of the Senate Foreign Relations Committee. It held extensive hearings in the United States and in the countries concerned before reporting in 1922. The committee favored a relatively rapid withdrawal from the Dominican Republic but cautioned that intervention in Haiti had to be continued for some time.
1929	
December 6	More than 20 Haitian demonstrators were killed by marines at Les Cayes. Soon President Herbert Hoover received Congressional authorization to send a commission of inquiry to Haiti.
1930	
March 26	President's Commission on Conditions on Haiti, also known as the Forbes Commission from its chair W. Cameron Forbes, presented a report critical of the Haiti intervention and recommended withdrawal of the United States from Haiti, at least by 1936. The same year the Moton Commission, chaired by Tuskegee Institute president Robert Russa Moton

and consisting of African–Americans, reported on education in Haiti.

1933

August 7   An Executive Accord prepared the way for the removal of United States forces and for Haiti to resume its status as a sovereign nation.

1934

August 15   The occupation regime in Haiti ended with the withdrawal of United States forces.

A. Congressional hearings and official reports

637. U.S. Congress. Senate. Select Committee on Haiti and Santo Domingo. *Inquiry into Occupation and Administration of Haiti and Santo Domingo, Pursuant to S. Res. 112*. Washington, DC: Government Printing Office, 1922.

    Emphasizes Haiti and constitutes one of the most important printed primary sources on the United States occupation of Haiti. There is a huge amount of contemporary documentation in the hearings, including depositions relating to alleged atrocities by marines and constabulary officers.

638. U.S. State Department. *Report of the President's Commission for the Study and Review of Conditions in the Republic of Haiti*. Washington, DC: Government Printing Office, 1930.

    Presents conclusions that point toward United States withdrawal. The commission was appointed in part because of the riots of 1929. The commission notes the existence of opposition to the occupation. Several pages deal with the Marine Brigade and the constabulary. The *Report* includes a detailed plan for Haitianization of the constabulary or *Gendarmerie d'Haiti*.

B. General studies

639. Adams, Thomas K. "Intervention in Haiti: Lessons Relearned." *Military Review* 76, no. 5 (September–October 1996): 45–56.

    Able survey of all aspects of the occupation that demonstrates mastery of the secondary materials. Adams, an expert in low-intensity conflict, makes useful comparisons between the events from 1915 to 1934 with those of the mid-1990s. In his view, there is little prospect for democracy in Haiti.

640. Balch, Emily Greene, ed. *Occupied Haiti*. New York: The Writers Publishing Company, Inc., 1927.

Contains the results of a study for the Women's International League for Peace and Freedom. Focusing on civic action and economic and social conditions in Haiti, the report was generally critical of United States policies.

641. Boot, Max. *The Savage Wars of Peace: Small Wars and the Rise of American Power*. Chap. 7, "Lords of Hispaniola: Haiti, 1915–34; Dominican Republic, 1916–24," pp. 156–81. New York: Basic Books, 2002.

Presents a highly favorable view of the occupation but does not ignore the abuses that occurred. Boot contends (p. 181) that Haiti and the Dominican Republic were "inhospitable soil" for democracy. The section dealing with the Dominican Republic is not included in this bibliography because of its brevity.

642. Butler, Smedley D. *General Smedley Darlington Butler: The Letters of a Leatherneck*. Chap. 9, "Chief of Police of Haiti," pp. 149–202, ed. Anne Cipriano Venzon. New York: Praeger, 1992.

Venzon describes the events leading to the United States intervention in Haiti in 1915 and presents a number of Butler's letters, which provide a detailed picture of marine experiences in Haiti and discuss the political situation. These materials are a significant primary source for understanding the occupation. Venzon includes extensive notes that identify those people mentioned and that provide important explanatory information.

643. Calhoun, Frederick S. *Power and Principle: Armed Intervention in Wilsonian Foreign Policy*. Chap. 3, "The Power of Ideology: Santo Domingo and Haiti," pp. 86–103. Kent, OH: Kent State University Press, 1986.

Concisely discusses United States interests in Haiti. Calhoun argues that Wilson initially shaped his policy toward Haiti on his earlier approach to the Dominican Republic from 1913 to 1915, that is, before his intervention in the Dominican Republic.

644. Davis, H.P. *Black Democracy: The Story of Haiti*. New York: Dodge Publishing Co., 1936.

Praises United States involvement in Haiti, but notes the existence of much opposition from both the elite and the people. More than a third of the text deals with the occupation. One chapter summarizes the work of the congressional committee that investigated American operations in Haiti and the Dominican Republic.

645. Douglas, Paul H. "The American Occupation of Haiti." Parts 1, 2. *Political Science Quarterly* 42 (June, September, 1927): 228–58; 368–96.

Stresses political history and economic and financial topics. It is relatively well balanced, despite Douglas' opposition to the intervention. Douglas admits that little is known about the decisions made by United States political leaders concerning Haiti. There is very little on the military aspects.

646. Evans, Frank E. "Salient Haitian Facts." *Marine Corps Gazette* 15, no. 2 (February 1931): 14–17, 65, 67, 69.

Presents some highlights from a marine handbook that, for financial reasons, was never published. The article is a general treatment of the history, geography, topography, climate, historic sites, and governmental structure of Haiti.

647. Gruening, Ernest H. "Conquest of Haiti and Santo Domingo." *Current History* 16 (March 1922): 885–96.

Discusses the United States search for pliant Haitian politicians who would cooperate with the United States and sign the desired agreements. Gruening stresses what he regards as the high-handed behavior of United States authorities in Haiti. Despite the title, the bulk of the article deals with Haiti.

648. "Haiti and Its Regeneration by the United States." *National Geographic Magazine* 38 (December 1920): 497–511.

Praises United States efforts to improve living conditions in Haiti and to reform Haitian political life. Several photographs show marines and constabulary in camp and in the field. The text deals with the Caco revolt of 1918–20 and with other military topics to some extent.

649. Heinl, Robert Debs, Jr., and Nancy Gordon Heinl. *Written in Blood: The Story of the Haitian People 1492–1971*. Boston, MA: Houghton Mifflin Company, 1978.

Seeks to explode several myths about the occupation, notably that Haitian casualties were exceptionally heavy and that southern officers were sent purposefully to Haiti. The authors deal with the fighting in 1915–18 in detail. This popularly written study is an important, well-documented treatise. Portions of it appeared in the November and December, 1978, and January, 1979, issues of the *Marine Corps Gazette*.

650. Hoffman, Jon T. *Chesty: The Story of Lieutenant General Lewis B. Puller, USMC*. Chap. 2, "'Great Lessons of Warfare': Haiti, 1919–21," pp. 23–41. New York: Random House, 2001.

This portion of the biography of a man who probably more than anyone else represents the skill, dedication, and toughness of the marine corps, focuses on Puller's experiences and contributions, of course, but at the same time Hoffman provides a succinct account of the Haitian campaign of 1918–20. Hoffman notes the charges made of atrocities in Haiti, although Puller was never involved in such matters.

651. Kelsey, Carl. "American Intervention in Haiti and the Dominican Republic." *Annals of the American Academy of Political and Social Science* 100 (March 1922): 166–202.

Constitutes an important study that is highly favorable to the United States. There is little on the fighting in Haiti.

652. Logan, Rayford W. "James Weldon Johnson and Haiti." *Phylon* 32 (Winter 1971): 396–402.

Reviews Johnson's experiences in Haiti and his series of articles in the *Nation* in 1920 that helped arouse protest about the occupation. Logan summarizes the articles, which are sharply critical of United States involvement with Haiti.

653. Millett, Richard, and G. Dale Gaddy. "Administrating the Protectorates: The U.S. Occupation of Haiti and the Dominican Republic." *Revista/Review Interamericana* 6 (Fall 1976): 383–402.

Analyzes the nature of United States control in the two countries and the degree to which American methods of administration actually furthered Washington's goals. The authors are very critical of what they view as inadequate planning by the State Department.

654. Millspaugh, Arthur C. *Haiti Under American Control 1915–1930*. Boston, MA: World Peace Foundation, 1931.

Carries the story of the occupation through the Forbes Commission of 1930. Approximately 20 percent of the text consists of documents bearing on the intervention. Millspaugh served in the United States administration in 1927–29. His is a balanced treatment.

655. Montague, Ludwell. *Haiti and the United States, 1714–1938*. Durham, NC: Duke University Press, 1940.

Provides a diplomatic analysis of the Haitian occupation. Perhaps the most useful chapter deals with "reconstruction" and details civic action and financial reforms. Another chapter covers the Haitianization of the government.

656. Musicant, Ivan. *The Banana Wars*. Chap. 5, "Haiti 1915–34," pp. 157–234. New York: Macmillan Company, 1990.

After reviewing Haitian history before 1915 briefly, Musicant offers a detailed description of the initial landings and the two Caco wars. Although relying largely on printed primary sources and secondary accounts for this chapter, Musicant includes some material from archives.

657. Pamphile, Leon D. "The NAACP and the American Occupation of Haiti," *Phylon* 47, no. 1 (4th Quarter 1986): 91–100.

The relatively new but increasingly powerful National Association for the Advancement of Colored People began an effort in 1920 to end the occupation of Haiti. James Weldon Johnson, who had roots in Haiti himself and who was a Republican Party leader, investigated conditions in Haiti and contacted elite figures in the country. His published criticisms aroused considerable attention, and the occupation became a significant issue in the presidential campaign of 1920. The NAACP continued to watch the Haitian situation and to criticize the occupation until 1934.

658. Posner, Walter H. "American Marines in Haiti, 1915–22." *Americas: A Quarterly Review of Inter-American Cultural History* 20 (January 1964): 231–66.

Stresses the financial aspects of the occupation. The article is a good discussion of the early years of United States involvement, based on secondary materials. There is little on the military aspects except for the controversy over casualties among blacks during the Caco revolt of 1918–20.

659. Schmidt, Hans. *Maverick Marine: General Smedley D. Butler and the Contradictions of American Military History*. Chap. 7, "The Haitian General," pp. 74–95. Lexington: University Press of Kentucky, 1987.

Early on, Butler was highly critical of some United States decision-makers in Haiti for what he regarded as their overly cautious policies. He led operations in part of Haiti during the first campaign in Haiti, between 1915 and 1916, and later he commanded the Haitian constabulary for some time. He initiated a compulsory service system for involving Haitians in road-building, but after his departure the program became much more oppressive, leading to the second period of fighting in Haiti between 1918 and 1920.

660. ———. *The United States Occupation of Haiti, 1915–1934*. New Brunswick, NJ: Rutgers University Press, 1971.

Highlights the racial and cultural differences that impeded communications between the marines and the Haitians. This is a judicious treatment of the controversial protectorate. Several sections deal with the fighting in 1915–16, 1918–20, and the riots of 1929. A 1995 reprint by Rutgers University Press includes a foreword by Stephen Solarz that compares events between 1915 and 1934 with those that occurred in the 1990s. Solarz takes a generally favorable view of the 1994 intervention.

661. Schreadley, Richard Lee. "Intervention—The Americans in Haiti, 1915–34." Ph.D. diss., Tufts University, 1972.

Views the intervention as a humanitarian endeavor and takes a highly sympathetic tone regarding the problems of the occupation authorities. Schreadley compared Vietnam and Haiti, criticizing both interventions because they were unrelated to strategic requirements and occurred in the absence of precise policy goals.

662. Shannon, Magdaline W. *Jean Price-Mars, The Haitian Elite, and the American Occupation, 1915–1935.* London: Macmillan Press, 1996.

Important study of elite resistance to the United States intervention. Price-Mars was a major intellectual figure in Haiti, but this book goes beyond being a biography. It is really a history of political developments during the occupation, which contains much information that would be difficult to find in other publications.

C. The decision to intervene

663. Baker, C.S. "Some Colorful Haitian History." *United States Naval Institute Proceedings* 50, no. 5 (May 1924): 723–43.

Asserts that United States intervention was the only alternative to a French landing. Baker discusses the revolution of 1915, the first days of the United States occupation, and the presidential election of 1915.

664. Blassingame, John W. "The Press and American Intervention in Haiti and the Dominican Republic, 1904–20." *Caribbean Studies* 9 (July 1969): 27–43.

Asserts that the major newspapers and periodicals in the United States generally supported the interventions in Haiti and the Dominican Republic until 1919. Both these nations and their peoples were depicted in extremely negative ways in the media.

665. Plummer, Brenda Gayle. *Haiti and the Great Powers, 1902–1915.* Baton Rouge: Louisiana State University Press, 1988.

Essential book for understanding the United States occupation. Plummer explores Haitian civilization and examines Haiti's relations, broadly interpreted, with Britain, France, Germany, and the United States. She includes much information about and analysis of the foreign economic penetration of Haiti. The extensive movement of United States warships to influence Haitian political developments before 1915 is also discussed.

D. The campaigns

666. Belton, Fred, and John Rogers. "Unsung Heroes of the Marine Corps—No. 3." *Leatherneck* 14, no. 6 (June 1931): 14, 44.

Haiti, 1915–34 157

Describes the efforts of Marine Sergeant Louis N. Bertol to establish a protected hamlet. Bertol later became commander of the Haitian president's Palace Guard. His greatest challenge came during the 1929 riots when the president's safety was threatened.

667. Butler, Smedley D. *Old Gimlet Eye: The Adventures of Smedley D. Butler As Told to Lowell Thomas.* New York: Farrar & Rinehart, Inc., Publishers, 1933.

Describes the early Caco campaigns, 1915–17 and the difficult living conditions in the field. The high point is the capture of Fort Rivière. Butler recounts many amusing encounters with the Haitian political elite in his role as commander of the constabulary, the *Garde d'Haiti*.

668. Gray, John A. "Boucan Carre." *Marine Corps Gazette* 16, no. 11 (November 1931): 28–32.

Describes an attack on a Caco camp amid rumors of increasing Caco activity. Gray feels that Péralte, the Caco chief, was given too much time to recruit a force before an attack was mounted against him. This was one of the first battles in which the *Garde d'Haiti* participated.

669. ———. "Cul de Sac." *Marine Corps Gazette* 16, no. 2 (February 1932): 41–44.

Recounts Gray's experiences in leading a *Garde* detachment against Benoit Batraville, a major Caco leader, in June, 1919. A United States attack on a camp netted supplies, but did little other damage. The force was then attacked by the Cacos. A later ambush by the *Garde* brought some results, but Batraville escaped.

670. Greathouse, G.H. "King of the Banana Wars." *Marine Corps Gazette* 44, no. 6 (June 1960): 28–33.

Recounts the famous marine story of Sergeant Herman H. Hanneken, who slipped into bandit leader Charlemagne Péralte's camp and killed him.

671. Heinl, Robert Debs, Jr. *Soldiers of the Sea: The United States Marine Corps, 1775–1962.* Annapolis, MD: United States Naval Institute, 1962.

Provides a brief but very fine account of fighting in Haiti, pages 170–80, 234–47.

672. Hill, Walter N. "A Haitian Reconnoissance" [sic]. *Marine Corps Gazette* 2, no. 3 (March 1917): 31–36.

Offers a participant's view of a marine patrol to counter a suspected revolt by the Piquettes, a Haitian nationalist group. The article is primarily a description of the people encountered and the country as it appeared during the patrol along the southern coast from Les Cayes.

673. "L.D." "Marine Manoeuvres [sic] Through Hatien [sic] Eyes." Translated by W.P. Upshur. *Marine Corps Gazette* 8, no. 3 (March 1923): 21–23.

Reviews briefly the marine exercises. They included a mock attack on a small fort and target practice. These operations took place at Camp General Russell near Port-au-Prince. The anonymous Haitian newspaper reporter's account originally appeared in *L'Essor*, February 14, 1923.

674. Millett, Allan R. in *Semper Fidelis: The History of the Marine Corps*. Chap. 7, "Hispaniola," pp. 178–211. Revised, expanded edition. New York: Free Press, 1991.

Depicts the effects of World War I on the military government in a highly readable, analytical account of the occupation. The chapter includes a lengthy discussion of the *Garde d'Haiti*.

675. "One Man Armies of Haiti." *Leatherneck* 14, no. 9 (September 1931): 12–13.

Begins with a review of the fighting at Thomonde in December, 1920. Following the Caco raid, a marine sergeant, James M. Darmond, moved into the enemy camp with a few companions and captured the rebel leader. Another episode that the article discusses involved a very similar set of circumstances: an attack on a rebel camp by Darmond with only a few men.

676. Stahley, Carl E. "Bertol of the Black Republic." *Leatherneck* 19, no. 2 (February 1936): 5, 56–57.

Focuses on Louis N. Bertol's role in guarding the life of Haitian president Borno. Owing to the large number of enemies Borno had incurred, this was a difficult task. The article also discusses Bertol's service in the Caco campaigns and his later assignments in the marine reserve.

677. Thompson, L.E. "The Haitian Campaign of 1919." Parts 1, 2. *Leatherneck* 11, nos. 10, 11 (October, December, 1928): 3, 53; 37, 54.

Describes life in an outpost at Boucan-Carre and combat patrols in that area. The article consists of what may be termed vignettes from the Caco wars. Despite the rough writing, this is an interesting series. The last article in the series states that more parts would be published, but, apparently, they never were.

E. Battle of Fort Rivière

678. Buckner, David N. *A Brief History of the 10th Marines*. Washington, DC: Headquarters, U.S. Marine Corps, History and Museums Division, 1981.

Examines the role of marine artillery in Haiti, August–November, 1915, pages 5–10. The climax is the attack on Fort Rivière when three marines, including Smedley Butler, earned the Medal of Honor.

679. Snyder, H.W. "Butler at Fort Rivière." *Marine Corps Gazette* 64, no. 11 (November 1980): 83–87.

Stresses Butler's careful planning for his daring attack. Butler combined dash with careful thinking. Snyder gives due credit to the two enlisted men who accompanied Butler in the initial reconnaissance that led to the fall of the insurgent fort.

680. Thrasher, Thomas E. "The Taking of Fort Rivière." *Marine Corps Gazette* 15, no. 2 (February 1931): 31–33, 64.

Describes the fort in considerable detail, and briefly reviews its capture. A useful chart depicts troop movements during the assault. Thrasher presents an important participant's account of the battle.

F. The role of air power

681. Bubur, K.F. "Foreign Aviation Field an Ideal Marine Base." *Leatherneck* 7 (December 20, 1924): 3–4, 19.

Describes the physical surroundings, accommodations, and the routine duties of the marine aviation unit in Haiti. Bubur emphasizes support activities, such as carrying mail, rather than tactical operations.

G. Military advisory activities

682. Bride, Frank L. "The Gendarmerie d'Haiti." *Marine Corps Gazette* 3, no. 12 (December 1918): 295–98.

Discusses the initial recruitment and training of troops for the *Gendarmerie* and its early achievements in maintaining security and carrying out civic action programs.

683. "The Carbon Copy Marine Corps." *Leatherneck* 17, no. 9 (September 1934): 8–9, 48–49.

Emphasizes the law enforcement and civic action functions of the *Gendarmerie*. The article also describes the ceremony that took place on August 1, 1934, when the *Gendarmerie* became a Haitian force officially. It examines the situation in Haiti in 1915 and Admiral Caperton's response to it at some length.

684. "The Garde D'Haiti." *Leatherneck* 14, no. 9 (September 1931): 9–11, 53.

Emphasizes the early years. Few marines had competence in French, which complicated training. Enemy contacts began as early as 1916. Several paragraphs deal with the fighting in 1918–20. Another section discusses *Garde* reactions to the riots of 1929 and the subsequent fire-setting campaign.

685. "The Haitianization of the Garde d'Haiti." *Leatherneck* 14, no. 3 (March 1931): 12–14, 54.

    Describes the beginnings of Haitianization of the force and the preparation of regulations and training manuals, based in part on French Army publications. The officer training program is examined in detail. The author notes that the Moton Commission of 1930 assisted in developing appropriate training methods.

686. "The Haitien Gendarmerie." *Marine Corps Gazette* 11, no. 6 (June 1926): 73–81.

    Reviews the Haitian Army as it was in 1915 and its replacement by the marine-trained *Gendarmerie*. The article contains very detailed information on the composition and compensation of the *Gendarmerie*.

687. "History of the Gendarmerie D'Haiti." Parts 1, 2, 3, 4. *Leatherneck* 10, nos. 5, 6, 9, 11 (May, June, September, November, 1927): 29; 33; 8; 10.

    Stresses the critical function of marines, especially enlisted men, in training and supervising members of the *Gendarmerie*. There is an excellent description of the duties of the marines attached to the *Gendarmerie*. Several paragraphs are concerned with the fighting in Haiti during and shortly after World War I.

688. McCrocklin, James H., comp. *1915 Garde D'Haiti 1934*. Annapolis, MD: United States Naval Institute, 1956.

    Constitutes the most complete published account of the marine organization and training of the *Gendarmerie*. Many manuscript materials were apparently used for the first time during the preparation of this book. The intervention and civic action operations are discussed and the Haitianization process is explored in depth. Chapter IV examines the combat actions in Haiti.

689. Rogers, John J. "The Haitianization of the Garde." *Leatherneck* 16, no. 10 (October 1933): 3–5, 51.

    Begins with the text of the order directing completion of Haitianization by October 1, 1934. Rogers describes training and construction activities of the *Garde*, among other topics. By the fall of 1933, only 33 marines were still attached to the *Garde*.

690. "Schools Conducted in Haiti." *Leatherneck* 6 (September 8, 1923): 4–5.

Reviews the compulsory education program for *Gendarmerie* members. The courses, conducted by Haitians, included basic writing and mathematics and the rudiments of Haitian law.

691. "What Have the Marines Done in Haiti? The Gendarmerie d'Haiti." Parts 1, 2, 3, 4. *Leatherneck* 8 (March 7, 14, 21, 28, 1925): 1, 3, 6; 1–2; 2–3; 1.

    Analyzes the organization of the *Garde* at considerable length. The series also provides a rather full discussion of the Caco campaigns.

H. The occupation regime

692. "American Marines in Haiti Exonerated." *Current History* 16 (August 1922): 836–41.

    Summarizes Senate committee findings on Haiti.

693. Callcott, Wilfrid Hardy. *The Caribbean Policy of the United States, 1890–1920*. Baltimore, MD: Johns Hopkins University Press, 1942.

    Criticizes severely the United States administration during the early years of the protectorate. This account contains a good deal of information on military operations, as well as political history, pages 404–22, 476–84.

694. Coffey, R.B. "A Brief History of the Intervention in Haiti." *United States Naval Institute Proceedings* 48, no. 8 (August 1922): 1325–44.

    Examines the early months of the occupation at length. The text contains copies of some of the first operational orders for the occupation as well as several maps of Port-au-Prince and other parts of Haiti. One section deals with the fighting in northern Haiti during the fall of 1915.

695. Craige, John H. "The Haitian Situation." *Marine Corps Gazette* 15, no. 3 (March 1930): 16–20.

    Reviews the Haitian governmental structure and the record of President Borno and Borno's political prospects. Craige also describes the economic and political problems of Haiti on the eve of the 1930 elections and notes the absence of an independent legislative body.

696. Healy, David. *Gunboat Diplomacy in the Wilson Era: The U.S. Navy in Haiti, 1915–1916*. Madison: University of Wisconsin Press, 1976.

    Focuses on Admiral William B. Caperton, who commanded the Atlantic Fleet's Cruiser Squadron and who oversaw the early stages of the occupation. Healy emphasizes the importance of studying decisions by United States commanders in the field. The work is primarily a governmental and political study.

697. Munro, Dana G. *Intervention and Dollar Diplomacy in the Caribbean 1900–1921.* Chap. 8, "Intervention in Haiti," pp. 326–87. Princeton, NJ: Princeton University Press, 1964.

Suggests that after 1916, responsibility for governing Haiti was cloudy, although the marine commander remained the most important United States official in the country.

698. ———. *The United States and the Caribbean Republics 1921–1933.* Chap. 4, "The Treaty Regime in Haiti," pp. 71–115. Princeton, NJ: Princeton University Press, 1974.

Describes the evolution of the occupation regime. Much of the chapter might be termed a financial history of the occupation. Munro is quite critical of the United States authorities in terms of the efficiency with which they governed Haiti.

699. Perkins, Whitney T. *Constraint of Empire: The United States and the Caribbean Interventions.* Westport, CT: Greenwood Press, 1981.

Describes in detail the process by which the United States assumed control of Haiti. Perkins also discusses the relationships between the Navy and State departments' representatives in Haiti. For some time, little progress was made in developing the country. Perkins' analysis is particularly valuable because he includes a discussion of political resistance to the American regime.

700. Renda, Mary A. *Taking Haiti: Military Occupation and the Culture of U.S. Imperialism, 1915–1940.* Chapel Hill: University of North Carolina Press, 2001.

Outstanding study of the occupation. Renda finds that paternalism was the key element in United States attitudes and actions during the occupation. According to Renda, what she calls "the culture of paternalism" was the "fabric that helped to determine the material practices of the thousands of marines and the scores of sailors and nonmilitary personnel who ruled Haiti for nineteen years" (p. 303). Despite the cultural emphasis of the book, Renda includes much material about the military campaigns.

701. Russell, John H. "The Development of Haiti During the Last Fiscal Year." *Marine Corps Gazette* 15, no. 6 (June 1930): 77–115.

Expresses the view that the marines were essential in Haiti as a force to support the small, but loyal, *Gendarmerie.* Russell, the American high commissioner, recounts the events of the year 1928–29, with emphasis on the strikes and riots that occurred.

702. "The Year in Haiti: Strikes, Riots, Investigations, Elections." *Marine Corps Gazette* 15, no. 11 (November 1930): 23–25, 61–64.

Begins with an examination of the actions of President Borno late in 1929, including his refusal to sanction the election of a legislative body. The article describes the riots, quotes the martial law declaration of December 4, 1929, and concludes with a discussion of the Forbes Commission report and the Haitian congressional elections of 1930.

I. Civic action

703. Belton, Fred. "History and Development of the Haiti Fire Department." *Leatherneck* 14, no. 7 (July 1931): 9, 50–51.

    Explains that setting fires was a political and military tactic in Haiti. Fire was used as a dangerous weapon in 1930 in particular.

704. ———. "Police Department, Port au Prince, Haiti." *Leatherneck* 14, no. 6 (June 1931): 7–8, 48–49.

    Asserts that crime had been reduced a great deal during the United States occupation. Traffic safety, too, had improved, within the limitations imposed by the age of the city. The police had been controlled by the marines from the beginning of the occupation.

705. Conrad, Charles. "A Year in Haiti's Customs and Fiscal Service." *United States Naval Institute Proceedings* 49 (April, 1923): 605–24.

    Emphasizes the improvised nature of initial operations in Haiti. Conrad was one of the 11 Pay Corps officers selected to supervise the Haitian customs service in 1915–16.

706. Duncan, G.H. "The Corps of Civil Engineers, U.S. Navy, in Haiti." *United States Naval Institute Proceedings* 56, no. 3 (March 1930): 209–19.

    Records the fine contributions made by public works projects such as water supply systems, roads, parks, bridges, and lighthouses that were planned and coordinated by naval officers. Duncan and his fellow officers found this duty both challenging and enjoyable. Many photographs of the impressive work done precede and follow the article.

707. Engle, Eloise. "King Faustin II." *Marine Corps Gazette* 52, no. 11 (November 1968): 49–54.

    Describes the governing of La Gohave, a Haitian island, by Marine Master Sergeant Faustin Wirkus, 1924–28. He won so much respect from the local inhabitants that they made him their king. Objections by the Haitian government eventually led to his transfer.

708. McMillen, Fred E. "Some Haitian Recollections." *United States Naval Institute Proceedings* 62, no. 4 (April 1936): 522–36.

Describes the work of the customs service and living conditions in Petit, where McMillen served as Port Captain in 1915–16.

709. Myers, Robert H. "The Marine 'King.'" *Leatherneck* 27, no. 1 (January 1944): 33–34.

Discusses Wirkus' installation as "King Faustin II." It also deals with his later marine career.

710. Wirkus, Faustin, and Dudley Taney. *The White King of La Gohave*. Garden City, NY: Doubleday, Doran & Company, Inc., 1931.

Relates Wirkus' own story of his rule as "king" of the Haitian island of La Gohave. The second section refers to Wirkus' service on the Haitian mainland, where he participated in patrols against the guerrillas.

711. Wise, Frederick May. "The Occupation of a Haitian Town." *United States Naval Institute Proceedings* 57, no. 12 (December 1931): 1629–32.

Deals with the occupation of Jeremie in southern Haiti in 1915. Wise emphasizes the need to negotiate with the Haitians and asserts that marine discipline prevented disorder in Haiti.

J. Haiti as a duty station

712. Brady, John D. "Haiti." *Marine Corps Gazette* 9, no. 6 (June 1924): 149–56.

Provides background information for officers going to Haiti. It deals primarily with living conditions and facilities, even including a detailed guide to garages in Port-au-Prince.

713. Coyle, Randolph. "Service in Haiti." *Marine Corps Gazette* 1, no. 12 (December 1916): 343–48.

Lists the towns held by the marines, and provides a description of the climate and living conditions for marines.

714. "Living Conditions in the Marine Corps." *Marine Corps Gazette* 11, no. 6 (June 1926): 203–9.

Discusses accommodations, cost of living, and the availability of servants.

715. Smith, Emily Kaighn. "The Story of a Volunteer Hostess." *Marine Corps Gazette* 7, no. 12 (December 1922): 374–76.

Describes the recreation center Smith opened for marines at Port-au-Prince in 1920. Her objective was to prevent rowdiness and drunkenness on the part of the marines.

K. The decision to withdraw

716. Dodge, R.A. "Massacre in Haiti," *United States Naval Institute Proceedings* 120, no. 11 (November 1994): 60–64.

    Examines the reasons for a clash between Haitians and marines in 1929 that soon led the Hoover administration to reassess its stance toward the continuing occupation of Haiti.

717. Ellis, L. Ethan. *Republican Foreign Policy, 1921–1933*. New Brunswick, NJ: Rutgers University Press, 1968.

    Outlines the efforts made to end the intervention, pages 270–76. Ellis gives Hoover great credit for trying to eliminate all intervention in Latin America on principle.

718. Grieb, Kenneth J. *The Latin American Policy of Warren G. Harding*. Chap. 5, "A Lower Profile in Haiti." Fort Worth: Texas Christian University Press, 1976.

    Examines Harding's efforts to reduce the United States presence in Haiti. Among the topics discussed is the special Senate investigating committee of 1921–22. Grieb makes excellent use of manuscript sources.

719. Gruening, Ernest. "The Issue in Haiti." *Foreign Affairs* 11 (January, 1933): 279–89.

    Urges strongly that the United States withdraw from Haiti as soon as possible and cease involving itself in Haitian affairs.

720. ———. "The Withdrawal from Haiti." *Foreign Affairs* 12 (July, 1934): 677–79.

    Praises President Roosevelt for removing the marines from Haiti.

721. Holly, Alonzo Potter Burgess. "Our Future Relations with Haiti." *Annals of the American Academy of Political and Social Science* 156 (July, 1931): 110–15.

    Pleads eloquently for better relations between the United States and Haiti. Holly gives President Hoover credit for moving toward more freedom for the Haitians.

722. Munro, Dana G. *The United States and the Caribbean Republics 1921–1933*. Chap. 9, "Withdrawal from Haiti," pp. 309–41. Princeton, NJ: Princeton University Press, 1974.

    Describes the rioting of 1929 and the work of the Forbes Commission. Munro also reviews the State Department's perspective on the conclusions of the commission. There were substantial conflicts between

United States representatives in Haiti and the Haitian government until 1932, when relations began to improve.

723. Spector, Robert M. *W. Cameron Forbes and the Hoover Commissions to Haiti (1930)*. Lanham, MD: University Press of America, 1985.

Exhaustive analysis of the Forbes Commission, which led eventually to the withdrawal of United States forces. Chapter 5 discusses the Moton Commission on Education. This is a judicious treatment. The extensive research is reflected by the thorough bibliography.

724. ——. "W. Cameron Forbes in Haiti: Additional Light on the Genesis of the 'Good Neighbor Policy.'" *Caribbean Studies* 6 (July 1966): 28–45.

Asserts that the Forbes Commission convinced Hoover that the marines should be withdrawn from Haiti and thereby made a step toward the Good Neighbor Policy. Spector uses the Forbes papers, which had not been available before.

L. The marine withdrawal

725. Fegan, J.C. "After Nineteen Years We Leave Haiti." *Marine Corps Gazette* 19, no. 8 (August 1934): 21–24.

Indicates that United States successes in improving conditions in Haiti made withdrawal possible two years early. Among the successes noted was the attraction of foreign capital to Haiti. The article also reviews the general course of the occupation.

726. "Final Ceremonies in Haiti." *Marine Corps Gazette* 19, no. 11 (November 1934): 20–21.

Documents the departure of the marines and the completion of Haitianization of the *Gendarmerie*. The General Order establishing the *Gendarmerie* as a Haitian force is quoted, as is the official statement on the withdrawal by Secretary of State Cordell Hull.

XII

Dominican Republic, 1916–24

Several marine landings took place before 1916 because of the unsettled state of the country. As early as 1904–5, the United States assumed responsibility for customs operations. When another revolution broke out in 1916, American marines and sailors landed, and on November 29, 1916, a military government was proclaimed.

There was fairly heavy fighting in 1916 and again in 1921–22. As in Haiti, there were experiments with aircraft. Trucks also made an early appearance as supply vehicles and as primitive armored cars. In 1920, Warren G. Harding pledged that he would withdraw from Haiti and the Dominican Republic. Despite setbacks during the negotiations, a treaty was finally achieved, and the marines departed, leaving a constabulary headed by Rafael Trujillo.

Chronology

1905	The United States first intervened by forcing the Dominican Republic to agree to turn over the administration of its customs system to the United States. The purpose was to ensure that foreign creditors were repaid properly, thereby making military action by one or more European powers unnecessary.
1916	
May 15	United States forces began landing to deal with the political conflicts that had divided Dominicans for some time.
November 29	The United States proclaimed a military government in the Dominican Republic, and United States naval and marine officers exercised complete authority over the country for years. Measures such as rigorous press censorship alienated

many Dominicans. A complete occupation occurred because the United States was unable to persuade enough Dominican political leaders to form a government that would carry out the policies that the United States demanded. Substantial fighting occurred for years as Dominican nationalists maintained their resistance to the military government.

1917
April 7   The *Guardia Nacional Dominicana* (later called the *Policia Nacional Dominicana*) was established as a paramilitary police force to replace earlier military and law enforcement bodies, which were regarded by United States authorities as thoroughly politicized.

1919
April   The marines deployed an aviation squadron to assist in their extensive counterinsurgency efforts. This innovation improved communications between marine units, but, otherwise, air operations did not change the nature or the dimensions of the conflict much.

1920
February   *Union Nacional Dominicana* established to express the extreme opposition of Dominican nationalists to the continued United States occupation and the military government that had been imposed.

November   The election of Warren G. Harding as president of the United States brought significant changes to the United States policy toward the Dominican Republic. The Republicans, among them Harding, had been critical of Woodrow Wilson's many interventions in the Caribbean area.

1921
June   United States military governor Rear Admiral Samuel S. Robison proposed the "Harding Plan" for the withdrawal of direct United States control of the Dominican Republic. Dominican opinion flatly rejected the plan because of the numerous conditions attached to it.

July 19   Senate Select Committee on Haiti and Santo Domingo created as a subcommittee of the Senate Foreign Relations Committee. It held extensive hearings in the United States and in the countries concerned before reporting in 1922. The committee favored a relatively rapid withdrawal from the Dominican Republic but cautioned the intervention in Haiti had to be continued for some time.

September	Dominican nationalist fighters announced their determination to keep fighting until the United States relinquished control of the Dominican Republic. This resolute stand compelled United States decision-makers to move away from military measures toward a political settlement.
1922	
May	A series of surrenders of guerrilla bands took place during the spring, bringing conflict to an end. The prospect of a United States withdrawal and an amnesty plan encouraged these surrenders.
June 30	A number of prominent Dominican political leaders and United States Secretary of State Charles Evans Hughes signed an agreement providing for the withdrawal of the United States from the Dominican Republic. Negotiations had begun in March when Francisco J. Peynado began discussions with major United States figures, including Hughes. Therefore, the new instrument for freeing the Dominican Republic is often known as the "Hughes-Peynado Plan."
October	Provisional government established after a settlement was reached between Washington and Dominican political leaders.
1924	
March	National elections held in the Dominican Republic.
September 18	The occupation ended when the last United States forces left the Dominican Republic.

A. Congressional hearings and official reports

727. U.S. Congress. Senate. Select Committee on Haiti and Santo Domingo. *Inquiry into Occupation and Administration of Haiti and Santo Domingo, Pursuant to S. Res. 112*. Washington, DC: Government Printing Office, 1922.

Provides much contemporary documentation. Even though the emphasis is upon Haiti, the hearings constitute one of the most important printed primary sources on the United States administration of the Dominican Republic.

B. General studies

728. Calder, Bruce J. *The Impact of Intervention: The Dominican Republic During the U.S. Occupation of 1916–1924*. Austin: University of Texas Press, 1984.

This is the authoritative study of the Dominican Republic under United States rule. Calder deals expertly with the political, diplomatic, economic, and military aspects of the occupation.

729. Calhoun, Frederick S. *Power and Principle: Armed Intervention in Wilsonian Foreign Policy.* Chap. 3, "The Power of Ideology: Santo Domingo and Haiti," pp. 75–86, 104–12. Kent, OH: Kent State University Press, 1986.

Succinct discussion of political developments in the Dominican Republic from 1911 to 1916 and of the shaping of a policy toward that country by Wilson and Secretary of State William Jennings Bryan. Calhoun then discusses the decisions to intervene and then to impose a military government.

730. Callcott, Wilfrid Hardy. *The Caribbean Policy of the United States, 1890–1920.* Baltimore, MD: Johns Hopkins University Press, 1942.

Provides a generally critical account of the early years of the occupation, beginning with the ultimatum that the United States presented to the Dominican Republic in November 1915, pages 400–404, 484–89.

731. Gruening, Ernest H. "Conquest of Haiti and Santo Domingo." *Current History* 16 (March 1922): 885–96.

Interprets the intervention in the Dominican Republic as following from the refusal of that nation to accept an ultimatum from the United States in 1915. Gruening compares the occupation to the German occupation of Belgium during World War I.

732. Habib, Douglas F. "'Representatives of a Powerful but Righteous Nation': United States Military Occupations in the Dominican Republic and the German Rhineland in the Progressive Era." Ph.D. diss., Washington State University, 2003.

The intervention and occupation policies of the Wilson administration have received a good deal of attention from scholars, and some comparative analyses have been undertaken. Habib's dissertation is innovative in comparing the Rhineland occupation with a Third World intervention. Interestingly, Habib found many similarities between the approaches of the United States military leaders in the Rhineland and the Dominican Republic.

733. Kelsey, Carl. "American Intervention in Haiti and the Dominican Republic." *Annals of the American Academy of Political and Social Science* 100 (March 1922): 166–202.

Presents the "official" view that the intervention in the Dominican Republic is proceeding smoothly and that the problems of the country are primarily the fault of its inhabitants. There is relatively little on the fighting in the Dominican Republic.

734. Knight, Melvin M. *The Americans in Santo Domingo.* New York: Vanguard Press, 1928.

Deals primarily with the intervention of 1916–24, although earlier periods are discussed. It is a somewhat disjointed, but nevertheless valuable, discussion. Especially noteworthy is chapter X, which is concerned with press censorship.

735. Leschorn, Helen. "American Atrocities in the Dominican Republic." *Current History* 15 (February 1922): 881–82.

Discusses some of the testimony given by witnesses before the Senate committee to investigate conditions in the Dominican Republic and Haiti (item 727). She notes that the hearings were quickly concluded and predicts that much more will be said by the Dominicans when hearings resume in Washington.

736. MacMichael, David Charles. "The United States and the Dominican Republic 1871–1940: A Cycle in Caribbean Diplomacy." Ph.D. diss., University of Oregon, 1964.

Includes excellent material on relations between the United States and the Dominican Republic from 1913 to 1916. There is also much background on the proclamation of the military government. One chapter covers poor treatment of the Dominicans by individual marines.

737. Munro, Dana G. *Intervention and Dollar Diplomacy in the Caribbean 1900–1921.* Chap. 7, "The Military Occupation of the Dominican Republic," pp. 269–325. Princeton, NJ: Princeton University Press, 1964.

Asserts that Wilson's major objective in taking the Dominican Republic was to establish a democratic regime. Munro reviews the military government's successes.

738. Musicant, Ivan. *The Banana Wars.* Chap. 6, "Dominican Republic I 1916–24," pp. 235–84. New York: Macmillan Company, 1990.

Deeply researched study that is thoroughly based on primary materials related to the intervention. Musicant provides a fine presentation of United States–Dominican relations before 1916 that puts later events into historical perspective. Not much is included about the complex withdrawal of the United States from the occupation in the early 1920s.

739. Neymeyer, Robert James. "The Establishment and Operation of the Dominican Republic Customs Receivership, 1905–16." Ph.D. diss., University of Iowa, 1990.

Provides essential information about and perspectives on the United States penetration of the Dominican Republic before the formation of an occupation government late in 1916. The objectives of the receivership were to encourage stability in the country and to reduce the chance that European powers might use the failure of the Dominican Republic to pay its debts as an excuse to employ military force and, possibly, to establish military installations.

740. Welles, Sumner. *Naboth's Vineyard: The Dominican Republic 1844–1924*. 2 volumes. New York: Payson & Clarke Ltd., 1928.

Constitutes a classic account of the occupation. In addition to a lengthy narrative, Welles includes the texts of a great many important documents and a noteworthy bibliography on relations between the two countries.

C. Strategy and tactics

741. Ancker, W.M. "The Imperialistic Mercenaries." *Marine Corps Gazette* 60, no. 3 (March 1976): 60–62.

Recalls Ancker's youthful experiences as a mounted marine on patrol in the Dominican Republic during World War I. The article describes the organization and operation of patrols in detail.

742. Buckner, David N. *A Brief History of the 10th Marines*. Washington, DC: Headquarters, U.S. Marine Corps, History and Museums Division, 1981.

Describes and illustrates the employment of marine artillery in the Dominican Republic between 1915 and 1917, pages 10–16.

743. Fuller, Stephen N., and Graham A. Cosmas. *Marines in the Dominican Republic 1916–1924*. Washington, DC: Headquarters, U.S. Marine Corps, History and Museums Division, 1974.

Supplies a fully documented survey of the occupation. There are sections on the constabulary and civic action as well as combat operations. There is a long section of photographs.

744. Heinl, Robert Debs, Jr. *Soldiers of the Sea: The United States Marine Corps, 1775–1962*. Annapolis, MD: United States Naval Institute, 1962.

Provides an excellent account of marine operations, despite its brevity, pages 180–85, 247–51. Of particular importance is the detailed description of the marines' counterinsurgency tactics.

745. "In Memoriam to Captains Hirshinger and Low." *Marine Corps Gazette* 1, no. 12 (December 1916): 397–402.

Deals primarily with the careers of these marine officers, but also describes the circumstances surrounding their deaths in the Dominican Republic during the early fighting in 1916.

746. Kilmartin, Robert C. "Indoctrination in Santo Domingo." *Marine Corps Gazette* 7, no. 4 (December 1922): 377–86.

Consists of an address written by Kilmartin for the marine commander. He explains the marine role and discusses local opposition to the military government and warns that misconduct by marines can cause serious problems.

747. Ley, Wes. "A Recruit in Santo Domingo." Parts 1, 2, 3. *Leatherneck* 10, 11, no. 8, nos. 2, 3 (August 1927; February, March 1928): 7–8, 40–42; 8–9, 57–58; 5, 49–50, 59.

    Recounts in diary form a marine's experiences between December 3, 1919 and May 20, 1920. In lively fashion, Ley narrates events in a marine camp. He went on several patrols but reports few contacts with the "bandits." This is a roughly written but essential account.

748. McClellan, Edwin N. "Operations Ashore in the Dominican Republic." *United States Naval Institute Proceedings* 47, no. 2 (February 1921): 235–45.

    Provides an excellent outline history of the military aspects of the occupation from 1916 through 1919. The article begins with a brief political history of the Dominican Republic.

749. Metcalf, Clyde H. *A History of the United States Marine Corps.* Chap. 13, "Occupation of the Dominican Republic," pp. 339–70. New York: G.P. Putnam's Sons, 1939.

    Describes troop deployments and movements fully. Several pages review the military government, and another section is concerned with the constabulary. Metcalf notes that while the constabulary had been intended to preserve democracy, it became the mainstay of the Trujillo dictatorship.

750. Miller, Charles J. "Diplomatic Spurs: Our Experiences in Santo Domingo." Parts 1, 2, 3. *Marine Corps Gazette* 19, nos. 2, 5, 8 (February, May, August 1935): 43–50; 19–25, 52–55; 35–54.

    Discusses the disarmament of the people, intelligence operations, and the nature of the forces arrayed against the marines. In addition to examining counterinsurgency operations, Miller surveys military government services.

751. Millett, Allan R. *Semper Fidelis: The History of the United States Marine Corps.* Rev. ed. Chap. 7, "Hispaniola," pp. 178–211. New York: Free Press, 1991.

    Includes a lengthy discussion of the constabulary, and describes the effects of World War I on the military government. This is a highly readable, critical account of the occupation.

752. "Private Kemp Reports on Our War in Santo Domingo." *Literary Digest* 60 (February 22, 1919): 105–6, 108.

Responds to stories of dispirited United States troops in North Russia with a letter on marine experiences in the Dominican Republic. Kemp provides a lucid picture of living conditions in the field and describes patrol activity at some length.

D. The role of air power

753. Boyden, Hayne D. "Aerial Adventures in Hispaniola." Parts 1, 2. *Leatherneck* 8 (August 1, October 10, 1925): 3–6; 6–7, 20.

    Narrates several of Boyden's air crashes in a breezy fashion.

754. ———. "Some Forced Plane Landings in Santo Domingo." *Marine Corps Gazette* 7, no. 6 (June 1922): 175–84.

    Details Boyden's many air crashes during his 27 months of service in the Dominican Republic.

E. Military advisory activities

755. Fellowes, Edward A. "Training Native Troops in Santo Domingo." *Marine Corps Gazette* 8, no. 12 (December 1923): 215–33.

    Covers the operation of an officer candidate school and training facilities for enlisted men between 1920 and 1923.

756. Goldwert, Marvin. *The Constabulary in the Dominican Republic and Nicaragua: Progeny and Legacy of United States Intervention.* Gainesville: University of Florida Press, 1962.

    Examines the status of the Dominican Army before 1916, the development of the constabulary under United States rule, and its role in the rise of Trujillo, chapters I–III. Despite its brevity, this is an important study.

757. Williams, Charles F. "La Guardia Nacional Dominicana." *Marine Corps Gazette* 3, no. 9 (September 1918): 195–99.

    Refers to the initial difficulties of recruiting a constabulary because of opposition to the intervention. This is primarily an administrative study with little information about operations. It includes a full discussion of the origins of the *Guardia* in 1917.

F. Military government/civic action

758. Baughman, C.C. "United States Occupation of the Dominican Republic." *United States Naval Institute Proceedings* 51, no. 12 (December 1925): 2306–27.

Focuses on the first two-and-a-half years of the military government. Baughman had charge of public works, communications, agriculture, and immigration, but he touches on other areas as well. This is primarily a legal and financial study.

759. Lane, Rufus H. "Civil Government in Santo Domingo in the Early Days of the Military Occupation." *Marine Corps Gazette* 7, no. 6 (June 1922): 127–46.

Describes the formation of the military government. Lane also deals with the problems of coping with foreign residents and settling boundary problems with Haiti.

760. Lord, Rebecca Ann. "An 'Imperative Obligation': Public Health and the United States Military Occupation of the Dominican Republic, 1916–24." Ph.D. diss., University of Maryland, College Park, 2002.

Valuable contribution to the literature dealing with the occupation. Lord examines the extensive efforts the occupation government made to deal with such problems as prostitution, venereal disease, and smallpox. She also gives considerable attention to the factors that hindered these campaigns. This dissertation would repay careful reading by anyone studying the occupation.

761. Moran, Brian Patrick. "Prison Reform in the United States Navy and the Dominican Republic: The Military Occupation and Prisons, 1900–1930." Ph.D. diss., University of Illinois, Chicago, 2000.

Explores the short- and long-term effects of changes in the Dominican prison system that were either mandated or recommended by the United States occupation authorities. Before the occupation, the views of United States Marine Corps and Dominican prison specialists seemed to be converging. Despite the fact that the Dominicans later eliminated some changes dating from the occupation era, United States influence seems to have been significant.

762. Thorpe, George C. "Dominican Service." *Marine Corps Gazette* 4, no. 12 (December 1919): 315–26.

Attributes much of the unrest in the country to German instigation. The article stresses civic action and military government. Thorpe emphasizes the need for suitable marine officers to serve with the constabulary.

763. Wellman, Leslie H. "Mapping Activities and Compilation of Handbooks by the Second Brigade, U.S.M.C. in the Dominican Republic." *Marine Corps Gazette* 8, no. 9 (September 1923): 161–73.

Describes the compilation of a two-volume handbook on the Dominican Republic that was classified "confidential" upon completion. The first portion of the article is a highly technical report on mapping.

G. The decision to withdraw

764. Curry, Earl H. *Hoover's Dominican Diplomacy and the Origins of the Good Neighbor Policy.* New York: Garland Publishing, Inc., 1979.

    Argues that Hoover did little to disavow intervention, using the president's Dominican policy as a case study. The first two chapters contain a very detailed discussion of the negotiations that ended the occupation in the early 1920s.

765. Grieb, Kenneth J. *The Latin American Policy of Warren G. Harding.* Chap. 4, "Withdrawal from the Dominican Republic," pp. 61–83. Fort Worth: Texas Christian University Press, 1976.

    Considers negotiations between 1921 and 1924. One stumbling block was the United States insistence on keeping a military mission in the country.

766. Juarez, Joseph Robert. "United States Withdrawal from Santo Domingo." *Hispanic–American Historical Review* 42 (May 1962): 152–90.

    Discusses the many factors that led to withdrawal. Juarez's primary concern is with the public debate over intervention rather than decision-making within the State Department. Despite the author's heavy reliance on secondary materials, this is an important study.

767. Munro, Dana G. *The United States and the Caribbean Republics 1921–1933.* Chap. 3, "Getting Out of Santo Domingo," pp. 44–70. Princeton, NJ: Princeton University Press, 1974.

    Examines negotiations between the United States and Dominican leaders. The Dominican government wanted some military advisers but placed such limitations on their use that the marines rejected the Dominican proposal.

768. Perkins, Whitney T. *Constraint of Empire: The United States and the Caribbean Interventions.* Westport, CT: Greenwood Press, 1981.

    Reviews the customs receivership and the pre-intervention period in some detail. The ultimately successful efforts to end the military government and the occupation comprise large portions of chapters 3, 7, and 9.

H. The marine withdrawal

769. "The Second Brigade Marines, Santo Domingo, D.R." *Leatherneck* 7 (August 9, 1924): 1–2.

    Deals briefly with the situation that led to the intervention, but, more importantly, contains a description of the ceremonies that surrounded the final withdrawal.

I. The aftermath

770. Hardy, Earl B. "Friendship of Dominicans is Outgrowth of Our Occupation." *Leatherneck* 16, no. 8 (August 1933): 5–7.

    Surveys the final years of the occupation, 1921–24, with emphasis on the final marine offensives to crush the insurgents. Hardy provides a detailed account of major operations under Marine General Harry Lee and quotes tributes to Lee by President Vasquez and General Lejeune.

771. Reisinger, H.C. "Preserving the Marine Tradition in the Land Columbus Loved." *Marine Corps Gazette* 15, no. 3 (March 1930): 1–9.

    Asserts that the occupation was a great success. This is an interesting, if superficial, look at the Dominican Republic and its army just before Trujillo assumed control.

XIII
Cuba, 1917–22

Parts of Cuba were occupied by marines when a disputed election and a resultant civil war created concern in the United States that the sugar crop was threatened. Some American leaders believed that the Germans were using the unrest to work against Allied interests.

Relatively few marines participated in the intervention, and, after the initial landings, there was little fighting. There was, however, tension. In 1917, the United States considered sending a cavalry division to Cuba. The following year, the occupation of the entire island by the marines was discussed. Finally, in 1920, deployment of an army infantry division was contemplated.

Eventually, a new election in Cuba was held, and the situation stabilized until the early 1930s.

Chronology

1917

March	Parties of marines were landed from United States warships because of a perceived breakdown of order in eastern Cuba, primarily Oriente Province, an area of large-scale cultivation of sugar. The marines were withdrawn quickly, however.
August	7th Marines (that is, the 7th Marine Regiment) was landed at Guantanamo Bay to be prepared for contingencies.
October 24	The 7th Marines moved into Oriente Province from Guantanamo Bay. A number of detachments were stationed at different towns to counter the possible endangerment of the sugar crop.
December	9th Marines sent to Cuba because of continuing concerns about disorder and the possible destruction of the sugar crop.

	This regiment was transferred from Cuba to Texas in July 1918, however.
1918 December	1st Marines joined the 7th Marines in Cuba to form the 6th Marine Brigade.
1919 August	Most marines left Cuban territory. Two companies remained at Camagüey.
1922 February 15	The remaining marines were withdrawn from Camagüey.

A. Diplomatic background of the intervention

772. Callcott, Wilfrid Hardy. *The Caribbean Policy of the United States, 1890–1920*. Baltimore, MD: Johns Hopkins University Press, 1942.

   Gives a highly critical account of the intervention. Relatively little is said about purely military matters, pages 471–76.

773. Meyer, Leo J. "United States and the Cuban Revolution of 1917." *Hispanic–American Historical Review* 10 (May 1930): 138–66.

   Surveys the events leading to the disputed election. United States naval movements and marine landings are examined, although obviously not to the extent that Gibean (item 774) studies them. Like Gibean, Meyer discounts German influence on the Cuban rebels.

B. Marine landings and the intervention

774. Gibean, Victor Hugo, Jr. "Relations of Cuba with the United States, 1916–21." Ph.D. diss., University of North Carolina, 1953.

   Includes much information on ship movements and the landing of marines. Plans for more ambitious interventions are considered in detail. In addition, Gibean thoroughly reviews the election of 1916 and the crisis that alarmed the Americans. He believes that there was little or no German involvement in the revolutionary movement.

775. Pérez, Louis A. *Intervention, Revolution, and Politics in Cuba, 1913–1921*. Pittsburgh, PA: University of Pittsburgh Press, 1978.

   This is the authoritative account of the political, economic, and military aspects of this intervention. Pérez skillfully describes the continuing impact of the Platt Amendment on Cuban politics and on the events of the "Sugar intervention."

776. Smith, Robert F. *The United States and Cuba: Business and Diplomacy, 1917–1960*. New York: Bookman Associates, 1960.

Constitutes an important analysis of the intervention, pages 18–21, 82–102, and 104–7. The military aspects are covered in some detail. The last few pages noted cover the final withdrawal of the marines in 1922.

C. Special studies

777. Clagett, John H. "Skipper of the Eagle: Rehearsal for Greatness." *United States Naval Institute Proceedings* 102, no. 4 (April 1976): 58–65.

Relates the achievements of H. Kent Hewitt, captain of the *Eagle*, a patrol vessel, during the early days of the Cuban civil war. He put landing parties ashore several times to protect United States owned cane fields during February and March 1917. Hewitt also organized a cavalry unit and converted a Model T Ford into a patrol car to extend the reach of his sailors.

XIV

Russia, 1918–20

American operations in Russia during and after World War I remain some of the most controversial of the small wars of the United States.

Allied pressures for United States participation played a major role in Wilson's decision to intervene in North Russia and in Siberia. Helping to reconstruct an eastern front in 1918, assisting the Czechoslovak Legion to make its way to the Pacific coast, and guarding the huge stocks of Allied supplies in Russia were other motivations.

While there was certainly hostility in Washington toward the Bolsheviks, United States involvement in Russia remained halting and less than enthusiastic. As has often been the case with small wars, the armed forces themselves were not desirous of squandering their strength on what they regarded as diversions.

A. General studies

778. Bacon, Eugene Howard. "Russian–American Relations, 1917–21." Ph.D. diss., Georgetown University, 1951.

    Explores thoroughly the issue of American intervention and interprets it primarily as an effort to mollify the Allies.

779. Calhoun, Frederick S. *Power and Principle: Armed Intervention in Wilsonian Foreign Policy.* Chap. 6, "The Power of Collective Security: Russia," pp. 185–218; chap. 7, "The Limits of Force: Russia, Bolshevism, and the Paris Peace Conference," pp. 219–49. Kent, OH: Kent State University Press, 1986.

    Skillful examination of the interventions in North Russia and Siberia, which emphasizes the Siberian expedition. Calhoun gives a good deal of attention to differences between the North Russia and Siberia operations, which the Wilson administration viewed as separate commitments. This is a well-documented study resting on extensive archival research.

780. Fic, Victor M. *The Collapse of American Policy in Russia and Siberia, 1918: Wilson's Decision Not to Intervene (March–October 1918)*. Boulder, CO: East European Monographs, 1995.

Severely critical account of United States policy with respect to Russia during the intervention period. Unlike many students of the period, Fic believes that a "New Russia" was developing with the help of the Czechoslovak Legion that could have been a viable and a much-to-be-preferred alternative to the Soviet regime. One of the strengths of this book is the amount of attention given to the Czechs, whose troops constituted a major military factor in Russia during and immediately after World War I. There is a large bibliography that includes many Russian materials.

781. Foglesong, David S. *America's Secret War Against Bolshevism: U.S. Intervention in the Russian Civil War, 1917–1920*. Chapel Hill: University of North Carolina Press, 1995.

Authoritative study that is thoroughly documented and that contains an extensive bibliography, which is of great value to anyone interested in consulting the materials cited in this work. Foglesong highlights Wilson's complex and confused policies regarding the Russian Civil War. Wilson clearly hated communism, but he also believed Russians should be allowed to make their own decisions about the future. Wilson argued for open government both on the national and international levels, but he also felt that many questions had to be settled covertly by a decision-making elite.

782. Goldberg, Harold J., ed. *Documents of Soviet–American Relations: Volume 1. Intervention, Famine Relief, International Affairs 1917–1933*. Gulf Breeze, FL: Academic International Press, 1993.

Valuable collection of materials, mostly, but not entirely, official documents. Most of the items pertinent to the intervention appear in sections 1 and 2, pp. 1–189, but documents in other sections are also useful. One of the important features of this reference work is the inclusion of many items from the Soviet perspective. The first pages of the introduction are a helpful summary of events. There are extensive notes and splendid indexes.

783. Goldhurst, Richard. *The Midnight War: The American Intervention in Russia, 1918–1920*. New York: McGraw-Hill Book Company, 1978.

Provides a chronological account of the intervention. Both the military and political aspects of the campaigns are well covered. This popular work, based mainly on secondary materials, is detailed and highly readable.

784. Killen, Linda. *The Russian Bureau: A Case Study in Wilsonian Diplomacy.* Lexington: University Press of Kentucky, 1983.

The War Trade Board of the United States Russian Bureau was established to give economic aid to Russia, specifically the deposed Kerensky government, not to the Soviets. The United States assumed that further economic development in Russia would end the attraction of communism to the Russian people and thus would destroy the revolutionary regime. In fact, the amount of money dedicated to the project was small, and little was accomplished. Killen offers an insightful observation that seems to explain much about United States policies toward Russia during its civil war. She comments that "the remarkable thing about American policy was not how it dealt with the Bolsheviks, but rather how it attempted to deal with Russia in spite of the Bolsheviks" (p. 131).

785. Shapiro, Sumner. "Intervention in Russia (1918–19)." *United States Naval Institute Proceedings* 99, no. 4 (April 1973): 52–61.

Assesses the intervention judiciously, and analyzes Soviet assertions concerning United States responsibility for the Allied effort. Much of the article criticizes United States ambassador Francis for serving his country very poorly in Russia and for indulging in much wishful thinking.

786. Sivachev, Nikolai V., and Nikolai N. Yokovlev. *Russia and the United States.* Translated by Olga Adler Titelbaum. Chicago: University of Chicago Press, 1979.

Reviews changing Soviet and American interpretations of United States participation in the Allied intervention from a Soviet perspective, pages 42–62.

787. Somin, Ilya. *Stillborn Crusade: The Tragic Failure of Western Intervention in the Russian Civil War, 1918–20.* New Brunswick, NJ: Transaction Publishers, 1996.

Somin's book is based largely on secondary works and published primary sources. Somin is highly anti-Soviet and is disappointed that the Allied powers and the United States did not engage themselves more actively in the Russian Civil War. Chapter 3, "'Why Not Save Siberia?' The Development of U.S. Policy," pp. 79–133, is especially important for this bibliography. The introduction includes a brief historiography of the debate about intervention and the reasons for the intervention and the results of the failure of the intervention and the collapse of the White Russian movement.

B. Diplomatic and political studies

788. Berle, Adolf A. "Our Undeclared War." *New Republic* 23 (June 16, 1920): 92–94.

    Examines the problem of Soviet–American relations. The article emphasizes Allied intrigues and the desires of various countries to carve out spheres of influence in Russia.

789. Davis, Donald E. and Eugene P. Trani. *The First Cold War: The Legacy of Woodrow Wilson in U.S.–Soviet Relations.* Columbia: University of Missouri Press, 2002.

    The authors make a persuasive case that United States policy toward the Soviet Union after World War II was largely a return to the stance taken toward revolutionary Russia by the Wilson administration. A segment of the book modestly titled "An Essay on Notes and Sources," pages 207–32, is a highly successful effort to present a historiography of the United States involvement in the Allied intervention in Russia.

790. Filene, Peter G. *Americans and the Soviet Experiment, 1917–1933.* Chap. 2, "Intervention and Withdrawal," pp. 39–63. Cambridge, MA: Harvard University Press, 1967.

    Does not discuss military aspects of the interventions in any detail. However, the chapter is highly useful for its portrayal of the atmosphere in the United States that fostered intervention on the one hand and pro-Soviet attitudes on the other.

791. Kennan, George F. *Soviet–American Relations, 1917–1920: The Decision to Intervene.* Princeton, NJ: Princeton University Press, 1958.

    Constitutes the most valuable single source on United States involvement in the Allied intervention in Russia. The various pressures on Wilson are described in considerable detail, along with his responses to them.

792. Strakhovsky, Leonid I. *American Opinion About Russia 1917–1920.* Toronto: University of Toronto Press, 1961.

    Examines contemporary sources and gives extensive quotations from newspapers, periodicals, memoirs, and diplomatic and military messages. Two sections deal specifically with the period of the United States intervention in Russia.

C. Military analyses

793. Boot, Max. *The Savage Wars of Peace: Small Wars and the Rise of American Power.* Chap. 9, "Blood on the Snow: Russia, 1918–20," pp. 205–30. New York: Basic Books, 2002.

Detailed surveys of operations in North Russia and Siberia. Boot skillfully describes the complexities of United States policy. He laments the failure of the Western powers to intervene more forcefully, which he believes was a profound error that led to decades of Soviet rule.

794. Dupuy, R. Ernest. *Perish by the Sword: The Czechoslovakian Anabasis and Our Supporting Campaigns in North Russia and Siberia 1918–1920*. Harrisburg, PA: Military Service Publishing Co., 1939.

Furnishes a very thorough account of the adventures of the Czechoslovak Legion and of the experiences of the United States troops, which is well worth reading today. American assistance for the Legion is stressed, as the title suggests.

795. Dupuy, R. Ernest, and William H. Baumer. *The Little Wars of the United States*. Chap. 7, "Interventions in Russia, 1918–20," pp. 169–213. New York: Hawthorn Books, Inc., 1968.

Delineates clearly the extent of United States participation in the Allied conflict with the Soviet Union.

796. Gardiner, J.B.W. "Our Military Problem in Russia." *World's Work* 36 (October 1918): 603–9.

States categorically that intervention in Russia should be limited to helping the Russians establish a new army to fight Germany. At the same time, however, Gardiner advocates sending a large army to Russia and fighting the Soviets as much as necessary. Most of the article consists of detailed maps of various areas of Russia.

797. "Red Russia as Our Foe." *Literary Digest* 58 (September 21, 1918): 9–13.

Charges the Soviets with working with the Germans. Therefore, the United States must consider the Soviets to be enemies. The article quotes many newspaper editorials that declare the Soviets are "outlaws." A detailed map of European Russia and a gazetteer are included.

XV

North Russia, 1918–19

Approximately 5,000 United States soldiers joined British forces in the Archangel area of North Russia during the summer of 1918 to guard Allied supplies and to prevent the Germans from taking the city.

While there was no fighting with the Germans, conflicts broke out between the Allied forces and the Soviets. At times, heavy fighting took place. The Americans suffered from the severe climate and the lack of appropriate rations and supplies, and were delighted when they were allowed to withdraw in advance of the British in the summer of 1919.

Many questions were raised in the United States about the deployment of combat troops against a country with which the United States was not at war, marking one of the first debates over presidential war powers.

Chronology

1917

November 7 Russian government overthrown in the Bolshevik Revolution. Allied governments became convinced that troops had to be sent to Russia to maintain an eastern front against the Central Powers.

1918

March British troops landed at Murmansk. Owing to the lack of Allied reserves to expand the intervention in Russia, the Allies pressed the United States to send troops.

July 3 The Supreme War Council of the Allies formally determined to intervene in North Russia. British forces already in the area began an offensive against the Bolsheviks.

September 4 The American Expeditionary Force North Russia (AEFNR) arrived at Archangel, and the majority of its members went to the front.
September 26 The United States demanded that the offensive in North Russia end immediately. This decision had no real influence on events, and U.S. forces continued to fight the Bolsheviks.
November 11 The end of World War I had no immediate effect on events in North Russia, but opposition to the North Russia venture increased among both United States leaders and the United States populace.

1919
February 14 Wilson allowed a few United States reinforcements to go to North Russia but only if withdrawal of all United States forces as soon as possible was guaranteed.
April United States troops began to leave the front, and in June, they began to be evacuated.
July 28 United States participation in the North Russia intervention ended.

A. Contemporary general studies

798. Albertson, Ralph. *Fighting Without a War: An Account of Military Intervention in North Russia*. New York: Harcourt, Brace and Howe, 1920.

Stresses the complete control that the British had over the United States troops and the British failure to cooperate with the other Allies. About half the book refers to the period when United States troops were at Archangel. Albertson claims to have been the last American evacuated from North Russia.

799. Costello, Harry J. *Why Did We Go to Russia?* Detroit, MI: Harry J. Costello, 1920.

Provides a useful participant's view of the confusion and hardship associated with the North Russia campaign. Costello's account is generally very critical of the British high command, but it is more positive in tone than most narratives written by participants in the expedition. One chapter discusses a "mutiny" on the part of a United States infantry company.

800. Cudahy, John. ("A Chronicler") *Archangel: The American War with Russia*. Chicago: A.C. McClurg & Co., 1924.

Puts the expedition into the broader perspective of World War I and the Russian Revolution. This is one of the most important military studies of the North Russia expedition. The fighting in which United States forces engaged is discussed in considerable detail. Cudahy goes beyond his personal experiences, making this a valuable contribution.

801. Moore, Joel R. "The North Russian Expedition: The 85th Division's Participation." *Infantry Journal* 29 (July 1926): 1–21.

    Asserts that the intervention was begun solely to revive the eastern front against Germany. Ultimately, the expedition had no real influence on the outcome of the war. The article includes a number of illustrations and a map showing the distribution of United States forces in North Russia.

802. Moore, Joel R., Harry H. Meade, and Lewis E. Johns. *The History of the American Expedition Fighting the Bolsheviki: Campaigning in North Russia 1918–1919*. Detroit, MI: Polar Bear Publishing Co., 1920.

    Includes a great deal on the fighting. The treatment is episodic, but very valuable. This richly illustrated account contains chapters on a variety of topics, such as the work of the Y.M.C.A.

803. Richardson, W.P. "America's War in North Russia." *Current History* 13 (February 1921): 287–94.

    Defends the dispatch of the expeditionary force. This survey of the expedition and its combat operations ignores the serious morale problems of the United States troops; otherwise, it is a full account.

B. Later analyses

804. Gordon, Dennis. *Quartered in Hell: The Story of the American North Russia Expeditionary Force 1918–1919*. Missoula, MT: Doughboy Historical Society/G.O.S., Inc., 1983.

    Recreates vividly the world of the expeditionary force. This magnificent book is based almost entirely on the recollections of participants and is heavily illustrated with contemporary photographs. The contributors are bitterly critical of the decision to keep the force in Russia after the Armistice.

805. Halliday, E.M. *The Ignorant Armies*. New York: Harper & Brothers, Publishers, 1960.

    Provides a fine introduction to the North Russia expedition. Halliday makes excellent use of published sources and interviews with many participants. He presents nothing from archival sources. A special feature is his discussion of several earlier works on the North Russia intervention.

806. Rhodes, Benjamin D. *The Anglo–American Winter War with Russia, 1918–1919: A Diplomatic and Military Tragicomedy.* New York: Greenwood Press, 1988.

Well-written and heavily researched account that emphasizes the military facets of the intervention in North Russia. The book does not neglect the views of the United States troops involved. There is a good deal of detail about the eventual withdrawal of Allied and United States forces. Although the book is based primarily on archival sources, it includes a short, but valuable, bibliography.

807. Strakhovsky, Leonid I. *The Origins of American Intervention in North Russia (1918).* Princeton, NJ: Princeton University Press, 1937.

Focuses on the political and military background of the intervention rather than the events that occurred. Strakhovsky believes that United States participation had the benefit of deterring the British and French from seizing Russian territory.

C. Combat narratives

808. "Archangel in Winter Makes No Hit with Michigan Troops." *Literary Digest* 60 (February 8, 1919): 99–108.

Records some of the many complaints made by United States troops at Archangel, as quoted in letters to the editor in various Michigan newspapers. Some of the letters are long and quite interesting. Many of the comments deal with living conditions.

809. "Facing the Bolsheviki with Frozen Guns." *Literary Digest* 60 (March 1, 1919): 62, 64.

Testifies to the bitter fighting in which United States troops were involved during December, 1918. Much of the material comes from the *New York Evening Post.* One American soldier is quoted as discounting Soviet military capability, but noting that the Soviets are present in overwhelming numbers.

810. "Fighting the Bolsheviki South of Archangel." *Literary Digest* 60 (February 8, 1919): 108–12.

Describes United States participation in the fighting near the Waga River in January 1919. One section of the article deals with the heroic actions of Captain M.J. Donoghue, who won the Distinguished Service Order in September 1918.

811. "What Happened at Archangel." *Independent* 97 (March 1, 1919): 279–80.

Discusses the Allied retreat under Soviet pressure. The author asserts that the Allies did not realize the Soviet capability for offensive operations. There is a brief description of the Soviet forces fighting the Allies. The *Independent* praises Wilson for being reluctant to intervene in Russia.

812. York, Dorothea. *The Romance of Company "A" 339th Infantry A.N.R.E.F.* Detroit, MI: McIntyre Printing Co., 1923.

Stresses the experiences of the enlisted men in what was a front-line company. There is a great deal about the character of the fighting, along with much criticism of the practice of scattering the Americans in tiny outposts. The book is also extremely critical of the British for their handling of the North Russia expedition.

D. Naval operations

813. Jackson, Chester V. "Mission to Murmansk." *United States Naval Institute Proceedings* 95, no. 2 (February 1969): 82–89.

Recalls the operations of the United States cruiser *Olympia* in the Murmansk area during the summer of 1918. Landing parties went ashore several times both at Murmansk and Archangel. Jackson was not in action, but he does offer an interesting narrative of a forgotten aspect of the Archangel venture.

814. Tolley, Kemp. "Our Russian War of 1918–19." *United States Naval Institute Proceedings* 95, no. 2 (February 1969): 58–72.

Provides an excellent survey of American naval activities in North Russia. Much of the narrative centers around the experiences of the men detached from the cruiser *Olympia* for land service early in the intervention. Among the more interesting episodes is the use of United States motor launches on Lake Onega for a short time in 1919.

E. Political critiques

815. Albertson, Ralph. "The Debacle of Archangel." *New Republic* 20 (November 14, 1919): 342–46.

Argues that the basic mission of the force was not realized because the Soviets had removed all the Allied supplies before the Allied troops arrived. Albertson praises United States soldiers in North Russia, but he charges that the leaders of the expedition were unaware of political problems.

816. Hibschman, Harry J. "Why American Soldiers in Russia?" *New Republic* 19 (July 30, 1919): 417–19.

Relates discussions with members of the Archangel force on the question in the title. Hibschman mentions rumors about a "mutiny." While discounting them he asserts that there was acute dissatisfaction about being kept at Archangel after the Armistice. The author agrees with some United States soldiers who believed that the intervention was intended to open Russia to British and French businessmen.

XVI
Siberia, 1918–20

The United States involvement with the Allied intervention in Siberia was primarily a matter of maintaining an American presence rather than fighting. Nevertheless, United States soldiers skirmished with both Soviet and White Russian troops at times.

Officially, the American objective was to help guard the route along which the Czechoslovak Legion made its way through Siberia to the Pacific. While aiding the Legion was important, the Americans also hoped that by maintaining troops there the Japanese would be deterred from gaining a foothold in Siberia.

The Allies, White Russians, and United States diplomats on the scene were adamant that American troops join the drive against the Soviets. Following his instructions from President Wilson and the War Department, General William S. Graves kept his force as neutral as possible and minimized losses while performing his mission of keeping a United States presence in the Allied army.

Chronology

1917
April — Russian Railway Advisory Mission established to assist Russia in maintaining the vital Trans-Siberian Railway. The mission recommended formation of an operational unit that became the Russian Railway Service Corps.

1918
March — Russian Railway Service Corps reached Vladivostok, after having spent some time in Japan because Vladivostok harbor

Siberia, 1918–20 193

	was frozen and because political dissension hampered effective commitment of the organization.
March 15	Supreme War Council of the Allies approved Japanese plans to put some of their troops into Siberia to guard Allied supplies and to lend support to anti-Communist elements.
August 3	United States Army's 27th and 31st Infantry Regiments directed to move from the Philippines to Vladivostok. Various other organizations joined them, and the whole force was designated the American Expeditionary Forces, Siberia.
September	Wilson froze United States forces in Siberia at 7,000.
November	Admiral Aleksandr V. Kolchak, the leader of anti-Communist elements, began a large-scale advance west, but his offensive failed late in 1919, leaving Allied forces with little reason to continue in Siberia.
1919	
January	Inter-Allied Advisory Technical Commission of Railway Experts established to assist in coordinating railway service in Siberia.
November	British troops left Siberia.
1920	
April	American Expeditionary Forces, Siberia completely withdrawn.
1922	
October	Japan ended its intervention in Siberia.
November	U.S.S. *Sacramento*, which had been monitoring Japanese activities, departed from Vladivostok after the Japanese withdrawal.

A. General studies

817. Curtis, Donald John, Jr. "'Hard Times Come Again No More': General William S. Graves and the American Intervention in Siberia, 1918–20." Ph.D. diss., Texas A&M University, 2000.

    Curtis rightly calls the United States intervention in Siberia "one of the most bizarre and confused occurrences in the nation's history." This is a comprehensive study of the highly complex situation that confronted the United States commander, General William S. Graves, and the skillful fashion in which he tried to fulfill his mandates from President Wilson and Secretary of War Newton D. Baker.

818. Dunscomb, Paul E. "U.S. Intervention in Siberia as Military Operations other than War." *Military Review* 82, no. 6 (November–December 2002): 98–102.

As the title indicates, this is an innovative effort to view the United States involvement in Siberia as a "military operation other than war," a theoretical construct now much used in military analysis. Despite its brevity, this is a major study that applies current methods to an often discussed and often puzzling episode. Dunscomb includes a welcome historiography of the intervention.

819. Ekbladh, David. "'Wise as a Serpent and Harmless as a Dove': John F. Stevens and American Policy in Manchuria and Siberia, 1918–24." *Prologue* 27, no. 4 (Winter 1995): 318–33.

After General Graves, John F. Stevens, as the leader of efforts to reconstruct and maintain the vital railways of Siberia, was the most powerful United States citizen to play a role in this intervention. This article is a major contribution to the literature about United States involvement in Siberia.

820. Maddox, Robert J. *The Unknown War with Russia: Wilson's Siberian Intervention*. San Rafael, CA: Presidio Press, 1977.

Examines not only the military aspects of the intervention, but also, more generally, relations between the United States and the various White Russian governments. Maddox asserts that hostility toward communism was Wilson's primary motive in authorizing the intervention in Siberia.

821. Manning, Clarence A. *The Siberian Fiasco*. New York: Library Publishers, 1952.

Tries too hard to see the beginnings of both World War II and the Cold War in the Siberian situation. Manning stresses the role of the United States in the intervention and severely criticizes Graves for his neutral stance in the Russian civil war. Manning also feels that the Allies should have done more to encourage separatist movements in Russia.

822. Melton, Carol Willcox. *Between War and Peace: Woodrow Wilson and the American Expeditionary Force in Siberia, 1918–1921*. Macon, GA: Mercer University Press, 2001.

Views the United States forces in Siberia as largely successful in the sense that they adhered closely to Wilson's wishes in their operations. Melton argues that Wilson employed the armed forces in many parts of the world, including Siberia, in new ways to promote peace rather than wage war. Chapter 4, "America's Siberian Policy Defined," is of special importance because of the contemporary and subsequent controversies about the mission of the expeditionary force. The book includes a number of unusual photographs, many of them from the author's collection. Appendices contain the texts of several key documents relating to the United States involvement in Siberia. The bibliography provides brief annotations for the various primary source collections utilized.

823. O'Connor, Richard. "Yanks in Siberia." *American Heritage* 25 (August 1974): 10–17, 80–83.

    Discusses General Graves' dilemma of having to try to follow Wilson's orders in the face of Allied pressure to intervene more actively in Russian affairs. This richly illustrated article praises Graves for his intelligence and moderation.

824. Smith, Gibson Bell. "Guarding the Railroad, Taming the Cossacks: The U.S. Army in Russia, 1918–20." *Prologue* 34, no. 4 (Winter 2002): 294–305.

    Fine overview of this intervention. Smith includes much detail about United States Army relations with Gregori Semenoff, a ruthless Cossack leader, who was working closely with the Japanese, who were bent on extending their influence over Siberia. This is a richly illustrated study.

825. Westall, Virginia Cooper. "AEF Siberia—The Forgotten Army." *Military Review* 48, no. 3 (March 1968): 11–18.

    Deflates completely most of the reasons given for United States involvement in the Siberian intervention. The papers of General Robert L. Eichelberger, upon which Westall's article is based, reveal the problems that faced Graves. In retrospect, Eichelberger commented that the United States should have cooperated more fully with the Allies, including Japan.

826. White, John Albert. *The Siberian Intervention*. Princeton, NJ: Princeton University Press, 1950.

    Gives essential background information on the intervention, although the amount of space given to purely military matters is quite limited. This is a searching examination of Siberia during the last years of czarist rule and the beginning of the Soviet era. White comments at length on the motivations of the intervening powers.

B. Political and diplomatic studies

827. Ackerman, Carl W. *Trailing the Bolsheviki: Twelve Thousand Miles with the Allies in Siberia*. New York: Charles Scribner's Sons, 1919.

    Sees the world confronted by a contest between communism and the League of Nations concept. Much of the book discusses the political situation in Siberia during 1917 and 1918. Ackerman notes the tension between General Graves and some other United States representatives and Allied leaders in Siberia.

828. Fic, Victor. *The Rise of the Constitutional Alternative to Soviet Rule in 1918: Provisional Governments of Siberia and All-Russia—Their Quest for Allied Intervention.* Boulder, CO: East European Monographs, 1998.

Provides a much-needed Russian perspective on the Allied Intervention in Siberia. Fic offers a richly documented survey of events in Russia, especially in Siberia, before the intervention and then proceeds to devote several chapters to the early period of Allied control of the area. Not too much attention is given specifically to the role of the United States, but, nevertheless, the book is required reading for anyone wishing to understand the background of this complex and controversial episode.

829. Giffin, Frederick C. "An American Railroad Man East of the Urals, 1918–22." *Historian* 60, no. 4 (Winter 1998): 812–30.

Examines the experiences of Benjamin O. Johnson, who was one of the major figures in the Russian Railway Service Corps (RRSC), a paramilitary organization attached to the United States army in Siberia. Johnson eventually received command of the corps, which was designed to help the Russians improve service on the Trans-Siberian Railway. He carried out the orders he received, which often involved giving considerable assistance to White Russian elements, but he had a conciliatory attitude toward the Soviets, believing that the old Russian government was worse than the Communists.

830. Hayes, Harold B., III. "The Iron(ic) Horse from Nikolsk." *Military Review* 62, no. 5 (May 1982): 18–28.

Reviews the intervention in Russia and then concentrates on a situation in which the author's grandfather, Harold B. Hayes, arranged a train and safe conduct for a Soviet delegation that negotiated the final surrender of White Russian forces in eastern Siberia.

831. Luckett, Judith A. "The Siberian Intervention: Military Support of Foreign Policy." *Military Review* 54, no. 4 (April 1984): 54–63.

Analyzes Graves' conflicts with the State Department and the other Allied leaders in Siberia. Luckett compares United States policies in Siberia with those pursued during the Korean and Vietnam wars in a highly useful fashion.

832. "Out of Siberia." *Independent* 101 (February 28, 1920): 323–25.

Endorses the United States policy of neutrality in the Russian civil war in Siberia. The article also discusses the end of the Kolchak regime and asserts that the Allied intervention policy was a mistake.

833. U.S. Congress. House. Committee on Military Affairs. *American Troops in Siberia, Hearings on H. Con. Res. 30 Making Rules to Return All American Soldiers from Countries with which We Are at Peace.* 66th Cong., 1st sess., 1919.

Emphasizes issues related to presidential war powers, with Secretary of War Baker making sweeping claims as to their extent. Some information is provided on United States military operations and the replacement of draftees by volunteers for Siberia.

834. Unterberger, Betty Miller. *America's Siberian Expedition, 1918–1920: A Study of National Policy.* Durham, NC: Duke University Press, 1956.

Examines the diplomacy of the Siberian expedition thoroughly. Several chapters deal with Allied efforts to embroil the United States in Siberia. The rest of the book focuses primarily on relations between the Allies and the White Russians.

835. ——. *The United States, Revolutionary Russia, and the Rise of Czechoslovakia.* Chapel Hill: University of North Carolina Press, 1989.

Given the fact that the Wilson administration frequently justified its part in the intervention in Siberia by referring to the need to assist the Czechoslovak Legion, which was made up of former members of the Austro-Hungarian Army who were fighting with the Russian Army for the independence of Czechoslovakia, anyone seriously interested in events in Siberia needs to consult this superior study. This is an important political and diplomatic analysis rather than a work on the military side of the intervention.

C. Personal narratives

836. Graves, William S. *America's Siberian Adventure 1918–1920.* New York: Peter Smith, 1931.

Focuses on the reasons for having United States troops in Siberia and Graves' efforts, as commander of the United States forces, to maintain neutrality during the Russian civil war. Not too much is said about troop movements and other operational matters. Nevertheless, it is an indispensable source.

837. Kindall, Sylvian G. *American Soldiers in Siberia.* New York: Richard R. Smith, 1945.

Constitutes a most important personal narrative of the Siberian expedition. Kindall recounts skirmishes between American troops and both communist forces and White Russian soldiers backed by the Japanese. He expresses extreme hostility toward the Japanese and a great deal of sympathy for the Russian people.

838. Smith, Herbert E. "Christmas in Siberia." *Leatherneck* 20 (December 1937): 5, 54.

Describes Christmas, 1919, as celebrated by Company M, 31st Infantry Regiment, near Razdolnoe, Siberia. Smith stresses the excellent morale of the troops and mentions the good relations they had with the Russian inhabitants. Fortunately, mail, including gift parcels, arrived during the day.

D. Naval and marine operations

839. Braisted, William Reynolds. Book IV, "The Siberian Intervention." pp. 343–406, in *The United States Navy in the Pacific, 1909–1922*. Austin: University of Texas Press, 1971.

Emphasizes the political ramifications of United States naval operations at Vladivostok, 1918–22. American naval officers came to favor Bolshevik over Japanese expansion in Siberia by the early 1920s.

840. McClellan, Edwin N. "American Marines in Siberia During the World War." *Marine Corps Gazette* 5 (June 1920): 173–81.

Examines the role of the marines who landed from the cruiser *Brooklyn* in June 1918, and who stayed at Vladivostok until October. McClellan quotes heavily from the *Brooklyn*'s log and relates several amusing stories about the marines' adventures.

XVII
Nicaragua, 1927–33

The Second Nicaraguan Campaign was one of the most important counterinsurgency wars waged by the United States. Its duration and intensity forced the marines to think harder about the nature of such conflicts, a process that led to the *Small Wars Manual* in 1940 (see item 123).

The controversial nature of the war pushed the State Department and the Hoover administration toward a more conciliatory policy regarding Latin America and opened the way to the Good Neighbor Policy of Franklin D. Roosevelt.

There were similarities between Nicaragua and Vietnam. United States policy-makers asserted that the worldwide Communist movement was behind the rebels. The Mexican government was also believed to be aiding the insurgents. Increasing use of air bombardment brought criticism from both Nicaraguans and American opponents of the war.

Chronology

1912	After the Nicaraguan election of 1912, the United States stationed a small body of marines, a "legation guard," in Managua to demonstrate its continuing commitment to ensuring the maintenance of what it viewed as order in the country.
1925 August 1	The legation guard was withdrawn once Nicaragua began making an effort to replace its army with a constabulary that was intended to combine military and police capabilities and thereby to maintain order without the need for foreign troops.

October 25	Conservatives, led by General Emiliano Chamorro Vargas, ousted the elected Liberal government. The United States objected strongly to the coup.
1926	
March	Emiliano Chamorro became president, bringing a Liberal attempt to retake power. Significant fighting occurred. An initial mediation effort by the United States failed, and fighting continued.
1927	
January 6	Marines returned to Managua in the face of the disputed election.
February 27	First marine aviation element arrived in Nicaragua, and it was followed by other units. Aviation significantly complemented ground operations by the marines and Nicaraguan national guard through direct combat and through support activities.
May	United States Marine Corps Lieutenant Colonel Robert Y. Rhea became the first commander of the *Guardia Nacional de Nicaragua*, a paramilitary police organization, to be trained and organized by the United States. The establishment and continuance of this force was a major issue in Nicaraguan politics for years. Both the Conservatives and some Liberals, such as Sandino, were highly critical of at least aspects of the new security organization.
May 4	A special United States envoy, Henry L. Stimson, persuaded most of the warring factions in Nicaragua to agree to the continuance in office of the incumbent president and to the supervision of the 1928 elections by the United States. Some Liberal elements, under the leadership of Augusto C. Sandino, refused to accept these terms.
July 16–17	Sandinistas attacked marines and members of the Nicaragua national guard at Ocotal. Marine aircraft supported the besieged marines and Nicaraguan security personnel, defeating their attackers with heavy losses. From then on, the Sandinistas fought as guerrillas rather than making frontal attacks.
December 30	Fighting at Camino Real demonstrated that the Sandino movement was stronger than assumed by United States authorities, and the decision was made to strengthen the marine contingent in Nicaragua.
1928	
January	Marines undertook a combined air and land offensive against Sandino's stronghold of El Chipote. Eventually, marines reached the camp on January 26.

November 4   Liberals won the national election, which was held under close United States control, and José Moncada became president. Sandino had advised Nicaraguans to boycott the election. Clearly, his appeal had an effect, but interpretations of the election results continue to vary.

1929
January   Newly elected president Moncada proposed the establishment of a military unit, the *Voluntario* Force, to supplement operations by the national guard. Despite considerable opposition by some United States representatives in Nicaragua, the unit was formed, but it was neither as successful nor as economical as anticipated, and it was terminated in August.

1930
December 31   Eight marines killed in a battle at Achuapa. The fighting prompted further criticism of the intervention in the United States and moved the Hoover administration to plan the withdrawal of all United States forces from Nicaragua after the next Nicaraguan election in 1932.

1932
November 6   Liberals won the national elections, and Juan B. Sacasa became president-elect. In a few years he was ousted by the *Guardia* and its leader Anastasio Somoza.

1933
January 2   Evacuation of United States marine and naval personnel from Nicaragua completed. Anastasio Somoza was appointed head of the *Guardia* through his political connections rather than a military record.

February 2   Sandino agreed to stop military operations and to withdraw from politics.

1934
February 21   Sandino and several associates were murdered by followers of Anastasio Somoza, who was rapidly moving toward complete domination of Nicaragua. Somoza's forces attacked Sandinista elements, but scattered fighting continued for many years.

A. The decision to intervene

   841. Ellis, L. Ethan. *Frank B. Kellogg and American Foreign Relations, 1925–1929.* "Nicaraguan Relations," Chap. 3, pp. 58–85. New Brunswick, NJ: Rutgers University Press, 1961.

Interprets American concern with Nicaragua in terms of the United States' perception of the possibility of a Nicaraguan canal. The chapter includes a good deal of information on the background of the Stimson mission.

842. Munro, Dana G. *The United States and the Caribbean Republics 1921–1933*. Chap. 7, "The Second Intervention in Nicaragua," pp. 187–254. Princeton, NJ: Princeton University Press, 1974.

Includes a thorough account of the events that led to intervention. Fears of Mexican and Communist involvement played a role, as did concern about protecting the alternative canal route.

843. Stimson, Henry L. *American Policy in Nicaragua*. New York: Charles Scribner's Sons, 1927.

Defends vigorously the United States intervention. The most significant passages deal with the Stimson agreement in 1927, which provided for a constabulary and American supervision of the 1928 elections.

844. U.S. Congress. House. Committee on Foreign Affairs. *Conditions in Nicaragua and Mexico, Hearings on H. Res. 373, H. Res. 372, H. Res. 368, H. Res. 388, H. Res. 389, H. Res. 394, H. Res. 376, H. Res. 371, H. Res. 357*. 69th Cong., 2nd sess., 1927.

Focuses on Coolidge's policy, but also deals with presidential war powers, dollar diplomacy, and the applicability of the Monroe Doctrine to situations in which one Latin American country (Mexico in this case) is allegedly intervening in the affairs of another.

B. General surveys

845. Baylen, Joseph L. "American Intervention in Nicaragua, 1909–33: An Appraisal of Objectives and Results." *Southwestern Social Science Quarterly* 35 (September 1954): 128–54.

Asserts that the United States intervened to protect United States property, to prevent intervention by other nations, and to protect the alternative canal route. Baylen attacks the unilateral character of the intervention in the 1920s.

846. ———. "Sandino: Patriot or Bandit?" *Hispanic–American Historical Review* 31 (August 1951): 394–419.

Emphasizes political events, but also discusses Sandino's military strategy and notes major battles. It is a valuable piece, based in part on Nicaraguan newspapers.

847. Boot, Max. *The Savage Wars of Peace: Small Wars and the Rise of American Power.* Chap. 10, "Chasing Sandino: Nicaragua, 1926–33," pp. 231–52. New York: Basic Books, 2002.

Adequate, but perhaps too brief account of this important episode. Boot misunderstands the attitude of the Communists toward Sandino and his movement.

848. Cox, Isaac Joslin. "Nicaragua and the United States 1909–27." *World Peace Foundation Pamphlets* 10 (1927).

Presents the facts about United States interventions in a relatively objective fashion, taking account of the accusations by critics of dollar diplomacy. About half the substantial pamphlet consists of documents reprinted from other sources.

849. Crawley, Eduardo. *Dictators Never Die: A Portrait of Nicaragua and the Somoza Dynasty.* New York: St. Martin's Press, 1979.

Includes a fairly detailed review of Sandino's campaigns. The book is based in part on interviews with figures from the Sandino era. Crawley clearly delineates the elder Somoza's role during the United States intervention.

850. Delgadillo, Roberto Carlos. "The Last Banana War: United States Policy and the Second United States Intervention in Nicaragua, 1927–33." Ph.D. diss., University of California–Los Angeles, 2004.

Fine analysis of United States policy-making with respect to Nicaragua. Delgadillo draws heavily upon a variety of sources beyond State Department records, including reports from the Office of Naval Intelligence and the Federal Bureau of Investigation. In good part, he argues that the decisions to intervene in Nicaragua and then to continue a United States military presence in the country arose from intelligence failures. The concerns about Nicaragua that existed among United States decision-makers and even the public in the United States were simply not realistic.

851. Denny, Harold Norman. *Dollars for Bullets: The Story of American Rule in Nicaragua.* New York: Dial Press, Inc., 1929.

Focuses on the intervention of the 1920s. Denny's book is very sympathetic to the United States, although it is not uncritical. Several chapters deal specifically with the fighting between the marines and constabulary and Sandino's men.

852. Frazier, Charles E., Jr. "Augusto Cesar Sandino: Good Devil or Perverse God?" *Journal of the West* 3 (October 1964): 517–38.

Asserts that Sandino's life and career will remain potent symbols in Nicaragua. Most of the article deals with Sandino's personal life, personality, political views, and the events surrounding his murder. A small section deals directly with his military operations.

853. Hall, Steven R. "Glimpses of Wilsonianism: United States Involvement in Nicaragua During the Coolidge Era." M.A. thesis, West Virginia University, 2007.

Intriguing reinterpretation of the second major United States intervention in Nicaragua. Most studies have accepted the statements of the Coolidge administration that it was involving the United States in Nicaraguan affairs because of United States interests and because of "Communist" influence on Nicaragua from Mexico. Hall asserts that the Coolidge policy was a continuation of the paternalistic approach of the Wilson administration to several countries in the Caribbean area.

854. Kamman, William. *A Search for Stability: United States Diplomacy Toward Nicaragua 1925–1933*. Notre Dame, IN: University of Notre Dame Press, 1968.

Contains much on the military aspects of the intervention, despite the title. Several chapters are devoted to a thorough examination of the gradual United States withdrawal. This is one of the most important studies of the Nicaraguan intervention.

855. Livingston, Michael John. "Officers and Diplomats: A Study in Relationships and Foreign Policy, Nicaragua, 1926–33." M.A. thesis, University of Southern California, 2004.

Pertinent study of relationships between United States diplomats and their United States military counterparts in Nicaragua. Given the length of the intervention, the controversy surrounding United States involvement, and the complexity of affairs in Nicaragua, this is a useful look at an important facet of the episode. The thesis contains quite a bit of information about the United States supervision of Nicaraguan elections, which is another of its positive features.

856. Musicant, Ivan. *The Banana Wars*. Chap. 7, "Nicaragua II 1927–34," pp. 285–361. New York: Macmillan Company, 1990.

Discusses events in Nicaragua from 1912 to 1926 briefly and then examines the fighting that broke out in 1926 and the controversial United States intervention that followed. In contrast to most parts of Musicant's fine book, this segment is based on published primary sources and secondary materials rather than archival holdings. This approach may well have been sufficient because of the quality of literature about the second intervention in Nicaragua.

857. Nalty, Bernard C. *The United States Marines in Nicaragua*. Rev. ed. Washington, DC: Headquarters, U.S. Marine Corps, Historical Branch, G-3 Division, 1962.

Provides an extensive account of military operations in Nicaragua, from the perspective of the marines. While earlier episodes are included, the bulk of the study deals with the fighting in the 1920s and 1930s. This is one of the key books on Nicaragua.

858. Navarro-Génie, Marco Aurelio. *August "César" Sandino: Messiah of Light and Truth.* Syracuse, NY: Syracuse University Press, 2002.

Comprehensive analysis of Sandino's complex ideology, compounded of political, religious, and philosophical elements. This is more than a book about theories, important as they are for understanding Sandino and his movement, though. The author provides a considerable amount of information about political and military developments in Nicaragua, especially in the areas of the country under Sandino's control. There is an extensive bibliography, which appears in expanded form at www.sandino.org

859. Schroeder, Michael J. *Close Encounters of Empire: Writing the Cultural History of U.S.–Latin American Relationships.* "Civil War, Imperialism, Nationalism, and State Formation Muddied Up Together in the Segovias of Nicaragua, 1926–34," pp. 208–68, ed. Gilbert M. Joseph, Catherine C. LeGrand, and Ricardo D. Salvatore. Durham, NC: Duke University Press, 1998.

Essential reading for anyone concerned with Nicaragua at the time of the intervention against Sandino. Schroeder discusses interrelationships between Nicaraguan actors and factions that are hardly touched upon in other works or, indeed, are completely ignored. The essay is more than a cultural study. It includes highly useful information about the military aspects of the conflict. Schroeder also provides an important historiography of the period. The sophistication of Schroeder's research can simply be hinted at in an annotation.

860. Selser, Gregorio. *Sandino.* Translated by Cedric Belfrage. New York: Monthly Review Press, 1981.

Attacks bitterly the United States air war against the guerrillas. Written originally in 1955, this is a readable and highly sympathetic treatment of Sandino's struggle. There is much on the campaigns in Nicaragua and on Sandino's tactics.

861. Tierney, John Joseph, Jr. "The United States and Nicaragua, 1927–32: Decisions for De-Escalation and Withdrawal." Ph.D. diss., University of Pennsylvania, 1969.

Compares the Nicaraguan conflict to the Vietnam war, emphasizing the political aspects of United States intervention. Tierney develops an interesting set of propositions about American participation in small wars, pages 312–28. Especially noteworthy is Tierney's ability to correlate military and political events.

862. U.S. State Department. *The United States and Nicaragua: A Survey of the Relations from 1909 to 1932.* Washington, DC: Government Printing Office, 1932.

Deals primarily with the 1920s and 1930s. This is a lucid statement of American–Nicaraguan relations from the official United States standpoint. The section on the military aspects stresses the role of the National Guard.

863. Whisnant, David E. *Rascally Signs in Sacred Places: The Politics of Culture in Nicaragua.* Chap. 9, "Ancestral Feats and Future Dreams: Sandino and the Politics of Culture," pp. 344–82. Chapel Hill: University of North Carolina Press, 1995.

Discusses the way in which Sandino and his followers have been portrayed by various first-hand observers and others, emphasizing the years in which Sandino led an army against the marines in Nicaragua. Interpretations and even the presentation of basic facts have differed considerably. Whisnant notes that "reliable details on Sandino's early years are not abundant" (p. 348). During his fighting years, Sandino was disturbed and angered by the persistent efforts of his opponents to picture him and his men as "bandits."

864. Williams, Dion. "The Nicaraguan Situation." *Marine Corps Gazette* 15, no. 11 (November 1930): 18–22, 53, 55–57.

Counters criticisms of the marine presence in Nicaragua. Williams denies that the United States is an imperialistic nation or that it is intervening in Nicaragua in the customary sense. He refutes assertions that Nicaragua is bankrupt.

C. Strategy and tactics

865. Broderick, J.M. "The Science of Jungle Patrols." *Leatherneck* 16 (August 1933): 8, 56–57.

Describes Broderick's patrol methods in Nicaragua. He emphasizes the need to divide patrols into two sections in order to flank ambushes, if necessary. This account displays much mistrust of the Nicaraguans because of widespread sympathy with Sandino.

866. Brooks, David C. "US Marines, Miskitos and the Hunt for Sandino: The Rio Coco Patrol in 1928." *Journal of Latin American Studies* 21, no. 2 (1989): 311–42.

Praises Merritt Edson's approach to fighting the Sandinistas. Edson actively recruited Miskitos (indigenous Nicaraguans) to assist the marines. Brooks analyzes Miskito experiences through history that help explain the willingness of the group to cooperate with the marines. Brooks usefully compares this episode to the United States military cooperation with some ethnic mountain peoples in Indochina in the 1960s and 1970s.

867. Clark, George B. *With the Old Corps in Nicaragua*. Novato, CA: Presidio Press, 2001.

Includes a useful chapter on the intervention and subsequent fighting in 1912 but emphasizes the events of the late 1920s and early 1930s. Clark did a considerable amount of research, using some important materials that had not been consulted previously. This is an important study, especially for anyone interested in the nature of combat in Nicaragua at the time. The numerous maps are a bit rough, yet they are clear and are extremely informative.

868. Daniels, John A. "Don't Plan These Battles." *Marine Corps Gazette* 25, no. 9 (September 1941): 19–20, 43–46.

Argues that guerrilla fighting and counterinsurgency cannot be planned and that it is dangerous to make the attempt. This is primarily an anecdotal record of fighting in Nicaragua during the late 1920s.

869. "Events in Nicaragua Since February 28, 1928." *Marine Corps Gazette* 13, no. 6 (June 1928): 143–46.

Contains a detailed list of engagements between marines and Sandinista forces. There are one or two sentences for each combat contact up to May 18, 1928.

870. Gray, John A. "The Second Nicaraguan Campaign." *Marine Corps Gazette* 17, no. 2 (February 1933): 36–41.

Emphasizes that the jungle in Nicaragua made the fighting more difficult than in Haiti and the Dominican Republic. Gray focuses on the early combat contacts in northwest Nicaragua during the winter and spring of 1928. The article cites many examples of coordination between marine aviation and ground forces.

871. Heinl, Robert Debs, Jr. *Soldiers of the Sea: The United States Marine Corps, 1775–1962*. Annapolis, MD: United States Naval Institute, 1962.

Discusses marine involvement in Nicaragua in considerable detail, pages 260–90. More attention is paid to political ramifications of military operations than in most of Heinl's narratives of marine expeditionary duty.

872. Hodges, Donald C. *Intellectual Foundations of the Nicaraguan Revolution*. Austin: University of Texas Press, 1986.

Part of chapter 4 of this book is devoted to Sandino's troops, "The Defending Army of National Sovereignty", and chapter 7 to Sandino's evolutionary military strategy.

873. ——. *Sandino's Communism: Spiritual Politics for the Twenty-First Century.* Chap. 3, "People's War, People's Army," pp. 41–55. Austin: University of Texas Press, 1992.

Views Sandino's army as similar to Bakunin's efforts to build a revolutionary army in France at the time of the Franco–German War in 1870. Hodges also relates the Sandinista force to revolutionary developments in Mexico.

874. Hoffman, Jon T. *Chesty: The Story of Lieutenant General Lewis B. Puller, USMC.* New York: Random House, 2001.

For two periods, Puller commanded M Company of the *Guardia* and developed it into a mobile force (hence the "M" designation) that was much feared by Sandino's troops because of its almost unceasing and highly aggressive patrol activity. Working with William A. Lee, a marine sergeant who served as a *Guardia* officer, Puller moved his company like guerrillas, traveling lightly and quickly. He spurned the use of animal transport during his first assignment as company commander but later used animals, because he realized that carrying more supplies would permit longer patrols. Hoffman makes interesting comparisons between the fighting in Haiti and Nicaragua.

875. ——. *Once a Legend: "Red Mike" Edson of the Marine Raiders.* Chaps 4 and 5, pp. 47–94. Novato, CA: Presidio Press, 1994.

Edson quickly realized the importance of the Coco River, in eastern Nicaragua, to the insurgents as an avenue of transportation and, potentially, to the marines, who sought to deny mobility to Sandino's forces. He carried out several "Coco patrols," using small units first to study the relatively unknown area of the country and later to drive insurgents away from the river and, if possible, to push them into the way of other marine or *Guardia* units. This effort was plagued by unfavorable weather conditions, a chronic lack of supplies, and other challenges. Nevertheless, Edson and his men developed effective ways to fight the guerrillas that included coordination with marine aviation.

876. Holmes, Maurice G. "With the Horse Marines in Nicaragua." *Leatherneck* 14, no. 4 (April 1931): 5–7, 50–55.

Recounts the woes of a mounted patrol with a long pack train. Holmes includes a valuable section on mounted patrol tactics and points out that the United States insistence on a patrol carrying large amounts of supplies with it reduces mobility. He also describes the problems of mounting and keeping packs on animals.

877. McClellan, Edwin North. "He Remembered His Mission." *Marine Corps Gazette* 15, no. 11 (November 1930): 30–32, 51–52.

Recounts marine efforts to protect a Liberal governor who had been appointed in 1927 as a result of the Stimson agreement. The escort mission brought contacts between the marines and Sandino's men, but there was no real fighting at that early stage.

878. ———. "The Saga of the Coco: Adventures Along Nicaragua's Most Dangerous Waterway." *Marine Corps Gazette* 15, no. 11 (November 1930): 14–17, 71, 73, 75, 77, 79.

Describes the establishment of a regular boat patrol. Geyer's patrol, September–October 1928, and Edson's patrol in the spring of 1928 are covered in chronological fashion.

879. Macaulay, Neill. "Counterguerrilla Patrolling." *Marine Corps Gazette* 47, no. 7 (July 1963): 45–48.

Reviews "Chesty" Puller's patrols in Nicaragua. Macaulay notes the problems of dealing with a population that is sympathetic to the rebels. He makes some comparisons between the Sandino campaigns and counterinsurgency warfare in Malaya and Vietnam.

880. ———. "Guerrilla Ambush." *Marine Corps Gazette* 47, no. 5 (May 1963): 41–44.

Analyzes two successful and two unsuccessful patrols in Nicaragua between 1927 and 1932. He distinguishes between combat and other patrols. Normally, patrols in the latter category should break off action, but fighting back may be fruitful and may often be the only possibility.

881. ———. *The Sandino Affair*. Chicago: Quadrangle Books, 1967.

Compares the Nicaraguan campaign with various other counterinsurgency wars, including the United States intervention in the Dominican Republic in the 1960s. This is an essential work for anyone who wants to understand the military aspects of the intervention in Nicaragua. Macaulay was a former officer in both the United States Army and Castro's forces.

882. Peard, Roger W. "The Tactics of Bush Warfare." *Infantry Journal* 38 (September–October 1931): 408–15.

Analyzes guerrilla warfare and counterinsurgency, using examples from Nicaragua to illustrate the stages that are identified. Several battles, including Ocotal, are examined in some detail. This is an extremely important study, one of the most precise examinations of the subject during the interwar period.

883. Proctor, Clarance B. "The Nicaraguan Expedition." *Leatherneck* 12, no. 6 (June 1929): 8–9, 55–56.

Provides a useful discussion of the problems of fighting guerrillas in the jungle. Proctor notes the difficulty in gaining accurate intelligence about Sandino's forces and in avoiding ambushes. He has high praise for the navy medical personnel attached to the marines and comments at length on the usefulness of marine aircraft.

884. U.S. Congress. Senate. Committee on Foreign Relations. *Use of the United States Navy in Nicaragua, Hearings Pursuant to S. Res. 137.* 70th Cong., 1st sess., 1928.

Constitutes an important source for the early years of the intervention. The navy seemed to respond fully to congressional demands for information. Among the witnesses was Admiral Julian Lane Latimer, Special Service Squadron commander. A chronology covers the period from December, 1926, through early February, 1928.

885. U.S. Marine Corps. Headquarters. Division of Operations and Training. "Protection of American Interests." *Marine Corps Gazette* 12, no. 9 (September 1927): 175–83.

Reviews in detail troop movements for the Nicaraguan intervention in 1926–27. The remainder of the article deals with the situation in China.

886. Walraven, J.G. "Typical Combat Patrols in Nicaragua." *Marine Corps Gazette* 14, no. 12 (December 1929): 243–53.

Begins with an examination of the climate and vegetation, and then proceeds to a review of patrol activity in Nueva Segovia Province. Walraven's discussion of the deployment of personnel on the march includes detailed sketches that pinpoint the position of every man and animal on typical patrols.

887. Williams, Dion. "Captain Richard Bell Buchanan, United States Marine Corps." *Marine Corps Gazette* 12, no. 6 (June 1927): 73–75.

Eulogizes Buchanan, who was killed in a fire fight with Nicaraguan "revolutionary guerrillas" on May 16, 1927. Williams also outlines marine operations in Nicaragua during the early part of 1927.

D. Major battles and campaigns

888. Beasdale, Victor F. "La Flor Engagement." *Marine Corps Gazette* 16, no. 2 (February 1932): 29–40.

Analyzes an engagement with Sandinista troops, February 13–14, 1928. This is an excellent examination of the battle. Two fine charts enhance the article. The marine/constabulary patrol's organization and equipment are discussed in considerable detail.

889. "Captain Robert Stuart Hunter, U.S. Marine Corps." *Marine Corps Gazette* 14, no. 3 (March 1929): 4–5.

Describes the fighting on May 13, 1928, in which Captain Hunter died. He received the Navy Cross for his actions that day. There is an unnumbered page of photographs.

890. Edson, Merritt A. "The Coco Patrol: Operations of a Marine Patrol Along the Coco River in Nicaragua, 1928–29." Parts 1, 2, 3. *Marine Corps Gazette* 20, 21, nos. 8, 11; 2 (August, November 1936; February 1937): 18–23, 38–48; 40–41, 60–72; 35–43, 57.

Constitutes one of the most detailed published narratives on warfare in Nicaragua during the Second Intervention. It is an eloquent statement on the hardships that the marines endured.

891. Harris, Harold D. "A Skirmish in Nicaragua." *Leatherneck* 21, no. 11 (November 1938): 7–8, 60–61.

Presents a series of problems for discussion, based on a battle when some of Sandino's troops attacked a pack train. Two charts illustrate the problems.

892. Lejeune, John A. "The Nicaraguan Situation." *Leatherneck* 11, no. 2 (February 1928): 10, 52.

Reviews current operations against Sandino. General Lejeune expressed the desire to finish the campaign before June 1928. Much of the article deals with the battle of Quilali. This engagement revealed that the rebels were much better armed and trained than had been thought.

893. ———. "The Situation in Nicaragua." *Leatherneck* 11, no. 4 (April 1928): 10, 52, 54.

Welcomes the decline in Sandino's forces and narrates the air attacks on El Chipote, his headquarters. General Lejeune also discusses a mutiny within the constabulary and its suppression by the marines. He expresses surprise that more episodes of this sort have not occurred in view of the fact that detachments are often left without much supervision.

894. McClellan, Edwin North. "The Nueva Segovia Expedition." Parts 1, 2. *Marine Corps Gazette* 16, nos. 5, 8 (May, August 1931): 21–25; 8–11, 59.

Recounts marine efforts to retake an American-owned mine captured by rebels. The second part deals primarily with a battle at Jicaro in 1927. The article contains extensive extracts from the official report and Sandino's correspondence.

895. McKenzie, Scott W. "Tiger of the Mountains." *Marine Corps Gazette* 55, no. 11 (November 1971): 38–42.

Narrates "Chesty" Puller's and William A. Lee's exploits in leading constabulary units against Sandino's men. Several skirmishes are described. The article also gives a good deal of information about the tactical innovations Puller and Lee introduced, which were later incorporated into the *Small Wars Manual* of 1940 (item 123).

896. "Marines and Machine Guns in Nicaragua." *Independent* 118 (January 22, 1927): 90–91.

Discusses the deployment of marines to Nicaragua and notes the number of naval personnel stationed there. The political situation is also summarized.

897. "A Patrol Makes Contact." *Leatherneck* 21, no. 12 (December 1938): 8, 56–57.

Presents the reader with a series of questions on how to react to certain patrol situations, with an actual patrol in Nicaragua used as an example. Part of the article discusses attacks on buildings held by guerrillas.

898. Rentflow, Frank Hunt. "And in Sunny Tropic Scenes." *Leatherneck* 14, no. 3 (March 1931): 8–9, 47–49.

Reviews the fighting in Nicaragua, starting with an ambush in December 1930, but moving back to 1927. Rentflow describes the major battles between 1927 and the end of 1930, with little regard to tactics. The article is enhanced by two sketches of marine operations in Nicaragua.

899. "Sandino and His 'Glorious' Exploits." *Leatherneck* 11, no. 9 (September 1928): 8, 54.

Accuses rebel leader Sandino of having marine corpses disinterred and held until the families of the slain men made efforts to secure the withdrawal of all the marines from Nicaragua. The writer includes a letter, said to have been written by Sandino, to substantiate the charges.

900. U.S. Marine Corps. Headquarters. Division of Operations and Training. "Combat Reports of Operations in Nicaragua." Parts 1, 2, 3, 4. *Marine Corps Gazette* 13, 14, nos. 12; 3, 6, 9 (December 1928; March, June, September 1929): 241–47; 16–30; 81–94; 170–79.

Deals with the battles of La Paz Centro, Ocotal and Telpaneca, and El Chipote and the suppression of guerrilla bands after El Chipote. This very important series is accompanied by a number of photographs.

901. "Will Fight Until We Die, Marine Major Tells Rebel." *Leatherneck* 10, no. 8 (August 1927): 22–23.

Provides a rather detailed account of the siege of Ocotal by Sandino's forces. The marines and constabulary held out despite the lack of water. The article, which is reprinted from the *Washington Post*, gives due credit to the marine airplanes that drove the rebels away from Ocotal by using aerial bombs.

902. Wood, John C., and J.D. Wilmeth. "Sandino Strikes Again." *Leatherneck* 22, no. 2 (February 1939): 8, 55–57.

Presents a series of tactical problems based on Wood's experience in Nicaragua in 1930. Wood complains that more responsive action by the navy would have dealt Sandino a serious blow in one situation.

E. The role of air power

903. "The Aerial Rescue of Ocotal." *Leatherneck* 10, no. 10 (October 1927): 9–11, 48.

Tells the story of both the besieged marines and constabulary and their rescuers from the air. The unnamed writer was at Ocotal and gives a vivid picture of the fighting. His account of the activities of the marine flyers as they sought to relieve the Ocotal garrison is also quite full.

904. "Air Operations in Nicaragua." *Leatherneck* 11, no. 9 (September 1928): 9, 51.

Reviews marine air operations during May and June 1928. A variety of activities are discussed, and, in addition to references to several fights, a number of amusing occurrences are noted. Despite the choppiness of the narrative, this is a useful article.

905. Brainard, Edwin H. "Marine Corps Aviation." *Marine Corps Gazette* 13, no. 3 (March 1928): 25–36.

Uses operational experiences in China and Nicaragua in a lecture to marine schools. Brainard emphasizes the variety of combat and support services and the need for aviation in a country where transportation is primitive and the terrain rugged.

906. ———. "The Marines Take Wings." *Leatherneck* 11, no. 8 (August 1928): 28–33.

Focuses on the battles at Ocotal and El Chipote. In addition to recounting the contribution of marine aviation to these victories, Brainard discusses the first use of marine airplanes in Nicaragua, early in 1927. They performed useful functions, particularly supplying isolated marine units and maintaining communications between them.

907. Heritage, Gordon W. "Forced Down in the Jungles of Nicaragua." *Leatherneck* 15, no. 5 (May 1932): 13–15, 48–49.

Focuses on the long march that Heritage and another noncommissioned officer endured when their aircraft was damaged during an attack they made on a rebel force. Fortunately, they encountered no enemy troops on their way to the coast. Both men received high decorations for their accomplishments.

908. Jennings, Kenneth A. "Sandino Against the Marines: The Development of Air Power for Conducting Counterinsurgency Operations in Central America," *Air University Review* 37, no. 5 (July–August 1986): 85–95.

Important survey that begins with air operations in 1926 and early 1927 during the Nicaraguan civil war. Jennings reviews the wide range of uses of aircraft by the marines between 1927 and 1933, but he readily admits the limitations of aviation in small wars. Although this is a worthwhile study, the author curiously ignored numerous relevant articles in the *Marine Corps Gazette* and *Leatherneck*, as well as a major analysis in the *United States Naval Institute Proceedings* (item 910).

909. Major, H.C. "Bringing the 'Ducks' from Nicaragua." *United States Naval Institute Proceedings* 59, no. 12 (December 1933): 1727–31.

Reviews the flight of marine aircraft from Nicaragua to the United States in January 1933. Much of the article deals with poor flying conditions and encounters with officials and soldiers in several Latin American countries.

910. McGee, Vernon E. "The Genesis of Air Support in Guerrilla Operations." *United States Naval Institute Proceedings* 91, no. 6 (June 1965): 48–59.

Provides the most complete published account of the innovations made by marine aviation in Nicaragua. General McGee notes that the marines had the right aircraft for their work, something which may not have been the case in Vietnam.

911. Montross, Lynn. "The Marine Autogiro in Nicaragua." *Marine Corps Gazette* 37, no. 2 (February 1953): 56–61.

Stresses marine experiments in Nicaragua in 1932, although it also gives something of the history of helicopters generally. The mechanical problems that plagued the OP-1 are described in detail, along with the reactions of the Nicaraguans to the strange new type of airplane.

912. Mulcahy, Francis P. "Marine Corps Aviation in the Second Nicaraguan Campaign." *United States Naval Institute Proceedings* 59, no. 8 (August 1933): 1121–40.

Covers all phases of aviation. In addition to combat support, aircraft proved useful for flying personnel and ballots during the 1928 elections. Mulcahy stresses the dangers facing aviators in Nicaragua. These included weather and terrain.

913. "Over the Halls of Montezuma: The Aerial Evacuation of Nicaragua." *Leatherneck* 16, no. 3 (March 1933): 10–11, 51–53.

Deals with Captain Mulcahy's flight of marine aircraft from Nicaragua to the United States. The article is presented in the form of a diary from January 2 to January 12, 1933.

914. Roper, William L. "Air Action in Nicaragua." *Marine Corps Gazette* 56, no. 11 (November 1972): 54–57.

Provides a valuable, contemporary account of the air relief of the marine/constabulary garrison of Ocotal. Roper was a newspaper reporter who interviewed Major Ross E. Rowell shortly after he and his fellow flyers carried out the first dive bombing attack, which effectively broke the siege of Ocotal.

915. Rowell, Ross E. "Aircraft in Bush Warfare." *Marine Corps Gazette* 14, no. 9 (September 1929): 180–203.

Focuses on Nicaragua, although reference is also made to the use of aircraft against insurgents in North Africa and the Middle East. Much of the article deals with ground support attacks. The author emphasizes the need for appropriate air base security.

916. ———. "Annual Report of the Aircraft Squadron, Second Brigade, U.S. Marine Corps, July 1, 1927, to June 30, 1928." *Marine Corps Gazette* 13, no. 12 (December 1928): 248–65.

Describes support activities primarily, but does contain a long section on aviation in counterinsurgency operations.

917. ———. "Marine Air in Nicaragua." *Marine Corps Gazette* 69, no. 5 (May 1985): 80–91.

Renowned marine flyer recalls his experiences in the air war on the Sandinistas.

918. Schroeder, Michael J. "Social Memory and Tactical Doctrine: the Air War in Nicaragua During the Sandino Rebellion, 1927–32," *International History Review* 29, no. 3 (September 2007): 508–49.

Detailed account of the air campaign and the ways in which it was portrayed by media in Latin America. Schroeder includes a number of striking editorial cartoons from Latin American newspapers. He argues, too, the air offensive had a lasting impact on Nicaragua. Schroeder asserts that "[t]error was built into the architecture of the aerial war in Nicaragua" (p. 548).

919. "Second Lieutenant Earl Albert Thomas, U.S. Marine Corps." *Marine Corps Gazette* 13, no. 12 (December 1928): 222–25.

Praises Thomas' bravery for his fighting desperately against guerrillas after his airplane crash and before his death.

920. Turner, Thomas C. "Flying with the Marines in Nicaragua." *Leatherneck* 14, no. 3 (March 1931): 5–7, 52–54.

Stresses the innovative nature of marine aviation operations in combatting guerrilla bands. Eyewitness reports document the importance of both air-to-ground attacks and supply by air. Much of the article deals with the Sandinista siege of Ocotal and the raising of the siege by marine flyers. The article includes statistics on air operations during 1929 and 1930.

F. Military advisory activities

921. Carlson, Evans F. "The Guardia Nacional de Nicaragua." *Marine Corps Gazette* 21, no. 8 (August 1937): 7–20.

Emphasizes the pioneering efforts of the marines in developing the National Guard. This is a thorough summary of Guardia combat actions with special attention to "Chesty" Puller's patrol on the Coco River.

922. Carter, C.B. "The Kentucky Feud in Nicaragua." *World's Work* 54 (July 1927): 312–21.

Describes Carter's abortive attempt to form a constabulary in 1925 and the opposition he received. Nicaraguans were simply too divided politically to support a nonpartisan military or police force. Carter asserts that some Mexicans were involved in the Nicaraguan civil war, but not necessarily the Mexican government.

923. Denig, Robert L. "Native Officer Corps, Guardia Nacional De Nicaragua." *Marine Corps Gazette* 17, no. 11 (November 1932): 75–77.

Locates the origins of the National Guard in the Stimson agreement of May 11, 1927. Denig gloomily, and accurately, predicts that the National Guard will soon lose its nonpartisan character. He also discusses various auxiliary forces attached to it and the problem of procuring higher officers for it.

924. Goldwert, Marvin. *The Constabulary in the Dominican Republic and Nicaragua: Progeny and Legacy of United States Intervention.* Gainesville: University of Florida Press, 1962.

Reviews the Nicaraguan military tradition, the United States desire to see a nonpartisan force established, and the conflicts between the Americans and Nicaraguans during the formation of the Guard.

925. Hanneken, Herman H. "A Discussion of the Voluntario Troops in Nicaragua." *Marine Corps Gazette* 26, no. 11 (November 1942): 120, 247–48, 250–54, 256, 258, 260, 262, 264, 266.

Explains the Voluntario units as necessary expedients because the Nicaraguan government was reluctant to commit the National Guard to field operations against Sandino. Two units of 90 men each served as Voluntario troops in 1929.

926. Keyser, Ralph Stover. "Constabularies for Central America." *Marine Corps Gazette* 11, no. 6 (June 1926): 87–97.

Describes State Department efforts to establish nonpartisan constabularies in the Central American countries to discourage military coups. At least half the article focuses on the Nicaraguan National Guard.

927. Millett, Richard. *Guardians of the Dynasty: A History of the U.S. Created Guardia Nacional de Nicaragua and the Somoza Family.* Maryknoll, NY: Orbis Books, 1977.

Examines the United States role in the formation of the constabulary in great detail. About half of this interesting study relates to the 1920s and 1930s. Chapter IV deals most specifically with combat operations against Sandino.

928. Reisinger, H.C. "La Palabra del Gringo! Leadership in the Nicaraguan National Guard." *United States Naval Institute Proceedings* 61, no. 2 (February 1935): 215–21.

Recounts a number of anecdotes demonstrating the loyalty of National Guard men. The author also cites several cases of serious indiscipline and explains the reasons for them. He praises highly marine officers and National Guard enlisted men in general.

G. Other military topics

929. Anderson, Charles R. "The Supply Service in Western Nicaragua: April, 1927 to April, 1929." *Marine Corps Gazette* 17, no. 5 (May 1932): 41–44.

Lists the many kinds of transportation (14 types) used to supply marines. Trucks were rarely utilized because of poor roads. Air drops were very helpful. Stockpiling of supplies was necessary in some areas.

930. Baker, C.S. "The Spanish Language and the Marine Corps." *Marine Corps Gazette* 14, no. 12 (December 1929): 254–59.

Highlights the problems of marine officers in Nicaragua who lacked fluency in Spanish. They often pay excessive fees for services because they cannot bargain. During the municipal elections of 1927, both Liberal and Conservative interpreters had to be used to avoid election fraud.

931. Cole, C.E. "Marine Corps Communications." *Marine Corps Gazette* 14, no. 9 (September 1929): 221–25.

Reviews communications problems that beset expeditionary forces. Cole deals at some length with Nicaragua, where army radios proved too heavy for convenient use. He also discusses the broader issue of retaining the necessary technical personnel.

932. Denig, Robert L., Jr. "Use of Chemical Agents in Guerrilla Warfare." *Marine Corps Gazette* 20, no. 8 (August 1936): 36–38.

Decries the United States inability to use chemical weapons in counterinsurgency because of public opinion. The article begins with a description of an incident in 1931, in Nicaragua, where marines might easily have broken up an ambush with white phosphorous grenades.

933. Pagano, Dom Albert. *Bluejackets*. Boston, MA: Meador Publishing Company, 1932.

Tells about the naval service in Nicaragua during 1926 and early 1927, when sailors were used to guard installations and to support Stimson's negotiations, prior to the arrival of marines in strength. This was before the advent of Sandino, and, therefore, the book has little on the fighting.

934. Peard, Roger W. "Bull Cart Transportation in the Tropics." *Marine Corps Gazette* 15, no. 2 (February 1931): 29–30, 47–49.

Examines closely the handling by marines of bull carts in Nicaragua during the late 1920s. Load levels and river crossing are among the topics discussed. About half the article relates to a marine patrol that used such carts during the summer of 1927 while disarming guerrillas in accordance with the Stimson agreement.

935. Schubert, Richard H. "Communications in Bush Warfare." *Marine Corps Gazette* 14, no. 6 (June 1929): 114–16.

Describes the major types of radio equipment used in Nicaragua and the problems associated with them. Modifications had to be made in army equipment, primarily to adapt it to mule transport.

H. Civic action

936. Bacevich, A.J. *Diplomat in Khaki: Major General Frank Ross McCoy and American Foreign Policy, 1898–1949*. Chap. 8, "Mission to Nicaragua," pp. 114–37. Lawrence: University Press of Kansas, 1989.

McCoy headed the commission that the United States established to supervise the 1928 election in Nicaragua in accordance with an agreement made in 1927 between the Conservatives and some Liberals to accept this supervision. The election was conducted in a highly successful manner, except for some abstentions because of Sandino's plea to boycott the process. McCoy's stay in Nicaragua was hardly peaceful, though, owing to significant friction between McCoy, an army officer, and the navy and marines.

937. Buell, Raymond Leslie. "American Supervision of Elections in Nicaragua." *Foreign Policy Association Information Service* 6 (December 24, 1930): 385–402.

Covers the 1928 presidential election and the congressional election of 1930. Buell discusses the issues associated with United States involvement. The article concludes with a detailed examination of the possible policies that the United States might pursue toward Nicaragua in the future.

938. Dodd, Thomas J. *Managing Democracy in Central America: A Case Study: United States Election Supervision in Nicaragua, 1927–1933*. New Brunswick, NJ: Transaction Publishers, 1992.

Thoroughly researched study based on various primary sources, including extensive interviews with key figures in Nicaragua during the period examined. Dodd's work has permanent value, in part because some of the documents he examined were destroyed in an earthquake. Dodd analyzes not only the 1928 and 1932 national elections but also municipal elections in 1930 and 1931.

939. Dodds, Harold W. "American Supervision of the Nicaraguan Election." *Marine Corps Gazette* 14, no. 6 (June 1929): 117–24.

Describes the legal mechanisms by which the United States came to oversee the Nicaraguan election of 1928, and discusses the many problems that developed during the campaign and the election.

940. Hardy, Earl B. "Supervision of Nicaraguan Elections." *Leatherneck* 15, no. 4 (April 1932): 7–9, 58.

Details in depth information on the organization and deployment of the Nicaraguan Electoral Missions of 1928, 1930, and 1932. A veteran of earlier missions, Hardy writes on the preparations for the 1932 election. This important article argues that the Electoral Missions did much good.

941. Linsert, E.E. "The Marine Corps in Nicaragua." *Leatherneck* 15, no. 9 (September 1932): 7, 59–60.

Stresses the political background of the intervention, and describes the Stimson agreement. Linsert contends that all Nicaraguans with any respect for law and order approved of the intervention. There is a brief section on the constabulary.

942. McClellan, Edwin North. "Supervising Nicaraguan Elections of 1928." *United States Naval Institute Proceedings* 59, no. 1 (January 1933): 25–38.

Gives a detailed account of United States involvement with the 1928 election and something of the background of the intervention. McClellan notes Sandino's efforts to block the election. He defends American intervention because of the need to protect the security of Central America.

943. "The Managua Disaster." *Marine Corps Gazette* 16, no. 8 (August 1931): 12–17, 35.

Examines marine relief efforts, including the use of aircraft to transport supplies, in the wake of the earthquake of March 31, 1931. The article lists marines and marine dependents killed.

944. Munro, Dana G. *The United States and the Caribbean Republics 1921–1933*. Chap. 8, "The Hoover Administration, Central America, and the Dominican Republic, 1929–33," pp. 255–308. Princeton, NJ: Princeton University Press, 1974.

Deals generally with United States intervention in Nicaragua between 1928 and 1933. United States supervision of the congressional election of 1930 and the presidential election of 1932 prevented violence by the guerrillas.

945. Phillips, Clinton A. "Earthquake in Managua." *Marine Corps Gazette* 74, no. 2 (February 1990): 60–67.

The author discusses the earthquake and the efforts that the marines and the Red Cross made to assist the Nicaraguans after this serious episode.

946. "Service in Nicaragua." *Field Artillery Journal* 19 (May–June 1929): 289–95.

Considers the experience of army officers who assisted other Americans in the supervision of the election of 1928. Much of the article consists of photographs, including some dealing with military operations such as the surrender of insurgents at Ocotal.

I. Political developments during the intervention

947. Howell, Charles F. "Neighborly Concern: John Nevin Sayre and the Mission of Peace and Goodwill to Nicaragua, 1927–28," *Americas* 45, no. 1 (July 1988): 19–46.

Thoroughly researched study of Sayre's efforts to bring peace to Nicaragua through mediation between the United States and the Sandinistas. Howell includes much material about the views of State Department officials about Nicaragua and Sayre's peace crusade.

948. Langley, Lester D. *The Banana Wars: An Inner History of American Empire 1900–1914.* Chap. 4, pp. 169–216. Lexington: University Press of Kentucky, 1983.

Emphasizes Nicaraguan manipulation of United States decision-makers through clever use of the Mexican menace. Some attention is given to military as well as political events. Langley dates the end of the campaign as April, 1931, when the navy warned United States citizens seeking protection to go to the coast.

949. Perkins, Whitney T. *Constraint of Empire: The United States and the Caribbean Interventions.* Westport, CT: Greenwood Press, 1981.

Focuses almost entirely on political events during the intervention. Perkins' treatment is generally more sympathetic to United States policies than many other analyses, chapters 6 and 10.

950. Salisbury, Richard V. "Mexico, the United States, and the 1926–27 Nicaraguan Crisis," *Hispanic American Historical Review* 66, no. 2 (May 1986): 319–39.

Notes Mexico's continued interest in Central America and its policies toward that area after the Mexican Revolution of 1910. Nicaraguan Liberals looked to Mexico as early as 1916 in the face of United States involvement with their country. Mexico and Nicaragua made a formal agreement for active Mexican support for the Nicaraguans. Salisbury is highly critical of the Coolidge administration's stance toward Nicaragua.

951. Schroeder, Michael J. "Bandits and Blanket Thieves, Communists and Terrorists: The Politics of Naming Sandinistas in Nicaragua, 1927–36 and 1979–90." *Third World Quarterly* 26, no. 1 (2005): 67–86.

Describes the systematic efforts of the United States to discredit the Sandinistas by branding them as "bandits." Sandinista tax gathering was used as one piece of "evidence" for this characterization.

952. Vincent, Ted. "The Harlem to Bluefields Connection: Sandino's Aid from the Black American Press," *Black Scholar* 16, no. 3 (May–June 1985): 36–42.

Documents the strong support given to the Sandinistas by the African-American press. Vincent notes that the African minority in Nicaragua was highly interested in Marcus Garvey's Universal Negro Improvement Association, which created a tie between Nicaragua and the

United States. Garvey's *Negro World* and the *Crisis* were the newspapers that maintained a continuing interest in Nicaragua.

J. The decision to withdraw

953. Buell, Raymond Lewis. "Getting Out of Central America." *Nation* 135 (July 13, 1932): 32–34.

    Reviews both United States policy toward Nicaragua and the current Nicaraguan political scene.

954. Garcia, Rogelio. "Opposition Within the Senate to the American Military Intervention in Nicaragua, 1926–33." Ph.D. diss., Columbia University, 1973.

    Discusses the Senate's effort to interest the public in the Nicaraguan situation. Gradually, its campaign was successful, and the administration lost popular support for the intervention.

955. "Recalling the Marines." *Outlook* 157 (February 25, 1931): 288–89.

    Discusses the partial withdrawal of marines that has been announced by Secretary of State Stimson. The article notes that training of the constabulary by the marines has not yet been completed.

956. Tarbuck, Ray D. "The Nicaraguan Policy of the United States." *United States Naval Institute Proceedings* 56, no. 2 (February 1930): 113–20.

    Describes the politics, society, and economy of Nicaragua. Tarbuck argues that the United States must help Nicaragua to achieve national unity. There seems to be no consistent United States policy, however.

957. Wood, Bryce. *The Making of the Good Neighbor Policy.* Chap. 11, "The Nicaraguan Experience," pp. 13–47. New York: Columbia University Press, 1961.

    Traces the beginnings of the Good Neighbor Policy to the Nicaraguan intervention. This is an excellent survey of United States policies regarding Nicaragua and the military and political problems confronting the United States in that country.

K. The evacuation

958. "The Evacuation of Nicaragua." *Leatherneck* 16, no. 1 (January 1933): 5–7, 47.

    Describes the redeployment of both air and ground forces. The article predicts that the United States has made a lasting impact on Nicaragua. All the battles in which marines were killed are listed.

959. "The Marines Return from Nicaragua." *Marine Corps Gazette* 17, no. 2 (February 1933): 23–27.

Reviews marine activity from January 1927, through the end of 1932, emphasizing the supervision of several elections. The article does not stress the fighting. It does, however, contain extremely detailed information on casualties and gives a description of the evacuation.

XVIII

Post–World War II doctrines and general studies

The Greek civil war and the Korean War kindled a concern with counterinsurgency. In the early 1960s President Kennedy's commitment brought a tidal wave of literature, only part of which can be examined in this section. A few of the articles and books cited here refer to Vietnam, but most of that literature has been omitted.

This section of the bibliography contains subsections on counterinsurgency in the 1950s, doctrines in the 1960s, theoretical approaches, the Kennedy counterinsurgency initiative, the role of Special Forces, the naval contribution, counterinsurgency training in the 1960s, systems analysis and counterinsurgency, counterinsurgency tactics, the weapons of counterinsurgency, intelligence and counterinsurgency, other facets, and the related topic of civic action.

A. Counterinsurgency in the 1950s

960. Beebe, John E. "Beating the Guerrilla." *Military Review* 35, no. 12 (December 1955): 3–18.

   Cites a number of instances in which Communist guerrillas have been defeated. This is done to stress the ability of the West to fight counterinsurgency wars. Beebe deals primarily with the Korean War but generalizes from this experience.

961. Booth, Waller B. "The Pattern That Got Lost." *Army* 31 (April 1981): 62–65, 67.

   Discusses "Porrex," a counterinsurgency exercise in 1950, in which aggressor troops simulated being Puerto Rican insurgents. Due to the Korean War and the immediate need for anti-guerrilla units, Booth contends, the lessons of the exercise were not fully learned.

962. Decker, William T. "Anti-Guerrilla Warfare." *Marine Corps Gazette* 30, no. 8 (August 1951): 22–25.

Emphasizes the need to treat civilians in guerrilla areas humanely, citing the disastrous experiences of the Germans in the Soviet Union during World War II. Decker also discusses the usefulness of helicopters in anti-guerrilla campaigns.

963. Downey, Edward F. "Theory of Guerrilla Warfare." *Military Review* 39, no. 5 (May 1959): 45–55.

Examines Chinese Communist guerrilla doctrine, and advocates that all United States military personnel, including reservists, receive extensive guerrilla warfare training.

964. Harris, Albert E. "Partisan Operations." *Military Review* 30, no. 8 (August 1950): 10–20.

Reviews anti-Axis guerrilla movements, and warns that similar operations might be a factor in future conflicts. Harris examines the problems of both partisan and anti-partisan forces.

965. Heinl, Robert D., Jr. "Small Wars—Vanishing Art?" *Marine Corps Gazette* 29, no. 4 (April 1950): 22–25.

Warns that marines have become so engrossed with amphibious warfare that they have forgotten how to fight guerrillas. These lessons should be relearned. Moreover, Heinl suggests that marines may have to fight as guerrillas, if the Soviets overrun Eurasia.

966. Johnson, J.R. "Antiguerrilla Exercise." *Marine Corps Gazette* 34, no. 12 (December 1955): 20–22.

Describes a rigorous company-level exercise at Vieques, Puerto Rico. One result was the positive evaluation given to the relevant field manual, FM 31-20, *Operations Against Guerrillas*.

967. Oppenheimer, Harold L. "Command of Native Troops." *Marine Corps Gazette* 30, no. 10 (October 1951): 50–59.

Reviews American experiences with the constabularies in the Caribbean. The author commanded Samoan troops and discusses them briefly. He examines the ways in which United States leadership can be provided to "native" troops but recommends that more research be conducted on this subject.

968. Paddock, Alfred H., Jr. *U.S. Army Special Warfare: Its Origins*. Rev. ed. Lawrence: University Press of Kansas, 2002.

Emphasizes psychological and unconventional warfare rather than counterinsurgency, reflecting the concerns of the army in the 1940s and 1950s. Nevertheless, this lucid and richly documented study is important to an understanding of the milieu in which the Special Forces concept developed.

969. "Readiness for the Little War: Optimum Integrated Strategy." *Military Review* 37, no. 4 (April 1957): 14–26.

Surveys the kinds of situations that can lead to American intervention, especially in the Middle East and South Asia. Several pages outline the characteristics of the force needed to fight small wars.

970. "Readiness for the Little War: A Strategic Security Force." *Military Review* 37, no. 5 (May 1957): 14–21.

Elaborates on the conclusion of the previous article. It discusses army, navy, air force, and other components of a counterinsurgency force. It briefly describes the role of Military Assistance Advisory Groups and Special Forces units.

971. "Special Warfare." *Leatherneck* 41, no. 2 (February 1958): 24–29.

Discusses special warfare training provided for marines by Special Forces troops at Fort Bragg. The article notes that although the primary mission of Special Forces is guerrilla warfare in the Soviet bloc in case of a general war, it is also responsible for counterinsurgency operations.

972. Wilkins, Frederick. "Guerrilla Warfare." *United States Naval Institute Proceedings* 80, no. 3 (March 1954): 311–18.

Reviews guerrilla operations through history in a very sophisticated article. Wilkins warns that counterinsurgency will probably be much more difficult in Asia than it was in Greece during the late 1940s, for example.

B. Counterinsurgency doctrine

973. Birtle, Andrew J. *U.S. Army Counterinsurgency and Contingency Operations Doctrine 1942–1976*. Washington, DC: United States Army Center of Military History, 2006.

Authoritative official study of the development of doctrine by the United States Army. This is a massive work with a large although "select" bibliography. Numerous maps and photographs add to the worth of this major achievement in the field of counterinsurgency research.

974. Bowman, Stephen Lee. "The Evolution of United States Army Doctrine for Counterinsurgency Warfare: From World War II to

the Commitment of Combat Units in Vietnam." Ph.D. diss., Duke University, 1985.

Argues that the army had little in the way of counterinsurgency doctrine at the beginning of its large-scale involvement in Vietnam. Thus, doctrinal development was extremely hurried. Bowman tries to determine whether this rapidly prepared doctrine actually assisted the forces in Vietnam to operate more effectively. Based on the sources he studied, Bowman finds that newly developed doctrine had little value for fighters in Vietnam, in good part because of the nature of the Vietnam war, which was not simply an insurgency.

975. Brask, James J. "Counterinsurgency as an Instrument of American Foreign Policy for the Analysis of Vietnam Counterinsurgency Evaluations." Ph.D. diss., Northern Illinois University, 1990.

Makes the provocative argument that the United States began fighting a counterinsurgency campaign in Vietnam before the armed forces had developed adequate doctrine for such an effort. Brask devotes a long chapter to French Indochina and then proceeds to discuss ways to evaluate the progress in a counterinsurgency environment. Because evaluating success in Vietnam was a highly important and highly controversial issue, this dissertation makes a real contribution to the study of counterinsurgency.

976. Cable, Larry E. *Conflict of Myths: The Development of American Counterinsurgency Doctrine and the Vietnam War.* New York: New York University Press, 1986.

Important study that reviews major post–World War II counterinsurgencies, including Malaya and South Korea. Chapter 6, "The Banana Wars, 1915–34," deals with the evolution of the *Small Wars Manual, 1940.* Cable does not examine the evolution of United States military doctrine as the Vietnam War continued, but he does offer a valuable picture of the United States stance toward insurgency and counterinsurgency before and during the very early stages of the Vietnam conflict.

977. Cassidy, Robert M. "Why Great Powers Fight Small Wars Badly," *Military Review* 82, no. 5 (September–October 2002): 41–53.

Thoughtful examination of the problem. Cassidy reviews a number of counterinsurgencies through history, although his emphasis is on the post–World War II years. One of the key factors in his view is the respective strength of will of the insurgents and counterinsurgents. For the insurgents, their war is a "total" conflict, and an important task for them is to increase the scope and intensity of the conflict to make the counterinsurgents commit more resources and to sustain more and more casualties. This tactic raises the political costs to the counterinsurgents and generates criticism of the war.

978. Corcoran, Edward Francis. "United States Defense Policy and the Third World—The Development of Doctrine, 1962–85." Ph.D. diss., University of South Carolina, 1986.

Sees significant continuities between the ways in which several United States presidential administrations, from Kennedy to Reagan, have tried to deal with insurgencies in Third World countries. Corcoran argues that the Reagan administration largely returned to the approaches that were developed during the Kennedy years.

979. Fairchild, David Ross. "Armies of Compassion? United States Military Training in Counterinsurgency Doctrine at the School of the Americas." M.A. thesis, University of Northern Colorado, 2004.

Vigorous attack on the programs undertaken at the School of the Americas, an institution for training military and police personnel from Latin America. Fairchild criticizes the school and its teachings, because he believes that at the least they encouraged political instability in Latin America. He asserts that there has been little real change in the mission and functioning of the school, using the anti-narcotics campaign in Colombia as evidence of his thesis.

980. Grant, Thomas Alexander. "Little Wars, Big Problems: The United States and Counterinsurgency in the Postwar World." Ph.D. diss., University of California–Irvine, 1990.

Asserts that the United States has been largely ineffective in prosecuting counterinsurgency since World War II. Grant thinks that possible explanations are "hidebound leadership," "bureaucratic rigidity," and "organizational fragmentation." The three case studies that he examines are the campaigns in the Philippines against the Huks, the Vietnam War, and United States involvement in El Salvador. Chapter V deals with the Kennedy administration's effort to change the counterinsurgency approach of the United States.

981. Johnson, Wray Ross. "From Counterinsurgency to Stability and Support Operations: The Evolution of United States Military Doctrine for Foreign Internal Conflict, 1961–96." Ph.D. diss., Florida State University, 1997.

This is a stimulating review of the ways in which the armed forces have changed their concepts of insurgencies and low-intensity conflicts. Johnson uses El Salvador, Somalia, and Bosnia–Herzegovina as case studies. The dissertation gives a useful view of military thinking about these matters, and, thereby, it is helpful background for anyone studying the massive commitment to low-intensity conflict in Afghanistan and Iraq.

982. Rose, Donald Gregory. "Peace Operations and Counterinsurgency: The United States Military and Change." Ph.D. diss., University of Pittsburgh, 2000.

Interesting study that compares military adjustment to the prospect and actuality of its greater involvement with low-intensity and peacekeeping operations in the 1960s and 1990s. Rose examines the Vietnam War, Operation Provide Comfort (aid to the Kurds in northern Iraq after the Gulf War), and Operation Restore Hope (humanitarian efforts in Somalia) and then discusses changes in military doctrine and military training programs in view of new emphases in mission.

983. Shafer, D. Michael. *Deadly Paradigms: The Failure of U.S. Counterinsurgency Policy.* Princeton, NJ: Princeton University Press, 1988.

Essential, highly sophisticated study. In the process of his search for the reasons for counterinsurgency successes and failures, Shafer explores the continuity in attitudes and assumptions of United States decision-makers. The cases he examines are the French in Indochina and Algeria and the United States in Greece, the Philippines, and Vietnam. Shafer asserts that "counterinsurgency as understood in theory has never been practiced" and goes further to state that counterinsurgency is "inapplicable as well as unapplied" (pp. 5, 6). This annotation just suggests the value of this book.

984. Sheehan, Kevin Patrick. "Preparing for an Imaginary War? Examining Peacetime Functions and Changes of Army Doctrine." Ph.D. diss., Harvard University, 1988.

General examination of the nature and purpose of army doctrine from the 1950s through the 1980s. Much of the dissertation deals with plans for large-scale warfare, the employment of the "Pentomic Division," "Active Defense," and "AirLand Battle." Chapter 6, "Doctrinal and Operational Change During Wartime—The U.S. Army and Vietnam, 1965–68," makes Sheehan's work a necessary reference for this bibliography. Sheehan is critical of the army for what he views as its failure to change its counterinsurgency stance in the Vietnam War.

985. U.S. Headquarters, Department of the Army. Department of the Army Field Manual FM 31–16. *Counterguerrilla Operations.* Washington, DC: Department of the Army, 1967.

Prescribes doctrine for forces of brigade strength and below. Much tactical information, including charts, is provided, along with some material on civil affairs. The manual is an important source of military thinking during the Vietnam conflict.

986. ——. *Department of the Army Field Manual FM 31–21. Guerrilla Warfare and Special Forces Operations.* Washington, DC: Department of the Army, 1961.

Focuses almost entirely on offensive guerrilla warfare. Only a short section relates to "Counter Guerrilla Operations."

987. U.S. Headquarters, Marine Corps. *Operations Against Guerrilla Forces.* FMFM-21. Washington, DC: Headquarters, United States Marine Corps, 1962.

Provides a detailed set of instructions for carrying out missions against guerrillas. A large section deals with "small unit operations." The Marine Corps considers the manual an update of the *Small Wars Manual, 1940* (item 123); however, it is not nearly as detailed as the earlier publication.

C. Small wars and counterinsurgency: general studies

988. Astor, Gerald. *Presidents at War: From Truman to Bush, the Gathering of Military Power to Our Commanders in Chief.* Hoboken, NJ: John Wiley & Sons, 2006.

Useful perspective on how presidents from Woodrow Wilson to George W. Bush have conceived their powers and duties in respect to military matters. There is much material on presidential decisions to enter or to initiate a given conflict. Astor argues that through the twentieth century presidents have increased ability their to begin military operations on their own authority, despite efforts Congress has made to assert itself in such decisions.

989. Barnet, Richard J. *Intervention and Revolution: The United States in the Third World.* New York: World Publishing Company (New American Library), 1968.

Attacks what Barnet regards as the United States effort to police the world. Vietnam and Greece are among the countries studied. The chapters on Lebanon and the Dominican Republic are listed and annotated in the appropriate sections of this bibliography.

990. Barrett, Raymond J. "The Problem of Lower Spectrum Violence." *Military Review* 46, no. 2 (February 1966): 90–93.

Warns that insurgency is stimulated by various sources, not simply Communist agitation. A variety of programs must be employed to deter insurgency. One key element is effective police work.

991. Beaumont, Roger A. "The Military Utility of Limited War." *Military Review* 47, no. 5 (May 1967): 53–57.

Disputes the common view that small wars are useful for testing new weapons and tactics. In fact, Beaumont argues, participating in such wars is likely to narrow the perspectives of the professional officer and do more harm than good.

992. Bjelajac, Slavko N. "Unconventional Warfare in the Nuclear Era." *Orbis* 4 (Fall 1960): 323–37.

Examines Soviet and Chinese doctrines of unconventional warfare. Bjelajac assails continued Western reliance on "massive retaliation" and argues for increasing the counterinsurgency capabilities of the Western nations.

993. Black, Edwin F. "The Problems of Counterinsurgency." *United States Naval Institute Proceedings* 88, no. 10 (October 1962): 22–39.

Discusses communist insurgency and United States countermeasures. Much of the article sets out principles for successful operations by country teams of United States military and diplomatic representatives in Third World countries. Greater integration of United States commands and greater use of technology in jungle environments are Black's chief recommendations.

994. Blaufarb, Donald S. *The Counterinsurgency Era: U.S. Doctrine and Performance from 1950 to the Present.* New York: The Free Press, 1977.

Deals primarily with Vietnam, Laos, and Thailand, but does consider other areas. This is a highly sophisticated, readable study that is essential for an understanding of American counterinsurgency developments. As an officer in the CIA, Blaufarb was a participant in many of the events he describes.

995. Bloomfield, Lincoln P., and Amelia C. Leiss. *Controlling Small Wars: A Strategy for the 1970's.* New York: Alfred A. Knopf, 1969.

Presents an analytical scheme for examining "local conflicts." The authors argue that the United States must develop alternatives to military intervention and discuss many of them at length.

996. Brands, H.W. "Decisions on American Armed Intervention: Lebanon, Dominican Republic, and Grenada." *Political Science Quarterly* 102, no. 4 (Winter 1987): 607–24.

Succinct but valuable analysis of these decisions. Lebanon in 1958 was included in the study, not Lebanon in the 1980s, to avoid an overemphasis on the Reagan administration. Brands discusses a number of facets of each intervention. He notes that policy elites, including members of both parties, were almost unanimous in supporting the president's decision to intervene and that the influence of

"Munich" appeasement continued to be a persuasive factor in deliberations about each crisis.

997. Bratton, Joseph K. "Regional War Strategy in the 1970's." *Military Review* 47, no. 7 (July 1967): 3–10.

Emphasizes that concern about escalation of the conflict is a dominant part of making policy for a limited or small war. Therefore, strategies for an all-out war and limited or small wars must be fully integrated. The article enunciates a number of strategic principles.

998. Buchanan, William J., and Robert A. Hyatt. "Building a Counterinsurgency Political Infrastructure." *Military Review* 48, no. 9 (September 1968): 25–41.

Emphasizes the integration of various counterinsurgency programs in a given country. This is a sophisticated sociological analysis, which uses group theory extensively.

999. ———. "Capitalizing on Guerrilla Vulnerabilities." *Military Review* 48, no. 8 (August 1968): 3–40.

Constitutes a very valuable essay on counterinsurgency. While the Vietnam conflict is mentioned, the principles put forth are generalized beyond that war. Section III deals specifically with the military side of counterinsurgency.

1000. Conley, Michael C. "The Strategy of Communist-Directed Insurgency and the Conduct of Counterinsurgency." *Naval War College Review* 21 (May 1969): 73–93.

Analyzes Communist insurgency, emphasizing the importance of the revolutionary cadre and the significance of "political integration" for societies that are fighting insurgency.

1001. Crandall, Russell. *Gunboat Democracy: U.S. Interventions in the Dominican Republic, Grenada, and Panama.* Lanham, MD: Rowman & Littlefield, 2006.

As the title suggests, Crandall believes that these interventions successfully restored democracy to the countries mentioned. He also defends the United States decisions to intervene on the ground that United States policy-makers feared security threats and dangers to United States nationals. The book is based on a few interviews, a number of federal publications, and a large number of books and articles.

1002. Delaney, Robert Finley. "A Case for a Doctrine of Unconventional Warfare." *United States Naval Institute Proceedings* 87, no. 9 (September 1961): 66–70.

Warns that the United States is neglecting guerrilla warfare, counterinsurgency, and psychological warfare. Delaney also discusses the potential naval role in a future guerrilla conflict.

1003. Drinnon, Richard. *Facing West: The Metaphysics of Indian-Hating and Empire-Building*. Minneapolis: University of Minnesota Press, 1980.

Traces American counterinsurgency theory and practice to traditions of racism and conquest established early in American history. Drinnon deals with such episodes as the Moro wars but, curiously, ignores marine ventures in the Caribbean. This is a highly provocative book that cuts across a number of academic disciplines to make its points.

1004. Gann, Lewis H. "Guerrillas and Insurgency: An Interpretative Survey." *Military Review* 46, no. 3 (March 1966): 44–59.

Gives a general perspective on insurgency through history. Gann warns against investing guerrillas with an aura of invincibility, pointing out that many guerrilla movements have failed.

1005. Gillert, Gustav J. "Counterinsurgency." *Military Review* 45, no. 4 (April 1965): 25–33.

Reviews army doctrine on counterinsurgency and the role of the Special Group (counterinsurgency) that was established by Kennedy. This article is an important examination of army organization during the height of the commitment to counterinsurgency.

1006. Greenberg, David Robert. "The U.S. Response to Philippine Insurgency." Ph.D. diss., Fletcher School of Law and Diplomacy (Tufts University), 1994.

This is a comparative study between the campaigns against the Huks and the unrest during the Aquino years. Greenberg credits the United States with exerting effective pressure on the Philippine government to reduce the attractiveness of insurgency programs, as well as providing extensive military aid and advice.

1007. Gurtov, Melvin. *The United States Against the Third World: Antinationalism and Intervention*. New York: Praeger Publishers, Inc., 1974.

Constitutes one of the most significant critiques of United States interventionism during the Eisenhower, Kennedy, Johnson, and Nixon administrations. Gurtov believes that intervention is frequently ineffective in the long run. Some of the small wars are considered, but the treatments are too brief to be listed separately.

1008. Guttieri, Karen. "Toward a Useable Peace: United States Civil Affairs in Post–Conflict Environments." Ph.D. diss., University of British Columbia, 1999.

"Civil affairs" refers to United States military efforts to restore or reform the institutions of societies that have been torn with either foreign or internal war. This dissertation is an impressive comparative study of civil affairs activities in the Dominican Republic (1965–66), Grenada (1983), and Panama (1989). Civil affairs is a topic of continuing significance, and, therefore, Guttieri's study has much to offer the student of counterinsurgency and post-conflict situations.

1009. Haffner, L.E. "Guerrilla War and Common Sense." *Marine Corps Gazette* 46, no. 6 (June 1962): 20–23.

Attacks much of what the author regards as conventional wisdom about guerrilla conflicts. He argues, for example, that infantry units can handle counterinsurgency operations without extensive reorganization or retraining. This is an excellent, if controversial, study.

1010. Harris, Richard Boyd. "Conflict and Reform: The Evolution of Special Operations in U.S. National Security Strategy." M.A. thesis, American University, 1988.

Traces the varying fortunes of special operations forces elements within the United States armed services in the post–World War II years. After a period of growth, especially during the Vietnam era, commitment to special operations forces and special operations generally declined, but then in the late 1970s interest arose again. Harris touches on several special operations, such as the Mayaguez and Iran rescue attempts. He believes that many in the United States see value in special operations forces to counter terrorism and other threats, but other military and political leaders continue to have reservations about them.

1011. Hartness, William M. "Social and Behavioral Sciences in Counterinsurgency." *Military Review* 46, no. 1 (January 1966): 3–10.

Argues that social science data and techniques need to be used more effectively in American counterinsurgency operations. In particular, some of the methods are useful in identifying insurgency movements in their early stages.

1012. Heilbrunn, Otto. "A Doctrine for Counterinsurgents." *Marine Corps Gazette* 48, no. 2 (February 1964): 30–38.

Analyzes various categories of insurgents and offers the theory that with the advent of air weapons in counterinsurgency there is less need to be concerned with political warfare to win over the local population.

1013. ———. "When Counterinsurgents Cannot Win." *Military Review* 49, no. 10 (October 1969): 36–43.

Warns that the West has not formulated a strategy for counterinsurgency that guarantees success. Heilbrunn criticizes American doctrine for ignoring terrorist campaigns. He uses various post–World War II conflicts to illustrate his point that Malaya and Greece were special cases and that not too much should be generalized from them.

1014. Hess, Gary R. *Presidential Decisions for War: Korea, Vietnam, the Persian Gulf, and Iraq.* 2nd ed. Baltimore, MD: Johns Hopkins University Press, 2001, 2009.

Well-written examination of policy-making in respect to post–World War II United States military ventures. The second edition of this book includes two chapters about George W. Bush and the Iraq War. There is an extensive "Bibliographic Essay," but the almost total absence of documentation limits the scholarly importance of an otherwise commendable work.

1015. Hessler, William H. "Guerrilla Warfare is Different." *United States Naval Institute Proceedings* 88, no. 4 (April 1962): 34–47.

Analyzes very ably the nature of guerrilla warfare, especially the motivations of guerrilla fighters. Hessler reminds readers of the differences between "bandits" and "guerrillas" and discusses the need for positive social and economic programs to counter guerrilla efforts. He also examines the role of sea power in small wars.

1016. Holliday, Sam C. "Warfare of the Future." *Military Review* 49, no. 8 (August 1969): 12–17.

Stresses the conflict of wills in an insurgency war. Therefore, counterinsurgency must focus on destroying the guerrilla infrastructure even more than the movement's armed forces.

1017. Jacobs, Walter Darnell. "Insurgency." *Marine Corps Gazette* 50, no. 2 (February 1966): 43–46.

Recommends further study of Communist insurgency doctrine. Jacobs asserts that the Communists do not concern themselves with the possible functions of regular troops in insurgency operations. Rather than employ the army in this field, the United States may want to use the marines or delegate the function to local forces.

1018. ———. "This Matter of Counterinsurgency." *Military Review* 44, no. 10 (October 1964): 79–85.

Ridicules many varieties of counterinsurgency "experts" in a witty and iconoclastic way. Jacobs suggests turning counterinsurgency over to the marines, leaving the army free to concentrate on deploying for a nuclear war with the Soviet Union.

1019. Karabell, Zachary. *Architects of Intervention: The United States, the Third World, and the Cold War, 1946–1962.* Baton Rouge: Louisiana State University Press, 1999.

Provides a number of case studies that Karabell uses to argue that elites in the countries where some form of intervention occurred often lured United States decision-makers to take a hand. The cases discussed include Greece, Guatemala, Laos, and, notably, Lebanon in 1958. If the interests of the United States and the foreign elites that urged intervention were reasonably compatible, then the involvements can be counted as successes.

1020. Kecskemeti, Paul. "Insurgency as a Strategic Problem." Rand Memorandum RM-5160-PR, February 1967.

Discusses Vietnam, but aims primarily at establishing principles for avoiding insurgency situations. The author argues that root causes must be removed to prevent insurgency and that the United States usually lacks the means to eliminate them.

1021. Kelly, George A. "Footnotes on Revolutionary War." *Military Review* 42, no. 9 (September 1962): 31–39.

Discusses the paradox that the United States would ordinarily welcome the emergence of anti-colonial movements but must use caution because of the Cold War and its priorities. Kelly stresses that the United States needs to act early in a revolutionary situation to prevent a movement from being taken over by the Communists.

1022. Kent, Irvin M., and Richard A. Jones. "The Myth of the Third Man." *Military Review* 46, no. 5 (May 1966): 48–56.

Notes that experts tend to focus on one aspect of counterinsurgency. In a given country, however, the insurgent movement is based on conditions and appeals that may vary from region to region. Military operations, civic action, and what the authors call "population and resource control" are interrelated ways to fight insurgency.

1023. King, Edward L. *The Death of the Army: A Pre-Mortem*. New York: Saturday Review Press, 1972.

Describes the development of counterinsurgency in acid terms, believing that it served the army's rather than the nation's needs. King asserts that the United States is in great danger unless it dismantles the means for waging counterinsurgency campaigns.

1024. Koburger, Charles W. "Komer's War: The Indirect Strategy in Action." *Military Review* 49, no. 8 (August 1969): 18–26.

Uses the example of a confrontation in Arabia in 1962–63. A radical coup in Yemen upset the local balance of power. Various options included use of limited American air power. A strong American guarantee to Saudi Arabia stopped Yemen from attacking. The article analyzes the decision-making in Washington during the crisis.

1025. ——. "Lower Spectrum War." *Military Review* 48, no. 7 (July 1968): 3–9.

Asserts that military planning must be increased for small-scale applications of military force in direct support of diplomatic initiatives. Several paragraphs specifically discuss reactions to the outbreak of insurgency warfare in an allied Third World nation.

1026. Ladd, Jonathan F. "Some Reflections on Counterinsurgency." *Military Review* 44, no. 10 (October 1964): 72–78.

Begins with a definition of the term, and argues that as a concept there is nothing new about it. Ladd emphasizes, however, that the United States lacks current experience with counterinsurgency and must be ready to learn from other nations.

1027. Lindsay, Franklin A. "Unconventional Warfare." *Military Review* 42, no. 6 (June 1962): 53–62.

Tries to explain the sources of communist success in waging insurgency wars. The ability of the communists to mobilize the population is given special emphasis. Lindsay suggests that Western countries will have to use well-supported political organizers in the villages of the Third World to counter communist tactics.

1028. Long, William F. "Counterinsurgency Revisited." *Naval War College Review* 21 (November 1968): 4–10.

Gives six very useful warnings about illusions concerning the nature of counterinsurgency. Long draws from the Vietnam experience to suggest that the United States cannot fight a counterinsurgency war for the inhabitants of a nation threatened by a revolutionary movement.

1029. ——. "Counterinsurgency—Some Antecedents for Success." *Military Review* 43, no. 10 (October 1963): 90–97.

Suggests that the United States should learn from European colonial campaigns in such areas as Malaya, Indochina, and Kenya, even though the United States will wish to distinguish its programs from the policies of the former colonial powers.

1030. ——. "A Perspective of Counterinsurgency in Three Dimensions—Tradition, Legitimacy, Visibility." *Naval War College Review* 22 (February 1970): 19–25.

Provides a sophisticated analysis of counterinsurgency in relation to American traditions. Examining the Vietnam experience, Long discusses the role of the legislative and executive branches in the United States, public opinion, and the media.

1031. McCuen, John J. *The Art of Counter–Revolutionary War: The Strategy of Counter–Insurgency*. Harrisburg, PA: Stackpole Books, 1966.

Provides a lucid analysis of post–World War II counterinsurgency tactics in countries such as Malaya, Greece, and Indochina. McCuen tends to ignore the political aspects of these campaigns and United States experiences in counterinsurgency, except for a few words on the Philippine Constabulary.

1032. MacCloskey, Monro. *Alert the Fifth Force: Counterinsurgency, Unconventional Warfare, and Psychological Operations of the United States Air Force in Special Warfare.* New York: Richard Rosen Press, Inc., 1969.

Contains a great deal of information on the development of air force doctrine and the types of aircraft used for counterinsurgency. The author reviews Communist Chinese guerrilla strategy and the United States Air Force experience in World War II.

1033. Meernik, James David. *The Political Use of Military Force in US Foreign Policy.* Aldershot, U.K.: Ashgate Publishing Limited, 2004.

This is not a study of counterinsurgency, but it is a major examination of many of the small wars of the United States. Meernik closely analyzes a number of theories that have been used to explain the use of military power by the United States. Although the book discusses United States military involvements before 1945, the bulk of the research seems to relate to the post–World War II years. Of the four explanations discussed, Meernik finds that security and economic concerns are the major reasons for generating military action.

1034. Miller, Roger J., and James A. Cochrane. "Counterinsurgency in Perspective." *Air University Review* 14 (September–October, 1963): 64–74.

Constitutes an interesting comparative study of the American, French, and Mexican revolutions and communist theories of insurgency. The major post–World War II counterinsurgency campaigns are reviewed.

1035. Mrazek, James E. "The Creativity of the Guerrilla." *Army* 14 (June, 1964): 86, 88, 90.

Emphasizes the spiritual and intellectual facets of guerrilla warfare. Interestingly, Mrazek argues that communist guerrilla movements are likely to take an independent line in foreign affairs once they are successful in their own country; thus, they need not become pawns of the Soviets.

1036. Paret, Peter, and John W. Shy. *Guerrillas in the 1960's.* 2nd ed. New York: Frederick A. Praeger, 1962.

Surveys the history of guerrilla warfare, and examines the implications for United States policy-makers. The authors stress the political character of guerrilla warfare. This is one of the most important studies of guerrilla operations written during the heyday of counterinsurgency.

1037. Procter, David E. "The Rescue Mission: Assigning Guilt to a Chaotic Scene." *Western Journal of Speech Communication* 51, no. 3 (Summer 1987): 245–55.

Focuses on presidential explanations of certain United States incursions into foreign countries. Procter views "rescue missions" as phenomena that are sharply delineated from other types of military operations, partly because of their limited scope. Presidents Eisenhower (Lebanon, 1958), Johnson (Dominican Republic, 1965), and Reagan (Grenada, 1983) argued that chaotic situations in those three nations required immediate action by the United States.

1038. Pustay, John S. *Counterinsurgency Warfare.* New York: Free Press, 1965.

Formulates doctrine for both Third World countries and for the United States. Communist theories of guerrilla warfare are examined. Chapter 8 deals specifically with the United States role in counterinsurgency. Pustay includes a great many proposals for structural and organizational changes needed to facilitate effective counterinsurgency.

1039. Raymond, William Montgomery, Jr. "Collision or Collusion? The Congress, the President, and the Ambiguity of War Powers." Ph.D. diss., University of Michigan, 1993.

Innovative research that covers numerous small wars and armed encounters ranging from Lebanon in 1958 to the covert campaign against the Sandinista regime in Nicaragua. Raymond argues that presidents can largely ignore the provisions of the War Powers Resolution of 1973 when deciding whether or not to take military action in a given situation. Congress does try to use the resolution from time to time, but, ultimately, it is satisfied to let the executive branch take the initiative. The resolution has had some influence on the use of military force, but not much. Congress is unlikely to make major moves against presidential power, however. Too much congressional involvement would mean unwanted congressional responsibilities.

1040. Record, Jeffrey. *Making War, Thinking History: Munich, Vietnam, and Presidential Uses of Force From Korea to Kosovo.* Annapolis, MD: Naval Institute Press, 2002.

As the title suggests, this book analyzes post–World War II United States interventions and possible interventions, such as Indochina in 1954, in terms of analogies suggested by the Munich appeasement agreement of 1938 and by the United States intervention in and ultimate withdrawal from Vietnam. This is a brief study, and, therefore, material from the book is not listed and discussed in other chapters of this bibliography. Nevertheless, this is a valuable analysis of an important element in United States decision-making about intervention.

1041. Sanders, Ralph. "The Human Dimension of Insurgency." *Military Review* 44, no. 4 (April 1964): 38–47.

Proposes a psychological approach to counterinsurgency. Sanders discusses the problems facing the people of the Third World and the ways in which the communists have utilized these problems for their own ends. He concludes with a few suggestions on how counterinsurgents can fight back in this area of conflict.

1042. Schlesinger, Robert J. "The Elusive Guerrilla." *Army* 14 (January 1964): 42–45.

Discusses the stages in the development of insurgency movements. Schlesinger emphasizes the extreme difficulty of locating guerrilla units in their early stages. Guerrilla attacks are completely unpredictable. This fact underlines the need for special aircraft to supplement ground patrols.

1043. Stanton, Louise. "The Civilian–Military Divide: Obstacles to the Integration of Intelligence in the United States." Ph.D. diss., Rutgers, The State University of New Jersey, 2007.

This important analysis views a counterinsurgency strategy as the answer to the low-intensity campaign being waged against the United States. Stanton describes the nature of such a strategy that must involve both civilian and military components at all levels of government in the United States. She argues, however, that United States laws and traditions militate strongly against the development of the kind of plan developed in the dissertation. Chapter 6 includes an important analysis of counterinsurgency, intelligence, and homeland security.

1044. Steel, Ronald. *Pax Americana.* New, rev. ed. New York: Viking Press, 1970.

Criticizes what the author regards as the confused objectives of United States policy makers. Chapters 15 and 16 are of special importance to the student of United States involvement in small wars.

1045. Stern, Ellen P., ed. *The Limits of Military Intervention.* Beverly Hills, CA: Sage Publications, 1977.

Contains a number of essays that consider possible limits to the effective use of military force by great powers, primarily the United States. These include the public antipathy to military intervention, smaller numbers of troops available, and other factors.

1046. Tillema, Herbert K. *Appeal to Force: American Military Intervention in the Era of Containment.* New York: Thomas Y. Crowell Company, 1973.

Compares interventions in Korea, Lebanon, South Vietnam, and the Dominican Republic. The Lebanon and Dominican Republic sections are examined in the appropriate sections of this bibliography. Chapter 5 deals with interventions that might have occurred but did not.

1047. Waddell, Ricky Lynn. "The Army and Peacetime Low Intensity Conflict, 1961–93: the Process of Peripheral and Fundamental Military Change." Ph.D. diss., Columbia University, 1994.

Deals in detail with the ways in which the United States Army can be induced to give more attention to low-intensity conflict. Waddell suggests that if civilian decision-makers work through military leaders that are termed here as "traditionalists," that is, those who believe the army must continue to focus on the possibility of large-scale military operations, then their efforts are likely to fail. Partnerships between civilian leaders and respected army leaders who have a commitment to extensive preparations for low-intensity conflict will prove more productive. Waddell also describes conflicts within the army between "traditionalists" and those who are termed "visionaries."

1048. Ward, Compton E. "Subversive Insurgency: An Analytical Model." *Naval War College Review* 20 (September 1967): 3–25.

Discusses post–World War II insurgencies. Section V makes recommendations on counterinsurgency strategy and tactics. Ward states clearly that the primary counterinsurgency offensive must be in the political, social, and economic spheres, with military operations being of only secondary importance.

1049. Weller, Jac. *Fire and Movement: Bargain-Basement Warfare in the Far East.* New York: Thomas Y. Crowell Company, 1967.

Examines strategy, tactics, and weapons used against communist guerrillas by United States, British, and Philippine forces. Much of the focus is on Indochina, but several chapters are on a more general level.

1050. Winslow, Richard Lawrence. "American Military Intervention in the Middle East: Its Political Prudence, Military Feasibility, and Moral Permissibility: An Analysis of Four Cases." Ph.D. diss., Georgetown University, 1989.

Impressive study of the interventions in Lebanon in 1958 and the 1980s, air attacks on Libya during the Reagan administration, and the use of United States naval vessels to protect merchant vessels in the Persian Gulf in the late 1980s. Winslow provides only brief discussions of "military feasibility," but, nevertheless, this is an important research effort, especially in light of the continuing massive involvement of the United States in the Middle East.

D. Small wars, counterinsurgency, and the media

1051. Mermin, Jonathan. *Debating Peace and War: Media Coverage of U.S. Intervention in Post–Vietnam Era*. Princeton, NJ: Princeton University Press, 1999.

   Advocates more active participation by the media in decisions about whether or not to intervene in particular situations. Mermin seeks to show that the media have not played a significant role in the process unless open rifts have developed among governmental decision-makers. He argues that the media need to search for views that may question or dispute the positions taken by the government.

1052. Thrall, A. Trevor. *War in the Media Age*. Cresskill, NJ: Hampton Press, 2000.

   Discusses the media treatment of the Vietnam conflict and official United States responses to it. Thrall analyzes the role of the media and United States media policies in Grenada, Panama, and the Gulf War, asserting that the government is increasingly able to transmit the messages it favors and to restrain alternative or critical messages.

1053. Western, Jon W. *Selling Intervention and War: The Presidency, the Media, and the American Public*. Baltimore, MD: Johns Hopkins University Press, 2005.

   This sophisticated analysis discusses an instance in which intervention was finally decided against, and four United States military interventions: Lebanon in 1958, Grenada, Somalia, and Iraq in 2003. Western focuses on policy-making elites. In his view, there are categories of elite positions with respect to intervention that persist in administration after administration. The members of the elite who hold such views are described by such terms as "Selective Engagers" or "Liberals."

E. The Kennedy counterinsurgency initiative

1054. Aliano, Richard A. *American Defense Policy from Eisenhower to Kennedy: The Politics of Changing Military Requirements, 1957–1961*. Athens: Ohio State University Press, 1975.

   Places the counterinsurgency effort into the broader perspective of changes in military policy. Aliano underscores the serious view Kennedy took of "wars of liberation" and his almost feverish commitment to increasing conventional force levels.

1055. Amory, John Forth. *Around the Edge of War: A New Approach to the Problems of American Foreign Policy*. New York: Clarkson N. Potter, 1961.

Predicts that American efforts to wage guerrilla warfare in communist countries will fail because they lack popular support. Amory claims that United States foreign policy is dictated by business interests.

1056. Deitchman, Seymour J. *Limited War and American Defense Policy.* Cambridge, MA: MIT Press, 1964.

Examines limited conventional wars, such as the Korean conflict, but focuses much of the discussion on unconventional warfare situations. This is a sophisticated statement of the Kennedy strategy of counterinsurgency.

1057. Gray, Colin S. *Strategic Studies and Public Policy.* Lexington: University Press of Kentucky, 1982.

Presents a critique of United States thinking about counterinsurgency and its relationship to the modernization process in the Third World, pages 113–16. Gray believes the Kennedy administration embraced counterinsurgency without understanding the problems associated with it. Despite its brevity, this is an important study.

1058. Norman, Lloyd, and John B. Spore. "Big Push in Guerrilla Warfare." *Army* 12 (March 1962): 28–37.

Describes at length the army's effort to carry out President Kennedy's mandate to improve its counterinsurgency capability.

1059. Walter, Richard J. *Cold War and Counterrevolution: The Foreign Policy of John F. Kennedy.* New York: Viking Press, 1972.

Assails Kennedy for his commitment to increased armaments and his propensity for confrontations with the Soviet Union and Communist China. Much of the book deals with Laos, Vietnam, and Berlin, but there is material on the two Cuban crises. The book is also useful for putting Kennedy's counterinsurgency drive into context.

1060. Weigley, Russell F. *The American Way of War: A History of United States Military Strategy and Policy.* New York: Macmillan Publishing Co., Inc., 1973.

Describes the Kennedy administration's concern with counterinsurgency. Interestingly, early studies of the Vietnam situation called for a possible attack on North Vietnam rather than the development of a counterinsurgency strategy, pages 445, 455–60.

1061. Wood, John S. "Counterinsurgency at the National and Regional Level." *Military Review* 46, no. 3 (March 1966): 80–85.

Recommends a number of ways for improving coordination. The problem is especially acute at the top, since country teams coordinate operations in foreign countries. The article reflects the imperfect implementation of Kennedy's drive for developing an effective counterinsurgency program, cutting across agency and service lines.

F. Special Forces and related units

1062. Adams, Raymond E. "Special Forces Aidman." *Army Information Digest* 18 (October 1963): 57–59.

    Deals primarily with the training of aidmen.

1063. Ahern, Neal J. "What It Is That's so Special About Special Forces." *Army Reservist* 9 (November 1963): 5–7.

    Reviews a visit to a United States Army Reserve Special Forces unit at Fort Bragg.

1064. Asprey, Robert R. "Special Forces: Europe." *Army* 12 (January 1962): 56–61.

    Discusses training for Special Forces personnel generally, despite the title.

1065. Blair, Edison T. "The Air Commandos." *Airman* 6 (August 1962): 18–23.

    Describes the activities of the 6th Fighter Squadron (Commando) in training Vietnamese in counterinsurgency techniques. Blair stresses the versatility of the commandos and briefly describes their aircraft.

1066. Blanchard, George S. "Special Warfare—Now!" *Army Information Digest* 19 (January 1964): 20–26.

    Sees special warfare as including counterinsurgency, psychological warfare, and civic action. Blanchard reviews the stages of insurgency: (1) propaganda and organizational development, (2) guerrilla operations, and (3) conventional warfare.

1067. Dodson, Charles A. "Special Forces." *Army* 11 (June 1961): 56–61.

    Discusses Special Forces recruitment and the day-to-day life of Special Forces men.

1068. Hamlett, Barksdale. "Special Forces: Training for Peace and War." *Army Information Digest* 16 (June 1961): 2–9.

    Discusses Special Forces training at Fort Bragg and the counterinsurgency role of Special Forces.

1069. Kinard, William H. "This is Special Warfare: U.S. Army Style." *Army Information Digest* 15 (June 1960): 2–11.

    Considers psychological warfare as well as Special Forces operations.

1070. Kleinman, Forrest K. "Under the Green Beret." *Army* 13 (October 1962): 41–47, 50.

Distinguishes between the guerrilla and counterinsurgency capabilities of Special Forces. Kleinman comments on the difference between Special Forces and the early Ranger units.

1071. McGlasson, W.D. "Special Forces: The New Elite." *National Guardsman* 18 (June 1964): 2–7.

Follows an Army National Guard Special Forces unit through a parachute jump.

1072. Peers, W.R. "Meeting the Challenge of Subversion." *Army* 15 (November 1964): 95–97.

Reviews the development of the army's concern with counterinsurgency.

1073. Rosson, William B., and others. "The U.S. Army in Special Warfare." *Army* 13 (November 1962): 59–68.

Stresses Vietnam operations, but deals with other topics as well.

1074. Seigle, John W., John E. Ralph, and Americo A. Sardo, eds. *Readings in Counterinsurgency 1963*. West Point, NY: United States Military Academy, Department of Social Sciences, 1963.

Supplements earlier material given to cadets in 1962.

1075. Spark, Michael. "Marines, Guerrillas, and Small Wars." In *The Guerrilla—And How to Fight Him: Selections from the Marine Corps Gazette*. Ed. T.N. Greene. New York: Frederick A. Praeger, 1962.

Argues that the marines have a unique capability for counterinsurgency warfare.

1076. U.S. Army. Special Warfare School. *Readings in Counter–Guerrilla Operations*. Rev. ed. Fort Bragg, NC: U.S. Army Special Warfare School, 1961.

Focuses primarily on foreign experiences.

1077. U.S. Department of the Army. Office, Chief of Information. *Special Warfare U.S. Army*. Washington, DC: Department of the Army, 1962.

Collects a number of articles, many of them reprinted from *Army, Military Review*, and other military periodicals. The book is heavily illustrated.

1078. Winkler, Melvin L. "The Army's Special Forces." *Army Information Digest* 21 (October 1966): 26–31.

Discusses Special Forces very generally. Some of the text refers to Vietnam.

1079. Yarborough, William P. "What Every Conventional Force Commander Should Know About Guerrillas." *Army* 13 (July 1963): 22–24, 26.

Analyzes the motivations for individual guerrillas. Yarborough was one of the leaders in the counterinsurgency program.

1080. "'Young Moderns' Are Impetus Behind Army's Special Forces." *Army* 12 (March 1962): 38–39.

Consists of brief biographical sketches of generals Yarborough and Rosson.

G. The navy and counterinsurgency

1081. Nelson, Andrew G., and Norman G. Mosher. "Proposed: A Counterinsurgency Task Force." *United States Naval Institute Proceedings* 92, no. 6 (June 1966): 36–45.

Proposes formation of naval groups to wage counterinsurgency warfare, in a very interesting and persuasive article. The authors discuss the various ways in which naval forces have been used to counter guerrillas, such as blockading supply routes.

1082. Seim, H.B. "The Navy and 'Fringe' War." *United States Naval Institute Proceedings* 77, no. 8 (August 1951): 834–41.

Presents an early discussion of communist insurgency. While low-intensity conflicts are "local," they form part of a strategic plan and have to be countered by strategic planning. Seim examines the ways in which the navy can fight communist threats.

H. Counterinsurgency training in the 1960s

1083. Anderson, Eddie. "Hawaiians Train for Jungle Warfare." *National Guardsman* 18 (February 1964): 20–21.

Describes maneuvers that emphasized counterinsurgency and jungle operations, including the avoidance of ambushes and the problems of dealing with primitive peoples.

1084. Barrow, Clay. "Hit Back Hard." *Leatherneck* 45, no. 7 (July 1962): 16, 18–23, 79.

Describes the "Anti-Guerrilla Warfare Course" conducted by the First Marine Division for junior officers and senior noncommissioned officers. Instructors emphasize the need for studying the political and economic roots of insurgency. Barrow stresses the need for everyone in a military organization to be ready for counterinsurgency, not just the rifleman.

1085. Bashore, Boyd T. "Organization for Frontless Wars." *Military Review* 44, no. 5 (May 1964): 3–16.

Asserts that counterinsurgency is a new kind of warfare and that conventionally trained military forces must be systematically retrained to fight guerrillas.

1086. Coble, Donald W. "They Called It 'Highland Fling.'" *National Guardsman* 19 (June 1965): 8–12.

Reviews the training of the 130th Air Commando Group of about 500 men at the United States Air Force Tropical Survival School in Panama. In addition to survival exercises, the unit provided medical assistance to local inhabitants and performed other missions.

1087. Easterbrook, Ernest F. "Realism in Counterinsurgency Training." *Army Information Digest* 17 (October 1962): 12–21.

Describes efforts to train the 25th Infantry Division for effective counterinsurgency operations. The exercises conducted were very imaginative.

1088. Edwards, T. C. "3d MarDiv Counterguerrilla Training." *Marine Corps Gazette* 47, no. 5 (May 1963): 45–48.

Discusses the separate training courses that were developed for infantry and staff and communications personnel.

1089. Goldoni, John E. "The Jungle: Neutral Adversary." *Army Information Digest* 16 (May 1961): 3–9.

Focuses on the United States Army Jungle Warfare Training Center in Panama which trains both individuals and small units.

1090. Lang, Daniel. "Set a Thief ... " *Leatherneck* 45, no. 1 (January 1962): 32–35, 78.

Reviews counterinsurgency training for new officers at Quantico. After receiving the training, these men are to pass it on to other marines. Lang argues that conventionally trained troops can adapt to counterinsurgency tactics fairly easily. Much of the article deals with living in the field and the employment of ambushes.

1091. McGlasson, W.D. "Jungle and Rice-Paddy Air Force!" *National Guardsman* 18 (February 1964): 2–6, 48.

Discusses joint training between Air Commandos and Special Forces in Alabama. Much of the text concerns the old, but highly useful, aircraft used for guerrilla supply and liaison. There is a short section on Air Commando activity in Vietnam.

1092. "MEDLANDEX 5:63." *Leatherneck* 47, no. 3 (March 1964): 26–31, 77.

Describes a practice landing to provide counterinsurgency assistance in "artifice." The heavily illustrated article emphasizes the need to combine United States and foreign forces quickly to fight an insurgency movement.

1093. Page, George. "Trials on the Wilderness Trail." *Army Information Digest* 17 (August 1962): 56–59.

Discusses training of two six-man Special Forces teams in West Virginia and a band of simulated "partisans." Aggressors were supplied by the 82nd Airborne Division. The exercise lasted three weeks.

1094. Palmer, Bruce, Jr., and Roy K. Flint. "Counter–Insurgency Training." *Army* 12 (June 1962): 32–39.

Describes maneuvers in West Virginia between elements of the 82nd Airborne Division and Special Forces troops. At the end of the article there is a brief overview of the United States experience with guerrilla warfare, including the Native American wars.

1095. "Swampex '63." *Leatherneck* 47, no. 1 (January 1964): 18–23, 65.

Reviews exercises at Croatan National Forest in North Carolina. The 2nd Marine Division and other forces had to deal with heavy "insurgent" attacks and had to learn to deal with the local "inhabitants." The article includes many illustrations.

1096. "Under the Canopy." *Leatherneck* 45, no. 8 (August 1962): 16–21.

Reports on United States Army Jungle Warfare Training Center operations in Panama. The article, which is filled with photographs of training activities, emphasizes the nature of the various types of jungle terrain and the ways in which they can help or hinder soldiers.

I. Systems analysis and counterinsurgency

1097. Martino, Joseph P. "Systems Analysis and Counter–Insurgency." *Air University Quarterly Review* 17 (September–October 1966): 23–33.

Comes close to arguing that counterinsurgency can only be mastered through the application of systems analysis. This is a discussion of village life in an underdeveloped country and an analysis of the policy of development and suppression of insurgency that must be followed.

1098. Rhyne, Russell F. "Operations Research and Counterinsurgency." *Military Review* 44, no. 5 (May 1964): 26–34.

Argues for joint military/civilian studies that will apply operations research to counterinsurgency. Rhyne emphasizes that more than statistical analysis is needed for this program.

1099. Saund, Dalip. "On the Limitations of Systems Analysis for Counterinsurgency Programs." *Air University Review* 19 (May–June 1968): 42–51.

Provides a critique of Martino's article (item 1097) and supplies an alternative model. Overall, Saund concludes that systems analysis does not work well with a realistic model and can be downright dangerous if done improperly.

J. Counterinsurgency tactics

1100. Evans, David L., III. "Lessons from Counterinsurgency Operations." *Air University Review* 15 (March–April, 1964): 48–54.

Stresses nonmilitary knowledge as fundamental to waging successful counterinsurgency. Evans warns that insurgency must be dealt with in its early stages. At the same time, United States personnel must refrain from interfering with local military forces, who have to come to their own decisions.

1101. Fisher, Albert L. "To Beat the Guerrillas at Their Own Game." *Military Review* 43, no. 12 (December 1963): 81–86.

Argues that small-unit tactics are the key to success. Random patrols reduce the mobility of guerrillas and prevent ambushes. Conventional heavy weapons must be de-emphasized.

1102. Leach, C.R., and C.J. O'Shea. "Artillery vs. Guerrillas." *Marine Corps Gazette* 46, no. 9 (September 1962): 48–51.

Distinguishes between guerrillas fighting in support of conventional forces and insurgent forces operating independently. Sporadic but precise use of artillery can disorient and harass guerrillas.

1103. Lyon, Harold C., Jr. "Cancer Action." *Army* 13 (August 1962): 50–53.

Expounds a theory of saturating guerrilla areas with small patrols. Lyon views this as a new approach and describes a hypothetical operation in detail, including a useful chart. Both units assigned and the tactics they employ must be switched frequently to maintain the effectiveness of such a strategy.

1104. Miller, Walter L., Jr. "Chemicals vs. Guerrillas." *Marine Corps Gazette* 48, no. 7 (July 1964): 37–39.

Argues for greater use of chemical agents in counterinsurgency operations, especially for defoliation, marking guerrillas, and preventing storm wave attacks from succeeding.

1105. Paret, Peter, and John W. Shy. "Guerrilla Warfare and U.S. Military Policy: A Study." *Airman* 6 (May 1962): 2–11.

Asserts that the key to successful counterinsurgency is dividing the guerrillas and the people supporting them. The authors note the problems facing the guerrilla leader, especially the dilemma of balancing political versus strictly military objectives.

1106. Steward, H. Douglas. "How to Fight Guerrillas." *United States Naval Institute Proceedings* 88, no. 7 (July 1962): 23–37.

Analyzes insurgency warfare. Steward emphasizes the need to orchestrate the total system of counterinsurgency tactics, which he asserts was done in the Philippines and Malaya. Destroying supply lines and impeding guerrilla movements are critical steps in fighting insurgency.

K. Counterinsurgency weapons

1107. Isaacs, Alvin C. "Weapons for Internal Defense 1970–80." *Military Review* 47, no. 4 (April 1967): 87–91.

Describes many futuristic weapons for counterinsurgency operations that were in various stages of development. Many of them were used later in Vietnam.

1108. Norman, Lloyd. "War Without Gadgets." *Army* 16 (December 1966): 53–59.

Reviews some of the weapons developed for counterinsurgency during the early 1960s. Norman is critical of the emphasis on technology that has not proved itself in Vietnam.

L. Intelligence and counterinsurgency

1109. Kent, Irvin M., and Ruth A. Caldwell. "A Stitch in Time." *Military Review* 48, no. 6 (June 1968): 69–74.

Discusses the need for better intelligence to detect the early signs of insurgency, which the authors see as threatening more than 50 Third World nations. The article recommends that the United States provide more intelligence training to foreign soldiers.

1110. LeMay, Curtis E. "Counterinsurgency and the Challenge Imposed." *Airman* 6 (July 1962): 2–9.

Discusses Communist guerrilla organization and tactics. General LeMay emphasizes the role of intelligence in counterinsurgency because of the relative invisibility of the enemy.

M. Other facets of counterinsurgency

1111. Blair, Edison T. "SAC's Role in Counterinsurgency." *Airman* 6 (December 1962): 2–3.

    Considers the importance of Strategic Air Command capability in keeping Soviet and Communist Chinese aggression at the insurgency level.

1112. Ozaki, John. "Defector Operations." *Military Review* 49, no. 3 (March 1969): 71–78.

    Presents a model plan to persuade enemy guerrillas to defect to the government side. Ozaki bases his work on the Huk campaigns in the Philippines and United States experiences in Vietnam. His plan revolves around the effective use of defectors in psychological warfare and direct military operations.

1113. Pence, Harvey J. "Signal Support of Special Warfare." *Army Information Digest* 18 (May 1963): 64–68.

    Describes the communications challenges of counterinsurgency warfare and the extensive efforts being made by the army to develop equipment that is suitable for use in such combat environments.

1114. Powers, Robert D., Jr. "Guerrillas and the Laws of War." *United States Naval Institute Proceedings* 89, no. 3 (March 1963): 82–87.

    Examines distinctions between legitimate combatants and criminals who take advantage of unsettled conditions. Powers reviews the history of international law pertaining to guerrillas and concludes that they should be treated as prisoners of war, if captured, unless they have committed war crimes.

1115. Schmid, Clarence H. "Funding and Equipping for Low Intensity Warfare." *Marine Corps Gazette* 50, no. 12 (December 1966): 37–41.

    Argues that armed services budgeting and management systems are designed for full-scale wars and preparing for them. Schmid recommends that changes be made to accommodate the needs created by counterinsurgency campaigns. For example, military units may need new types of supplies for civic action projects and fewer combat supplies.

1116. White, Wolfred K. "Interpreter—or Filter?" *Military Review* 48, no. 2 (February 1968): 78–83.

    Outlines the problems of American officers' reliance on interpreters in counterinsurgency and civic action operations in the Third World. The article stresses that the interpreter is often a man in the middle, politically and administratively, and notes that his effectiveness is often limited by his relatively low status.

N. Civic action

1117. Auletta, Anthony J. "Ten–Nation Progress Report." *Army* 13 (July 1963): 53–59.

   Discusses civic action programs all over the world. Educational and medical activities are stressed. Auletta traces the current interest in civic action to the reconstruction of South Korea with the help of the United States armed forces after the Korean War.

1118. Barber, Willard F., and C. Neale Ronning. *Internal Security and Military Power: Counterinsurgency and Civic Action in Latin America.* Columbus: Ohio State University Press, 1966.

   Presents not only an excellent examination of the subjects noted in the title, but also analyzes the development of counterinsurgency thinking in the United States. The authors note briefly the role of the marines and the navy during the 1920s and 1930s.

1119. Bellinger, John B., Jr. "Civilian Role in Anti-guerrilla Warfare." *Military Review* 41, no. 9 (September 1961): 91–94.

   Discusses the role of the Theater Army Civic Affairs Command (TACAC). Its effectiveness depends on generating appropriate reforms, ensuring that counterinsurgency is the responsibility, primarily, of local forces, preventing terrorism, and providing technical assistance.

1120. Bentz, Harold F., Jr. "Psychological Warfare and Civic Action." *Army* 13 (July 1963): 62–65.

   Emphasizes the need for coordinating military efforts and United States Information Agency programs with local psychological warfare operations. Bentz asserts that psychological warfare is the primary element; military operations and civic action must support it.

1121. Buskirk, Lawrence E. "Nationbuilding." *Military Review* 49, no. 11 (November 1969): 10–17.

   Discusses the agricultural nature of the Third World economies and the ways in which the armed forces can contribute to improving agricultural production. This can be accomplished mainly by United States personnel teaching mechanical skills to foreign soldiers, who can then train the farmers.

1122. Duffy, John J. "Signpost: Success in the Philippines." *Army* 13 (July 1963): 60–62.

   Consists of an interview with Colonel Harry Lambert, a United States adviser in the Philippines. Civic action programs began there after the renewal of Huk activity in 1962.

1123. Glick, Edward Bernard. "Military Civic Action." *Army* 17 (September 1967): 67–70.

   Cautions that much is unknown about the nature and effectiveness of civic action or the extent to which it is accepted by the Congress and by United States public opinion. There is also a real danger of creating expectations in host countries and then not being able to live up to them.

1124. Krasin, Chaiyo. "Military Civic Action in Thailand." *Military Review* 48, no. 1 (January 1968): 73–77.

   Describes both American and Thai military civic action projects. These involve the improvement of communications, including roads and radio broadcasting. The improved radio communication system lends itself to psychological warfare along the Thai border, where Communists from Indochina have some influence.

1125. Lansdale, Edward G. "Civic Action Helps Counter the Guerrilla Threat." *Army Information Digest* 17 (June 1962): 50–53.

   Points to the skillful use of civic action by communist guerrillas. People must be won over by both spontaneous acts of kindness by counterinsurgency forces and by well-planned projects for improving life through the use of military personnel and resources.

1126. Leacacos, John P. "The Search for and Development of Soldier-Statesmen." *Army* 13 (April 1963): 48–54.

   Discusses Kennedy's push for more politically sensitive military leaders, in part because of the need for effective civic action programs. Leacacos stresses the importance of stopping an insurgency movement in its first stages.

1127. Livingston, Hoyt R., and Francis M. Watson. "Civic Action: Purpose and Pitfalls." *Military Review* 47, no. 12 (December 1967): 21–25.

   Warns that without proper planning, some civic action programs can be harmful. Civic action must be carried out with the same degree of organization as other military tactics. The planners, moreover, must not get too far away from the society and the environment.

1128. Rostow, Walt W. "Guerrilla Warfare in Underdeveloped Areas." *Airman* 6 (May 1962): 15–19.

   Stresses the role of military personnel as modernizers during a passing stage of world history and asserts that "Communism is best understood as a disease of the transition to modernization" (p. 16). This famous address has been widely published. It is one of the most important statements of the Kennedy strategy of counterinsurgency and civic action.

1129. Roush, Maurice D. "Master Plans for Nationbuilding." *Military Review* 49, no. 2 (February 1969): 77–85.

Discusses the problems of Third World nations and the difficulties they face in organizing to cope with them. Roush suggests that the United States establish a National Institute for Progress as part of the Agency for International Development and examines the possible American military role in this field.

1130. Slover, Robert H. "This is Military Civic Action." *Army* 13 (July 1963): 48–52.

Emphasizes training local nationals in civic action. Slover notes the problem of procuring appropriate military personnel and states that the level of involvement of United States service personnel varies with the nature of the project.

1131. Smith, Laun C. "Military Civic Action in Central America." *Military Review* 49, no. 1 (January 1969): 64–71.

Reviews civic action projects in Guatemala, Nicaragua, El Salvador, and Honduras. Smith emphasizes that the local military is learning to help the people rather than repress them and notes the beneficial United States influences that are bringing this about.

1132. Walterhouse, Harry F. "Civic Action—A Counter and Cure for Insurgency." *Military Review* 42, no. 8 (August 1962): 47–54.

Emphasizes the need for civic action programs, answers possible criticism of such programs at considerable length, and suggests the human resources that the army could tap to assure effective implementation of civic action projects.

1133. ———. *A Time to Build, Military Civic Action: Medium for Economic Development and Social Reform*. Columbia: University of South Carolina Press, 1964.

Represents probably the most comprehensive treatise on this facet of counterinsurgency. Chapters examine foreign experiences, American precedents, such as the Civilian Conservation Corps (CCC), and civic action in various parts of the world. Curiously, nothing is said about navy and marine projects in the Caribbean before World War II.

XIX

North China, 1945–49

In September, 1945, approximately 50,000 marines landed in northern China to assist in the disarmament and repatriation of Japanese servicemen and civilians. By guarding coal mines and rail lines, the marines also aided the Chinese Nationalists during the beginning of their conflict with the Chinese Communists.

Several clashes occurred between marines and Chinese Communist forces. Communist attacks on marine-guarded installations seem to have foreshadowed the tactics that became more familiar during the Vietnam conflict.

When repatriation ended, the numbers of marines were speedily reduced. Marine operations were confined closely to the coast as time went on, and, finally, as Chinese Communist armies came closer, the marines were withdrawn along with remaining naval units.

Chronology

1945

August 10   United States Joint Chiefs of Staff ordered to assist Chinese National forces to move into territory formerly under Japanese occupation, to retrieve Allied prisoners of war, and facilitate the repatriation of remaining Japanese forces in China. Concern about an increase of Soviet influence in the Far East was also part of the reason for deploying United States forces to North China.

September 20   A small advance party of the Marine III Amphibious Corps, which had been assigned the tasks of aiding the Chinese Nationalists, locating and caring for Allied prisoners of war, and Japanese repatriation, arrived in Shanghai.

September 30	Initial landing of elements of the main force began at Tianjin.
October 6	Formal surrender of Japanese forces in the Tianjin area.
October 6	Initial skirmish between marines and Chinese Communist troops on the road from Tianjin to Peiping occurred as the marines attempted to remove roadblocks. With stronger support, including aircraft, the road was opened the next day.
October 7	An effort to land more marines at Chefoo failed, owing to vigorous warnings from the local Chinese Communist commanders that a landing would be resisted.
October 10	After a delay because of a typhoon, the Chefoo force was landed at Tsingtao
October 25	Formal surrender of Japanese forces in the Tsingtao area.
November 7	Army Lieutenant General Albert C. Wedemeyer, who headed United States forces in China, advised the marine commander that marines would be needed in North China for some time.
November 14	Chinese Communist forces attacked a train whose passengers included Major General DeWitt, who then commanded the Marine 1st Division.
November 16	Marines were assigned to guard a number of railway bridges, in order to free Chinese Nationalist troops to drive Chinese Communists away from the railways.
December 6	Marine aircraft, under certain conditions, were permitted to respond to Chinese Communist fire.

1946

January 10	Truce between Chinese Nationalist and Chinese Communist forces arranged through United States mediation. The level of fighting lessened considerably for a time, although some skirmishes occurred. The truce helped to facilitate the repatriation of Japanese troops and civilians. With the substantial progress made in repatriation, the decision was made to reduce the marine presence in North China.
June 28	A truce arranged through United States mediation to end heavy fighting in Manchuria, where the January 10 ceasefire had not been honored. Despite the truces there was an increasing number of military encounters between Chinese Communists and marines.
July 13	Chinese Communists seized seven marines who had strayed from their security duties. The marines returned to United States control on July 24.
July 15	The last repatriation vessel left China for Japan.
July 29	Chinese Communists made a vigorous attack on a marine highway convoy at Anping, with the allegation that the marines had been operating with Chinese Nationalist forces.

	Some marines were killed and others wounded. This episode led to a consolidation of marine elements.
September 30	Last marines removed from railway security duties.
October 3–4	Chinese Communist attack on a marine supply depot at Hsin Ho. The Chinese Communists were driven off, although they succeeded in carrying away some ammunition.

1947

April 5	Chinese Communists made a second vigorous, but unsuccessful attack on the Hsin Ho supply point. Some marines were wounded, and a number of Chinese Communists were killed or wounded before they retreated. The remaining munitions at the depot were transferred to the Chinese Nationalists, with the process ending in September.
December 25	Several marines who were hunting fell into the hands of Chinese Communists. One of them was mortally wounded in an initial exchange of fire. The survivors were not returned to United States control until April 1, 1948.

1949

May 16	The last remaining marines were withdrawn from Tsingtao.

This chronology has omitted a considerable number of military movements, force reductions, and skirmishes between 1945 and 1949.

A. General studies

1134. Churley, Robert A. "The North China Operation." Parts 1, 2. *Marine Corps Gazette* 31 (October, November 1947): 10–16, 17–22.

   Focuses on the activities of the First Marine Division between October 1945 and June 1947. Much of the article is devoted to the political and military situation in North China and Manchuria at that time.

1135. Clark, George B. *Treading Softly: U.S. Marines in China, 1819–1949*. Chap. 5, "Back to China, 1945–49," pp. 130–65. Westport, CT: Praeger, 2001.

   Provides a useful survey of events in North China and gives fairly detailed accounts of the numerous skirmishes and other clashes between marines and the Chinese Communist forces. The human interest stories that Clark relates nicely supplement the more formal presentation of the same period in Frank and Shaw (item 1138).

1136. Colbourn, Colin. "Caught in the Crossfire: Marines in North China, 1945–49." *Leatherneck* 91, no. 4 (April 2008): 42–46.

Fine overview of marine service in North China and the fighting that occurred between the marines and the Chinese Communists. The numerous carefully selected photographs add to the worth of this article.

1137. Dobbs, Charles M. "American Marines in North China, 1945–46." *South Atlantic Quarterly* 76, no. 3 (Summer 1977): 318–31.

Emphasizes the problems that confronted the marines in North China. The Chinese Communists became openly hostile, and the Chinese Nationalists were dissatisfied with United States policies. Sharp criticism of the United States involvement in China by several senators added to the complexity of the situation. By the beginning of 1947 most of the marines had been withdrawn.

1138. Frank, Benis M., and Henry I. Shaw, Jr. "North China Marines." Part V, vol. 5, of *Victory and Occupation: History of U.S. Marine Corps Operations in World War II*. Washington, DC: Government Printing Office, 1968.

Constitutes the most complete official account of the North China intervention. The narrative is supplemented by photographs and useful maps and charts. There are extensive notes to official sources and a few other items.

1139. Guo, Xixiao. "The Anticlimax of an Ill-Starred Sino-American Encounter," *Modern Asian Studies* 35, no. 1 (February 2001): 217–44.

Much of the article deals with the Anping clash between marines and Chinese Communist troops in July 1946 and the Shen Clung rape controversy. Guo argues that these and other incidents generated significant anti-American feeling and that the anti-American movement did not occur simply because it was instigated by the Chinese Communists.

1140. ———. "Paradise or Hell Hole? U.S. Marines in Post-World War II China," *Journal of American–East Asian Relations* 7, nos. 3–4 (Fall/Winter 1998): 157–85.

Solidly based research about the life of the marines in North China and their role in the Chinese Civil War. In addition to many health hazards, the marines faced sharp encounters with Chinese Communist troops and friction with the Chinese Nationalists, including the police.

1141. Houchins, Lee Stretton. "American Naval Involvement in the Chinese Civil War, 1945–49." Ph.D. diss., American University, 1971.

Describes a number of navy and marine operations in the early postwar period in China, including the evacuation of Allied prisoners of war and the transporting of Nationalist forces to North China. The initial marine landings are examined in Chapter III, which includes a detailed discussion of the Chefoo incident, in which Chinese Communists refused to allow a marine landing.

1142. *Marshall's Mission to China, December 1945–January 1947: The Report and Appended Documents.* Vol. I. Arlington, VA: University Publications of America, Inc., 1980.

Provides a chronological and fairly detailed picture of the situation in North China during 1945 and 1946, chapters XVIII and XXXI. Clashes between the marines and Chinese Communist forces are only touched upon very briefly.

1143. Shaw, Henry I. *The United States Marines in North China 1945–1949.* Washington, DC: Headquarters, U.S. Marine Corps, Historical Branch, G-3 Division, 1960.

Seems to have been superseded effectively by Frank and Shaw (item 1138), except that it contains appendices on casualties in North China. This information does not appear in the later volume.

1144. U.S. Congress. House. Committee on International Relations. *Selected Executive Session Hearings of the Committee, 1943–50. Volume VII. United States Policy in the Far East. Part I.* Washington, DC: Government Printing Office, 1976.

Discusses the employment of marines in North China, pages 105–6, 117, 134, 135–37, 139–40, 151. The confusion over objectives assigned to the marines is clearly apparent from the testimony. The marines are to remain neutral in the civil war, but they are helping the Nationalists.

1145. U.S. Department of State. *United States Relations with China.* Washington, DC: Government Printing Office, 1949.

Contains material relevant to marine service in North China. The lack of an index forces the researcher to search carefully for information.

1146. Yang, Zhiguo. "U.S. Marines in Quindao: Military–Civilian Interaction, Nationalism, and China's Civil War, 1945–49." Ph.D. diss., University of Maryland, 1998.

Argues that the marine presence in North China had a significant impact on China during the early years after World War II. After an initial welcome by the Chinese, the marines were subjected to increasing criticism. Various incidents occurred that the Communists could employ to discredit United States involvement in China. Yang's extensive use of Chinese archives and his effort to put the marine episode into perspective by discussing the earlier German and Japanese impacts on Quindao make this an especially valuable study.

B. Diplomatic and political perspectives

1147. Chern, Kenneth S. *Dilemma in China: America's Policy Debate, 1945.* Hamden, CT: Archon Books, 1980.

    Discusses the issue of United States military involvement in the Chinese civil war, including the deployment of marines in North China, at some length, chapters IV and V. Chern strongly disapproves of American aid to the Nationalists.

1148. Feis, Herbert. *The China Tangle: The American Effort in China from Pearl Harbor to the Marshall Mission.* Princeton, NJ: Princeton University Press, 1953.

    Describes very briefly United States landings in North China and the refusal of Communist forces to permit debarkation at Chefoo in September, 1945. Feis also reviews negotiations with the Soviets for the removal of both Soviet and United States forces from China.

1149. "GI Welcome Wears Out in China." *Amerasia* 10 (December 1946): 173–74.

    Quotes complaints in Chinese newspapers about misbehavior by United States service personnel, especially drunk and disorderly charges. *Amerasia* suggests that the problem is that United States forces should not be in China. There is nothing worthwhile for American troops to do; therefore, they are rebelling through antisocial actions.

1150. Pepper, Suzanne. *Civil War in China: The Political Struggle, 1945–1949.* Berkeley: University of California Press, 1978.

    Discusses political attacks by Chinese Communists and other opponents of the Nationalists on the presence of American troops in China, pages 52–58. Pepper focuses on student unrest after an alleged rape in 1946.

C. Repatriation of the Japanese in China

1151. Klein, Edwin. "Back to Japan." *Marine Corps Gazette* 30, no. 3 (March 1946): 17–19.

    Examines repatriation activities of the marines. Several photographs illustrate the repatriation process. This was one of the major responsibilities of the marines in the first year after the war.

1152. ———. "Situation in North China." *Marine Corps Gazette* 30, no. 4 (April 1946): 11–14.

    Defends United States participation in the repatriation program, and carefully delineates the problems of negotiating with the Nationalists and the Communists. The article includes a map of events in North China during September and October 1945.

D. Marines as railway guards

1153. Aplington, Henry II. "China Revisited." *Marine Corps Gazette* 57, no. 7 (July 1973): 24–31.

Emphasizes the newness of low-intensity conflict to World War II-era marines. They also had to adapt to the changes in China due to World War II. The marines were very thinly spread over railroad and other guard posts. Without helicopters, communication and transportation were difficult.

1154. Hittle, James D. "On the Peiping–Mukden Line." *Marine Corps Gazette* 31, no. 6 (June 1947): 18–26.

Describes marine services on guard duty at several coal mines and along the railroads of North China. There is much information on the organization and armament of guard units. Hittle emphasizes that most officers are quite inexperienced, but they are performing well in a very touchy situation.

1155. Moore, Lynn. "Fuel for the Fires of China." *Leatherneck* 29, no. 8 (August 1946): 9–12.

Discusses guard duty along railroads between coal mines and the coast. One page deals with life at Bridge 96. The text is supplemented by interesting photographs.

E. Clashes with Chinese Communist forces

1156. Aplington, Henry II. "North China Patrol." *Marine Corps Gazette* 33, no. 6 (June 1949): 46–55.

Describes efforts to recover several marines captured by Chinese Communists in July 1946. Aplington gives an excellent analysis of the advantages guerrillas often have over regular troops. These advantages may be overcome in part by extensive air support.

1157. Palmer, Thomas A. "The First Confrontation: U.S. Marines in North China, 1945–47." *Marine Corps Gazette* 54, no. 4 (April 1970): 22–28.

Puts the North China operation into the broader context of the Communist–Nationalist conflict. Palmer emphasizes the strength of the initial force in China and asserts that on a tactical level the marines were successful. Their landings were a setback to the Chinese Communists. Misunderstandings about the nature of Chinese Communism in the United States, it is said, helped bring pressure for the withdrawal of the marines.

1158. Polete, Harry. "China Expedite." *Leatherneck* 30, no. 3 (March 1947): 22–25.

Describes air supply operations in support of the marine contingent in China. Polete focuses on a special flight to carry medicine to protect the marines from a sleeping sickness epidemic that swept North China in the summer of 1946. A number of photographs of air activities, especially in Hawaii, are included.

1159. Schuon, Karl. "Shanghaied in an Oxcart." *Leatherneck* 30, no. 1 (January 1947): 2–7.

Emphasizes the marine search for a few men captured by the Chinese Communists in the summer of 1946 and the experiences of the prisoners. The capture itself is briefly described. The article includes many photographs of the negotiations that were required for the return of the marines.

F. Other topics

1160. Polete, Harry. "China Police." *Leatherneck* 30, no. 4 (April 1947): 3–6.

Describes mixed police patrols in Peiping, consisting of one man each from the marines, the army, the Chinese Army, and the Peiping police. Because Chinese personnel were included, the patrols could make searches effectively. Liaison between the two services and the Chinese authorities at higher levels is discussed. Polete notes marine cooperation with Nicaraguan police in the 1920s and 1930s.

1161. Tsou, Tang. "Civil Strife and Armed Intervention: Marshall's China Policy." *Orbis* 6 (Spring 1962): 76–101.

Studies exhaustively possibilities for committing United States troops to support the Nationalists. No one advocated it, partly because of the weakness of the United States armed forces. Tang Tsou cites misconceptions about the Chinese Communists and the weakness of the Nationalist government as other factors.

XX
Lebanon, 1958

For several months in the summer and fall of 1958, United States forces held parts of Lebanon to prevent a coup or attacks from Syria. Intervention took place under the Eisenhower Doctrine of 1957, which assured Middle Eastern powers of United States aid if they were threatened with internal subversion or foreign invasion.

Fighting in Lebanon, the formation of the United Arab Republic, and a revolution in Iraq led the United States to land marines and soldiers. After it was realized that the new Iraqi government and the United Arab Republic were not as leftist as had been thought, the crisis alleviated.

The Lebanon operation was one of the most dramatic evidences of the significance of United States naval power in the Mediterranean Sea after World War II.

Chronology

1953
April — Early in the Eisenhower administration, Secretary of State John Foster Dulles and other United States leaders became concerned about anti-Western trends in the Middle East in the wake of a successful revolution in Egypt that overthrew the monarchy. Closely associated as it was with the former major colonial powers in the Middle East, the United States and its policies were objects of suspicion in many parts of the region.

1955
September 27 — Egypt signed an agreement with Czechoslovakia, an Eastern bloc country, to secure military items. This event caused

considerable concern among United States decision-makers. Relations between Egypt and the United States improved temporarily because the United States opposed the British, French, and Israeli attack on Egypt after that country took full control of the Suez Canal. Nevertheless, the United States continued to be identified with the British and French.

1957
January 5   In a speech Eisenhower recommended that Congress pass the Middle East Resolution, which offered support, including tangible military and economic aid, and, if necessary, United States military forces, to Middle East countries that were being attacked by the forces of Communism, whether from another country or through internal subversion. The resolution was passed, after congressional modifications, and it became known as the Eisenhower Doctrine. This policy was the basis for the 1958 intervention in Lebanon, and, indeed, President Camille Chamoun quickly endorsed the doctrine. The United States policy did not by any means stem the increasing anti-colonialism in the Middle East and consequent ill-feeling toward Western bloc countries.

April   Opponents of Chamoun established a United National Front, consisting of Muslims and Christians

1958
February 1   Egypt and Syria joined to form the United Arab Republic. The United States did not attack the establishment of a new state, but it was troubled by another sign of increasing militancy in the Middle East, especially among the Arabs.

May   Fighting broke out in Lebanon as the result of months of tension between the Chamoun government and its critics. Chamoun began asking the United States for military support. Chamoun, it should be noted, wanted to continue his presidency despite a constitutional provision that made it impossible.

June 30   Chamoun announced to the Lebanese that he would no longer attempt to continue as president. There seemed to be some prospect of a settlement in Lebanon.

July 14   A Pan-Arab uprising toppled the Iraq monarchy and government that had been aligned with the Western bloc in the Cold War. This was a major setback for Western policies in the Middle East.

July 14   Chamoun requested United States military support for his government, arguing that there might be a coup by Pan-Arab elements in Lebanon in the aftermath of the revolution in Iraq.

July 15	Initial landing of some United States Marines near Beirut. More marines arrived within the next few days. Lebanon and the United States believed that unrest in Lebanon was being supported by the United Arab Republic.
July 19	Elements of the United States Army's 24th Infantry Division began arriving, beginning with an airborne battle group that was a component of the division.
July 24	Major General Paul D. Adam, an army officer, assumed command of the force after a brief period when the marines and soldiers operated independently of one another.
July 31	Fuad Shabib was elected president of Lebanon by the legislative chamber of that country. Shabib was a compromise candidate agreed upon by supporters and opponents of Chamoun.
September 24	Shabib was inaugurated as president of Lebanon.
October 17	A newly formed Lebanese cabinet received a vote of confidence, thereby further stabilizing the political situation.
October 25	The last United States forces departed Lebanon.

A. Political surveys of the Lebanon crisis

1162. Agwani, M.S. *The Lebanese Crisis, 1958: A Documentary Study.* New York: Asia Publishing House, 1965.

Contains most of the pertinent documents. Chapters 5 through 7 deal specifically with the United States intervention and the reactions to it.

1163. Al-Aiban, Bandar Mohammed. "United States Policy in the Middle East and Its Intervention in Lebanon, 1955–58." Ph.D. diss., Johns Hopkins University, 1996.

Its broad scope helps make this dissertation a useful part of the literature about the 1958 intervention. Coverage of the landings in Lebanon and Jordan is quite limited, but there are valuable chapters about such topics as the Eisenhower Doctrine and the amendments made to it, the Camille Chamoun administration in Lebanon, and the temporary easing of tensions in Lebanon after the intervention.

1164. Alin, Erika G. *The United States and the 1958 Lebanon Crisis: American Intervention in the Middle East.* Lanham, MD: University Press of America, 1994.

Heavily researched study that puts the events of 1958 into the broader framework of United States–Lebanese relations during the 1950s. In addition to presenting an impressive political analysis, the book gives due attention to the military aspects of the 1958 landings. Alin views the Eisenhower administration's behavior during the crisis as having been determined almost entirely by fears that events in the Middle

East were directly related to the Cold War and to the Soviet Union's effort to expand its influence in the region.

1165. Barnet, Richard J. *Intervention and Revolution: The United States in the Third World*. Chap. 7, "The Lebanese Civil War and the Eisenhower Doctrine," pp. 132–52. New York: World Publishing Company (New American Library), 1968.

Discusses the political background of the landings, but says relatively little about the military aspects. Barnet points to episodes in Jordan and Syria during 1957 that led to United States naval movements to support the Eisenhower Doctrine.

1166. Gendzier, Irene L. *Notes from the Minefield: United States Intervention in Lebanon and the Middle East, 1945–1958*. New York: Columbia University Press, 1997.

Essential reading for anyone interested in the 1958 intervention. This massively researched study provides much information about the diplomacy preceeding the 1958 intervention and the negotiations about the scope and the length of the intervention once it had begun. Gendzier believes that Iranian and Israeli recommendations that the United States react to the Lebanon crisis militarily influenced the decisions made in Washington.

1167. Korbani, Agnes G. *U.S. Intervention in Lebanon, 1958 and 1982: Presidential Decisionmaking*. New York: Praeger, 1991.

Sees broad similarities between the behavior of presidents Eisenhower in 1958 and Reagan in the 1980s in respect to Lebanon. Although the 1958 intervention can be viewed as a partial success, the second major United States involvement with Lebanon was a clear failure. Korbani skillfully analyzes the backgrounds to the two interventions, including the complexities of Arab nationalism and the role of the Arab–Israeli conflict.

1168. Lambrakis, George Basil. "Perception and Misperception in Policy-Making: The U.S. Relationship with Modern Lebanon, 1943–76." Ph.D. diss., George Washington University, 1989.

Puts the events of 1958 into perspective by providing an overview of Lebanese–United States relationships for approximately the first 30 years of the recent independence of Lebanon. Lambrakis suggests that many Lebanese leaders, like major political figures in many Third World countries, believed that United States support for their nations meant a higher degree of commitment than was intended by United States decision-makers.

1169. Meo, Leila M.T. *Lebanon—Improbable Nation: A Study in Political Development*. Bloomington: Indiana University Press, 1965.

Describes the political scene in Lebanon and the lessons of the Eisenhower Doctrine. Only a few pages deal directly with the United States intervention, but the book is indispensable for an understanding of the 1958 situation.

1170. Murphy, Robert. *Diplomat Among Warriors.* Garden City, NY: Doubleday & Company, Inc., 1964.

Focuses on Ambassador Murphy's visits to various Middle Eastern nations to explain the United States intervention. Murphy also discusses the movement of troops to Lebanon and his negotiations with Lebanese leaders.

1171. Quandt, William B. "Lebanon, 1958, and Jordan, 1970," chap. 7, pp. 222–88, in Barry Blechman, Stephen S. Kaplan, David K. Hall, William B. Quandt, Jerome N. Slater, and Robert M. Slusser. *Force Without War: U.S. Armed Forces as a Political Instrument.* Washington, DC: Brookings Institution Press, 1978.

Examines the variety of motives behind the intervention. Quandt asserts that the Iraq coup triggered the intervention. For a time, there were plans for further interventions, but these were quickly scrapped when it became apparent that the situation was not as serious as thought initially.

1172. Qubain, Fahim I. *Crisis in Lebanon.* Washington, DC: Middle East Institute, 1961.

Analyzes the Lebanon crisis and discusses the gradual relaxation of tension between United States and Lebanese forces that led to the formation of joint military patrols. The book is based on United Nations documents, Middle Eastern newspapers, radio broadcasts, and other materials.

1173. Rostow, Walt W. *The Diffusion of Power: An Essay in Recent History.* Chap. 9, "Crisis in the Middle East," pp. 94–98. New York: Macmillan Publishing Co., 1972.

Recounts the facts of the Lebanon episode and comments that Eisenhower acted to stabilize an ambiguous situation even though the extent of Soviet involvement was uncertain. Half the chapter deals with Eisenhower's speech of August 13, 1958, which represented a coming to terms with new nationalist groups in the Middle East.

1174. Steeves, W.D., Jr. "Crisis in Retrospect." *Marine Corps Gazette* 57, no. 12 (December 1973): 31–34.

Argues that Eisenhower's decision to implement the Eisenhower Doctrine in Lebanon was based on a solid grasp of conditions in the Middle East and of the military problems involved in committing

United States troops. Steeves discusses various interpretations of the Lebanon operations.

1175. Tillema, Herbert K. "The Four Interventions: Lebanon," pp. 45–52, in *Appeal to Force: American Military Intervention in the Era of Containment*. New York: Thomas Y. Crowell Company, 1973.

Suggests that almost as soon as troops were landed, the Eisenhower administration realized that the crisis was not serious and acted rather quickly to ease the tension and to withdraw its forces. The chapter is based on interviews with Ambassador Murphy and others, as well as secondary materials.

1176. U.S. Senate. Committee on Foreign Relations. *Situation in the Middle East*, 85th Cong., 2nd sess., 1958.

Provides much information about the decision to land troops. The legal basis for the landings, the developing military situation, and other topics are discussed amid spirited questioning, particularly of Under Secretary of State Herter. For reasons of secrecy, these hearings were not published until 1980.

1177. Wolf, John Berchmans. "An Interpretation of the Eisenhower Doctrine, Lebanon 1958." Ph.D. diss., American University, 1967.

Furnishes an excellent survey of the background of the Lebanon crisis through chapters on the rebellion in Lebanon and the coup in Iraq. Chapter 5 deals specifically with the intervention and includes information on plans made in 1957 for a similar operation that was to have been a joint Anglo–American landing. When intervention actually occurred, however, the United States forces went into Lebanon and the British into Jordan. There is an extensive bibliography.

B. General military studies

1178. Bodron, Margaret M. "U.S. Intervention in Lebanon, 1958." *Military Review* 56, no. 2 (February 1976): 66–76.

Analyzes the political background at some length and briefly describes the landings. Bodron believes the United States move was risky but successful.

1179. Gray, David W. *The U.S. Intervention in Lebanon: A Commander's Reminiscence*. Fort Leavenworth, KS: U.S. Army Command and General Staff College, Combat Studies Institute, 1984, reprinted 1986.

This is an extremely valuable account of the operation. Gray is quite candid in describing problems associated with the organization and implementation of the Lebanon venture, including his own mistakes.

Gray gives considerable attention to contingency plans for possible Middle Eastern crises well before the 1958 episode. A significant portion of the document consists of specific remarks that correct or supplement the Spiller study (item 1182). Gray makes the interesting comment that Lebanon's President Chamoun "did not fear an external as much as he did an internal attempt on his life by members of the Palace Guard, which was part Arab and part Christian" (p. 41).

1180. Saunders, Richard M. "Military Force in the Foreign Policy of the Eisenhower Presidency," *Political Science Quarterly* 100, no. 1 (1985): 97–116.

Usefully places the Lebanon landings within the context of Eisenhower's use of the great military strength of the United States. Saunders views Eisenhower as having been extremely conscientious about employing force, a stance that was in line with Eisenhower's conservative economic policies. Thus, the Lebanon operation was undertaken only after extremely careful consideration of the Middle East situation.

1181. Smith, Lynn D. "Lebanon—Professionalism At Its Best." *Military Review* 39, no. 6 (June 1959): 36–46.

Praises the conduct of United States servicemen and asserts that their professional behavior contributed significantly to accomplishment of their mission. Their behavior resulted from the excellence of their training.

1182. Spiller, Roger J. *"Not War But Like War": The American Intervention in Lebanon.* Fort Leavenworth, KS: Combat Studies Institute, U.S. Army Command and General Staff College, 1981.

Views the Lebanon landing as a model of effective intervention, but does not ignore the problems that occurred. This is the only thorough study from the perspective of the army other than Gray and (item 1179). There is much material on planning for contingencies in the Middle East and the logistical aspects of the operation. Maritime aspects of the landings are by no means overlooked.

C. The landings

1183. "Dateline—Lebanon." *Leatherneck* 41, no. 10 (October 1958): 14–23.

Depicts the daily life of marines in Lebanon, and discusses their contacts with the Lebanese people. The article includes a useful compilation of photographs.

1184. "Lebanon Landing." *Leatherneck* 41, no. 9 (September 1958): 24–33.

Emphasizes the deployment of marines in readiness for possible combat and the utilization of patrols to avoid possible attacks. The article in large part consists of photographs.

1185. Pierce, Philip N. "Lebanon." *Leatherneck* 45, no. 9 (September 1962): 30–37.

Deals specifically with the first landing as seen through the eyes of a gunnery sergeant and other participants. The tone is half humorous. One of the most interesting aspects of the article is the description of the harassment tactics used by rebels against the marines, including the driving of animals into marine positions.

1186. Wade, S.S. "Operation Bluebat." *Marine Corps Gazette* 43, no. 7 (July 1959): 10–23.

Defends forcefully United States action in Lebanon. Wade, who commanded United States land forces in Lebanon briefly, identifies July 16, 1958, as the critical day of the intervention and relates his own experiences during a confrontation with the Lebanese Army. The article stresses the lessons of the operation in logistics and intelligence.

D. Political conflicts during the landings

1187. Ashton, N.J. "A Great New Venture? Anglo–American Cooperation in the Middle East and the Response to the Iraqi Revolution, July 1958." *Diplomacy and Statecraft* 4, no. 1 (March 1993): 59–89.

Like many other students of the Middle East, Ashton believes that the Eisenhower administration misinterpreted Egyptian and Pan-Arab moves in the mid-1950s. For a while the United States viewed efforts toward Arab unity as parts of a Soviet-sponsored strategy. The British had a better understanding of events in the Middle East, and, therefore, Ashton believes that Great Britain and the United States worked along somewhat separate lines in 1958.

1188. Bishku, Michael B. "The 1958 American Intervention in Lebanon: A Historical Assessment." *American–Arab Affairs* no. 31 (Winter 1989–90): 106–19.

Emphasizes that the Eisenhower Doctrine arose as a result of United States fears of Soviet penetration of the Middle East and United States hostility toward the Nasser regime in Egypt, which was taking a Pan-Arab stance. The author gives considerable attention to the attitudes of many leading Eisenhower administration figures, especially Eisenhower himself and Secretary of State John Foster Dulles. He does not neglect CIA involvement in Lebanon. Like other students of the Lebanon episode, Bishku believes the United States learned to be more cautious about opposing Arab nationalism and especially to be careful not to take sides in political conflicts within Lebanon. Bishku goes on to write that the Reagan administration forgot the latter lesson in the 1980s.

1189. Hadd, H.A. "Orders Firm But Flexible." *United States Naval Institute Proceedings* 88, no. 10 (October 1962): 80–89.

Explains the reasons for his timing of operations that Ambassador McClintock objected to. As commander of the marines in Lebanon under General Wade's general authority, Hadd insisted that military decisions must be made through the normal chain of command.

1190. Hahn, Peter L. "Securing the Middle East: The Eisenhower Doctrine of 1957," *Presidential Studies Quarterly* 36, no. 1 (March 2006): 38–47.

Highly informative analysis of the origins, development, and effects of the relatively short-lived Eisenhower Doctrine. Hahn views Eisenhower's stance as an effort to project United States power into the Middle East in the wake of the failed Suez operation undertaken by the old colonial powers, Great Britain and France, along with Israel. Some Middle Eastern countries, including Iran and Pakistan, welcomed Eisenhower's initiative, but other countries were reserved or outright hostile. Hahn offers relatively little material about the 1958 Lebanon crisis, but this is a significant source for anyone who wants a full understanding of the episode.

1191. Koburger, Charles W., Jr. "Morning Coats and Brass Hats." *Military Review* 45, no. 4 (April 1965): 65–74.

Portrays the intervention as an example of the need to be ready to wage limited warfare. Koburger briefly describes the early landings and emphasizes the problems in coordination that arose when military commanders and diplomats clashed.

1192. Little, Douglas. "His Finest Hour? Eisenhower, Lebanon, and the 1958 Middle East Crisis," *Diplomatic History* 20, no. 1 (Winter 1996): 27–54.

Argues in a sharply critical account of the application of the Eisenhower Doctrine in Lebanon that although the intervention in 1958 was successful in the short run, it could have had serious consequences, such as the spread of conflict in the Middle East and further United States military interventions. Little asserts that "the success of Eisenhower's experiment with 'limited war' owed as much to good luck as to wise crisis management" (p. 46). Unrest in the Middle East was the result of increased Arab nationalism, but the Eisenhower administration asserted that it was the product of Soviet intrigue, knowing that its statements were false. Little claims that this example of misrepresenting events to justify United States military involvement was a technique that was used in respect to Vietnam. In his view, too, President Camille Chamoun skillfully exploited United States fears of communism to gain the support needed to bolster his shaky regime.

1193. McClintock, Robert. "The American Landing in Lebanon." *United States Naval Institute Proceedings* 88, no. 10 (October 1962): 65–79.

Praises the landing for demonstrating United States power and for maintaining the independence of Lebanon. Ambassador McClintock is critical of what he views as the inflexible orders given to local marine commanders. This circumstance, he asserts, retarded the effective use of force at the right time.

1194. Ovendale, Ritchie. "Great Britain and the Anglo–American Invasion of Jordan and Lebanon in 1958," *International History Review* 16, no. 2 (June 1994): 284–303.

Stresses that Great Britain no longer had the strength to lead Western efforts to obstruct the Soviet Union's efforts to expand its influence in the Middle East. Ovendale discusses the Middle East incursions of 1958 largely in terms of relations between Great Britain and the United States, and asserts that Great Britain encouraged the United States to take the lead in dealing with Middle Eastern crises.

1195. Pearson, Ivan. "The Syrian Crisis of 1957, the Anglo–American 'Special Relationship,' and the 1958 Landings in Jordan and Lebanon," *Middle Eastern Studies* 43, no. 1 (January 2007): 45–64.

Valuable study that provides new information about United States and British planning for possible military action to support the governments of Jordan and Lebanon. The precipitating factor for such planning was the hostile stance of Syria toward the West, not the Iraq revolution that did not occur until the summer of 1958. Pearson demonstrates that the United States and Lebanon were seriously considering intervention in Lebanon and Jordan earlier than has been stated in most studies of the 1958 crisis.

E. The navy's role in Lebanon

1196. Baldridge, Elward F. "Lebanon and Quemoy—The Navy's Role." *United States Naval Institute Proceedings* 87 (February 1961): 94–100.

Reviews briefly the naval operations that were ordered in support of the Lebanon landings. Baldridge emphasizes that the navy initially played the paramount role in the crisis.

1197. Bryson, Thomas A. *Tars, Turks, and Tankers: The Role of the United States Navy in the Middle East, 1800–1979*. Chap. 7, "The Eisenhower Doctrine and the Landing in Lebanon, 1958," pp. 123–40. Metuchen, NJ: Scarecrow Press, 1980.

Asserts that while a United States response to Middle Eastern developments in the summer of 1958 was needed, the unilateral character of the Eisenhower Doctrine was a mistake. The chapter presents a compact but highly detailed account of the landing.

1198. Calhoun, Christopher S. "Lebanon: That Was Then." *United States Naval Institute Proceedings* 111, no. 9 (September 1985): 74–80.

Applauds the deployment of the 6th Fleet in support of United States policies in the Middle East. Calhoun emphasizes the strength of the fleet and its ability to provide supplies to land elements of the intervention and asserts that without the fleet "it would have been virtually impossible for the United States to land and support troops in the Middle East" (p. 80).

1199. Jones, J.K., and Hubert Kelley, Jr. "Crisis in the Middle East." Chap. 1, pp. 1–23, in *Admiral (31-Knot) Burke: The Story of a Fighting Sailor*. Philadelphia, PA: Chilton Books, 1962.

Follows Arleigh A. Burke through the height of the crisis, July 14–15, 1958. The authors describe the steps Burke took to counter apparent threats and strengthen the 6th Fleet.

F. The marines in Lebanon

1200. Shulimson, Jack. *Marines in Lebanon 1958*. Washington, DC: Headquarters, U.S. Marine Corps, Historical Branch, G-3 Division, 1966.

Demonstrates the effectiveness of marines in pacifying Lebanon and achieving United States policy goals. The pamphlet gives a lengthy description of the confrontation between Lebanese troops and the marines. This richly documented study is based on research in the archives of all the armed forces and of the State Department. Shulimson's text is enhanced by fine maps and photographs.

G. Special military studies

1201. Olinger, Mark A. "Airlift Operations During the Lebanon Crisis." *Army Logistician* 37, no. 3 (May–June 2005): 30–33.

Detailed description of the air movement of a marine battalion to Lebanon from the United States. Although United States forces did not encounter significant hostile forces in Lebanon, the operation was an important test of airlift capability.

1202. Wade, Gary H. *Rapid Deployment Logistics: Lebanon, 1958.* Fort Leavenworth, KS: U.S. Army Command and Staff College, 1984.

Detailed analysis of the role of logistics in Lebanon. Wade describes the logistics doctrine that governed operations in 1958, the deployment, and lessons from this experience. Numerous maps and diagrams enhance an understanding of the complex structure of the operation and the forces that carried it out.

H. The influence of the Lebanon operation on the armed forces

1203. Braestrup, Peter. "Limited War and the Lessons of Lebanon." *Reporter* 20 (April 30 1959): 25–27.

Concludes that the United States lacks appropriate limited war capability. Aging equipment threatened the success of the Lebanon operation. Braestrup's findings are based on interviews with military officers and other national security specialists.

1204. Hadd, Harry A. "Who's a Rebel? The Lesson Lebanon Taught." *Marine Corps Gazette* 46, no. 3 (March 1962): 50–54.

Argues that political information must be made available not only to command personnel, but also to others when landings are made. Hadd describes the early days of the intervention and the dangers of confrontation with the Lebanese Army.

1205. Sights, Albert P., Jr. "Lessons of Lebanon: A Study in Air Strategy." *Air University Review* 16 (July–August 1965): 28–43.

Discusses the strategic considerations underlying the Lebanon intervention and the role of airpower in the landings. Sights gives attention to both air force and navy participation in air operations. The airlift to Lebanon takes up several pages. The conclusion discusses the lack of interservice coordination and the poor preparation, generally.

XXI

Dominican Republic, 1965-66

United States military forces returned to the Dominican Republic in order to prevent the possibility of a communist seizure of power. When former president Juan Bosch's followers tried to overthrow the military regime, United States officials believed that his movement was communist-infiltrated, if not communist-dominated.

The Dominican episode resembled the Lebanon landings of 1958 in several respects. There was little fighting, and United States troops were withdrawn fairly quickly. The Organization of American States became involved in the intervention just as the United Nations supervised the Lebanon situation as United States forces were evacuated.

A large literature developed on the Dominican intervention. While most military items have been included in this bibliography, coverage of diplomatic and political materials has been selective.

Chronology

1962
December 20 Juan Bosch became the first democratically elected president of the Dominican Republic in many years. He led the Dominican Revolutionary Party, a left-liberal organization.

1963
February 27   Bosch assumed the presidency of the Dominican Republic.
September 25 Bosch was overthrown by the Dominican military and forced back into exile. He intended to resume the presidency of his country and began working toward that goal. Although the United States government mistrusted Bosch, it did not approve

of the military coup. It did recognize the new Dominican regime eventually, however.

1965

April 24 — A pro-Bosch faction within the military and others of his supporters began an uprising to return Bosch to power.

April 27 — Many United States citizens were evacuated from Santo Domingo by a United States naval force that had been ordered to the area. At this point, the pro-Bosch movement appeared to be on the point of total collapse, and the situation seemed to be stabilizing.

April 28 — President Lyndon B. Johnson ordered a marine landing to assist United States citizens and to protect their property.

April 29 — Johnson ordered further marines ashore to prevent a possible Communist seizure of the Dominican Republic. Other forces, including elements of the army followed, and by mid-May there were about 23,000 United States troops in the country. The marines had left by June 6, leaving the 82nd Airborne Division, an army force, as the largest United States military element.

May 6 — At the urging of the United States, the Organization of American States established an Inter-American Force, which was subsequently renamed the Inter-American Peace Force, to participate in the further pacification of the Dominican Republic.

May 23 — Hugo Penasco Alvim, a Brazilian general, arrived in the Dominican Republic to command the Inter-American Force, which by then included the United States components in the country. The Latin American contingents of the Inter-American Force were quite small compared to the United States contribution, but they had considerable political importance. Their presence helped reduce criticism of the initial United States military moves and aided in giving the intervention a greater sense of legality.

June 15–16 — After a tense period with a number of confrontations between members of the peace force and members of the rebel movement, significant fighting occurred in Santo Domingo. The rebels sustained considerable damage during the period of combat. Quiet was restored, but sporadic incidents continued to occur not only in Santo Domingo and elsewhere in the Dominican Republic.

September 3 — A provisional government was formed, which ruled the Dominican Republic for nearly a year.

1966

June — National elections held. By the time of the elections there had been significant changes in the configuration of the

	Dominican political parties. Although a Bosch victory had been widely predicted, the new president was Joaquín Balaguer, who had been associated with the Trujillo dictatorship.
July 1	Foreign troops began leaving the Dominican Republic in accordance with an Organization of American States resolution.
September 14	Two United States soldiers were killed, the last combat casualties of the operation.
September 19	Final withdrawal of United States forces from the Dominican Republic. The Inter-American Peace Force ended the next day.

A. General studies

1206. Barnet, Richard J. *Intervention and Revolution: The United States in the Third World.* Chap. 8, "The Dominican Republic," pp. 153–80. New York: World Publishing Company (New American Library), 1968.

Discusses the political background rather than the military aspects of the intervention. Barnet is highly critical of United States decision-makers. The chapter also briefly examines the United States threat of a marine landing to force the remnants of the Trujillo regime to leave the island in 1961.

1207. Bass, Jeff D. "The Appeal to Efficiency as Narrative Closure: Lyndon Johnson and the Dominican Crisis, 1965." *Southern Speech Communication Journal* 50, no. 2 (Winter 1985): 103–20.

Suggests that while there was some elite opposition to the intervention in the Dominican Republic, the general public in the United States approved of Johnson's actions because the operation was accomplished with dispatch.

1208. Brown, Seyom. *The Faces of Power: Constancy and Change in United States Foreign Policy from Truman to Johnson.* Chap. 19, "The Vietnam and the Dominican Interventions," pp. 278–306. New York: Columbia University Press, 1968.

Emphasizes President Johnson's fear of repercussions worldwide, if he did not act decisively in the Dominican Republic. Brown accepts Johnson's explanation for the landing of marines and then examines the justifications for the larger intervention.

1209. Carey, John, ed. *The Dominican Republic Crisis 1965: Background Paper and Proceedings of the Ninth Hammarskjold Forum.* Dobbs Ferry, NY: Oceana Publications, Inc., 1967.

Constitutes a treatise on the international law aspects of the intervention. The proceedings are very interesting, partly because Adolf A. Berle discusses the decision to land troops from the perspective of a participant. The book includes the best early bibliography on the subject.

1210. Chester, Eric Thomas. *Rag-Tags, Scum, Riff-Raff, and Commies: The U.S. Intervention in the Dominican Republic, 1965–1966.* New York: Monthly Review Press, 2001.

A largely political analysis, emphasizing the protracted negotiations between United States representatives and Dominican leaders. Chester assails United States policies and actions before and during the intervention. One chapter and an appendix deal in detail with the Dominican election of 1966, which Chester views as a demonstration election.

1211. Draper, Theodore. *The Dominican Revolt: A Case Study in American Policy.* New York: Commentary, 1968.

Suggests that the mental contortions of the Johnson administration in justifying the landings were almost more damaging than the intervention itself. This is a highly sarcastic dissection of United States explanations for the Dominican decisions.

1212. Felten, Peter Gerhard. "The 1965–66 United States Intervention in the Dominican Republic." Ph.D. diss., University of Texas, Austin, 1995.

Downplays the role of the United States in Dominican political development during the intervention. The Johnson administration was alarmed by the perceived prospect of the Dominican Republic following the path of Castro's Cuba. Elements of the Dominican political elite favored the intervention and encouraged the United States to keep some forces in the country. Eventually, elections were held that brought about a resolution of political conflict along lines favored by the United States, but internal political negotiations brought a settlement, not United States involvement.

1213. ———. "The Path to Dissent: Johnson, Fulbright, and the 1965 Intervention in the Dominican Republic." *Presidential Studies Quarterly* 26, no. 3 (Fall 1996): 1009–18.

By 1965, Senator J. William Fulbright had become critical of the Vietnam commitment. Initially, he supported the Dominican Republic intervention, but then he began questioning that move. When he made a relatively conciliatory speech about the Dominican Republic, the Johnson administration made a vehement counterattack, thereby intensifying the split between the two political leaders.

1214. Georgetown University. Center for Strategic Studies. *Dominican Action—1965: Intervention or Cooperation?* Washington, DC: Center for Strategic Studies, Georgetown University, 1966.

Justifies the United States landings on the basis of a danger to foreigners and of communist subversion. The heart of the book is a day-by-day analysis of events, April 24–30, 1965. Appendices identify major personalities and provide other background data.

1215. Gleijeses, Piero. *The Dominican Crisis: The 1965 Constitutionalist Revolt and American Intervention.* Baltimore, MD: Johns Hopkins University Press, 1978.

Provides the best scholarly treatment of the intervention from the Bosch point of view. The book is based in part on many lengthy interviews. Among those consulted were former president Juan Bosch and General Bruce Palmer, Jr., commander of United States forces in the Dominican Republic.

1216. Haley, P. Edward. "Comparative Intervention: Mexico in 1914 and Dominica in 1965," pp. 40–59, in *Intervention or Abstention: The Dilemma of American Foreign Policy*, ed. Robin Higham. Lexington: University Press of Kentucky, 1975.

Stresses the environment in which United States interventions take place and the goals that they have. In both Mexico and the Dominican Republic, the objective was to bring to power a moderate, democratic regime. Haley proposes a sequence of demands that the United States is likely to make upon participants in a civil war before it intervenes.

1217. Herman, Edward S. and Frank Brodhead. *Demonstration Elections: U.S.-Staged Elections in the Dominican Republic, Vietnam, and El Salvador.* Chap. 2, "The Dominican Republic," pp. 55–92. Boston, MA: South End Press, 1984.

The book deals largely with El Salvador and to a lesser extent with Vietnam. The single chapter about the Dominican Republic is a severely critical account of the election held there during the United States intervention. The authors assert that elections are only one element in a democratic political system. They believe, too, that the elections in the three countries discussed were manipulated.

1218. Kane, William Everett. *Civil Strife in Latin America: A Legal History of U.S. Involvement.* Baltimore, MD: Johns Hopkins University Press, 1972.

Concentrates on United States efforts to use the Organization of American States after the landings had occurred, pages 213–20. Kane also briefly discusses Juan Bosch, the beginning of the Dominican revolt, and American support for Trujillo.

1219. Keith, LeeAnna Yarbrough. "The Imperial Mind and U.S. Intervention in the Dominican Republic, 1961–66." Ph.D. diss., University of Connecticut, 1999.

A biting critique of the intervention. Like other authors, Keith believes that the Johnson administration greatly overestimated the possibility of a Communist coup. She asserts that the Kennedy administration and, initially, the Johnson administration, wanted a democratic political system introduced to the Dominican Republic. As time passed, the United States government reverted to long-standing views that Dominicans were incapable of maintaining democracy, and, therefore, the United States had to step in to restore order and direct Dominican political development.

1220. Kelso, Quinten Allen. "The Dominican Crisis of 1965: A New Appraisal." Ph.D. diss., University of Colorado, 1982.

Utilizes material declassified in the late 1970s to suggest that although the initial landings may have been justified, fears of a communist coup were never substantiated. In any event, a more decisive posture by United States authorities might have brought about a settlement without a large-scale intervention.

1221. Lowenthal, Abraham F. *The Dominican Intervention.* Cambridge, MA: Harvard University Press, 1972.

Focuses on the decision-making process in a sophisticated study, based on many interviews and classified sources. Washington was undecided at many points in the crisis, but there was a determination that there must be no "second Cuba." The bibliography is especially important.

1222. McPherson, Alan. "Misled by Himself: What the Johnson Tapes Reveal About the Dominican Intervention of 1965," *Latin American Research Review* 38, no. 2 (Winter 2003): 127–46.

Emphasizes that President Lyndon B. Johnson was extremely concerned about the possible effects of a leftist coup in the Dominican Republic on political debates and voter perceptions in the United States. Rather than depending as heavily on advisers and United States diplomats as had been assumed by most students of the Dominican intervention, Johnson seems to have made the final decisions on his own.

1223. ———. *Yankee No! Anti–Americanism in U.S.–Latin American Relations.* Chap. 4, "Dominican Republic, 1965: Episodic Anti–Americanism and U.S. Containment," pp. 116–62. Cambridge, MA: Harvard University Press, 2003.

The chapter begins with a detailed examination of the Kennedy administration's efforts to study and to counter anti–Americanism in Latin America, including the Dominican Republic. In fact, the United States was viewed positively by the Dominicans in the early 1960s, but that situation changed. McPherson argues that through the intervention of 1965, "Lyndon Johnson's government ... virtually created anti–Americanism out of nothing and stirred up a largely avoidable resentment" (p. 131). The author skillfully traces the currents of anti–Americanism in the Dominican Republic during the intervention.

1224. Martin, John Bartlow. *Overtaken by Events: The Dominican Crisis—From the Fall of Trujillo to the Civil War.* Garden City, NY: Doubleday & Company, Inc., 1966.

Emphasizes the period from 1962 to 1964, but part 4 deals with the events of 1965. Martin gives a good picture of the hectic negotiations and describes many incidents involving United States forces. He staunchly defends Johnson's course of action and argues that a revolutionary movement must become communist.

1225. Perkins, Whitney T. *Constraint of Empire: The United States and the Caribbean Interventions.* Chap. 16, "The Dominican Republic," pp. 211–29. Westport, CT: Greenwood Press, 1981.

Emphasizes the acute fears in the Johnson administration that "another Cuba" would develop in the Dominican Republic. Perkins is more sympathetic to Johnson and the State Department than most commentators. He also touches on the end of the Trujillo era.

1226. Rabe, Stephen G. "The Johnson Doctrine." *Presidential Studies Quarterly* 36, no. 1 (March 2006): 48–58.

The "Johnson Doctrine" dating from 1965 was a principle that the United States would not tolerate the establishment of another communist regime in the Western Hemisphere. Rabe argues that Johnson's pronouncement was simply a continuation of earlier United States efforts to exercise a high degree of control over Central and South America. President John F. Kennedy had made a similar pronouncement in 1963, moreover.

1227. Slater, Jerome N. "The Dominican Republic, 1961–66," chap. 8, pp. 289–342, in Barry Blechman, Stephen S. Kaplan, David K. Hall, William B. Quandt, Jerome N. Slater, and Robert M. Slusser. *Force Without War: U.S. Armed Forces as a Political Instrument.* Washington, DC: Brookings Institution Press, 1978.

Asserts that the United States intervened to prevent a communist takeover rather than to stop Bosch from regaining the presidency. Military aspects of the operation are discussed in some detail. Slater feels that measures other than intervention would have been effective.

1228. ———. *Intervention and Negotiation: The United States and the Dominican Republic.* New York: Harper & Row, 1970.

Argues that the United States aimed at bringing democratic government to the Dominican Republic rather than more military rule. Slater deals at some length with the operations of the Inter-American Peace Force.

1229. Stevenson, Adlai E. *The Papers of Adlai E. Stevenson: Volume VIII, Ambassador to the United Nations 1961–1965*. Ed. Walter Johnson. Boston, MA: Little, Brown and Company, 1979.

Contains two of Stevenson's United Nations speeches on the intervention. A number of personal letters written in 1965 are also relevant. There is a sharp difference in his public and private assessments of the situation.

1230. Szulc, Tad. *Dominican Diary*. New York: Delacorte Press, 1965.

Provides a detailed running account of events, April 24–May 27, 1965. There is a good deal of information on United States military movements and the scattered fighting that occurred after the intervention. Szulc warns against confusing communist and non-communist radical movements that attack established governments.

1231. Thompson, Denis William. "The Decisional Milieu of the Dominican Crisis of 1965: A Comparative Decision-Making Framework." Ph.D. diss., University of Kansas, 1970.

Focuses on decisions by the United States government, the Organization of American States, and the United Nations. The decisions to land the marines and later to expand the landings are of major significance. The sequence of events in the landings and the expansion of the intervention forces are recounted in considerable detail.

1232. Tillema, Herbert K. *Appeal to Force: American Military Intervention in the Era of Containment*. New York: Thomas Y. Crowell Company, 1973.

Provides an able, if brief, outline of events leading to the landings, pages 60–68. Another section, pages 87–91, asserts that the motives of the United States were complex. The safety of United States citizens and fear of anarchy played a role in United States thinking along with other factors.

1233. West, Richard. *The Gringo in Latin America*. Chap. 3, "The Dominican Republic and Haiti: LBJ's Big Stick—Another Part of the Island." London: Jonathan Cape, 1967.

Attributes United States opposition to Bosch to his desire to reduce the Dominican armed forces and to his reluctance to impose sanctions on Cuba. West believes that both sides in the civil war initially welcomed United States intervention.

1234. Woods, Randall B., with commentary by Petra Goede and Andrew Rotter. "Conflicted Hegemon: LBJ and the Dominican Republic." *Diplomatic History* 32, no. 5 (November 2008): 749–66.

Asserts that the Dominican Republic was President Lyndon B. Johnson's top priority for several months in 1965. In part, Woods attributes Johnson's reaction to the crisis to Johnson's life experiences. Johnson's motivations seem to have been highly mixed. He was deeply concerned about having "another Cuba" during his term of office and about how an indecisive response to an alleged Communist move in the Caribbean might reflect on his Vietnam policy. At the same time, the president was not unaware of how still another United States intervention in the Caribbean would be received in Latin America.

B. General military studies

1235. Musicant, Ivan. *The Banana Wars.* Chap. 8, "Dominican Republic II 1965," pp. 362–69. New York: Macmillan Company, 1990.

Brief review based largely on published primary sources. Relatively little about the political background or Dominican history from the first intervention through the early 1960s is included. Nevertheless, this is a good introduction to the military facets of the episode.

1236. Palmer, Bruce, Jr. *Intervention in the Caribbean: The Dominican Crisis of 1965.* Lexington: University Press of Kentucky, 1989.

Palmer commanded the intervention forces in the Dominican Republic, a fact that makes this book a major primary source for the episode. Chapter 8 is a long and interesting "Assessment" of the intervention. Chapter 9, "Caribbean Realities for the U.S.," includes a comparative review of the Dominican and Grenada operations. Palmer believes that "the two island interventions have much in common" (p. 172).

1237. Schoonmaker, Herbert G. *Military Crisis Management: U.S. Intervention in the Dominican Republic, 1965.* New York: Greenwood Press, 1990.

Concise but thorough examination of the intervention. Schoonmaker discusses the political aspects of the Dominican situation and analyzes cooperation between the United States armed forces and civilian elements of the federal government, but his emphasis is on the military facets of the episode. Schoonmaker believes the operation was an important example of the use of military power to achieve specific political objectives, and asserts that the intervention has continuing lessons for the future.

1238. Yates, Lawrence A. *Power Pack: U.S. Intervention in the Dominican Republic, 1965–1966*. Fort Leavenworth, KS: Combat Studies Institute, U.S. Army Command and General Staff College, 1988. http://purl.access.gpo.gov/GPO/LPS58624

Probably the most detailed account from an official United States perspective. The study is thoroughly based on primary sources, including interviews with participants and participants' oral histories. Maps, an extensive chronology, and other features make this an indispensable examination of the Dominican Republic episode.

C. Naval operations

1239. "Answering a Call in a Crisis." *All Hands*, no. 582 (July 1965): 2–5.

Discusses naval contributions to the Dominican Republic intervention. Much of the article revolves around the carrier *Boxer*. Little is said about marine encounters with the rebels. Rather, the emphasis is on the care given the evacuees. The article includes many photographs of the evacuees and naval activities.

1240. Dare, James A. "Dominican Diary." *United States Naval Institute Proceedings* 91, no. 12 (December 1965): 36–45.

Reviews the experiences of a naval task force that had just completed exercises off Puerto Rico when orders to proceed to the Dominican Republic arrived. This is a very useful study of the initial landings and the evacuation of refugees.

1241. Young, Gene. "Boxer's First Round." *Leatherneck* 49, no. 7 (July 1965): 26–31.

Portrays in diary form, April 25–May 3, 1965, helicopter support for landings in the Dominican Republic from the carrier *Boxer*. Young examines at some length the evacuation of United States nationals. He emphasizes the intense pressure on the air crews.

D. Marine operations

1242. Berger, Paul A. "The Corridor." *Leatherneck* 49, no. 8 (August 1965): 24–29.

Discusses the corridor stretching through Santo Domingo City between the marines and the 82nd Airborne Division. There was sporadic sniping. Many photographs depict the routines of the United States servicemen and the Dominican civilians.

1243. ———. "Peace Force: On Line." *Leatherneck* 49, no. 8 (August 1965): 18–23, 70.

Describes sniper attacks on marines during the early part of the occupation and the formation of the international force under the Organization of American States. The snipers are viewed as poorly disciplined, disorganized, and inexperienced in using weapons. Berger points out the excellence of the M-79 grenade launcher as an antisniper weapon.

1244. ——. "Rebels Fire First." *Leatherneck* 49, no. 7 (July 1965): 32–35, 87.

Contrasts the pleasant features of Santo Domingo City in ordinary circumstances with the atmosphere that prevailed during the fighting between United States forces and the rebels. Noting the tense situation, Berger discusses the restrictions on returning fire that were imposed on the United States service personnel.

1245. Tompkins, R. McC. "Ubique." *Marine Corps Gazette* 49, no. 9 (September 1965): 32–39.

Furnishes a wealth of information about the initial landings and the evacuation of United States nationals. Tompkins carries the story of the occupation to June 6, 1965, when the last marines left. This is a thorough and highly readable account. The narrative is supplemented by a detailed two-page map of Santo Domingo City.

1246. Tucker, Phil, and Paul A. Berger. "Embassy Beachhead." *Leatherneck* 49, no. 8 (August 1965): 68–69.

Describes the fighting in the vicinity of the United States embassy in Santo Domingo City. Shooting was heavy at times and almost constant. One building was cleared six times by the United States marines. In the two days of firing before the marines were relieved, they advanced only 150 to 200 yards beyond the embassy wall.

E. Army operations

1247. Borch, Frederic L. *Judge Advocates in Combat: Army Lawyers in Military Operations from Vietnam to Haiti*. Chap. 8, "Judge Advocates in Operations Other Than War, 1965–94 Operation Power Pack, 1965–66" pp. 268–72. Washington, DC: Government Printing Office, 2001.

Much of the judge advocates' work was with courts martial and damage claims. The extremely restrictive rules of engagement caused problems and influenced drafting of better rules in the future interventions.

1248. Chew, Peter T. "On a Lonely Point." *Army* 15 (June, 1965): 92–94.

Relates the experiences of a rifle platoon in the Dominican Republic early in May 1965. Much of the article deals with contacts with Dominican snipers. One of these episodes ended quickly when the

Americans were allowed to respond to enemy fire with a 106-mm recoilless rifle.

1249. Clingham, James H. "'All American' Teamwork." *Army Information Digest* 22 (January 1967): 19–23.

Stresses the civic action work of the 82nd Airborne Division in the Dominican Republic. Soldiers established medical and food services, maintained the water supply, and improved the drainage system of Santo Domingo City. One paragraph discusses a fire fight with rebels on June 15, 1965.

1250. Greenberg, Lawrence M. *United States Army Unilateral and Coalition Operations in the 1965 Dominican Republic*. Washington, DC: Analysis Branch, U.S. Army Center of Military History, 1987.

Fine case study of a United States intervention. Greenberg skillfully meshes military and political events in his narrative. His many sources are fully documented, and the narrative is enhanced by an impressive bibliography and several specialized maps. Lawrence's analysis is balanced. He recognizes that the intervention damaged relations between the United States and segments of Latin American opinion and caused divisions among political elites in the United States.

1251. Moskos, Charles C., Jr. "Grace Under Pressure: The U.S. Soldier in the Dominican Republic." *Army* 16 (September 1966): 41–42, 44.

Gives considerable praise to members of the 82nd Airborne Division for their discipline and endurance. Despite provocation from extremists, United States service personnel maintained order and did not overreact to frequently tense situations.

1252. Moulis, Wallace J., and Richard M. Brown. "Key to a Crisis." *Military Review* 46, no. 2 (February 1966): 9–14.

Discusses psychological warfare during the Dominican operation. The Dominican media were for the most part in rebel-held areas, which complicated the United States operation. Many expedients were used to get the message over, and, apparently, they were successful.

1253. Palmer, Bruce, Jr. "The Army in the Dominican Republic." *Army* 15 (November 1965): 43–44, 136, 138.

Reviews the movement of United States forces to the Dominican Republic. General Palmer, who commanded the United States troops employed, emphasizes the demands put on the soldiers by the need to avoid incidents and to minimize the use of force. His article includes many statistics on the personnel and supplies used in the operation.

1254. ———. "XVIII Airborne Corps—All the Way." *Army Information Digest* 22 (January 1967): 12–18.

Mixes early history of the corps with material on its contribution in the Dominican Republic. Most of the accompanying photographs refer to Dominican service. The article includes a letter from the commander of the international forces that commends the Americans.

1255. "Swift as Eagles." *Army Information Digest* 20 (July 1965): 5–11.

Chronicles the rapid air deployment of paratroopers to the Dominican Republic and Vietnam in May 1965. Most of the text relates to the Dominican operation. Early developments are summarized, and notice is taken of civic action projects that developed.

1256. Turner, Frederick C., Jr. "Experiment in Inter-American Peace-Keeping." *Army* 17 (June 1967): 34–39.

Explores the integration of United States forces into the international force that ultimately came to occupy the Dominican Republic. The training that so many Latin American officers had received in the United States greatly facilitated the cooperation that emerged. The United States had responsibility for supporting the entire force.

F. Air force operations

1257. "Reservists in Crises." *Air Reservist* 17 (July 1965): 4–7.

Discusses Air Force Reserve involvement with operations in the Dominican Republic, primarily by providing additional airlift capacity, but also in other ways. The article lists reserve units that participated and gives a good deal of statistical data on their operations. On some days, as many as 100 reserve aircraft were flying missions related to the Dominican Republic.

XXII

From the Nixon Doctrine through the end of the Cold War, 1970–90

For a time, in the aftermath of the United States withdrawal from Vietnam, there was a reduced emphasis on studying small wars and counterinsurgency. The Nixon Doctrine, which dated from 1969, indicated that the United States would assist friendly countries with military and other aid but would not commit large numbers of troops.

During the 1970s the phrase "low-intensity conflict" began to supersede the word "counterinsurgency" in discussions of combat below the thresholds experienced in the world wars and Korea. Many writers embraced the concept of "low-intensity conflict" enthusiastically, but other analysts were critical of the phrase and sometimes were outright sarcastic in their remarks about it. Some argued that "low-intensity conflict" was and is simply an attempt to replace the word "counterinsurgency" with its memories of the effort in Vietnam. Others asserted that the "low-intensity conflict" includes challenges, such as anti-terrorism operations, that cannot readily be subsumed under "counterinsurgency."

A. Doctrines

1258. U.S. Department of the Army. Headquarters. *Low Intensity Conflict*. FM 100–120. Washington, DC: Department of the Army, 1981.

> Reflects the Vietnam experience very clearly. The Army Institute for Military Assistance prepared this exposition of current counterinsurgency doctrine. Chapter 7 deals specifically with "counterguerrilla operations." Extensive appendices include a bibliography of relevant military publications, primarily field manuals, and a glossary.

1259. United States Army Training and Doctrine Command. Joint Low-Intensity Conflict Project. *Joint Low-Intensity Conflict Project. Final Report. Volume I: Analytical Review of Low-Intensity Conflict.* Fort Monroe, VA: United States Army Training and Doctrine Command, 1986.

Although issued by the army, this important publication was the result of extensive inter-service cooperation. It was a major study that discusses virtually every aspect of low-intensity conflict, including intelligence, civil affairs, and civic action. The second volume of the report includes supporting material. It was classified.

B. Small wars: general studies

1260. Carney, John T., Jr. and Benjamin F. Schemmer. *No Room for Error: The Covert Operations of America's Special Tactics Units from Iran to Afghanistan.* New York: Ballantine Books, 2002.

Firsthand discussion of the evolution of special operations forces after the Vietnam War. The volume includes extensive notes, a bibliography, and a list of acronyms with explanations. Most chapters of this significant book are separately cited and annotated in various segments of this bibliography.

1261. Cole, Ronald H. "Grenada, Panama, and Haiti: Joint Operational Reform." *JFQ*, no. 20 (Autumn/Winter 1998/1999): 57–64.

Discusses the improvements continually being made between operations, especially between Grenada and Panama. Nevertheless, further changes need to be made to make such efforts fully joint enterprises.

1262. Granstaff, Bill. *Losing Our Democratic Spirit: Congressional Deliberation and the Dictatorship of Propaganda.* Westport, CT: Praeger, 1999.

Provides detailed analyses of the congressional debates about the intervention in Lebanon in the 1980s, the Gulf War, and the humanitarian assistance effort in Somalia. Although Granstaff's concern is with patterns of political communications rather than the three conflicts as such, his book is a valuable study of congressional actions and reactions in respect to them.

1263. Hall, David Locke. *The Reagan Wars: A Constitutional Perspective on War Powers and the Presidency.* Boulder, CO: Westview Press, 1991.

Measured but ultimately vigorous defense of the Reagan administration's major military moves: the involvement in Lebanon, the invasion of Grenada, the attack on Libya, and naval operations in the Persian Gulf in 1987–88. This is an in-depth legal analysis of the powers of both the president and the Congress in respect to waging war. Hall believes that Congress erred by establishing what is an ineffective procedural route, the War Powers Resolution, to influence executive branch behavior. A better approach would be for Congress to debate the substantive issues associated with a military operation. The chapters concerned with Lebanon and Grenada are included in the appropriate sections of this bibliography.

1264. Huchthausen, Peter. *America's Splendid Little Wars: A Short History of U.S. Military Engagements: 1975–2000.* New York: Viking, 2003.

Well-written introduction to the subject. The texts of the chapters are supplemented by highly useful, uncluttered maps, and the bibliography is rather extensive. Most chapters have only a few notes, but, clearly, Huchthausen has made good use of the many items he cites in the bibliography. Most chapters of this book are listed separately and annotated in the appropriate sections of this bibliography.

1265. Ifedi, John-Patrick Afamefuna. "United States Hegemony and Foreign Policy in the Third World: Continuity and Change in the Use of Force as a Strategy of Containment, 1980–90." Ph.D. diss., Howard University, 1996.

Focuses largely on United States assistance to anti-Communist insurgencies in Angola, Nicaragua, and Afghanistan, but Ifredi also examines the United States interventions in Grenada and Panama to explain why a "proinsurgency strategy" was not employed in those countries.

1266. Jones, Robert Keith. "Limited Warfare as a Pragmatic Concern: The Bounds of Domestic Consensus, the Controlled Use of Force, and United States Security." Ph.D. diss., Louisiana State University and Agricultural and Mechanical College, 1995.

Emphasizes the challenge of limited war for United States defense. Jones explores the theory of limited war and closely examines Secretary of Defense Caspar Weinberger's approach to the waging of limited warfare. He applies Weinberger's criteria for a United States commitment to a given conflict to the United States experience with limited war. There are extensive chapters about the Korean, Vietnam, and Gulf wars and the interventions in Grenada and Panama.

1267. Lehman, John. *Making War: The 200-Year-Old Battle Between the President and Congress Over How America Goes to War.* New York: Scribners, 1992.

Despite the title, Lehman's book emphasizes the Carter, Reagan, and H.W. Bush administrations. The author, who was Reagan's navy secretary, is highly critical of the role of Congress and is a vigorous advocate of presidential primacy in foreign affairs.

1268. Manwaring, Max G., ed. *Uncomfortable Wars: Toward a New Paradigm of Low Intensity Conflict.* Boulder, CO: Westview Press, 1991.

Interesting group of essays written largely by active duty or retired career army officers and several highly experienced government officials. The backgrounds of the authors and their essays in this volume reflect a considerable concern with Latin America, especially Central America just at the time the Cold War was ending. The introduction by Edwin G. Corr recognizes the reduction in East–West tensions, but he asserts that the importance of this collection "may grow" as a result of that development (p. 4). Although the authors are clearly in favor of United States intervention if such a move seems appropriate, they recognize the significance that "legitimacy" plays in low-intensity conflict and realize that host nations may well need to make substantial changes to win legitimacy.

1269. Morales, Waltraud Q. "US Intervention and the New World Order: Lessons From the Cold War and Post–Cold War Cases." *Third World Quarterly* 15, no. 1 (March 1994): 77–101.

Discusses the justifications offered by the Reagan and H.W. Bush administrations for their interventions. Morales notes that the War Powers Resolution has had little impact on decision-making. Nevertheless, the public in the United States has accepted the actions of the government in Grenada, Panama, and the Persian Gulf. "The obvious lesson of all three interventions was the persuasive power of quick victory" (p. 86). Morales does not deny that some interventions may be justified and may be beneficial, but maintains that making appropriate decisions is extremely difficult. He also points out that in the operations noted and in other instances, one objective was "to disarm and control militarized groups that the United States had overarmed for nearly a decade" (p. 95).

1270. Szafranski, Richard. "Thinking About Small Wars." *Parameters* 20, no. 3 (September 1990): 39–49.

Warns that the military capabilities of many small nations are increasing. They have more and better weapons and their military leaders are likely to be highly competent. Intelligence about such countries may be limited, moreover. Szafranski recommends considerable care in deciding to intervene in another country and reminds decision-makers that intervention "begets violence, and violence is never subject to absolute control" (p. 44). He makes a number of pertinent suggestions if the decision is made to intervene.

C. Small wars and the media

1271. Bolling, Landrum R., ed. *Reporters Under Fire: Media Coverage of Conflicts in Lebanon and Central America.* Boulder, CO: Westview Press, 1985.

Valuable collection of essays that developed from a seminar held at the Institute for the Study of Diplomacy of Georgetown University's School of Foreign Service. Much of the content deals with reporting the fighting in Central America, but there is material about the civil war in Lebanon, as well. This is a brief volume, but it covers important issues related to bias and possible manipulation of the media by foreign enemies.

1272. Fox, Terrance Maurice. "The Media and the Military: An Explanatory Theory of the Evolution of the Guidelines for Coverage of Conflict." Ph.D. diss., Florida State University, 1995.

Asserts that efforts to set guidelines for media coverage of military events have generally been examples of "cultural lag," that is, techniques and, especially, technology for gathering news have advanced more rapidly than the ability of government to develop methods for exerting control. The coverage of this dissertation is broad, from the American Revolutionary War through the Gulf War.

1273. Otsuki, Hiromi. "The Role of the Media in American Foreign Policy." M.A. thesis, University of Western Ontario, 1995.

Argues that strategies for dealing with the media have become significant elements in pursuing and achieving United States foreign policy goals. Otsuki studies the media with respect to Grenada and the Gulf War, noting the effective cooperation of the media with the United States government.

1274. Rid, Thomas. *War and Media Operations: The U.S. Military and the Press from Vietnam to Iraq.* London: Routledge, 2007.

Analyzes the ways in which the United States armed forces have learned to deal with the media and with the criticism that often emanates from them. Rid deals with all the interventions from the 1960s through Iraq, including such episodes as Grenada and Somalia, and seeks to determine how and to what degree the military services' experiences with media handling have influenced their policies and behavior in Iraq.

1275. Sonoski, Patrick J. "Facilitating the Diversionary Theory of War: The President and the Media in United States Foreign Policy." M.A. thesis, University of Western Ontario, 1997.

Suggests that in order to use a foreign policy crisis to divert public attention from problems within the United States, a president must handle the media skillfully to secure its support for the current military effort. The case studies examined are Grenada and the Gulf War. Sonoski found that media support for administration policy in these conflicts did assist presidents Reagan and H.W. Bush to raise their standing with the public in the United States considerably.

1276. Westphal, D'Val J. "From the Mai Lai Massacre to the Slaughter of Sarajevo: A Deconstruction of Media Coverage of Contemporary United States Military Events." Ph.D. diss., University of New Mexico, 1995.

Severely critical account of the media response to the military undertakings of the United States in Vietnam, Grenada, Panama, the Persian Gulf, Somalia, and Bosnia. The author argues that the ability of the media to report news independently of government direction and control has steadily diminished since the Vietnam War. He explores a number of factors to explain the changes that have occurred.

D. Counterinsurgency and low-intensity conflict theory after Vietnam

1277. Armstrong, Charles L. "Decline of Low-Intensity Conflict and the Rebirth of Small Wars." *Marine Corps Gazette* 73, no. 4 (April 1989): 37–38.

Rejects the phrase "low-intensity conflict" and advocates use of the term "small wars," pointing out that the *Small Wars Manual, 1940* continues to have value.

1278. ———. "Making the Insurgent Quit." *Marine Corps Gazette* 74, no. 12 (December 1990): 50–55.

Interesting discussion of the language used in low-intensity conflict situations as part of psychological warfare efforts.

1279. Bahnsen, Peter. "Protracted Warfare and the Role of Technology: The United States," pp. 201–10 in *Guerrilla Warfare and Counterinsurgency: U.S.–Soviet Policy in the Third World*, ed. Richard H. Shultz, Jr., Uri Ra'anan, Robert L. Pfaltzgraff, Jr., William J. Olson, and Igor Lukes. Lexington, MA: Lexington Books, 1989.

Careful explication of the problems of developing and applying technology to low-intensity conflict. Bahnsen warns that "[h]elicopters, artillery, and aircraft often provide the illusion of activity that looks like military effectiveness, but without the substance" (p. 204). This brief essay contains many insights that cannot be summarized in an annotation.

1280. Barnes, Rudolph C., Jr. "Diplomat Warrior." *Military Review* 70, no. 5 (May 1990): 55–63.

Discusses the appropriate mix of diplomatic and military operations. "Where the threat is ambiguous and indirect ... diplomacy must play the dominant role, and military force must be subordinated to political objectives" (p. 55). Changes in the international political scene, the nature of current security challenges, and the increasing salience of low-intensity conflict necessitate the cultivation of diplomatic skills by United States military leaders.

1281. ———. "Politics of Low Intensity Conflict." *Military Review* 68, no 2 (February 1988): 2–10.

Argues that the complexity of low-intensity conflict lies in "the lack of cohesiveness or clear lines of demarcation between opposing sides" (p. 3). Barnes writes, too, that the United States usually becomes involved in a low-intensity conflict only after a strong insurgency has developed. He applauds the creation of the U.S. Special Operations Command (USSOCOM) but warns that the structure alone is not enough to prepare for low-intensity conflicts.

1282. Beverage, Harold G. "Armor Support in Low-to Mid-Intensity Conflict." *Armor* 99, no. 5 (September–October 1990): 15–16.

Criticizes United States capability for employing armor in low-intensity conflicts, especially those in which the enemy may have some significant armor. Beverage praises the M551 Sheridan for its effectiveness, but he points out that the army has only a few Sheridans.

1283. Bond, Peter A. "In Search of LIC (Low-Intensity Conflict)." *Military Review* 66, no. 8 (August 1986): 79–88.

Emphasizes the breadth of the low-intensity conflict concept and explains the four major groups of duties, such as "peacekeeping operations," that are bundled into it. These groups of duties often overlap, however.

1284. Bruton, James K. and Wayne D. Zajac. "Cultural Interaction: The Forgotten Dimension of Low-Intensity Conflict." *Special Warfare* 1 (April 1988): 29–33.

Discusses deficiencies in preparing for what the authors term "the socio-political arena of low-intensity conflict" (p. 29), and offers a plan for providing orientation and indoctrination for troops deploying abroad. The authors believe that the army has the people with suitable knowledge and experience to be good trainers, but they must be utilized.

1285. Casper, Lawrence E. "Attack Aviation Battalion in Low-Intensity Combat." *Army* 40, no. 8 (August 1990): 59–60.

Looks at the helicopter battalions that support light infantry divisions and briefly reviews the field manuals that present current low-intensity conflict doctrine and helicopter tactics. Casper argues that new doctrine is needed to guide company-level support in low-intensity conflict.

1286. Codevilla, Angelo M. "Can the United States Mount Political and Psychological Operations in Protracted Warfare?" pp. 309–19, in

Guerrilla Warfare and Counterinsurgency: U.S.–Soviet Policy in the Third World, ed. Richard H. Shultz, Jr., Uri Ra'anan, Robert L. Pfaltzgraff, Jr., William J. Olson, and Igor Lukes. Lexington, MA: Lexington Books, 1989.

Argues that shortcomings in United States efforts to carry out "political and psychological operations in protracted warfare" are entirely the fault of decision-makers in this country who "have viewed political and psychological operations not as a means to achieve success but as half-measures intended to avoid commitment" (p. 311). Codevilla derides such leaders and points to a number of situations in which the United States could have attained successes by using bold and imaginative tactics.

1287. Costello, Thomas J. "Counterfire Concept for Light Divisions." *Field Artillery* no. 2 (April 1989): 25–31.

Light divisions intended for rapid deployment in emergencies face problems owing to their limited artillery and communications capabilities. Costello discusses augmentations that are intended to increase their ability to operate in low-intensity conflict environments. These augmentations include the use of radar. Costello is especially concerned about the possibility that low-intensity conflict can easily give way to higher-intensity warfare.

1288. Currey, Cecil B. "Edward G. Lansdale: LIC (Low-Intensity Conflict) and the Ugly American." *Military Review* 68, no. 5 (May 1988): 44–57.

Reviews the United States Air Force counterinsurgency leader's career from his days in the advertising field to his highly influential role in developing responses to insurgencies as a result of his careful studies and observations. Currey asserts that Lansdale's teachings can be of considerable value to the United States in establishing strategies to deal with insurgencies in Central America.

1289. Darling, Roger. "A New Conceptual Scheme for Analyzing Counterinsurgency." Parts 1, 2. *Military Review* 54, nos. 2, 6 (February, June 1974): 27–38; 54–66.

Details the process by which social conflict becomes insurgency, using examples from Communist China and pre-Castro Cuba. Various types of movements need to be distinguished. Darling suggests that counterinsurgency should be studied further and not neglected.

1290. Dean, David J. *The Air Force Role in Low-Intensity Conflict*. Maxwell Air Force Base, AL: Air University Press, 1986.

Despite the title, much of this useful book is devoted to a recent counterinsurgency effort by Morocco, from which Dean seeks to draw lessons. There is also a chapter about British use of aviation in Middle Eastern campaigns against guerrillas. Another chapter deals with the Special Air Warfare Center of the United States Air Force, which underwent a name change and which was eliminated in the 1970s. The author concludes with a proposal for future air force contributions to low-intensity conflicts in which the United States is involved.

1291. ——. ed. *Low-Intensity Conflict and Modern Technology.* Maxwell Air Force Base, AL: Air University Press, 1986.

Highly impressive collection of essays by leading specialists in low-intensity conflict. The topics covered include aircraft for low-intensity conflicts and the importance of intelligence and medical services in low-intensity conflicts. The marines and the navy, as well as the air force and the army, are represented by authors. There are several civilian contributors, as well. A number of the essays are listed separately in this bibliography and annotated.

1292. Dreyer, June. "Subnational Groups, Unconventional Warfare, and Modern Security Planning," pp. 123–44 in *The Future of Conflict: Six Papers and Summaries of Discussion Presented during a Series of Seminars, conducted by the National Security Affairs Institute 16 November 1978–4 April 1979*, ed. John J. McIntyre. Washington, DC: National Defense University Press, 1979.

Analyzes the roots of guerrilla and terrorist movements, suggesting some remedies for reducing the risks of violence. A number of proposals are made for improving United States preparations for incidents and low-intensity warfare. Dreyer warns that "policy decisions on attitudes toward subnational conflict must be made with consideration for legitimate aspirations for social change" (page 136).

1293. Evans, Ernest. *Wars Without Splendor: The U.S. Military and Low-Level Conflict.* New York: Greenwood Press, 1987.

Brief, but important effort to explore the nature of what is more generally termed "low-intensity conflict" and the views taken by the United States military of such conflicts, and to propose a method for organizing the United States armed forces for service in such environments. Evans carefully documents the aversion of the armed services to low-intensity conflict and to developing the kinds of components that might be helpful in coping with it.

1294. Foster, Gregory D. and Karen A. McPherson. "Mobilization for Low Intensity Conflict." *Naval War College Review* 38 (May–June 1985): 49–64.

Reviews World War II experience briefly but emphasizes that the environment and mobilization requirements are quite different today.

The authors then examine steps in mobilization and make a sophisticated analysis of the advantages and disadvantages of pursuing a flexible approach to mobilization when attempting to meet different types of threats.

1295. Fulton, John S. "Debate About Low-Intensity Conflict." *Military Review* 66, no. 2 (February 1986): 60–67.

Argues that "low-intensity conflict's definition is too broad and the category is too large" (p. 61). Fulton further asserts that each conflict is unique, which militates against the formulation of doctrine. He explores the nature of conflict in Third World countries and expresses the view that a single force is unlikely to be able to maintain capabilities for waging both high- and low-intensity conflict.

1296. Gates, John M. "Dialogue: On Words or Reality? The Humpty Dumpty Approach to Doctrine Development." *Military Review* 68, no. 5 (May 1988): 58–63.

Scores what Gates views as the ambiguity of the phrase "low-intensity conflict" and points out that a number of "commonly cited examples of low-intensity conflict have definitely not been at the low level of violence the term implies" (p. 60). Gates does not think the phrase "military operations short of war" is appropriate either, because it seems to minimize the dangers of such involvements. These phrases are misleading euphemisms employed because of the unhappy memory of Vietnam.

1297. Girling, John L.S. *America and the Third World: Revolution and Intervention*. London: Routledge & Kegan Paul, 1980.

Uses pluralist and Marxist perspectives in studying revolution in the Third World and United States military responses to it. Girling urges that discrimination be used in determining the appropriate response for a particular situation. Chapters 7 and 8 focus, respectively, on limited war and counterinsurgency.

1298. Goodman, Allan E. "The Future of United States Development Assistance in the Insurgency Environment," pp. 283–93 in *War, Strategy, and Maritime Power*, ed. B. Mitchell Simpson. New Brunswick, NJ: Rutgers University Press, 1977.

Sees the United States concern with security and anti-Communism as obstructing good relations with Third World nations. Goodman asserts that because many American officials believe Vietnam was a special case, they have learned little from the war.

1299. Gregson, Wallace C. "Sea-Based Indirect Warfare." *Marine Corps Gazette* 74, no. 5 (May 1990): 43–45.

Brief description of the tactics likely to be employed by guerrilla forces, which will often try to pressure counterinsurgents to take measures that will antagonize the local population.

1300. Grinter, Lawrence E. "Nation Building, Counterinsurgency, and Military Intervention." pp. 237–56 in *The Limits of Military Intervention*, ed. Ellen P. Stern. Beverly Hills, CA: Sage Publications, 1977.

Applies the lessons of Vietnam to future United States involvement in similar conflicts. Grinter emphasizes the difficulty of fighting such a war with complete television coverage and with the CIA the subject of public scrutiny.

1301. Hamby, Larry B. "A Realignment of US Army Doctrine," pp. 313–34, in *Low-Intensity Conflict and Modern Technology*, ed. David J. Dean. Maxwell Air Force Base, AL: Air University Press, 1986.

Notes the army's extensive low-intensity conflict commitments through the years and then records the lack of interest in low-intensity conflict after the Vietnam War before a revival of concern in the early 1980s. Hamby then proceeds to a fine statement and analysis of current army doctrine.

1302. Hitchcock, Norman E. "Lessons for U.S. involvement in Future Civil Wars." *Marine Corps Gazette* 74, no. 7 (July 1990): 54–56.

The "lessons" Hitchcock analyzes are drawn largely from Vietnam.

1303. Hunt, John B. "Emerging Doctrine for LIC (Light Intensity Conflict)." *Military Review* 71, no. 6 (June 1991): 51–60.

Despite the high-level nature of the Gulf War, low-intensity conflict continues to be a major concern of the United States armed forces. Hunt discusses in some detail the current primary doctrinal statement, Field Manual FM 100–120/Air Force AF Pamphlet 3–20 and other sources of doctrine and policy. The key is now "balanced development–political, economic, and social" (p. 54). Hunt asserts that in a low-intensity conflict environment the "aim is to avoid escalating to war and to solve the problems in the political sphere" (p. 57).

1304. Hunter, Horace L. "Military Operations in Support of Counterinsurgency: A Primer for U.S. Officers." *Marine Corps Gazette* 74, no. 7 (July 1990): 40–45.

Examines the challenges of fighting an insurgency in a Third World country by training and equipping government forces for counterinsurgency.

1305. Hurley, David W. "LIC (Low-Intensity Conflict)—The Cultural Component." *Marine Corps Gazette* 73, no. 12 (December 1989): 19–21.

Warns that low-intensity conflict is decidedly different from other kinds of warfare, and, therefore, planning for it must take these distinctions into account.

1306. Ivey, Robert A. "Tough Lessons Learned at the JRTC (Joint Readiness Training Center)." *Military Intelligence* 16, no. 2 (April–June 1990): 15–18.

Reports on an exercise conducted by a battalion of the 29th Infantry Division (Light) and some Virginia National Guard personnel. Ivey notes the significant communications problems that were the result of the limited experience of many radio operators and equipment shortages or deficiencies. This is a straightforward account with many important points.

1307. Just, Ward. "Futures," in *Military Men*, chap. 7. New York: Alfred A. Knopf, 1970.

Discusses the change from military intervention to improved and expanded programs for preparing United States advisers to operate in Third World nations believed to be threatened by insurgency. The chapter deals primarily with the MAOP (Military Assistance Officers Program) at Fort Bragg.

1308. Kirkpatrick, Jeane. "Protracted Conflict and U.S. Policy," pp. 5–11 in *Guerrilla Warfare and Counterinsurgency: U.S.–Soviet Policy in the Third World*, ed. Richard H. Shultz, Jr., Uri Ra'anan, Robert L. Pfaltzgraff, Jr., William J. Olson, and Igor Lukes. Lexington, MA: Lexington Books, 1989.

Predicts that the United States will necessarily be much concerned with low-intensity conflict in Third World countries because of Soviet involvement and instigation of unrest. Kirkpatrick contends that the Soviet Union is as determined as ever to build a socialist world order and portrays that country as highly skillful in its Third World operations.

1309. Klare, Michael T. *Beyond the 'Vietnam Syndrome': U.S. Interventionism in the 1980's*. Washington, DC: Institute for Policy Studies, 1981.

Examines efforts to rebuild United States capability to use military action for gaining policy goals in the Third World. These involve destroying the "Vietnam Syndrome," that is, the belief that the United States should not intervene militarily in other countries. Chapter VII covers current counterinsurgency planning.

1310. ——. *War Without End: American Planning for the Next Vietnams*. New York: Alfred A. Knopf, 1972.

Deals extensively with the development of counterinsurgency from the early 1960s through the Vietnam War. This important study

1311. Lambeth, Carl L. "C3I for Special Operations Forces: On War's Threshold." *Army* 40, no. 4 (April 1990): 49–52, 55–56.

includes a useful bibliography and a "Research Guide" to aid further study.

With the end of the Cold War, United States defense policy will be concerned more with low-intensity conflict. The authors discuss communications needs in low-intensity conflict and recent development in communications equipment.

1312. Lehman, Christopher M. "Protracted Insurgent Warfare: The Development of an Appropriate U.S. Doctrine," pp. 121–33 in *Guerrilla Warfare and Counterinsurgency: U.S.–Soviet Policy in the Third World*, ed. Richard H. Shultz, Jr., Uri Ra'anan, Robert L. Pfaltzgraff, Jr., William J. Olson, and Igor Lukes. Lexington, MA: Lexington Books, 1989.

Assesses the effectiveness of the "containment" policy that the United States has followed during the Cold War and then discusses the need to develop appropriate doctrine to meet the continuing danger of low-intensity conflict to national interests. Overall, the containment strategy has worked. The Reagan administration has made much progress in developing a stance regarding low-intensity conflict, but Lehman writes that more needs to be done. One continuing problem is that the United States looks at individual countries rather than regions, despite the fact that many problems are regional rather than strictly national.

1313. Long, William F., Jr. "Counterinsurgency: Corrupting Concept." *United States Naval Institute Proceedings* 105, no. 4 (April 1979): 56–64.

Reviews critically the development of counterinsurgency strategy. Long assails the way in which the Vietnam War was waged and asserts that Special Forces units should have been used for fomenting revolt in North Vietnam, not for training South Vietnamese troops.

1314. Longoria, Alfred, Jr. "Ethics in Low-Intensity Conflict." *Marine Corps Gazette* 73, no. 4 (April 1989): 64–65.

Discusses the challenge of working with dysfunctional or even oppressive regimes in the face of insurgencies.

1315. "Low-Intensity Conflict." *Marine Corps Gazette* 72 (March 1988): 32–34.

Analyzes the roots of those movements that instigate low-intensity conflicts. Although such conflicts are found most often in Third World countries, they can occur in the developed world, as well.

1316. McInnis, Charles W. "Dialogue: On Words or Reality: A Nonsense Phrase (Military Operations Short of War)." *Military Review* 68, no. 5 (May 1988): 58, 64–69.

Analyzes the concept of "military operations short of war." McInnis strongly disputes the need for separate low-intensity conflict doctrine. For McInnis, combat is combat whatever the level of intensity. Doctrine can be usefully developed for aspects of low-intensity conflict, such as peacekeeping, but there is no need for an "overall catch word" (p. 69).

1317. Maechling, Charles. "Insurgency and Counterinsurgency: The Role of Strategic Theory." *Parameters* 14 (Autumn 1984): 32–41.

Notes the renewed concern in the United States about low-intensity conflict threats and examines in some depth revisions in Communist insurgency theory, which involve attacks on the economic system. Maechling stresses the often indiscriminate United States support for oppressive regimes that cannot quell insurgencies in the long run. He advises, too, against unilateral interventions.

1318. Manwaring, Max G. "Toward an Understanding of Insurgent Warfare." *Military Review* 68, no. 1 (January 1988): 28–35.

Reviews the characteristics of insurgencies in detail. Manwaring emphasizes the importance of United States military service members as trainers of host country security forces. He also warns that counterinsurgency is more than "a simple war of attrition" (pp. 32–33). Countries beset by insurgencies must be analyzed as entire structures in order to develop suitable strategies to combat those insurgencies. Those concerned with waging counterinsurgency campaigns must avoid focusing on a single facet of an ongoing conflict.

1319. Marsh, John O., Jr. "Comments on Low-Intensity Conflict by Secretary of the Army John O. Marsh Jr." *Military Review* 69, no. 2 (February 1989): 2–5.

Marsh views low-intensity conflict as a major threat to the United States, observing that our nuclear weapons have not deterred opponents operating at the low-intensity level. He warns that hostile nations may be able to play on fears and doubts in the United States to achieve their aims.

1320. Metz, Steven. "Counterinsurgent Campaign Planning." *Parameters* 19, no. 3 (September 1989): 60–68.

Emphatic warning that the ability of the United States to wage counterinsurgency is not keeping up with the importance of this kind of conflict. The problem is that "we have no strategic vision for the

Third World and thus no global strategy for low-intensity conflict" (p. 62). Metz presents a description of the planning process for counterinsurgency.

1321. Molloy, Ivan. *Rolling Back Revolution: The Emergence of Low Intensity Conflict*. London: Pluto Press, 2001.

Close analysis of low-intensity conflict and its evolution as a concept and as a policy during the Reagan years. After describing the nature of low-intensity conflict, Molloy discusses its application to Central America, especially Nicaragua, and to the Philippines.

1322. Moore, John Norton. "Global Order, Low Intensity Conflict and a Strategy of Deterrence." *Naval War College Review* 39 (January–February 1986): 30–46.

Expresses great concern about Soviet subversive efforts and offers ways to counter them. Other regimes, including those not of a Communist nature, can be aggressive, but the Soviet Union remains the main threat. The focus of the article is Communist activity in Central America.

1323. Morelli, Donald R. "Low-Intensity Conflict: An Operational Perspective." *Intelligence Defense Review* 17, no. 9 (1984): 1219–20; *Military Review* 62, no. 11 (November 1984): 2–16.

Describes ways in which the United States Army can improve its low-intensity conflict capabilities. Morelli identifies the needs of the army in this field and stresses the interlocking nature of military and non-military elements in a low-intensity environment. Military power and its application cannot be the full answer to low-intensity threats.

1324. Mott, Charles P. "Realistic LIC (Low-Intensity Conflict) Strategy in Latin America." *Military Review* 69, no. 5 (May 1989): 16–23.

Roundly criticizes the results of extensive attempts to develop a strategy for meeting current military challenges to the United States in Latin America. Mott describes those attempts at some length. The resources committed to the area are extremely limited, and in the face of serious economic, social, and political problems, the United States armed forces can only accomplish so much.

1325. Olson, William J. "Low Intensity Conflict—The Institutional Challenge." *Military Review* 69, no. 2 (February 1989): 6–17.

Highly sophisticated assessment of the barriers to the effective conduct of low-intensity conflict by the United States. Olson notes the failure to learn lessons from earlier conflicts and to apply those lessons where appropriate to current situations. This analysis fully retains its value today.

1326. ———. "Organizational Requirements for LIC (Low-Intensity Conflict)." *Military Review* 68, no. 1 (January 1988): 8–16.

Examines recent developments in the United States strategy and organization for low-intensity conflicts. Olson emphasizes the problems that continue to obstruct effective United States responses to such conflicts. One of Olson's prescriptions is an education effort aimed at both the military and civilian elements of the federal government that would facilitate better understanding of the problem and more agreement about how to meet insurgencies and other types of military confrontations.

1327. ———. "U.S. Objectives and Constraints: An Overview," pp. 13–42 in *Guerrilla Warfare and Counterinsurgency: U.S.–Soviet Policy in the Third World*, ed. Richard H. Shultz, Jr., Uri Ra'anan, Robert L. Pfaltzgraff, Jr., William J. Olson, and Igor Lukes. Lexington, MA: Lexington Books, 1989.

Compares the contemporary United States with Great Britain in the late nineteenth and early twentieth centuries because both nations, Olson believes, faced "accumulating challenges and [a] declining capability" (p. 15). Olson questions the usefulness of the "low-intensity conflict" concept, but he nevertheless attempts to define it. He discusses various models of activities that can be considered as low-intensity conflict and examines the many problems the United States confronts in attempting to fashion an adequate approach to dealing with low-intensity threats.

1328. Osborn, George K., and William J. Taylor, Jr. "The Employment of Force: Political-Military Considerations," in *U.S. Policy and Low-Intensity Conflict*, ed. Sam C. Sarkesian and William L. Scully, 17–47. New Brunswick, NJ: Transaction Books, 1981.

Reviews the various regions of the world to ascertain the likelihood of low-intensity conflicts that might involve the United States. The essay also discusses the concept of "low-intensity conflict" generally.

1329. Osgood, Robert Endicott. *Limited War Revisited*. Boulder, CO: Westview Press, 1979.

Stresses the influence of the Vietnam War on limited war doctrines formulated by United States military theorists. A short section discusses counterinsurgency concepts in the 1960s, and another deals with current views of limited war.

1330. Ottaway, David, "What Are the Constraints on the U.S. Role in Third World Protracted Conflict?" pp. 47–58 in *Guerrilla Warfare and Counterinsurgency: U.S.–Soviet Policy in the Third World*, ed. Richard H. Shultz, Jr., Uri Ra'anan, Robert L. Pfaltzgraff, Jr., William J. Olson, and Igor Lukes. Lexington, MA: Lexington Books, 1989.

Severely criticizes the Reagan administration for not developing doctrines and forces to cope with challenges in the sphere of low-intensity conflict. Ottaway reviews a number of what he terms "constraints." Much of this essay examines "moral restraints." Ottaway warns that the executive branch needs to develop skills in ensuring "political management of [commitments in Third World countries] to ensure sustained public support and for far better government coordination of whatever effort any administration decides to make" (p. 57).

1331. Paschall, Rod. "Low-Intensity Conflict Doctrine: Who Needs It?" *Parameters* 15 (Autumn 1985): 33–45.

Although there is a need to develop doctrine that covers various types of low-intensity conflict, not just counterinsurgency, insurgency, nevertheless, remains a much more serious problem than terrorism. Paschall reviews patterns of insurgency in Asia, primarily in Vietnam, in order to suggest several countermeasures.

1332. Quester, George H. "The Guerrilla Problem in Retrospect." *Military Review* 56, no. 8 (August 1976): 44–55.

Compares Chinese Communist insurgency theory and United States doctrines on counterinsurgency.

1333. Rittgers, Courtney M. "The Hard Corps Soldier." *Soldiers* 33 (January, 1978): 6–11.

Describes Special Forces in the post–Vietnam period, including its current organization and varied missions.

1334. Sarkesian, Sam C. "Low-Intensity Conflict: Concepts, Principles, and Policy Guidelines," pp. 9–32 in *Low-Intensity Conflict and Modern Technology*, ed. David J. Dean. Maxwell Air Force Base, AL: Air University Press, 1986.

Discusses the problems the United States faces in coping with low-intensity conflicts in the post–Vietnam period. Sarkesian views low-intensity conflict as involving a security phase in which the threatened government and its allies take protective measures, and a second phase in which they hit back at the insurgents. The United States finds it difficult to adjust to the nature of low-intensity conflicts because they "do not conform to democratic notions of strategy or tactics" (p. 28). At the same time, some low-intensity campaigns may have to be fought.

1335. ———. "Myth of US Capability in Unconventional Warfare." *Military Review* 68, no. 9 (September 1988): 2–17.

As the title of the article suggests, Sarkesian, a leading authority on unconventional warfare, believes the United States has not devoted enough analysis to low-intensity conflict. He assails much of the research and writing about ways to deal with this type of environment and observes that low-intensity conflict demands a wide variety of non-military approaches. Sarkesian further warns that the East–West confrontation in Europe is still central to United States security and that fact needs to be kept in mind by low-intensity conflict theorists.

1336. ———. *The New Battlefield: The United States and Unconventional Conflicts.* New York: Greenwood Press, 1986.

This astute analysis by one of the United States experts on unconventional warfare retains its value, even though it was written in the last years of the Cold War, and, necessarily, much of Sarkesian's attention was given to the Soviet role in Third World conflicts that might involve the United States. Sarkesian emphasizes the obstacles the United States faces in developing appropriate postures regarding unconventional conflicts and warns that it "has yet to develop a long-range political-military policy and strategy to respond to the new battlefield" (p. 37). There is a brief and somewhat dated but still useful bibliographic essay.

1337. Sarkesian, Sam C., and William L. Scully, eds. *U.S. Policy and Low-Intensity Conflict.* New Brunswick, NJ: Transaction Books, 1981.

Presents the results of a conference at Loyola University in November, 1979. The introduction assesses the relevance of the Vietnam experience to future wars.

1338. Scales, Robert H. "Calling Down Thunderbolts in Small Wars." *Army* 39, no. 7 (July 1989): 71–74, 76–77, 79.

Analyzes the ways in which firepower of all kinds can be used by the United States armed forces in low-intensity conflict. There are special challenges in employing firepower in low-intensity conflict because of the difficulty of pinpointing the fall of projectiles. Better communications are needed between commanders needing fire and those providing it.

1339. Schlaak, Thomas M. "The Essence of Future Guerrilla Warfare: Urban Combat." *Marine Corps Gazette* 60, no. 12 (December 1976): 18–26.

Warns that guerrilla warfare can be a major threat in Third World cities and even in the United States. The nature of urban guerrilla warfare is analyzed at length. Several pages discuss the means for fighting back and are based primarily on Commonwealth experience. Discusses the nature of low-intensity conflict, reviews the outlook for such conflicts in major regions of the world, and the configuration of United States forces for meeting Soviet challenges in those areas.

1340. Shultz, Richard H. "Low-Intensity Conflict and US Policy: Regional Threats, Soviet Involvement, and the American Response," pp. 71–102 in *Low-Intensity Conflict and Modern Technology*, ed. David J. Dean. Maxwell Air Force Base, AL: Air University Press, 1986.

Discusses the nature of low-intensity conflict, reviews the outlook for such conflicts in major regions of the world, and examines the configuration of United States forces for meeting Soviet challenges in those areas.

1341. Shultz, Richard H., Jr., and Alan Ned Saborsky. "Policy and Strategy for the 1980's: Preparing for Low Intensity Conflicts," pp. 191–227 in *Lessons from an Unconventional War: Reassessing U.S. Strategies for Future Conflicts*, ed. Richard A. Hunt and Richard H. Shultz, Jr. New York: Pergamon Press, 1982.

Warns that the Soviets are continuing to foment insurgency in the Third World. The authors advise that providing more military assistance, strengthening United States counterinsurgency capability, and developing the Rapid Deployment Force are all necessary.

1342. Sloan, Stephen. "US Strategy for LIC (Low-Intensity Conflict): An Enduring Legacy or Passing Fad?" *Military Review* 70, no. 1 (January 1990): 42–49.

Describes the Reagan administration's increasing interest in expanding United States capabilities for low-intensity conflict but asks whether this commitment will persist in the H.W. Bush era. Sloan warns that low-intensity conflict is generally a slow affair, actions such as the Grenada invasion notwithstanding, and criticizes Congress's apparent unwillingness to acknowledge this fact. There are many obstacles to developing a suitable posture for waging successful low-intensity conflict.

1343. Stilwell, Richard G. "U.S. Counterinsurgency: Political and Psychological Dimensions," pp. 299–308, in *Guerrilla Warfare and Counterinsurgency: U.S.–Soviet Policy in the Third World*, ed. Richard H. Shultz, Jr., Uri Ra'anan, Robert L. Pfaltzgraff, Jr., William J. Olson, and Igor Lukes. Lexington, MA: Lexington Books, 1989.

Argues that psychological warfare is a crucial element in low-intensity conflict. Stilwell describes Ramon Magsaysay's successful offensive against Communist insurgents in the Philippines in the 1950s that used numerous political, economic, and social tactics. The United States lacks a commitment to psychological warfare, whether in support of its own operations or in response to enemy moves.

1344. Summers, Harry G. "On Joint Doctrine for Low-Intensity Conflict," pp. 365–81 in *Low-Intensity Conflict and Modern Technology*, ed. David J. Dean. Maxwell Air Force Base, AL: Air University Press, 1986.

Broad discussion of United States defense policies and United States national interests. This analysis makes a number of important points, such as the need to establish a unified command in a particular country or region. The seemingly straightforward questions that Summers asks identify issues that have often been resolved badly and that are likely to be difficult to resolve in the future.

1345. ——. "Principles of War and Low-Intensity Conflict." *Military Review* 65, no. 3 (March 1985): 43–49.

Argues that the army overreacted to the advent of nuclear weapons in World War II and ignored older principles that remain applicable to low-intensity conflict. One major problem has been the involvement in military operations abroad without proper concern with the principle of "unity of command." Summers observes that the Korean War and the Vietnam War were quite different in that respect. In Korea, all forces were firmly under one leader.

1346. Swain, Richard. "Removing Square Pegs from Round Holes—Low-Intensity Conflict in Army Doctrine." *Military Review* 67, no. 12 (December 1987): 2–15.

Sophisticated treatment of existing United States Army doctrine for low-intensity conflict. Swain notes the expanding definition of low-intensity conflict during the 1980s. He believes separate doctrinal works on warfare generally and low-intensity conflict are needed and discusses the nature of the field manuals he deems appropriate.

1347. Szymanski, Michael W. "Hoist with the LIC (Low-Intensity Conflict) Petard." *Military Review* 68, no. 9 (September 1988): 18–26.

This pessimistic assessment begins with the statement that the "US military will be the likely loser in a LIC [low-intensity conflict] operation" (p. 19). The conduct of low-intensity conflict is likely to stray from the political and legal principles of the United States, thus gradually undermining support for an operation within the United States. Szymanski predicts that United States administrations will find it "difficult to sell foreign military adventure to the public in the future" (p. 26).

1348. Taylor, Richard H. "What Are These Things Called 'Operations Short of War'?" *Military Review* 68, no. 1 (January 1988): 4–7.

Briefly explores the nature of potential or actual conflict situations. Taylor writes that operations short of war "are undertaken to carry out strategic or tactical tasks to attain political purposes and to frustrate those of an adversary in an environment of routine, peaceful competition or LIC [low-intensity conflict]" (p. 5).

1349. Taylor, Richard H. and John D. McDowell. "Low-Intensity Campaigns." *Military Review* 68, no. 3 (March 1988): 2–11.

Useful reminder that several levels of conflict can exist in a single situation, as shown by Vietnam, where large-scale heavy fighting might be occurring at the same time as guerrilla warfare might be conducted elsewhere in the country. Lower levels of conflict may escalate rapidly to higher levels, moreover. The authors carefully relate existing army doctrine to the demands of low-intensity conflict.

1350. Thompson, Loren B., ed. *Low-Intensity Conflict: The Pattern of Warfare in the Modern World*. Lexington, MA: Lexington Books, 1989.

Numerous important essays by recognized specialists discuss aspects of low-intensity conflict, especially, but not entirely, as understood and employed by the United States. Several essays stress low-intensity conflict as a response to terrorism. Others deal with topics, such as foreign aid to countries that need United States support in the face of insurgencies.

1351. Tocchet, Gary J. "Doctrine and LICs (Low-Intensity Conflicts)—Can the Gap be Bridged?" *Air Defense Artillery* (January–February 1990): 14–25.

Reviews the application of air defense in a variety of conflicts following World War II. Tocchet warns that the nature of conflict is constantly subject to change, thus necessitating constant adaptations. Third World countries have acquired substantial numbers of new aircraft and other weapons in the 1970s and 1980s, making operations in those lands more challenging. Thus, the United States needs to develop new air defense doctrine.

1352. Trager, Frank N., and William L. Scully. "Low-Intensity Conflict: The U.S. Response," pp. 175–98 in *U.S. Policy and Low-Intensity Conflict*, ed. Sam C. Sarkesian and William L. Scully. New Brunswick, NJ: Transaction Books, 1981.

Considers the concept of "low-intensity conflict" at length. The essay discusses the Rapid Deployment Force and concludes that the United States is still not prepared to cope with Soviet assertiveness.

1353. Ware, Lewis B., Stephen Blank, Lawrence E. Grinter, Jerome W. Klingaman, Thomas P. Ofcansky, and Bynum E. Weathers. *Low-Intensity Conflict in the Third World*. Maxwell Air Force Base, AL: Air University Press, 1988.

Impressive collection of essays written by regional specialists that examine all areas of the Third World. Many of the conflicts analyzed did not directly involve the United States. Therefore, for the purposes of this bibliography, the most important segment of the book is Jerome W. Klingaman's "US Policy and Strategic Planning for Low-Intensity Conflict," pp. 161–78, which is a careful examination of the "low-intensity conflict" concept. Klingaman notes that the Soviets have a number of advantages in supporting revolutionary movements that work against the interests of the United States.

1354. Weinrod, W. Bruce. "Counterinsurgency: Its Role in Defense Policy." *Strategic Review* 2 (Fall 1974): 36–40.

Discusses counterinsurgency policy in the Nixon era. In accordance with the Nixon Doctrine, less dependence is being put on United States forces. The emphasis has shifted to assisting friendly nations to develop their own defense capabilities.

1355. Wood, Samuel S. "Joint Fire Support in Low-Intensity Conflict." *Military Review* 71, no. 3 (March 1991): 14–18.

Deals with the problem of determining how much firepower to use in low-intensity conflict situations, primarily using Panama as an example of how artillery and air weapons can be employed effectively without inflicting unnecessary civilian casualties and damage. There were clear rules for using these types of weapons. Definite rules must always be stipulated and applied in low-intensity conflict.

1356. Yarborough, William P. "Counterinsurgency: The U.S. Role—Past, Present, and Future," pp. 103–14 in *Guerrilla Warfare and Counterinsurgency: U.S.–Soviet Policy in the Third World*, ed. Richard H. Shultz, Jr., Uri Ra'anan, Robert L. Pfaltzgraff, Jr., William J. Olson, and Igor Lukes. Lexington, MA: Lexington Books, 1989.

Describes efforts by the Kennedy administration to develop United States capability for waging counterinsurgency campaigns and then turns to efforts to learn from the Vietnam experience. Yarborough praises the formation of a special military command for unconventional warfare, the Special Operations Command, but he believes other steps are necessary. More efforts are needed to improve United States capabilities in conducting low-intensity conflict, but Yarborough is thoroughly cognizant of the danger of small operations becoming major efforts that may well fail to achieve United States objectives. Given Yarborough's career as a low-intensity conflict specialist and commander, this is an especially important essay.

1357. Yates, Lawrence A. "From Small Wars to Counterinsurgency: US Military Interventions in Latin America Since 1898." *Military Review* 69, no. 2 (February 1989): 74–86.

General survey based largely on the best secondary sources, as they existed in the late 1980s. Despite the title of the article, Yates discusses the Philippine–American War, 1899–c.1902, for several pages. He criticizes objections by the State Department to what he regards as effective anti-guerrilla measures. This is a thoughtful study that tries to apply the lessons of pre–World War II campaigns to present involvements.

1358. Zais, Mitchell M. "Dialogue on Low-Intensity Conflict. LIC (Low-Intensity Conflict): Matching Missions and Forces." *Military Review* 66, no. 8 (August 1986): 79, 89–99.

Criticizes the vagueness of the low-intensity conflict concept and points out, quite rightly, that "the level of intensity may be a matter of perspective" (p. 89). Zais astutely discusses the various types of missions that may be assigned to the armed forces and identifies the types of units that are most appropriate for the tasks that need to be accomplished. This is an exceptionally worthwhile contribution.

E. Counterinsurgency forces: general

1359. Koch, Noel C. "Special Operations Forces and Low Intensity Conflict: Crisis in U.S. Leadership," pp. 43–46 in *Guerrilla Warfare and Counterinsurgency: U.S.–Soviet Policy in the Third World*, ed. Richard H. Shultz, Jr., Uri Ra'anan, Robert L. Pfaltzgraff, Jr., William J. Olson, and Igor Lukes. Lexington, MA: Lexington Books, 1989.

Considers the problems of developing elements of the United States armed forces that can deal with low-intensity threats. Despite the widespread agreement within the United States government that low-intensity military capability is needed, the reality is that opposition to special operations components of the armed services remains vigorous. Because of the complexity of the issues involved and, even more, because of the strong feelings involved in the controversy, political leaders tend to avoid making firm decisions in support of special operations forces and insist that those decisions be implemented within the defense establishment.

1360. Perkins, Stuart L. "'Protracted Warfare': The U.S. Force Structure and C3," pp. 235–50 in *Guerrilla Warfare and Counterinsurgency: U.S.–Soviet Policy in the Third World*, ed. Richard H. Shultz, Jr., Uri Ra'anan, Robert L. Pfaltzgraff, Jr., William J. Olson, and Igor Lukes. Lexington, MA: Lexington Books, 1989.

Examines the pattern of "command, communications, and control" in the United States government with respect to low-intensity conflict. Like a number of other authors, Perkins is critical of the fact that although such conflicts "are usually regional phenomena, the Washington bureaucracy is single-country oriented" (p. 241). He recommends a number of ways to improve organizational structures and communications systems to cope with various problems, including low-intensity conflict.

1361. Van Haute, Edward B. "Combat Service Support in Low-Intensity Conflict." *Marine Corps Gazette* 75, no. 4 (April 1991): 36–39.

Discusses the use of particular types of forces and capabilities in low-intensity conflict in view of the duties that may have to be carried out.

F. Counterinsurgency forces: army and air force

1362. Darragh, Shaun M. "Rangers and Special Forces: 'Two Edges of the Same Dagger.'" *Army* 27 (December 1977): 14–19.

States that Special Forces is back to its original role—preparing to lead guerrillas in Europe, and perhaps the Middle East, against the Soviets. Ranger experiences in Vietnam, especially reconnaissance operations, are discussed briefly.

1363. Dix, Billy. "Long-Range Air Power and Standoff Weapons in Low-Intensity Conflict," pp. 139–54 in *Low-Intensity Conflict and Modern Technology*, ed. David J. Dean. Maxwell Air Force Base, AL: Air University Press, 1986.

As the title suggests, the article discusses the employment of long-range aircraft tactically in low-intensity conflicts. Because of his concern about trying to have a single force try to cope with nuclear and conventional air threats, Dix believes there should be an air element devoted to conventional assignments, including low-intensity conflict.

1364. Hines, Charles A. "Military Police in Contingency Operations: Often the Force of Choice." *Parameters* 20, no. 3 (September 1990): 11–18.

Discusses the characteristics of the army's Military Police Corps that make it extremely useful in low-intensity conflict environments, and points to the numerous times that military police have given indispensable services during the combat and humanitarian relief efforts.

1365. Innes, John O. "Low-Intensity Conflict in the 'Outback'." *Army Logistician* (May–June 1991): 38–40.

Discusses the logistical aspects of an exercise involving Hawaii-based troops who were moved to Australia temporarily for training. Both the water supply and excessive dust were major problems. The dispersion of operations intensified various supply and maintenance difficulties.

1366. Kafkalas, Peter N. "Light Divisions and Low-Intensity Conflict: Are They Losing Sight of Each Other?" *Military Review* 66, no. 1 (January 1986): 19–27.

Describes the formation of new divisions which were intended to play specialized roles within the army, but it seems as though they are moving toward the general infantry division pattern because the army continues to emphasize possible high-intensity conflict in Europe. Kafkalas argues that the light divisions should be tasked for low-intensity conflict.

1367. Shallcross, H.C. "The Other SOS." *Air University Review* 26 (May–June, 1975): 41–49.

Discusses the Special Operations School at Hurlburt Field, Florida. The author argues that, despite Vietnam, counterinsurgency operations are possible, and a cadre of personnel trained in counterinsurgency is essential. The article also examines the air force's response to Kennedy's counterinsurgency initiative in 1962.

1368. Stratton, Ray E. "US Air Force Special Operations and Technology: Time for a Reappraisal," pp. 335–64 in *Low-Intensity Conflict and Modern Technology*, ed. David J. Dean. Maxwell Air Force Base, AL: Air University Press, 1986.

Examines in some detail the deployment of air force elements in low-intensity conflict environments and the need to further develop and refine air force doctrine regarding such subjects as targeting and psychological operations.

G. Counterinsurgency forces: marines and navy

1369. Crossland, Roger L. "Rusty Hand of Steel: The Naval Raid." *United States Naval Institute Proceedings* 105, no. 12 (December 1979): 60–63.

Argues that the United States should develop commando raids from the sea as a possible tactic. Crossland notes that the *Pueblo* and *Mayaguez* episodes were ideal for surface naval rescue efforts. He also briefly covers naval participation in Operation Phoenix in Vietnam.

1370. Etzold, Thomas H. "Neither Peace Nor War: Employing Naval Forces Short of General War," pp. 269–88 in *Low-Intensity Conflict and Modern Technology*, ed. David J. Dean. Maxwell Air Force Base, AL: Air University Press, 1986.

Sophisticated treatment that begins by pointing out that the navy has little doctrine to govern its operations. Instead, it develops tactics to deal with possible specific situations. Etzold stresses the significance of the navy as a force that can project power as a means for deterring actions that clash with United States policies.

1371. Linn, Thomas C. "Marine Corps is Special Operations." *United States Naval Institute Proceedings* 114, no. 2 (February 1988): 48–51.

As the title suggests, Linn views the marines as especially well-suited for low-intensity conflict. One factor is their ability to land and withdraw quickly, with the support of the fleet. Linn goes on to list and to discuss briefly the many roles played by the marines.

1372. McGwire, Michael. "Changing Naval Operations and Military Intervention," pp. 151–78 in *The Limits of Military Intervention*, ed. Ellen P. Stern. Beverly Hills, CA: Sage Publications, 1977.

Identifies a number of factors that have contributed to a decrease in the importance of sea power. The increasing complexity of weapons systems and the greater assertiveness of many states about their territorial waters are among these factors.

1373. Melshen, Paul. "Taking on Low-Intensity Conflicts." *Marine Corps Gazette* 71, no. 1 (January 1987): 44–51.

Notes the development and continuing activity of the army's light infantry divisions and Special Forces groups but argues, nevertheless, that the marines are and should be the premier low-intensity conflict force.

1374. Palen, Don G. "Close-in Battle Plan." *United States Naval Institute Proceedings* 113, no. 1 (January 1987): 67–73.

Details the numerous ways in which warships can be protected from a wide variety of terrorist attacks.

1375. Pratt, Andrew N. "Low-Intensity Conflict and the United States Marine Corps," pp. 289–312 in *Low-Intensity Conflict and Modern Technology*, ed. David J. Dean. Maxwell Air Force Base, AL: Air University Press, 1986.

Recalls the extensive experience the marines have had with low-intensity conflict and usefully examines the kinds of missions to which the marines may be assigned.

1376. Reese, Howard C. "Low Level Warfare: What Role for the Navy?" *Marine Corps Gazette* 55, no. 8 (August 1971): 27–30.

Emphasizes the usefulness of the navy in an environment in which deployment of ground troops, especially the army, may not be possible for domestic political reasons. The army has been ambivalent about its commitment to low-intensity warfare, giving the marines and the navy a chance to fill the gap.

1377. Rylander, R. Lynn. "Future of the Marines is Small Wars." *Naval War College Review* 40 (Autumn 1987): 64–75.

Emphasizes the importance of non-military factors in fighting small wars. Rylander identifies small wars as "the most immediate threat to U.S. security" currently and in the long term. He discusses United

States Marine Corps roles in light of Secretary of Defense Caspar Weinberger's principles for United States military commitments and predicts a smaller role for the marines in small wars than in their earlier history.

1378. Wasielewski, G. "Revolutionary Warfare." *Marine Corps Gazette* 73, no. 9 (September 1989): 50, 52–54.

Argues that the marines are not ready to combat revolutionary movements in a low-intensity conflict environment. Under President John F. Kennedy the armed forces made a significant commitment to understanding and fighting revolutionaries, but in Vietnam more traditional, mid- or high-intensity methods were employed for the most part.

1379. Zurat, Michael L. "Training for Small Wars." *Marine Corps Gazette* 71, no. 1 (January 1987): 48–49.

Traces the decline of marine commitment to low-intensity conflict in the wake of the Vietnam War.

H. Evacuation operations

1380. Barrett, Raymond J. "Evacuation of American Citizens." *Marine Corps Gazette* 59, no. 1 (January 1975): 31–37.

Stresses that evacuations are likely to occur, even if other low-intensity warfare missions are not undertaken. Barrett notes the problem of Americans who do not want to be evacuated and the lack of legal authority to remove them. Of special interest is the discussion of evacuation exercises in 1973 and 1974.

1381. Parks, W. Hays. "Evacuation by Military Force." *Marine Corps Gazette* 62, no. 9 (September 1978): 24–33.

Emphasizes that the United States remains responsible for protecting its nationals. The article touches on a number of evacuations and rescue missions and recommends the preparation of a field manual to cover this kind of operation.

1382. Zinser, Larry R. "The BLT in Evacuation Operations." *Marine Corps Gazette* 57, no. 12 (December 1973): 23–30.

Presents proposed doctrines for evacuations, using the United States experiences in the Dominican Republic in 1965 and in Lebanon in 1958 as a basis. Zinser's work is based in part on interviews with marines who participated in these operations. He discusses such matters as working with United States diplomatic officials and organizing the evacuation itself.

I. The legal aspects of counterinsurgency

1383. Anderson, William T. "Legal Restraints in Low-Intensity Conflict." *Marine Corps Gazette* 73, no. 4 (April 1989): 43–44.

    Discusses possible application of another element of the Geneva Convention to United States operations.

1384. Mehl, S.P. "Civilian Combat Casualties: Our Moral and Legal Responsibility." *Marine Corps Gazette* 60, no. 10 (October 1976): 43–46.

    Notes the steadily rising civilian casualty rate in the wars of the twentieth century and provides some guidelines to soldiers for minimizing these casualties. The article relies in part on Vietnam experiences.

1385. O'Brien, William V. *The Conduct of Just and Limited War*. New York: Praeger Publishers, 1981.

    Analyzes the law of insurgency and counterinsurgency in light of the Vietnam War. The book is written from a Catholic perspective and is very sympathetic to the power waging a counterinsurgency campaign.

J. Intelligence and counterinsurgency

1386. Bevilacqua, A.C. "Intelligence and Insurgency." *Marine Corps Gazette* 60, no. 1 (January 1976): 40–46.

    Reminds readers that the United States can be involved in counterinsurgency again and stresses the need to understand guerrilla infrastructure. Intelligence should be used to eliminate guerrilla leaders and to avoid too heavy use of firepower.

1387. Davis, Robert B. "Changing Roles for MI (Military Intelligence)." *Military Intelligence* 16, no. 2 (April–June 1990): 32–35.

    Important examination of the problems and promise of military intelligence in the current environment. Davis asserts that low-intensity conflict demands a major intelligence commitment.

1388. Oseth, John M. "Intelligence and Low-Intensity Conflict." *Naval War College Review* 37 (November–December 1984): 19–36.

    Echoes President Reagan's call for a stronger focus on Third World countries threatened by subversion. Oseth usefully discusses varying attitudes in the United States toward low-intensity conflict, and argues at length for more emphasis on the use of "human intelligence" in contrast to technological methods for gathering intelligence. He makes the interesting point that "low-intensity conflict arenas must be a lively intelligence concern even when they are not a pressing policy concern" (p. 34).

1389. Purkiss, Charles T. "Intelligence Requirements for Low-Intensity Conflicts," pp. 175–89 in *Low-Intensity Conflict and Modern Technology*, ed. David J. Dean. Maxwell Air Force Base, AL: Air University Press, 1986.

Although the United States possesses significant intelligence resources and processes, its armed forces need to adapt them further to the challenges of low-intensity conflict. An especially important segment of the essays presents "objectives of US intelligence in low-intensity conflict." Purkiss outlines the resources, such as an appropriate number of linguistics experts, that are needed to achieve such objectives. He makes the important point that local military and civilian leaders need to be the ones to determine what kinds of intelligence are required to assist in making the proper decisions.

1390. Stewart, John F. "Military Operations in Low-Intensity Conflict: An Organizational Model." *Military Review* 68, no. 1 (January 1988): 17–27.

Argues that although military intelligence has developed encouragingly during United States operations in Central America, this vital element in low-intensity conflict needs more analysis. Stewart describes the various facets of military intelligence associated with low-intensity conflict and stresses the need to keep intelligence collection and analysis levels in close communication. This is an important study of perhaps the key element in low-intensity conflict operations.

K. Civil affairs/civic action

1391. Barnes, Rudolph C., Jr. "Civil Affairs—A LIC (Low-Intensity Conflict) Priority." *Military Review* 68, no. 9 (September 1988): 38–49.

Asserts that "civilian support is a principal element in LIC" (p. 39). Hence, civil affairs operations are needed to help rally citizens behind the government. Civil affairs components have even more importance in low-intensity conflict situations than in conventional combat. Barnes reviews the history of civil affairs functions in the United States Army at some length and notes that in both the United States Army and the United States Marine Corps almost all civil affairs personnel are assigned to reserve elements.

1392. Brady, Robert G. "Civil Affairs FID/UW (Foreign Internal Defense/Unconventional Warfare) Battalion and Its Implications for SOF (Special Operation Forces) in LIC (Low Intensity Conflict) Operations." *Special Warfare* 4, no. 1 (Winter 1991): 12–17.

Notes that civil affairs units have not been tasked to support special operations forces or to perform in low-intensity conflict environments. The author discusses the organization, functions, and training of a new battalion in considerable detail.

1393. Bruss, Donald J. "The Emerging Role of US Military Health Care in Low-Intensity Conflict," pp. 223–35 in *Low-Intensity Conflict and Modern Technology*, ed. David J. Dean. Maxwell Air Force Base, AL: Air University Press, 1986.

Points out that although the United States armed forces do provide assistance to foreign nations in many emergency situations, they have "little or no official responsibility for helping third world countries build viable health care systems" (p. 227). Thus, a strategy should be developed for providing military medical services on a more systematic basis and for assisting foreign military forces to move toward supplying similar services.

1394. Carns, Edwin H.J. and Michael F. Huebner. "Medical Strategy (in a Low-Intensity Conflict LIC situation)." *Military Review* 69, no. 2 (February 1989): 37–43.

Views the provision of medical services by United States military personnel as a two-way street. Such assignments provide important support for host countries and give the medical personnel useful experience. The authors focus primarily on medical aspects of low-intensity conflict in Latin America.

1395. Decker, David A. "Civil Affairs: A Rebirth or Stillborn?" *Military Review* 67, no. 11 (November 1987): 60–64.

Although the army is once again demonstrating an interest in low-intensity conflict, much needs to be done. Civil affairs specialists have been among the few members of the military who have maintained an interest in low-intensity conflict in the years after Vietnam. Decker emphasizes the need for a long-term program for the army to develop area or regional specialists for possible deployment in low-intensity situations.

1396. Flores, Susan J. "Marine Corps Employment in Low-Intensity Conflict." *Marine Corps Gazette* 73, no. 4 (April 1989): 30–34.

Civil affairs responsibilities are a vital part of an effective low-intensity campaign, but civil affairs capabilities in the marine corps are found largely in reserve, not active, units and individuals.

1397. Gaillard, Regina. "Case for Separating Civic Actions from Military Operations in LIC (Low Intensity Conflict)." *Military Review* 71, no. 6 (June 1991): 30–41.

Advocates the establishment of a specific military component to carry out civic action and various types of non-military assistance to host countries. Gaillard briefly reviews the development of civic action in the United States from the 1950s onward, including the revival of interest during the 1980s, and duly notes the criticisms sometimes leveled at the civic action concept. Both domestic and foreign mistrust of civic action programs might be reduced by clearly separating the structures for waging low-intensity conflict and for providing aid and training. In this connection, the author mentions the Civilian Conservation Corps of the 1930s approvingly.

1398. Gosnell, P. Wayne. "Low Intensity Conflict and the U.S. Military Reviewed." *National Guard* 43 (October 1989): 20–23.

Discusses the use of civic action programs by the military. Gosnell examines recent writings that urge a much greater use of members of the National Guard and the Army Reserve for executing such programs.

1399. ———. "Reserve Component Role in LIC (Low-Intensity Conflict)." *Military Review* 69, no. 2 (February 1989): 65–73.

Argues that Army Reserve Component personnel can contribute significantly to establishing an effective strategy for meeting military threats to the United States in Latin America. Support elements of the army reserve could provide construction, medical, and other forms of support to host countries. Unfortunately, low-intensity conflict commitments usually involve the deployment of combat forces first, only to be followed later by support elements. Gosnell believes that changing this sequence could substantially alleviate the problems that lead to insurgencies.

1400. Holmes, David R. "Some Tentative Thoughts After Indo–china." *Military Review* 57, no. 8 (August 1977): 84–87.

Asserts that civic action and psychological warfare were not used in Vietnam as much as they should have been. In future counter-insurgency operations, moreover, statistical analyses should not be used as the primary method for assessing the usefulness of programs.

1401. Taylor, James A. "Military Medicine's Expanding Role in Low-Intensity Conflict." *Military Review* 65, no. 4 (April 1985): 26–34.

Argues that medical services provided by the armed forces to foreign civilians can have a significant part in combating insurgencies. Taylor asserts that when giving some assistance military medical personnel "cease to function in the combat service role" (p. 30). He describes in some detail the kinds of medical services that can be offered.

XXIII

Lebanon, 1982–84

United States marines landed in August 1982, as part of a multi-national force to bolster the Gemayel government. United States military activities centered on guarding the Beirut airport. The deployment of marines in Lebanon aroused considerable concern in Congress from the beginning.

On October 23, 1983, a suicide truck driven by an Arab extremist killed a large number of marines and other United States servicemen. This episode was thoroughly investigated by the marines and by various other official agencies (as well as the media).

As fighting continued and criticism of the United States presence grew, the administration slowly moved toward a decision to move the marines to naval ships off the coast. This move was carried out in the spring of 1984, but not before navy planes and ships exchanged fire with Syrian forces.

Chronology

1982

June 6	Israeli forces entered southern Lebanon after exchanges of fire with members of the Palestine Liberation Organization (PLO) as a result of the attempted assassination of the Israeli ambassador to the United Kingdom. The Israelis wanted to push PLO forces farther away from the Lebanon–Israel border to prevent additional rocket attacks.
June 12	After Israeli forces advanced into Lebanon, reaching Beirut, a ceasefire was achieved between Israeli and PLO forces.
June 24	United States marines assisted in the evacuation of United States nationals from Juniyah, Lebanon, near Beirut.

July 1	Israeli troops besieged Beirut, where there was a large concentration of PLO troops.
August 25	United States marines joined British, French, and Italian forces in a Multi-National Force (MNF) to effect the withdrawal of PLO elements from Beirut.
September 10	The MNF was removed from Lebanon after the completion of its mission.
September 14	The president-elect of Lebanon, Bashir Gemayel, was killed.
September 16–18	Gemayel followers attacked refugee camps in revenge for his assassination, and hundreds of Lebanese and Palestinians were killed.
September 29	Marines returned to Lebanon, as part of a new MNF, joining French and Italian forces. The marines were stationed at the Beirut airport and were positioned to help shield Beirut from Israeli troops. Numerous skirmishes and other incidents occurred for months. Fighting took a steady toll of the marines.

1983

April 18	A bomb attack on the United States embassy in Beirut inflicted still more casualties, including some marines. Limited fighting, including the exchange of artillery fire, continued for months.
October 23	A car bomb attack on marine headquarters at the Beirut airport killed many marines and other United States personnel. The attack resulted in controversy about the adequacy of security measures at the installation.
November 9	Some marines transferred from Beirut to ships. Nevertheless, combat actions involving United States ground, air, and naval forces continued through the winter of 1983–84.

1984

January	Marine positions at the Beirut airport were strengthened. Scattered fighting persisted.
February 26	After a gradual evacuation not only of military personnel, but also of civilians, the marines relinquished responsibility for security duties at the Beirut airport.

A. Congressional hearings and official reports

1402. U.S. Congress. House. Committee on Armed Services. Subcommittee on Investigation. *Adequacy of U.S. Marine Corps Security in Beirut.* 98th Cong., 1st sess., 1983.

Gives a minute description of the truck attack on the marines. Another section reviews the marine mission in Lebanon. Much of the report deals with the problem of balancing the need to maintain security and to remain "highly visible."

1403. U.S. Congress. House. Committee on Foreign Affairs. *Statutory Authorization under the War Powers Resolution—Lebanon, Hearing and Markup on H.J. Res. 364 and H. Res. 315*. 98th Cong., 1st sess., 1983.

Makes a searching examination of the role of United States forces in Lebanon and of the dangers that threaten them. Military facets of the situation are thoroughly examined.

1404. U.S. Department of Defense. *Report of the DOD Commission on Beirut International Airport Terrorist Act, October 23, 1983*. Washington, DC: Government Printing Office, 1983.

Presents the results of an extensive investigation of the attack. In addition to probing all aspects of the incident, the report examines terrorism in Lebanon and makes recommendations to the armed forces for responding to terrorist assaults.

B. General studies

1405. Bell, J. Bowyer. *Dragonwars: Armed Struggle and the Conventions of Modern War*. Chap. 1, "Dragonworld: Lebanon," pp. 17–58. New Brunswick, NJ: Transaction Publishers, 1999.

Briefly contrasts the situation in Lebanon in 1983 with that in 1958 and then skillfully interweaves the Lebanon background with the day of the attack. As in other areas of conflict, Bell indicates, United States decision-makers lacked meaningful knowledge of the complexities of local politics and empathy with the local people.

1406. Frank, Benis M. *U.S. Marines in Lebanon, 1982–1984*. Washington, DC: History and Museums Division, Headquarters, U.S. Marine Corps, 1987. http://purl.access.gpo.gov/GOP/LPS98826

Authoritative account of the marines' experiences in Lebanon. Frank used oral history interviews, as well as detailed reports and other marine documents. A number of valuable appendices are included in the book, including a chronology. There are also numerous photographs and several important maps. This is a fine piece of work by the United States Marine Corps. It deals almost entirely with the military side of the episode; so those needing political background information will need to turn to other studies.

1407. Gaddo, Randy. "'The BLT [Battalion Landing Team] is Gone!' One Marine Remembers His Beirut Brothers 20 Years Later," *Leatherneck* 86, no. 10 (October 2003): 20–27.

Recollections of a survivor depict the attack in detail.

1408. Hall, David Locke. *The Reagan Wars: A Constitutional Perspective on War Powers and the Presidency.* Chap. 5, "Lebanon, 1982–84," pp. 135–67. Boulder, CO: Westview Press, 1991.

Reviews the Lebanon involvement at some length but emphasizes the question of the president's power to order the use of force or potential force under the circumstances in Lebanon. Hall stoutly defends the actions of the Reagan administration. He asserts that the "Lebanon case history serves to illustrate the point that the lawfulness of a presidential employment of military force does not depend on its popularity" (p. 147).

1409. Hallenbeck, Ralph A. *Military Force as an Instrument of U.S. Foreign Policy: Intervention in Lebanon, August 1982–February 1984.* New York: Praeger, 1991.

Major analysis by a respected defense researcher. This thoroughly documented study includes a detailed chronology, a list of acronyms and their meanings, and several fine maps. The book is based in part on Hallenbeck's massive 1987 dissertation at the Pennsylvania State University.

1410. Hammel, Eric M. *The Root: The Marines in Beirut, August 1982–February 1984.* San Diego, CA: Harcourt Brace Jovanovich, 1985.

Provides perspectives of marines who were in Beirut. Hammel's work is based on a large number of interviews. Many marines initially told Hammel that they had not anticipated a car bomb attack, but later interviews brought statements that requests to strengthen their position had been rejected.

1411. Herspring, Dale R. *The Pentagon and the Presidency: Civil–Military Relations from FDR to George W. Bush.* Chap. 10, "The Military and Reagan," pp. 278–85. Lawrence: University Press of Kansas. 2005.

Stresses the opposition of Secretary of Defense Caspar Weinberger and the Joint Chiefs of Staff to the dispatch of marines to Lebanon, owing to their highly indefinite mission and the extremely dangerous situation. Reagan's stance that the marines had to remain in Lebanon changed only after the disastrous attack of October 23.

1412. Huchthausen, Peter. *America's Splendid Little Wars: A Short History of U.S. Military Engagements: 1975–2000.* Chap. 4, "Intervention in Lebanon," pp. 45–63. New York: Viking, 2003.

Begins with a concise but thorough discussion of the recent history of Lebanon, emphasizing areas of conflict. A number of problems beset the marines, beginning with a significant lack of clarity about the nature of their mission in Lebanon. Huchthausen reviews in some detail the skirmishes involving the Multi-National Force both before and after the major attack on the marines. He asserts that United States authorities learned nothing from the Lebanon episode, thereby setting a course for a somewhat similar sequence of events in Somalia.

1413. Kemp, Geoffrey. "The American Peacekeeping Role in Lebanon," chap. 6, pp. 131–42 in *The Multinational Force in Beirut 1982–1984*, eds. Anthony McDermott and Kjell Skjelsbaek. Miami: Florida International University Press, 1991.

Careful examination of the disputes in Washington about the appropriate policy to be followed with respect to Lebanon. Reagan and many of his advisers advocated a strong stand by the United States that required almost immediate Israeli and Syrian withdrawal from Lebanon. That strategy probably would have required a much larger military commitment by the powers that had established the Multi-National Force, which was politically unacceptable, however. Kemp clearly believes that the United States lost an opportunity to bring peace to Lebanon in 1982. He briefly outlines a number of factors that led to the failure of the Lebanon policy of the United States.

1414. Kondracke, Morton. "Wading In." *New Republic* 189 (October 10, 1983): 11–13.

Faults the administration for failing to justify the change in the role of United States forces from neutral peacemakers to participants. Reagan feels that the United States and the Soviet Union are struggling in Lebanon and that the United States must not give way.

1415. "Lebanon, Bloody Lebanon." *New Republic* 189 (October 10, 1983): 7–10.

Emphasizes the importance of the Multi-National Force for the maintenance of the Gemayel regime. The *New Republic* rejects the view that the United States should walk away from what might appear to be senseless fighting and praises President Reagan and the Congress for working together to keep United States troops in Lebanon.

1416. "The Lebanon Strategy." *New Republic* 189 (December 26, 1983): 5–7.

Defends the United States objective of maintaining the existing government in Lebanon, but questions whether an appropriate strategy to achieve this goal has been worked out. Loss of United States aircraft suggests that the United States lacks the capability for the kind of air strikes it has been mounting.

1417. Mead, James M. "The Lebanon Experience." *Marine Corps Gazette* 67, no. 2 (February 1983): 30–40.

Provides a very full account of the marines during their early months in Lebanon. Colonel Mead discusses the security precautions and living conditions of the marines. Several maps and many action photographs accompany the article.

1418. "The 'Peacekeeping' Fraud." *New Republic* 189 (November 14, 1983): 7–9.

Criticizes the deployment of marines in Lebanon and the maintenance of the fiction that they are a neutral element. The marines are exposed to attack, but they are not allowed to patrol and thereby guard against an assault. The article faults the United States for failing to cooperate openly with Israel.

1419. Petit, Michael. *Peacekeepers at War: A Marine's Account of the Beirut Catastrophe*. Boston, MA: Faber and Faber, 1986.

Although to a considerable extent the book discusses the life of an enlisted marine generally, it is also a taut account of the 1983 bombing and the aftermath.

1420. Pintak, Larry. *Seeds of Hate: How America's Flawed Middle East Policy Ignited the Jihad*. London: Pluto Press, 2003.

Interprets current opposition to the United States in much of the Muslim world as a response to events in Lebanon in the 1980s. Pintak believes that instead of being a peacekeeper in Lebanon, the United States became an active enemy of Lebanese Muslims.

1421. Smyrl, Marc E. *Conflict or Codetermination: Congress, the President, and the Power to Make War*. Chap. 5, "Congress Defends Its Gains," pp. 97–117. Cambridge, MA: Ballinger Publishing Company, 1988.

Reviews the events in Lebanon that generated the establishment of a multi-national force to restore peace and United States policy with respect to Lebanon and to this force. The congressional response to the situation and executive and legislative branch bargaining about the role of the War Powers Resolution take up most of this chapter. Smyrl praises the Multi-National Force in Lebanon Resolution and asserts that the Reagan administration generally showed "an implicit understanding of the boundaries of acceptable military initiatives" (p. 115).

1422. Van Hassell, Agonstino. "Beirut 1983: Have We Learned This Lesson?" *Marine Corps Gazette* 33, no. 10 (October 2003): 33–36.

Argues that the United States armed forces need more sophisticated doctrine that would differentiate between peacekeeping operations that are unlikely to involve combat and those that may well require significant application of military power.

C. The withdrawal of the marines

1423. Church, George J. "Failure of a Flawed Policy." *Time* 123 (February 27, 1984): 16–20.

   Describes naval gunfire support, which had to be limited to retaliation and not to aid the Lebanese Army. The battleship *New Jersey*, powerful as it is, did not affect the fighting in Beirut. The Druse seem willing to allow the marines to depart peacefully, but the possibility of fighting still exists.

1424. Fromm, Joseph. "Lebanon Pullout—The Spreading Impact." *U.S. News and World Report* 96 (February 20, 1984): 28–32.

   Analyzes Reagan's dilemma of having to withdraw without totally undermining the Gemayel government. Fromm notes the controversy over naval bombardments. The coming presidential election is playing a role in decision-making, but even withdrawal may not keep Lebanon out of the campaign.

1425. Smith, William E. "The Marines Leave Lebanon." *Time* 123 (March 5, 1984): 26–28.

   Describes the evacuation as "almost furtive" in contrast to the formal withdrawal of the Italians. The marines are said to be discouraged about moving to the ships and about their inability to keep the peace in Lebanon.

1426. Whitaker, Mark. "Reagan's Surprise Retreat from Beirut." *Newsweek* 103 (February 20, 1984): 36–39, 42–43, 45, 47.

   Notes that the evacuation is to be carried out gradually, although the marines and navy have the capacity for a fast withdrawal. The State Department and other agencies long objected to Secretary of Defense Weinberger's plan to take the marines out. Many Lebanese are described as bitter over the United States departure.

1427. "Who Lost Lebanon?" *New Republic* 190 (February 27, 1984): 7–8.

   Compares Reagan's increasing air and naval action in Lebanon while removing land forces with Nixon's disengagement in Vietnam. The *New Republic* blames both Reagan and the Congress for the loss of ground to enemies of the United States. Reagan was not given enough support and finally had to give in to the Democrats in Congress.

D. Other aspects of the Lebanon intervention

1428. Hawkins, Tammy. "Under Fire in Lebanon." *Soldiers* 39 (March 1984): 8–9.

    Recounts the experiences of a small army detachment sent to Lebanon to assist the marines in locating artillery targets. Four of thirty-one men were killed in the terrorist attack of October 23, 1983.

XXIV
Grenada, 1983–85

The situation in Grenada began to concern the United States as early as 1979, when Maurice Bishop became prime minister of the newly independent nation. Grenada became more and more friendly with the Soviet Union and Cuba. A particular worry for the United States was the construction of an airport that might handle sophisticated warplanes.

An acute crisis arose when extremists overthrew and killed Bishop. Soon after, United States forces joined with troops from various Caribbean nations to invade Grenada on October 25, 1983. There were several days of rather sharp fighting with Cuban troops and some of the forces serving Grenada. United States medical students who seemed to be threatened by the unstable situation were rescued unharmed, and the United States avoided a hostage situation.

Interpretations of the intervention varied considerably, and the invasion remains a controversial topic. The United States announced a complete withdrawal of its forces in 1985.

Chronology

1974
February 7   Grenada granted independence by Great Britain.

1979
March 13   Maurice Bishop led a successful coup against Prime Minister Sir Eric Gairy and soon made friendly overtures to Cuba and Eastern bloc countries.
November 19 Grenada announced that Cuba would assist in the construction of a large airfield in the country. Several North African nations also provided financial support for the project.

1980

May Grenada gave the U.S.S.R. landing rights for reconnaissance aircraft.

1983

September 14 After a stormy meeting of Central Committee of the ruling party, the New Jewel Movement, Bishop's powers were significantly restricted.

October 14 Bernard Coard, deputy prime minister, seized power from Bishop with military support.

October 17 United States Navy task force left Norfolk, Virginia and later embarked marines at Camp Lejeune.

October 19 Bishop and other ousted members of the Bishop cabinet were reported killed.

October 21 General Hudson Austin assumed power in Grenada.

October 21 The United States announced that it had diverted a task force on route to Lebanon to Grenada to protect its citizens.

October 22 Cuba informed the Grenada government that additional Cuban forces could be sent to the island.

October 23 The United States was asked by the Organization of Eastern Caribbean States to intervene in Grenada.

October 25 United States and Caribbean area forces landed in Grenada. All branches of the United States armed forces participated. There was sharp fighting at some points.

December 15 Most United States forces left Grenada.

December 19 Elections were held in Grenada.

A. Congressional hearings and official reports

1429. "JCS Replies to Criticism of Grenada Operation." *Army* 34 (August 1984): 28–33, 36–37.

Full text of a report prepared to rebut the Military Reform Caucus' critique of the invasion. This is a highly detailed response to each major contention made in the Caucus analysis.

1430. U.S. Congress. House. Committee on Foreign Affairs. Subcommittee on Inter–American Affairs. *United States Policy Toward Grenada, Hearing.* 97th Cong., 2nd sess., 1982.

Provides useful background information for understanding United States concern about Grenada. Various viewpoints, including that of the government of Grenada, are presented. Of special interest is the statement by Sally A. Shelton, who was the United States ambassador to Grenada from 1979 to 1981.

1431. U.S. Congress. House. Committee on Foreign Affairs. Subcommittee on International Security, Scientific Affairs, and Western Hemisphere Affairs. *United States Military Actions in Grenada: Implications for United States Policy in the Eastern Caribbean, Hearings.* 98th Cong., 1st sess., 1984.

Explores the reasons for the invasion of Grenada and includes questions on current policies, especially about the date on which the occupation will be terminated. Not much is said about the military aspects of the invasion, but, nevertheless, this is an essential source.

1432. U.S. Congress. Senate. Committee on Foreign Relations. *The Situation in Grenada, Hearing.* 98th Cong., 1st sess., 1983.

Discusses the reasons for the invasion and its legal ramifications. Military matters such as the level of resistance encountered by United States forces are also considered.

1433. U.S. Information Service. *Grenada: Background and Facts.* Washington, DC: Government Printing Office, 1983.

Constitutes a kind of official United States reference book on the invasion. The short pamphlet includes a chronology, statements by many political leaders and newspapers that support American actions, and statements by the United States Government.

B. Documentary collections

1434. Seabury, Paul and Walter A. McDougall. *The Grenada Papers.* San Francisco, CA: Institute for Contemporary Studies, 1984.

Collection of documents seized in Grenada that is intended to justify the United States intervention. The documents deal with internal political developments and relations with Communist bloc countries. The book includes a chronology, a list of the persons named in the documents, two maps, and an explanation of the many acronyms employed.

1435. Valenta, Jiri and Herbert J. Ellison, eds. *Grenada and Soviet/Cuban Policy: Internal Crisis and U.S./OECS Intervention.* Boulder, CO: Westview Press, 1986.

About half the book consists of documents seized in Grenada that discuss Soviet and Cuban involvement with Grenada. There are also a number of essays discussing what are viewed as the broader implications of Communist influence in Grenada and the Caribbean and the impact that the Grenada episode had on Soviet policies and on relations between the Western and Eastern bloc countries. Probably of special importance for the purposes of this bibliography are contributions by Dov S. Zakheim and George Liska that discuss the Grenada invasion and its connection to relations between the United States and the Soviet Union.

C. General studies

1436. Burrowes, Reynold A. *Revolution and Rescue in Grenada: An Account of the U.S.–Caribbean Invasion.* New York: Greenwood Press, 1988.

   Largely a political analysis. Much of the book discusses foreign reactions to the attack. Chapter 4 deals briefly with the military side, including shortcomings of the United States performance, such as the lack of maps and an uncertainty about the location of the United States citizens who were to be rescued.

1437. Dunn, Peter M. and Bruce W. Watson, eds. *American Intervention in Grenada: The Implications of Operation "Urgent Fury."* Boulder, CO: Westview Press, 1985.

   Valuable collection of essays written by officers in the United States armed forces and recognized specialists in the fields of international relations and defense studies. Given the swiftness with which Operation Urgent Fury was implemented and the limited scope of military action, the book's chapters are largely devoted to political issues. One of the military segments is Chapter 8, "Amphibious Aspects of the Grenada Episode" by Frank Uhlig, Jr. A useful feature of the book is an extremely detailed chronology of events.

1438. "Grenada: Special Report," *Airman* 39 (January 1984): 37–44.

   Capsule descriptions of the experiences of air force personnel in a number of situations in Grenada or in situations related to the mission, including the handling of Cuban prisoners.

1439. Lewis, Gordon K. *Grenada: The Jewel Despoiled.* Baltimore, MD: Johns Hopkins University Press, 1987.

   Socialist analysis that discusses political developments rather fully. Lewis believes that much economic progress was made in Grenada before the United States invasion. He gives some attention to military operations in 1983, emphasizing the size and power of the United States military elements in the invasion. He contends that although the United States intended to demonstrate the capabilities of its special operations personnel, they generally failed in their missions.

1440. MacDonald, Scott B., Harald M. Sandstrom, and Paul B. Goodwin, Jr., eds. *The Caribbean After Grenada: Revolution, Conflict, and Democracy.* New York: Praeger, 1988.

Solid collection of essays. Every aspect of the Grenada episode is examined. Despite the title, various chapters discuss the Grenada revolution and the resulting United States invasion. Owing to space limitations, the chapters cannot be listed and annotated separately, but this is one of the most significant books about events before, during, and after the invasion.

1441. O'Shaughnessy, Hugh. *Grenada: Revolution, Invasion, and Aftermath.* London: Hamish Hamilton with The Observer, 1984.

Early study that thoroughly condemns the invasion. O'Shaughnessy asserts that the United States lied consistently about events in Grenada and largely controlled media coverage of the attack.

1442. Payne, Anthony, Paul Sutton, and Tony Thorndike. *Grenada: Revolution and Invasion.* New York: St. Martin's Press, 1985.

Part 1 examines such topics as the political evolution of Grenada and United States and Cuban policies in the Caribbean area. Chapter 7 analyzes the invasion. The authors contest most United States statements concerning the situation in Grenada and the attack. This is a relatively early study of the Grenada intervention, and it is thus based largely on newspaper stories and a few government publications.

1443. Schoenhals, Kai P. and Richard A. Melanson. *Revolution and Intervention in Grenada: The New Jewel Movement, the United States, and the Caribbean.* Boulder, CO: Westview Press, 1985.

Largely an overview of the revolutionary movement in Grenada and a general discussion of United States policies toward the Caribbean countries. Military aspects of the intervention are hardly mentioned, but this is an important study, which is extensively documented.

1444. Stack, Cecil. "Grenada." *Soldiers* 28 (February 1984): 37–42.

Extensively illustrated overview of participation in the Grenada operation.

D. The decision to intervene

1445. Allman, T.D. "The Doctrine That Never Was." *Harpers Magazine* 268 (January 1984): 14, 18, 20–22.

Argues against using the Monroe Doctrine as a justification for the Grenada invasion and similar operations. Allman deals primarily with the historical development of the Monroe Doctrine, particularly its role in United States expansion and interventions.

1446. "Anatomy of a Small War." *New Republic* 189 (November 21, 1983): 7–9.

Examines justifications for the invasion. The article rejects the view that the United States must always practice nonintervention. Generally, the *New Republic* approves of the invasion, because it will help the people of Grenada. The argument is put forth that liberals would have approved of intervention in Haiti.

1447. Ashby, Timothy. "Grenada: Soviet Stepping Stone." *United States Naval Institute Proceedings* 109, no. 12 (December 1983): 30–35.

Discusses the development of Grenada as a Soviet/Cuban base. Ashby applauds the Reagan administration for recognizing the danger of the Soviet buildup and acting to end it. Part of the article examines the strategic position of Grenada and its usefulness as a Soviet base in conjunction with Cuba and Nicaragua.

1448. Barnet, Richard J. "The Empire Strikes Back." *Progressive* 48 (January 1984): 16–17.

Asserts that the invasion is simply an effort by Reagan to flex United States military muscles and to impress other nations, an attempt that has failed badly, according to Barnet. The author expresses the fear that the United States will be encouraged by its success in Grenada to intervene openly in Nicaragua.

1449. Beck, Robert J. *The Grenada Invasion: Politics, Law, and Foreign Policy Decisionmaking.* Boulder, CO: Westview Press, 1993.

Seemingly exhaustive study of the reasons for the invasion and, especially, the sequence of events in the decision to mount the intervention. An entire chapter deals with the justifications put forth by the Reagan administration. As the title of the book suggests, Beck gives due attention to the legal aspects of the episode.

1450. ——. "The 'McNeil Mission' and the Decision to Invade Grenada," *Naval War College Review* 44, no. 2 (Spring 1991): 93–112.

Detailed examination of a secret mission by Ambassador Francis J. McNeil to Barbados just before the invasion of Grenada. McNeil's assignment was to determine whether or not the leaders of several Caribbean nations fully supported military action. This study of the decision to invade Grenada supports the Reagan administration's explanation of its deliberations. Beck also describes deception measures used to mask the intentions of the United States.

1451. Bennett, Ralph Kinney. "Grenada: Anatomy of a 'Go' Decision." *Readers Digest* 124 (February 1984): 72–77.

Presents a day-by-day capsule summary of events, October 17–25, 1983. According to Bennett, Reagan insisted on avoiding a situation in which Americans might become hostages. As the title suggests, the focus is on the actions of United States decision-makers.

1452. "Gunboats to Grenada." *New Republic* 189 (November 14, 1983): 10, 12.

Describes Secretary of State Shultz's statement on the invasion. The article concedes that some concern about Grenada's ties with the Soviet bloc was justified but rejects the view that Americans on the island were in danger. The invasion is attributed in part to the United States desire to intimidate Nicaragua and the rebels in El Salvador.

1453. Hooker, Richard D., Jr. "Presidential Decisionmaking and Use of Force: Case Study Grenada." *Parameters* 21, no. 2 (Summer 1991): 61–72.

Argues that events in Grenada and their implications forced the United States to act against a threat. Hooker usefully relates the Reagan administration's national security stance specifically to Grenada and asserts that the episode cannot be interpreted in isolation. Despite the success of the Grenada operation, the decision-making process showed flaws.

1454. Massing, Michael. "Grenada Before and After." *Atlantic* 253 (February 1984): 76–87.

Emphasizes the conflicts in the country, and asserts that the invasion won general approval among the inhabitants. Massing deals at length with United States policies toward Grenada under Carter and Reagan and the intense political competition on the island.

E. Military and naval operations

1455. Adkin, Mark. *The Battle for Grenada.* Lexington, MA: Lexington Books, 1989.

Extremely critical account of the United States handling of the operation by a British officer who participated in the invasion. The armed forces demonstrated a number of shortcomings.

1456. "Army Evaluating Lessons Learned in Grenada Fight." *Aviation Week and Space Technology* 120 (January 16, 1984): 131, 133–36.

Discusses the army's evaluation of its combat helicopters as shown by the Grenada experience. This is a highly technical article, supplemented by several action photographs, two of which relate to the Grenada action.

1457. "Beirut/Grenada: Special Report." *Air Reservist* 36 (November/December 1983): 6–7, 27, 31.

Stresses the Grenada experience, despite the title. Several reservists describe their mobilization and their work in Grenada. The last page consists of an interview with an air force staff sergeant who was part of a C-141 crew who helped to evacuate the United States medical students on Grenada.

1458. Bolger, Daniel P. "Operation Urgent Fury and Its Critics." *Military Review* 66 (July 1986): 57–69.

Answers many criticisms of the execution of Operation Urgent Fury. Bolger very briefly outlines the invasion and then deals with objections of specific individuals. He describes the island and notes that significant fighting occurred initially, casting doubt on assertions that the United States used too much force. Bolger emphasizes, too, that the operation had a vital rescue component that may often be given too little weight in analyses.

1459. Borch, Frederic L. *Judge Advocates in Combat: Army Lawyers in Military Operations from Vietnam to Haiti*. Chap. 2, "Grenada: Army Lawyers in Transition," pp. 59–85. Washington, DC: Government Printing Office, 2001.

Judge advocates were initially challenged by the unexpectedly large number of prisoners taken. Cubans were considered prisoners of war, and all of the personnel secured were given the protections of the Geneva Convention. Later, lawyers were much involved with damage claims submitted by Grenada citizens.

1460. Canan, James W. "Blue Christmas Coming Up." *Air Force Magazine* 67 (January 1984): 78–81.

Provides a full description of air force operations in the Grenada invasion. Much of the article discusses the accomplishments of individual air force officers and men during the attack. A good deal of material on combat in Grenada is included.

1461. Carney, John T., Jr. and Benjamin F. Schemmer. *No Room for Error: The Covert Operations of America's Special Tactics Units from Iran to Afghanistan*. Chap. 9–10, pp. 108–67. New York: Ballantine Books, 2002.

Highly critical account of the planning and execution of Operation Urgent Fury. The authors portray the effort as a confused operation that they compare to a "Japanese Kabuki dance created by three or four choreographers speaking different languages" (p. 116). This is an important discussion that should be read by those interested in this episode.

1462. Cole, Ronald H. *Operation Urgent Fury Grenada: The Planning and Execution of Joint Operations in Grenada 12 October–2 November*

1983. Washington, DC: Joint History Office, Office of the Chairman of the Joint Chiefs of Staff, 1997.

Briefly discusses the origins of the crisis and then provides a chronological account of planning activities. The narrative moves from October 12 through November 2, 1983. The study is heavily documented, largely from official sources. Cole also drew information from interviews with a number of those who were involved in planning the operation. Several maps and a highly detailed index contribute to the worth of this volume.

1463. Collins, James L. and David M. Casmus. "Air Defense in Grenada." *Air Defense Artillery* (Spring 1984): 13–15.

Discusses the phased deployment of three air defense batteries to support brigades of the 82nd Airborne Division. A number of problems emerged for the air defense artillery during the invasion, one of them being a relative lack of mobility owing to the absence of vehicles.

1464. Cusick, John J. and Mark J. Flavin. "'Golden Griffins' Pave the Way." *Army Logistician* 16 (July–August 1984); 28–31.

Describes the role of the 407th Supply and Service Battalion of the 82nd Airborne Division and the division's support command in Operation Urgent Fury. Cusick emphasizes that prior training of these elements made their achievements possible.

1465. Harper, Gene. "Duty on the Spice Island." *Soldiers* 39 (September 1984): 16–20.

Discusses the duties and experiences of the few United States troops remaining with the Caribbean Peace Force (CPF) in Grenada. A military police company is the largest element, and the article focuses largely on this unit.

1466. Harper, Gilbert S. "Logistics in Grenada: Support No-Plan Wars." *Parameters* 20, no. 2 (June 1990): 50–63.

Underlines the logistical problems encountered in Grenada, which was an unexpected commitment in contrast to the intervention in Panama in 1989. Poor communications and coordination dogged the initial effort. Harper is highly critical of the general failure to follow established doctrine.

1467. Herspring, Dale R. *The Pentagon and the Presidency: Civil–Military Relations from FDR to George W. Bush.* Chap. 10, "The Military and Reagan," pp. 285–90. Lawrence: University Press of Kansas, 2005.

Like some other interventions and special operations, Operation Urgent Fury was dogged by poor intelligence. Lack of planning, poor communications, and severe inter-service conflicts were also major problems.

1468. House, Art. "Grenada: Army Reserve Goes Into Action." *Army Reserve Magazine* 30 (Spring 1984): 19–21.

Tells the little-known story of army reserve civil affairs specialists who followed combat forces to Grenada. The article examines the experiences of a number of reservists and the contributions they made to bringing life back to normal in Grenada.

1469. Huchthausen, Peter. *America's Splendid Little Wars: A Short History of U.S. Military Engagements: 1975–2000.* Chap. 5, "Intervention in Grenada," pp. 65–85. New York: Viking, 2003.

Although not ignoring some of the shortcomings in the execution of the invasion, Huchthausen generally praises the operation and its results. He believes the United States government was concerned about the possibility of a hostage situation developing in Grenada that would resemble the seizure of hostages by the Iranians a few years earlier. Quite a bit of space in the chapter is necessarily given to political events in Grenada in the early 1980s.

1470. "Operation Island Breeze Was Not a Breeze," *Army Logistician* 16 (November–December 1984): 20–21.

Heavily illustrated account of logistical support given to the Caribbean Peace Force (CPF) that assumed responsibility for peacekeeping in Grenada from the United States armed forces after the invasion.

1471. "Operation Toylift to Grenada." *National Guard* 38 (March 1984): 20–23, 39.

Focuses on the flight of toys donated by a Rhode Island manufacturer to Grenada by a Rhode Island Air National Guard unit. The article also includes a fair amount of information on other aspects of the Grenada operation and life in the small nation after the United States occupation.

1472. Potter, Michael C. "Destroyers Deliver 'Urgent Fury.'" *United States Naval Institute Proceedings* 121, no. 10 (October 1995): 96–99.

Detailed account of destroyer actions during the invasion. These included gunfire support for United States ground forces ashore, including SEALS (navy commando troops). The navy also rendered considerable support to army helicopters during the heavy fighting.

1473. Radin, Robert M. and Raymond A. Bell. "Combat Service Support of Urgent Fury." *Army Logistician* 16 (November–December 1984): 16–19.

Reviews the work of the 1st Corps Support Command's material management center in connection with Grenada. The authors emphasize the phased deployment of the 2nd Support Center's material management center to the island.

1474. Schafer, Charles H. "Logistics Support of the Caribbean Peacekeeping Force." *Army Logistician* 17 (September–October 1985): 22–26.

Discusses the experiences of a "logistics technical assistance team" sent to Grenada to provide training and advice to the Caribbean Peace Force (CPF) that maintained stability on the island after the invasion.

1475. Schemmer, Benjamin F. "Grenada Highlighted One of DOD's Major C3 Problems, But Increased Funding Is Bringing Solutions." *Armed Forces Journal International* 121 (February 1984): 50, 52–53.

Examines communications and other problems associated with the Grenada operation. There are a number of rather sarcastic comments about communications difficulties in the air force. Even maps seemed to be a persistent problem for the marines.

1476. Thompson, Dale L. "The Guard in Grenada." *United States Naval Institute Proceedings* 110, no. 11 (November 1984): 65–69.

Coast Guard operations at Grenada were carried out by four vessels, three cutters and a buoy tender as a support ship. The Coast Guard did not participate in the invasion, but it aided in stability efforts from early December 1983 to May 1984. Despite logistics and communications challenges, the Coast Guard provided needed services and cooperated with the army, the Caribbean Peace Force, and the United States embassy.

1477. Tiwathia, Vijay. *The Grenada War: Anatomy of a Low-Intensity Conflict*. New Delhi: Lancer International, 1987.

Sharply critical account of the invasion by an experienced Indian Army officer. In Tiwathia's view, the operation was flawed from the planning stage through the execution of the attack, which cost many lives unnecessarily.

1478. "The U.S. Army in Grenada." *Army* 33 (December 1983): 29–31.

Survey of army operations in Grenada. This is largely a collection of photographs.

F. Political and legal aspects

1479. Davidson, Scott. *Grenada: A Study in Politics and the Limits of International Law*. Aldershot, U.K.: Avebury Gower Publishing Company Ltd., 1987.

Painstakingly researched and massively documented study of the legal aspects of the intervention. One chapter includes a description of the consideration of the Grenada matter by the United Nations.

1480. De Camp, William T. "Grenada: The Spirit and the Letter of the Law." *Naval War College Review* 38 (May–June 1985): 28–36.

Argues that the United States medical students in Grenada were really in danger. Moreover, De Camp vigorously rejects comparisons between the Grenada operation and the Soviet invasion of Afghanistan. This is an interesting defense of the actions of the Reagan administration, but it is not an in-depth analysis from the perspective of international law.

1481. Hall, David Locke. *The Reagan Wars: A Constitutional Perspective on War Powers and the Presidency.* Chap. 6, "Grenada, 1983," pp. 167–209. Boulder, CO: Westview Press, 1991.

Detailed examination of the invasion, emphasizing the constitutional ramifications and the effect of the War Powers Resolution. Hall defends the actions of the Reagan administration against objectors. He insists emphatically that the consultation of the president with Congress under the War Powers Resolution does not give Congress an unlimited amount of time to consider possible military action.

G. The media and the Grenada invasion

1482. Hughes, Wayne P. "Guarding the First Amendment—For and From the Press." *Naval War College Review* 37 (May–June 1984): 28–35.

Asserts that public opinion in the United States strongly approved of the armed forces' media policy in Grenada and that the exclusion of the media was good decision in this instance. There does need to be a better relationship between the media and the government. Hughes warns that "dollar values and egos are ... threats to, not defenders of, a free press" (p. 34).

1483. Zoll, Atwell. "The Press and the Military: Some Thoughts After Grenada." *Parameters* 14 (Spring 1984): 26–34.

Offers a brief historical review of the activities of correspondents in British and United States wars, and then examines the controversy that arose about the United States keeping the media out of the Grenada invasion. Zoll explores the complex issues and agrees that the media's unhappiness at being excluded was "understandable and, in most part [sic], justifiable" (p. 31). He asserts that the Department of Defense needs to improve its relationships with the media.

H. African-American perspectives on Grenada

1484. "Attack on Grenada." *Black Enterprise* 14 (January 1984): 11–12.

   Focuses on criticism of the invasion by many American blacks and by foreign countries. The article briefly describes the invasion and Reagan's hostility toward the government of Grenada since 1981. The accomplishments of Bishop's government are examined.

1485. Booker, Simeon. "Tour of Grenada After the U.S. Invasion Reveals Grim Past and Uncertain Future." *Jet* 65 (January 16, 1984): 30–33, 37–39.

   Criticizes the socialist experiments of the regime in Grenada bitterly and emphasizes the violence in the country. Booker applauds Reagan's actions and briefly describes the Soviet/Cuban military buildup in Grenada.

XXV
Panama, 1989

Operation Just Cause followed years of tension between the United States and Panama. After having had close relationships with United States intelligence and law enforcement agencies in their operations against leftist governments in the Caribbean area and against narcotics dealers, Manuel Noriega became an increasing embarrassment to the United States. After a long career in Panamanian security forces, Noriega effectively controlled Panama by 1987. The following year, he was indicted in the United States for various crimes related to his role in the illegal narcotics trade. The Reagan and H.W. Bush administrations took many steps against the Noriega regime at the same time that anti-Noriega sentiment increased in Panama. Noriega was skilled in maintaining his position, however, and, finally, on December 20, 1989, United States forces invaded Panama. After several days of fighting, the country was secured and Noriega surrendered after a dramatic confrontation. Subsequently, a new government was formed in Panama, and Noriega was taken to the United States to be tried and, eventually, convicted. Because the Panama crisis developed gradually, the United States was able to plan its operations carefully, and when it invaded Panama, the operation went more smoothly than the attack on Grenada.

Chronology

1970
August — Manuel Noriega appointed head of Panamanian military intelligence in recognition of the support he gave to the new dictator, Omar Torrijos. Nationalism and resentment about the continued existence of the Panama Canal Zone that divided Panama into two segments continued to grow. In the 1950s

and 1960s there had been sharp clashes between Panamanians and United States citizens, including United States military personnel, and there were some deaths on both sides.

1988

February 5	Federal grand juries indicted Manuel Noriega because of his involvement in the narcotics traffic, including money laundering.
March 11	The United States began what became a phased imposition of sanctions on Panama because of the Noriega dictatorship. Increases of United States military strength in the Panama Canal Zone began about that time, and these continued until the eventual invasion in 1989.

1989

May 7	Opponents of the Noriega regime won a national election in Panama, but Noriega refused to accept the results. The United States utilized portions of Panamanian territory for military exercises as another way of pressuring the government of Panama.
October 3	Some elements of the Panama Defense Force (PDF) rebelled against the regime in an effort to overthrow the Noriega dictatorship, but the coup attempt failed, partly because the United States did not go beyond encouraging the anti-Noriega movement. An extremely tense situation developed between the United States and Panama, and numerous incidents took place, some of which had serious consequences.
October–December	The United States revised and completed plans for military action against Noriega.
December 15	The Noriega regime seemed undeterred by United States actions, including the numerous sanctions imposed on Panama, and the Panamanian National Assembly declared that there was a "state of war" between Panama and the United States.
December 20	United States forces invaded Panama. The Panama Canal was closed for security reasons, the first closure since it was first opened in 1914. A new Panamanian government was established under United States auspices in the Panama Canal Zone shortly before the invasion.
December 26	Panamanian resistance to United States forces ended.
December 30	The United Nations General Assembly passed a resolution criticizing the United States invasion of Panama.

1990

January–April	Operation Promote Liberty undertaken by the United States to support the formation of a new and reformed government for Panama.

January 3   Noriega surrendered to United States authorities more than a week after he had sought refuge with the papal nuncio at Panama City.
February 7   Congress approved a special assistance program for Panama in the aftermath of the invasion, but continuing problems obstructed implementation of the aid.
February 13 The last United States forces sent to Panama for the invasion left the country, leaving approximately the same number of United States personnel that had been stationed in the Panama Canal Zone prior to the build-up for the invasion.

A. General studies

1486. Buckley, Kevin. *Panama: The Whole Story.* New York: Simon & Schuster, 1991.

This is a dramatically told story of Panamanian–United States relations from the mid-1980s through the early days in Panama after the invasion. Buckley's detailed account is based largely on newspaper stories and numerous interviews.

1487. Gilboa, Eytan. "The Panama Invasion Revisited: Lessons for the Use of Force in the Post Cold War Era." *Political Science Quarterly* 110, no. 4 (Winter 1995–96): 539–62.

Presents a long overview of the unsuccessful efforts by the Reagan and H.W. Bush administrations to unseat Noriega, and an examination of the reasons for Bush's decision to invade Panama. Repeated United States attacks on Noriega simply strengthened his position, eventually making his removal all the more important to the United States. Gilboa believes that a stronger United States position earlier on would have warned Noriega about the possible consequences of his actions. He asserts, moreover, that similar errors, especially the failure to indicate clearly the level of United States concern, were made in dealing with the Bosnia, Somalia, and Persian Gulf situations. This is a highly enlightening study, which includes information about the many times the United States ignored Noriega's misdeeds.

1488. Independent Commission of Inquiry on the U.S. Invasion of Panama. *The U.S. Invasion of Panama: The Truth Behind Operation "Just Cause."* Boston, MA: South End Press, 1991.

Consists largely of statements by people from Panama about the invasion and its effects. There are also some special reports and other materials. The book is extremely critical of the intervention and the policies that are said to have been implemented by the United States following the attack, which is termed "the brutal and racist U.S. invasion of Panama" (p. 1).

1489. Johns, Christina Jacqueline, and P. Ward Johnson. *State Crime, the Media, and the Invasion of Panama*. Westport, CT: Praeger, 1994.

Largely an analysis of the reasons for the deterioration of the relationship between the Noriega regime and the United States. The authors assert that the invasion was the "result of long-term policy goals" (p. 21). One objective was effectively to undo the Panama Canal treaties negotiated during the Carter administration, but the situation was highly complex. It is said that Noriega was insufficiently hostile to Nicaragua and Cuba to suit United States decision-makers, moreover.

1490. Scranton, Margaret E. *The Noriega Years: U.S.–Panamanian Relations, 1981–90*, Boulder, CO: Lynne Rienner Publishers, 1991.

This is a searching examination of the United States relationship with Noriega and the changing policies of the United States with respect to Panama. It is a vital source for anyone wishing to understand the fall of the Noriega regime.

B. Military studies: general

1491. Cole, Ronald H. *Operation Just Cause, Panama: The Planning and Execution of Joint Operations in Panama, February 1988–January 1990*. Washington, DC: Joint History Office, Office of the Chairman of the Joint Chiefs of Staff, 1995, 1996. http://purl.access.gpo.gov/GPO/LPS48960

Brief, but heavily documented, chronological account of the Panama intervention. An interesting segment deals with the role of the then recently passed Goldwater–Nichols Defense Reorganization Act of 1986 and its influence on planning and the command structure.

1492. Donnelly, Thomas, Margaret Roth, and Caleb Baker. *Operation Just Cause: The Storming of Panama*. New York: Lexington Books, 1991.

A leading study of the invasion of Panama. The book has many virtues, but documentation is limited to a few pages that cite newspaper stories for the most part. The authors assert that their work "is the product of hundreds of interviews with the participants in Just Cause," (xiv) but there is no listing of those who were interviewed. The book includes several useful features, such a glossary, fine illustrations, and a number of charts that depict major military events.

1493. Flanagan, Edward M., Jr. *Battle for Panama: Inside Operation Just Cause*. Washington, DC: Brassey's, Inc., 1993.

Fine study by a retired army lieutenant general. Flanagan interviewed a number of participants or corresponded with them, and the notes to the book clearly identify the sources of many statements. Several pages discuss the "Sources," moreover. An excellent segment documents the military events of the operation. Flanagan also gives some attention to the decline in relations between Panama and the United States.

1494. Huchthausen, Peter. *America's Splendid Little Wars: A Short History of U.S. Military Engagements: 1975–2000*. Chap. 8, "Storming Panama," pp. 113–26. New York: Viking, 2003.

Asserts that the Panama invasion became a "prototype for [future United States] military action" that illustrated the use of "small, highly mobile armed forces tailored to resolve policy disputes rapidly and cleanly with decisive force" (p. 113). Huchthausen discusses a number of what he believes are unique or at least innovative features of the invasion, such as the fact it "was the first American military intervention to completely integrate conventional and special operations forces" (p. 123).

1495. Watson, Bruce W. and Peter G. Tsouras, eds. *Operation Just Cause: The U.S. Intervention in Panama*. Boulder, CO: Westview Press, 1991.

Important group of essays written by seasoned specialists in military affairs, especially intelligence, and international relations. A number of chapters pertain directly to military facets of the operation. These include discussions of air power by Noris Lyn McCall, logistics by John W. Turner, and civil affairs by Edward F. Dandar, Jr. One of the helpful components of the book is an extensive chronology prepared by Bruce W. Watson and Lawrence S. Germain.

1496. Yates, Lawrenc A. "Panama, 1988–90: The Disconnect Between Combat and Stability Operations." *Military Review* 85, no. 3 (May/June 2005): 46–52.

Emphasizes the fact that fighting and stability operations occurred concurrently in Panama for a while. Yates discusses planning for Operation Just Cause in some detail, stressing that planning for combat and for stability efforts proceeded in relative isolation from one another. Although the Panama invasion was a success, serious problems could arise in the future if such uncoordinated planning occurred.

1497. ———. *The U.S. Military Intervention in Panama: Origins, Planning, and Crisis Management June 1987–December 1989*. Washington, DC: Center of Military History, United States Army, 1989.

Essential study of the development of United States planning for what became Operation Just Cause. This is a thoughtful and forthright book by a seasoned military historian. The documentation is thorough. Yates does not ignore the fact that there were ties between various elements of the United States government and Noriega for some time. Now retired, Yates is preparing a study of the execution of Operation Just Cause, which promises to be another fine contribution to United States military history.

C. Military studies: specialized

1498. Borch, Frederic L. *Judge Advocates in Combat: Army Lawyers in Military Operations from Vietnam to Haiti*. Chap. 3, "Panama: Operation Just Cause, 1989–90," pp. 87–119. Washington, DC: Government Printing Office, 2001.

Legal activities ran more smoothly than in Grenada, partly because of the experience gained in Grenada and partly because a number of judge advocates were already in Panama in the Canal Zone. The complicated rules of engagement established in Panama necessarily created work for army lawyers.

1499. Briggs, Clarence E. III. *Operation Just Cause: Panama, December 1989: A Soldier's Eyewitness Account*. Harrisburg, PA: Stackpole Books, 1990.

Constitutes an important part of the literature about Operation Just Cause, in part because of Briggs' emphasis on the "rules of engagement" that governed United States military operations. This book usefully complements Borch's chapter (item 1498) by giving the reactions of a well-educated combat officer who is not an attorney to the challenges of the rules that were imposed in Panama.

1500. Carney, John T., Jr., and Benjamin F. Schemmer. *No Room for Error: The Covert Operations of America's Special Tactics Units from Iran to Afghanistan*. Chap. 13, pp. 200–223. New York: Ballantine Books, 2002.

Detailed description of the mounting and execution of special operations during the Panama invasion. Although the authors carefully discussed several shortcomings of those special operations, they believed that Operation Just Cause proved "the first major success for American special operations forces" (p. 222).

1501. Reynolds, Nicholas E. *Just Cause: Marine Operations in Panama, 1988–1990*. Washington, DC: History and Museums Division, Headquarters, U.S. Marine Corps, 1996.

Succinct, but excellent, examination of the marines and their experiences in Panama. This study is solidly based on primary sources and is thoroughly documented. There are numerous superb maps and interesting illustrations.

1502. Shultz, Richard H., Jr. *In the Aftermath of War: US Support for Reconstruction and Nation–Building in Panama Following Just Cause.* Maxwell, AL: Air University Press, 1993. http://purl.access.gpo.gov/GPO/LPS47001

Discusses the Noriega regime, planning for Operation Just Cause, the establishment of a "Military Support Group" by the United States, and the reconstruction of a "security force" in Panama. The Military Support Group seems to have combined civil affairs, military government, and military advisory functions. Although this organization was helpful in dealing with the many problems found in post-conflict Panama, Shultz asserts that there is a continuing "need for an interagency mechanism that can plan for and respond to postconflict crises in an effective way" (xiii). Shultz also asserts that psychological operations should have played a larger role in Panama.

1503. Steele, Dennis. "Operation Just Cause." *Army* 40, no. 2 (February 1990): 34–37, 40–44.

Praises army forces for having accomplished most of their goals on the first day of the invasion. This is a helpful overview of military events. The article is richly illustrated, and it includes several helpful maps.

1504. Thurman, Maxwell B. and William W. Hartzog. "Simultaneity—The Panama Case." *Army* 43 (November 1993): 16–20, 22–24.

Thurman commanded United States forces in Panama, making this article a major primary source. The authors assert that Operation Just Cause demonstrated the correctness of current army doctrine, that is, the application of maximum force on hostile headquarters or strong points, often termed "centers of gravity." There is an important analysis of the factors that made the operation a success.

XXVI

From the end of the Cold War to the War on Terrorism, 1991–2001

This period began with a major military event, the Persian Gulf War and related operations. The war was followed by several controversial United States involvements in international peacekeeping efforts, which occurred in Somalia, Haiti, and Bosnia. Those peacekeeping operations and continuing concerns with counterinsurgency and low-intensity conflict produced significant contributions to the literature, but less attention was given to these subjects than during the Vietnam years or later after the 9/11 attack and the invasions of Afghanistan and Iraq.

A. Small wars, military humanitarian efforts, and peacekeeping: general studies

1505. Blechman, Barry M. and Tamara Cofman Wittes. "Defining Moment: The Threat and Use of Force in American Foreign Policy since 1989." *Political Science Quarterly* 114, no. 1 (1999): 1–30.

Examines a number of cases after the end of the Cold War in which military force was either employed or existed as a real possibility. The events examined were Somalia, Macedonia, North Korea, Taiwan, Panama, Iraq, Bosnia, and Haiti. The authors interpret only Somalia as an outright failure, although they view Panama, Iraq, Bosnia, and Haiti as limited successes. They note that threats of force were not particularly useful. Often, threats had to be replaced by military action. Several factors affect the credibility of threats by the United States. A key factor seems to be the reluctance of public opinion in the United States to accept military casualties. If there are casualties momentum builds for a withdrawal of military assets. Blechman and

Wittes believe that the United States needs to improve its ability to make threats credible in order to avoid being drawn into protracted conflicts.

1506. Brune, Lester H. *The United States and Post–Cold War Interventions: Bush and Clinton in Somalia, Haiti, and Bosnia, 1992–1998.* Claremont, CA: Regina Books, 1998.

Fine discussion of these three interventions. Brune provides a single, detailed chronology of these events, well-written chapters that are impressively documented, and an extensive bibliography. The chapters related to the interventions are listed and annotated in the appropriate segments of this bibliography. Brune, a seasoned researcher and writer, demonstrates a considerable ability to discuss complicated occurrences in a lucid fashion. The bulk of the book deals with Bosnia.

1507. Burk, James. "Public Support for Peacekeeping in Lebanon and Somalia: Assessing the Casualties Hypothesis," *Political Science Quarterly* 114, no. 1 (1999): 53–78.

Vigorously assails the widely held view that public opinion in the United States cannot and will not support military commitments abroad when United States forces sustain casualties. Much of the article examines that view at some length, dealing in particular with assertions made by security expert Edward Luttwak. Burks goes on to examine public opinion poll data in respect to casualties and decisions about continuing military operations. On the basis of his research, he does not hesitate to recommend that the "casualties hypothesis" be discarded entirely.

1508. Campbell, Kenneth J. "Once Burned, Twice Cautious: Explaining the Weinberger–Powell Doctrine," *Armed Forces and Society* 24, no. 3 (1998): 357–74.

A cautious view of the viability of the Weinberger–Powell doctrine. On the one hand, many governmental leaders have endeavored to stay away from large-scale, long-term entanglements that might turn into conflicts resembling the Vietnam War. On the other hand, though, concerns about maintaining the political and military reputation of the United States and about humanitarian crises may make avoiding major commitments difficult. Campbell surveys the decline in United States military capability during the Vietnam War as a result of declining morale within the armed forces, a factor that encouraged the evolution of a conservative, rather than an assertive, view of the appropriateness of military efforts in Third World countries.

1509. Cassidy, Robert M. *Peacekeeping in the Abyss: British and American Peacekeeping Doctrine and Practice After the Cold War.* Westport, CT: Praeger, 2004.

Well-researched and well-reasoned approach to a rather new aspect of military operations that may well increase in importance in the future. This is a searching comparative study of British and United States counterinsurgency and peacekeeping doctrine. There are also specific chapters that discuss the United States handling of Somalia and the British commitment in Bosnia.

1510. Connaughton, Richard. *Military Intervention and Peacekeeping: The Reality.* Aldershot, U.K.: Ashgate Publishing Ltd., 2001.

Although not keyed specifically to the United States experiences, this book is a useful overview of a number of peacekeeping efforts during the 1990s. Connaughton, a recognized international affairs analyst, writes much about the United States role in the Persian Gulf, Somalia, and Kosovo, thus putting it into international perspective. As suggested by the inclusion of the word "reality" in the title, this is a forthright and, indeed, highly informative study.

1511. Gentry, John A. "Complex Civil–Military Operations: A U.S. Military-centric Perspective," *Naval War College Review* 53, no. 4 (Autumn 2000): 57–76.

Examines limitations in United States military performance in a number of recent missions, notably peacekeeping efforts. For the most part, the United States military continues to be resistant to such missions and the requirements imposed by them. Planning operations of this sort requires civilian, as well as military, participation.

1512. Hendrickson, Ryan C. *The Clinton Wars: The Constitution, Congress, and War Powers.* Nashville, TN: Vanderbilt University Press, 2002.

Focuses on the significance of the War Powers Resolution in relation to United States military action abroad during the Clinton administration, especially in Somalia, Haiti, and Bosnia. Despite some efforts by the Republican-controlled Congress to assert itself regarding military commitments, Clinton generally had considerable freedom of action. Hendrickson also studies how political party loyalties influence opinions about the scope of congressional war powers and how United States commitments abroad under United Nations auspices are viewed by Congress.

1513. McCombie, Ryan J. "Special Operations Forces and Small Wars." *Parameters* 25, no. 3 (Autumn 1995): 108–13.

Discusses several important books about low-intensity conflict and United States special operations forces. The books reviewed cover a number of conflicts, most of which included United States commitments, and the training and organization of United States forces.

1514. MacKinnon, Michael G. *The Evolution of US Peacekeeping Policy Under Clinton: A Fairweather Friend?* London: Frank Cass, 2000.

Traces the Clinton administration's retreat from a relatively strong commitment to United Nations peacekeeping activities to a much more hesitant stance than was embodied in Presidential Decision Directive 25 (PDD-25). The change came after the Somalia involvement, largely as a result of pressure from Congress.

1515. Rosen, Leora N., Paul D. Bliese, Kathleen M. Wright, and Robert K. Gifford. "Gender Composition and Group Cohesion in U.S. Army Units: A Comparison of Five Studies." *Armed Forces and Society* 25, no. 3 (1999): 365–86.

Three of the five studies deal with the individual situations during the Somalia, Haiti, and Gulf War periods. Generally, women showed less unit cohesion than men, although such a finding is not made in respect to one garrison sample and to the Haiti sample. Various topics were examined, including commitment to the missions involved and social relationships outside duty hours.

1516. Weiss, Thomas G. *Military–Civilian Interactions: Humanitarian Crises and the Responsibility to Protect.* 2nd ed. Lanham, MD: Rowman & Littlefield Publishers, Inc., 2005.

Valuable survey of post–Cold War humanitarian interventions. Most episodes are given their own chapters, although East Timor and Kosovo and Afghanistan and Iraq have combined chapters. Weiss' research is thoroughly documented, and his analyses are thoughtful.

B. Counterinsurgency and low-intensity conflict theory

1517. Dougherty, Kevin J. "Fixing the Enemy in Guerrilla Warfare." *Infantry* 87, no. 2 (March–June 1997): 33–35.

Using the examples from the Greek Civil War in the late 1940s and behind the lines operations during the Korean War, Dougherty discusses bringing guerrillas into direct contact with regular forces, thereby preventing the guerrillas from using one of their classic tactics of disappearing when faced with large enemy troop units.

1518. Fastabend, David. "Categorization of Conflict." *Parameters* 27, no. 2 (Summer 1997): 75–87.

Using a dramatic incident that occurred during the peacekeeping operation in Bosnia as an example, Fastabend explores the debate over the value of the low-intensity conflict and military operations short of war concepts. Some authorities have suggested that employing the latter term could result in serious harm to peacekeepers in situations where they could face heavy attacks or other challenges.

1519. Harned, Glenn M. "Nature of Insurgency: Melting the Iceberg." *Special Warfare* 11, no. 3 (Summer 1998): 34–35.

Makes an intriguing comparison between an insurgency and an iceberg, because both have structures that are largely under the surface. This is a very brief but valuable contribution that reaffirms "the importance of the populace as the center of gravity" in low-intensity conflict (p. 34).

1520. Hunt, John B. "OOTW (Operations Other Than War): A Concept in Flux." *Military Review* 76, no. 5 (September–October 1996): 3–9.

Portrays "military operations other than war" as still vague. Terminology continues to be unsettled, and Hunt discusses a number of writings about the subject in search of better ways to conceptualize its nature and the response it requires from the United States armed forces.

1521. Hunter, Horace L., Jr. "Ethnic Conflict and Operations Other Than War." *Military Review* 73, no. 11 (November 1993): 18–24.

Proposes that the United States expand its low-intensity doctrine to discuss the possibility of encountering ethnic strife during either counterinsurgency campaigns or peacekeeping operations. In this connection, Hunter examines United Nations secretary-general Boutros Boutros-Ghali's *An Agenda for Peace.*

1522. "Interview: James R. Locher III, Assistant Secretary of Defense for Special Operations and Low Intensity Conflict." *Special Warfare* 6, no. 2 (May 1993): 33–35.

Locher discusses current United States low-intensity conflict planning and notes the complexity of such conflict. "Almost everything we do in the special-operations world depends upon effective interagency planning and coordination" (p. 33). Interestingly, Locher says that military force might have to be used to prevent a given country from developing significant chemical, biological, and radiological weapons.

1523. Kilgore, Joe E. "PSYOP (Psychological Operations) in Support of Low-Intensity Conflict." *Special Warfare* 5, no. 2 (October 1992): 26–31.

Analyzes the development of psychological operations to support particular missions abroad. Kilgore stresses the need to expand United States psychological operations capability and to ensure that members of elements of the armed forces not directly concerned with such operations recognize their value to their own missions.

1524. Metz, Steven. "Victory and Compromise in Counterinsurgency." *Military Review* 72, no. 4 (April 1992): 47–53.

Discusses the increasing emphasis the United States is giving to establishing the "legitimacy" of a foreign government as part of a strategy to defeat the insurgency in a low-intensity environment. Metz argues against this approach, which he views as rooted in Cold War thinking. Instead, he advocates bringing the government and the insurgents to the bargaining table. In Metz's view, the insurgents often have some reasonable demands to make, which should be accommodated.

1525. Metz, Steven and James Kievit. "Siren Song of Technology and Conflict Short of War." *Special Warfare* 9, no. 1 (January 1996): 2–10.

Wisely states that "immense military advantage does not automatically bring strategic success" (p. 2). After examining some important technologies of military value, the authors present a "revolution in military affairs" that could take place by 2010. It includes a "Conflict Containment Agency" that coordinates a wide variety of activities. The article concludes with three alternative future patterns of development for United States special operations forces.

1526. Odom, William E. *On Internal War: American and Soviet Approaches to Third World Clients and Insurgents.* Durham, NC: Duke University Press, 1992.

Assesses both United States and Soviet commitments to Third World countries as largely unsuccessful. Odom uses El Salvador, Guatemala, the Philippines, and the Middle East as what he terms "Case Assessments." On the basis of his experience and his research, the author recommends that the United States "eschew involvement in internal wars entirely" (p. 215), yet he recognizes that the United States cannot and will not follow such a policy. This is a key work that deserves careful study.

1527. Sullivan, Brian R. "Special Operations and LIC (Low-Intensity Conflict) in the 21st Century." *Special Warfare* 9, no. 2 (May 1996): 2–7.

Suggests the likelihood of "unconventional and irregular military challenges to U.S. interests," despite the unlikely prospect of a "major war" for some time (p. 2). Sullivan explores the many social, economic, and political problems that can generate conflicts. He warns that when the United States intervenes in a foreign country, it must avoid giving the impression that it is trying to change the host society to conform to a United States pattern.

XXVII

Persian Gulf War, 1990–91

The Persian Gulf War arose from Iraq's seizure of Kuwait on August 2, 1990, after a period of increasing tension between the two countries. Iraq had long claimed Kuwait as a "lost" province, but Iraq cited additional reasons for the invasion. The dramatic attack aroused condemnation that was almost worldwide. Iraq made numerous political and diplomatic moves to tie its attack on Kuwait to the Arab–Israeli dispute in order to win support for its actions from the Arab countries, but these efforts were largely unsuccessful.

The conflict was in three stages. A Coalition of nations was established under United Nations auspices and under United States leadership, which conducted a major military build-up in the Gulf region, primarily in Saudi Arabia. This was Operation Desert Shield, which was intended to protect Saudi Arabia and some smaller Arab oil-producing countries in the area.

After protracted negotiations and further political and diplomatic efforts, Iraq was warned to remove its forces from Kuwait by January 15, 1991. Because Iraq did not comply with this demand, Coalition forces began an effort to free Kuwait and later to invade Iraq. This was Operation Desert Storm, which lasted from January 17 to February 28, 1991.

The third segment of the Gulf conflict was Operation Provide Comfort I, which was an effort to provide humanitarian aid to Kurds in northern Iraq who had rebelled against the Saddam Hussein regime and to protect the Kurds from Iraqi air attacks by establishing a "no-fly zone" over the Kurdish portion of Iraq. Operation Provide Comfort I was largely a United States effort, although it was based on United Nations resolution 688. The operation lasted from April 6 to July 24, 1991, and it was followed by successive "no-fly zone" stipulations.

The Persian Gulf War demonstrated the ability of the United States to build a temporary alliance of great magnitude and the capability to mount and execute a huge military effort. Nevertheless, the war was not as successful as first had seemed to be the case. Despite hopes that the loss of the war would topple Saddam Hussein, he showed he was able to weather not only this crisis but also continuing tensions, including the imposition of severe sanctions by the United Nations. Moreover, Iraqi military forces were not as severely impacted by Coalition military action as had been thought.

Chronology

1990

August 2	Iraq invaded and occupied Kuwait. The United States reacted immediately by freezing both Iraqi and Kuwaiti assets. Meanwhile, the United Nations Security Council condemned the invasion by passing its resolution 660.
August 3	The Soviet Union ended shipments of armaments to Iraq and joined the United States in a statement attacking Iraq's occupation of Kuwait.
August 4	The European Community applied an array of economic sanctions on Iraq because of its seizure of Kuwait.
August 8	Iraq annexed Kuwait. The United States responded by beginning Operation Desert Shield, a massive force build-up, primarily in Saudi Arabia.
August 18	For the first time during Operation Desert Shield, United States forces fired their weapons. United States naval vessels used warning shots to compel some Iraqi tankers to return to port.
August 21	Iraq proposed discussions to end the Persian Gulf standoff, but the United States rejected this offer, insisting that Iraq must restore Kuwait's independence.
August 28	The United States stated that it was not seeking the ouster of the Saddam Hussein regime. Instead, it would be satisfied if Iraq ended its occupation of Kuwait.
August 29	Iraq announced that Kuwait had become one of its provinces.
September 13	The United States warned Iraq against sponsoring terrorist attacks.
October 16	Iraq offered to withdraw its force from Kuwait in return for some territorial concessions, but the United States flatly opposed the Iraqi move.
November 1	Saudi Arabia and the United States agreed on the conduct of a possible future war to free Kuwait.
November 8	As the crisis continued, the United States announced a significant increase in its forces assigned to Operation Desert Shield.
November 29	United Nations Security Council resolution 678 authorized the use of armed force to expel Iraq from Kuwait, if Iraq did not voluntarily leave Kuwait by January 15, 1991.
December 8	King Hussein of Jordan proposed a conference on Middle East affairs, including the Israeli–Palestinian conflict, to take

	place at the same time that Iraq removed its forces from Kuwait.
December 21	Iraq formally stated that it would not leave Kuwait by January 15, 1991.
December 26	The number of United States military personnel in the Middle East poised for an attack on Iraq stood at about 300,000. Nevertheless, there were disquieting statements from military leaders and rumors during December about the readiness of United States forces for war with Iraq.
1991	
January 5	President Bush warned Iraq once again that war would follow if Kuwait were not evacuated. Popular opinion in the United States generally supported Bush's stand on the Kuwait issue.
January 9	A meeting between the United States secretary of state and the foreign minister of Iraq failed to produce an agreement that would end the Gulf crisis.
January 17	Massive Coalition air strikes began against targets in Iraq, and the United States Navy added missiles to the offensive. Highly successful air attacks continued for the next several days, using sophisticated weapons that were far ahead of those employed during the Vietnam War. Iraq responded with missiles targeted against both Saudi Arabia and Israel.
January 22	Iraq began destroying Kuwait oil installations by setting them afire, but such actions had no real effect on Coalition military operations.
January 25	Iraq began to allow large quantities of oil to flow into the Persian Gulf with the apparent intent to damage desalinization facilities in Saudi Arabia.
January 27	Secretary of Defense Dick Cheney announced that a full land assault on Iraq was part of the war plan.
January 29	The United States and the Soviet Union announced that a firm statement by Iraq that it would end its seizure of Kuwait would bring an end to the war. Previously, the United States had contended that Kuwait must be free of Iraqi forces before there could be a return to peace. The following day, the United States said that there had been no change in policy, however.
February 3	Estimates indicated that Iraq had been the target of more tons of bombs than had been dropped in all of World War II.
February 6	The United States completed its deployment of forces for the Gulf War.
February 16	Iraq made a peace proposal that was immediately rejected by the Coalition governments. Nevertheless, diplomatic activity continued, and the United States appeared pleased to have the Soviet Union as a mediating power in the struggle.

February 21	The Coalition rejected a peace settlement proposed by Iraq through the Soviet Union.
February 24–27	Land offensive against Iraq carried out by Coalition forces after the failure of further peace efforts. Incessant air attacks had already significantly impaired Iraqi resistance capabilities. Coalition command of the air prevented the Iraqis from detecting ground offensives, moreover.
February 27	After Iraq had agreed to withdraw from Kuwait but tried to avoid accepting United Nations resolutions concerning its seizure of Kuwait, Iraq finally agreed both to remove its forces from Kuwait and to comply with all the United Nations resolutions.
February 28	Operation Desert Storm was concluded.

A. Bibliography

1528. Orgill, Andrew. *The 1990–91 Gulf War: Crisis, Conflict, Aftermath: An Annotated Bibliography.* London: Mansell, 1995.

Important compilation that emphasizes the military aspects of the war, although a good deal of material relating to political and other topics is included. No items with publication dates after September 1993 are contained in the bibliography; so the book is becoming quite outdated. Nevertheless, it is a resource that repays careful examination.

B. Dictionaries and encyclopedias

1529. Newell, Clayton R. *The A to Z of the Persian Gulf War, 1990–1991.* Lanham, MD: Scarecrow Press, 2007.

This work was originally published as the *Historical Dictionary of the Persian Gulf War, 1990–1991* by Scarecrow in 1998. Although necessarily concentrating on military topics, Newell's study provides an excellent overview of the war. An introduction describing the coming of war, a number of maps, a chronology, which provides a perspective on events preceding the crisis, and several appendices expand the usefulness of the book.

1530. Schwartz, Richard A. *Encyclopedia of the Persian Gulf War.* Jefferson, NC: McFarland & Company, Inc., 1998.

Every aspect of the conflict is expertly covered in this succinct but thorough reference source. The detailed index and the numerous cross references between entries considerably increase the value of the study. There is also a good chronology that extends from 1988 to

1991, but, despite its worth it cannot compare with the book-length chronologies included below in this bibliography.

C. Chronologies

1531. BBC World Service. *Gulf Crisis Chronology*. Harlow, U.K.: Longman Current Affairs, 1991.

    Massive day-to-day account of the war from September 1990 through March 1991 based on the extensive resources of the sponsoring agency, the British Broadcasting Corporation. The chronology also appeared as Frederick Stanwood, Patrick Allen, and Lindsay Peacock, *The Gulf War: A Day-by-Day Chronicle*. London: Heinemann, 1991.

1532. Blair, Arthur H. *At War in the Gulf: A Chronology*. College Station: Texas A&M University, 1992.

    Highly detailed chronology that extends from June 26, 1990, through February 28, 1991. The entries are thorough, often running to several paragraphs. There are several maps and a small section of photographs. This is a major reference source for the war.

1533. Mattox, Henry E. *A Chronology of United States–Iraqi Relations, 1920–2006*. Jefferson, NC: McFarland & Company, Inc., 2007.

    Although it includes a number of military events, this bibliography seems to emphasize political occurrences. All the entries are brief, owing to the long chronological coverage of the book. The book is useful to students of both the Gulf War and the Iraq War, although the BBC compilation (item 1531) remains the most extensive chronological reference for the Gulf conflict.

D. General studies

1534. Bacevich, Andrew J. "'Splendid Little War': America's Persian Gulf Adventure Ten Years On," pp. 149–64 in *The Gulf War of 1991 Reconsidered*, ed. Andrew J. Bacevich and Efraim Inbar. London: Frank Cass, 2003.

    Reviews the initially optimistic assessment of the results of the Gulf War in the United States. Later, despite President George H.W. Bush's assertion that the demonstration of United States power in the Gulf would deter destabilizing elements in the world, numerous interventions by the United States occurred in the 1990s. Thus, there has been a sweeping militarization of United States foreign policy. The Weinberger or Weinberger–Powell Doctrine has effectively been discarded.

1535. Bacevich, Andrew J. and Efraim Inbar, eds. *The Gulf War of 1991 Reconsidered*. London: Frank Cass, 2003.

Stimulating collection of essays that for the most part emphasize political rather than the military aspects of the conflict. Space constraints in this bibliography prevented inclusion of all the essays as separate entries, but several are listed and annotated.

1536. Bellamy, Christopher. *Expert Witness: A Defence Correspondent's Gulf War 1990–91*. London: Brassey's, 1993.

Early, but impressive, contribution to the literature about the war. Bellamy is a British correspondent, who provides a broad, highly thoughtful view of what happened in the Gulf. The fine graphics, including numerous maps, and the many unusual photographs make the book an important source by themselves.

1537. Bin, Alberto, Richard Hill, and Archer Jones. *Desert Storm: A Forgotten War*. Westport, CT: Praeger, 1998.

Sophisticated treatment that reviews all aspects of the conflict, including the political and economic facets. The authors' description of the campaign itself is outstanding. The authors make skillful use of statistical charts and sketch maps in explaining highly complex events. The wide range of the authors' research is also impressive, embracing, as it does, European, North African, and Middle East publications.

1538. DeAtkine, Norvell B. "Middle East Scholars and the Gulf War." *Parameters* 23, no. 2 (Summer 1993): 53–63.

Severely critical assessment of initial studies of the Gulf War by what DeAtkine terms "Arabist [sic] revisionists" (p. 53). Their predictions of a massive reaction in the Middle East against the Coalition partners were unrealized. The author also assails those critics who contended that the Bush administration should have totally defeated Iraq and asserts that an occupation of Iraq by the United States would have been disastrous for this country.

1539. Divine, Robert A. "Historians and the Gulf War: A Critique." *Diplomatic History* 19, no. 1 (Winter 1995): 117–34.

Examines eleven studies that deal with political or military facets of the war or both. A variety of points of view are represented in the accounts reviewed, but cynicism about the United States' intent and about the behavior of the United States is characteristic of many works. This is a valuable essay about a number of books that retain their importance years after the Gulf War.

1540. Friedman, Norman. *Desert Victory: The War for Kuwait*. Annapolis, MD: Naval Institute Press, 1991.

Early, but very detailed overview of the war. About a third of the book consists of appendices that discuss Coalition forces and their weapons and other equipment, Coalition and Iraqi air losses, and Iraqi naval units lost, and the use of Scud missiles by the Iraqis. There are extensive notes, but, for the most part, these expand statements in the text rather than offer references to sources.

1541. Gordon, Michael R. and Bernard E. Trainor. *The Generals' War: The Inside Story of the Conflict in the Gulf.* Boston, MA: Little, Brown and Company, 1995.

Surprisingly candid examination of the Gulf War, focusing, as the title suggests, on major commanders. Quite a bit is said about "inter-service rivalry," moreover. Interviews were carried out over a period of years as a war threatened and as it occurred. Moreover, further interviews were conducted after the end of hostilities. Interviewees included United States civilians and foreign nationals, as well as members of the United States armed forces. The book is admirably documented.

1542. Head, William and Earl H. Tilford, Jr., eds. *The Eagle in the Desert: Looking Back on U.S. Involvement in the Persian Gulf War.* Westport, CT: Praeger, 1996.

Leading defense analysts discuss all facets of the conflict, including logistics, and the roles of the air force, army, navy, and marines. The political and diplomatic aspects receive less attention, but there are two valuable chapters on those topics. The tables and graphics, some of which contain unusual information, are highly useful, as is the fine bibliography.

1543. Huchthausen, Peter. *America's Splendid Little Wars: A Short History of U.S. Military Engagements: 1975–2000.* Chap. 9, "The Gulf War: Desert Shield," pp. 127–41; Chap. 10, "The Gulf War: Desert Storm," pp. 142–51. Chap. 11, "The Rescue of the Kurds in Northern Iraq," pp. 152–58. New York: Viking, 2003.

Fine analyses of all three operations. Huchthausen does not neglect describing the weapons and tactics used against the Iraqis. He asserts that "Desert Shield was more demanding logistically and larger than any other operation in military history, including the Normandy landings in 1944" (p. 137). This statement seems debatable, though, given the size and complexity of some Soviet operations of World War II. Nevertheless, these three chapters are valuable introductions, which include highly useful maps.

1544. Kagan, Frederick W. and Chris Kubik, eds. *Leaders in War: West Point Remembers the 1991 Gulf War.* London: Frank Cass, 2005.

Generals and field grade officers of the United States Army wrote the chapters in this book, based on their experiences with respect to the Gulf War. These are graphic accounts of what occurred both in battle and in support activities. Two chapters discuss logistics, and other chapters describe such topics as military police operations and engineer support. The book is relatively short, but it contains valuable and highly readable statements.

1545. Khadduri, Majid and Edmund Ghareeb. *War in the Gulf, 1990–1991: The Iraq–Kuwait Conflict and Its Implications.* New York: Oxford University Press, 1997.

Valuable study for those interested in the origins of the war. The authors argue that there was more at issue than a simple attack by one country on another. They review the long history of disputed borders with respect to Kuwait and many peaceful, but unsuccessful, negotiations that had been undertaken long before 1990 to solve the problem of Kuwait. Thus, they come to the conclusion that various countries and international organizations were to blame, in part, for the war, not simply Iraq under Saddam Hussein.

1546. Leyden, Andrew. *Gulf War Debriefing Book: An After Action Report*, ed. Camille Akin. Grants Pass, OR: Hellgate Press, 1997.

Extremely useful reference book. The readable and seemingly quite objective volume is packed with helpful information about weapons and tactics and many other subjects. One intriguing table displays what is referred to as "ethnic breakdown" of the United States armed services and their forces in the Gulf War. The reference value of the book is diminished by the lack of an index or an extensive table of contents. The user is forced to thumb through the text, but that exercise is well worth the trouble.

1547. Mazarr, Michael J., Don M. Snider, and James A. Blackwell, Jr. *Desert Storm: The Gulf War and What We Learned.* Boulder, CO: Westview Press, 1993.

These experienced defense analysts argue that the United States pursued diplomatic courses extensively to avoid war in the Persian Gulf before resorting to force. The other half of the book reviews the war and ends with a realistic assessment of the pressures within the United States that militate against the preservation of effective armed forces in the post–Cold War era. The authors rightly emphasize the "jointness" of the effort of the armed services in the Gulf and the need to maintain that ability to operate almost as a single service.

1548. Mueller, John. "The Perfect Enemy: Assessing the Gulf War." *Security Studies* 5, no. 1 (Autumn 1995): 77–117.

Insightful study that focuses on Iraqi weaknesses, such as a high desertion rate well before the beginning of the Gulf War. Mueller analyzes the quality of Iraqi military leadership, which he finds quite deficient. He also examines the numbers of Iraqi military and civilian casualties in considerable detail, finding that those numbers have been much exaggerated.

1549. Nye, Joseph, Jr. and Roger K. Smith, eds. *After the Storm: Lessons from the Gulf War.* Lanham, MD: Madison Books, 1992.

Notable collection of essays written largely by leading defense analysts and journalists. Bernard E. Trainor, a retired marine general, is the sole senior military figure. Much attention is given to the economic and diplomatic results of the war, although several chapters examine its effects on the United States defense establishment. The book includes an extensive chronology and copies of many United Nations documents.

1550. Record, Jeffrey. *Hollow Victory: A Contrary View of the Gulf War.* Washington, DC: Brassey's (US), Inc., 1993.

Argues that Bush wanted to eliminate Saddam Hussein and the perceived military threat from Iraq. Conversely, the United States lost much by allowing Saddam to remain in control of Iraq. Thus, the conflict was not nearly as much of a success as it has often been viewed. One of the values of the book is Record's keen analysis of the diplomacy of the years just before the Gulf War. Record cautions that Coalition forces were so much stronger than those of Iraq that it may be unwise to draw many lessons from the conflict.

1551. United States. Department of Defense. *Conduct of the Persian Gulf War; Final Report to Congress.* Washington, DC: Department of Defense, 1992. All three volumes are available online through the Defense Technical Information Center located at Fort Belvoir, Virginia, at www.dtic.mil/dtic/search/tr/

Huge compilation of materials. One volume consists of eight chapters that run from the Iraqi seizure of Kuwait through air, naval, and ground campaigns. The two volumes of appendices are at least as important as the report. The first group of appendices discusses topics such as intelligence, the role of reserve forces, and the role of women in the conflict. Another volume was issued as "Appendix T, Performance of Selected Weapons Systems," which deals with a variety of land, sea, and air weapons.

1552. U.S. News and World Report. *Triumph Without Victory: The Unreported History of the Persian Gulf War.* New York: Times Books, 1992.

Compiled by the staff of a respected news journal, the book is an undocumented survey. U.S. News indicates that the "sources for the events described in this book have been the people directly involved" (x). About 600 interviews were undertaken with major figures, including President George H.W. Bush and Secretary of Defense Dick Cheney.

1553. Watson, Bruce W., ed. *Military Lessons of the Gulf War*. London: Greenhill Books, 1991.

Very early assessment of the war's "lessons" by authors from several countries, mostly from the United States or Great Britain and mostly coming from military intelligence backgrounds. References are largely to current news publications, but the studies are worthwhile owing to the authors' knowledge and experience. One section deals with diplomacy in the war, and the rest of the book examines military facets, including electronic warfare, logistics, and an interesting treatment of "Command and Control" by Joel H. Nadel.

1554. Yetiv, Steve A. *The Persian Gulf Crisis*. Westport, CT: Greenwood Press, 1997.

Exemplary guide to the subject. Yetiv, a seasoned specialist in Middle Eastern affairs, provides a carefully crafted discussion of the background to the war. An interesting chapter subtitled "A Classic Case of Individuals Driving History" focuses on George H. W. Bush and Saddam Hussein. An extensive annotated bibliography, a series of biographies, and other features complement the text.

E. Decision to go to war

1555. Chomsky, Noam. "The US and the Gulf Crisis," pp. 11–29 in *The Gulf War and the New World Order*, ed. Haim Bresheeth and Nira Yuval-Davis. London: Zed Books, 1991.

Sarcastic and highly critical analysis of United States policy regarding the Iraqi seizure of Kuwait. In this essay written on the eve of the Gulf War, Chomsky asserts that the United States has intentionally worked against any diplomatic solution of the crisis in favor of a war against Iraq, because of its desire to maintain control over oil resources in the region. He believes that Iraq would have agreed to a peaceful settlement.

1556. Dubois, Thomas R. "Weinberger Doctrine and the Liberation of Kuwait." *Parameters* 21, no. 4 (Winter 1991–92): 24–38.

Studies the process by which the United States government determined whether or not to commit the country to undoing Iraq's seizure of Kuwait. Dubois systematically analyzes the decision in terms of the principles for United States military action that had been enunciated by Secretary of Defense Caspar Weinberger.

1557. Francona, Rick. *Ally to Adversary: An Eyewitness Account of Iraq's Fall from Grace.* Annapolis, MD: Naval Institute Press, 1999.

Indispensable book for understanding United States–Iraqi relations before the Gulf War. Francona discusses a period between 1987 and 1988 when "the U.S. Department of Defense enjoyed a professional, cooperative relationship with the Iraqi Ministry of Defense" (xv), because concern about Iran put aside other considerations. Later, Francona was part of General Norman Schwarzkopf's staff as Schwarzkopf's interpreter, and thus he could record a valuable account of the planning and execution of the coalition effort to liberate Kuwait.

1558. Herspring, Dale R. *The Pentagon and the Presidency: Civil–Military Relations from FDR to George W. Bush.* Chap. 11, "The Military and George Bush," pp. 310–23. Lawrence: University Press of Kansas, 2005.

Relates at length the political moves that preceded the Gulf War. Bush and Secretary of Defense Dick Cheney wanted a decisive response once Kuwait had been invaded, but General Colin Powell counseled a more circumspect approach to the crisis. This segment deals largely with planning the war with a few paragraphs on ending it.

1559. Kirkpatrick, Jeane J. *Making War to Keep Peace.* Chap. 2, "Iraq Invades Kuwait," pp. 9–58. New York: HarperCollins, 2007.

Praises Bush's decisive behavior in the face of the first major crisis after the end of the Cold War. Despite his wish to take a determined stand against aggression, Bush took the time to build United Nations and congressional support for his actions, which was a wise course in Kirkpatrick's view, if he were to achieve the establishment of a "new world order."

1560. Nelson, W.H. "What's the Answer for the Gulf?" *United States Naval Institute Proceedings* 116, no. 12 (December 1990): 32–37.

Examines the geopolitical significance of the Persian Gulf region and the degree to which the United States has interests in the area. The author also discusses Soviet activity in the Middle East and reviews recent United States naval operations in the Persian Gulf.

1561. Renshon, Stanley A., ed. *The Political Psychology of the Gulf War: Leaders, Publics, and the Process of Conflict.* Pittsburgh, PA: University of Pittsburgh Press, 1993.

Early, but notable, collection of essays by leading scholars. A wide range of topics is covered, with significant material on public opinion in countries or areas other than the United States. All segments of the book are informative, but one of the most useful essays is Stephen J.

Wayne's "President Bush Goes to War: A Psychological Interpretation from a Distance." Renshon's "The Gulf War Revisited: Consequences, Controversies, and Interpretations" is an important analysis.

1562. Sherry, Michael S. *In the Shadow of War: The United States Since the 1930s.* Chap. 9, "A Farewell to Militarization? 1988–95," pp. 462–82. New Haven, CT: Yale University Press, 1995.

This is a highly important study that documents how warfare and the threat of war have molded culture in the United States and, more specifically, have helped shape decision-making on a wide range of issues in this country. In a segment of the chapter cited, Sherry provides an excellent review of the heated debate about the wisdom of initiating war against Iraq and an examination of the rapid decline of H.W. Bush's credibility soon after the end of the fighting. As in the other segments of his book, Sherry manages to offer numerous insights in a relatively few pages.

F. Air operations

1563. Andrews, William F. *Airpower Against an Army: Challenge and Response in CENTAF's Duel with the Republican Guard.* Maxwell Air Force Base, AL: Air University Press, 1998.

Important element in the literature about the air campaign. Andrews begins with a sophisticated discussion of the problems of writing about military events and then proceeds to describe in considerable detail both Iraq's Republican Guard and the air force response to the mission that it was given. Because the Republican Guard was viewed as the strongest component of the Iraqi armed forces, it was essential to destroy it. This assignment required the air force to make changes from its existing doctrine. Much of the book deals with the air force's ability to adapt. Andrews' valuable study includes a detailed chronology and an annotated bibliography.

1564. Clancy, Tom, with Chuck Horner. *Every Man a Tiger.* New York: G.P. Putnam's Sons, 1999.

Chuck Horner led the combined air campaign during the Gulf War, but Clancy and Horner went far beyond writing a biography of Horner in this book. It is largely a detailed, although undocumented, history of the preparation for the war and events that occurred during the fighting. The authors do not flinch from discussing the conflicts within the United States armed forces. Given Horner's air force career and Clancy's knowledge of the armed forces, it is hardly surprising that they give readers a "feel" for the strengths and weaknesses of the armed forces.

1565. Cohen, Eliot A. and Thomas A. Keaney, eds. *Gulf War Air Power Survey.* Washington, DC: Department of the Air Force, 1993.

Truly massive study consisting of a summary volume and five large specialized volumes. The detailed segments of the series deal with such topics as planning, operations, logistics, and weapons. There were some deletions of material for security reasons, especially information about "space operations." Volume 5 consists of statistical information and a thorough chronology. Although the *Monthly Catalog of U.S. Government Publications* does not cite an Internet source, the whole series is available through the Air Force Historical Studies Office (www.airforcehistory.hq.af.mil/Publications/Annotations/gwaps. htm). Simply typing in "Gulf War Air Power Survey" through a search engine should access this impressive work, moreover.

1566. Davis, Richard G. *Decisive Force: Strategic Bombing and the Gulf War.* Washington, DC: Air Force History and Museums Program, United States Air Force, 1996.

Succinct but important overview of the crucial role of the United States Air Force in the rapid victory of Coalition forces in the Gulf War. The book is largely text, but it includes a number of striking photographs and readable tables and diagrams.

1567. ———. *On Target: Organizing and Executing the Strategic Air Campaign Against Iraq.* Washington, DC: Air Force History and Museums Program, United States Air Force, 2002.

Impressive overview of the air force role in both Desert Shield and Desert Storm. Nearly half the book examines the Persian Gulf crisis, the movement of air force elements to the area, and planning for the war. Davis' research centered on "security classified USAF documents" and a considerable number of interviews of participants. Nevertheless, there is an extensive bibliography.

1568. Fabian, Robert. "Storm Warnings: Cruise Missile Lessons from the Gulf War." *United States Naval Institute Proceedings* 128, no. 10 (October 2002): 86–89.

Analyzes the effectiveness of air force and navy cruise missiles used against especially dangerous Iraqi targets. Despite their usefulness, cruise missiles cannot be used in isolation, but, rather, they need to be used only as part of a plan for an air offensive. Fabian also recommends a strategy for dealing with hostile cruise missiles in the future.

1569. Hallion, Richard P. *Storm Over Iraq: Air Power and the Gulf War.* Washington, DC: Smithsonian Institution Press, 1992.

Exemplary study of the United States air campaign in the Gulf War, focusing primarily on operations by the United States Air Force, although the marines and navy are by no means ignored. The superb maps and numerous charts and diagrams complement the text. Hallion often compares aspects of the Gulf War with other conflicts, such as World War II and Vietnam, a feature that puts the Gulf War into perspective. The documentation is excellent. Although this was an early effort and although more information is likely to appear, the book retains its importance.

1570. Jamieson, Perry D. *Lucrative Targets: The U.S. Air Force in the Kuwaiti Theater of Operations.* Washington, DC: Air Force History and Museums Program, United States Air Force, 2001.

As the title suggests, this study focuses on operations that occurred specifically in the Kuwait area, as opposed to Iraq generally. Jamieson describes the origins of the Kuwait crisis and the planning of the air campaign in some detail, but most of the book deals with operations. The study is massively documented, and there is a fine "Essay on Sources."

1571. Kennedy, William V. "The Tank is Dead but the Cavalry Lives On." *United States Naval Institute* Proceedings 120, no. 11 (November 1994): 50–53.

Examines in detail United States and French helicopter assaults on Iraqi armor. In view of the success of the helicopter in the Gulf War, Kennedy strongly questions United States commitment to its airborne and heavy armor forces.

1572. Mann, Edward C. *Thunder and Lightning: Desert Storm and the Airpower Debates.* Maxwell Air Force Base, AL: Air University Press, 1995.

This volume is part of a study that includes Richard T. Reynolds' *Heart of the Storm* (item 1576). Mann discusses the Gulf War air campaign in light of the development of United States air power since World War II. He believes that the air force performance in the Gulf vindicated air force doctrine, and he asserts that the air force effort was thoroughly in line with the parts played by the other services. In effect, he downplays inter-service rivalry concerning the Gulf War.

1573. Olsen, John Andreas. *Strategic Air Power in Desert Storm.* London: Frank Cass, 2003.

Thoroughly grounded in extensive research that included numerous interviews with participants, this is a major study of the subject. Much of the book is devoted to planning for the air war, although a long chapter analyzes the fighting. The author carefully discusses the availability of source material and his use of those items that are not classified. There is a seemingly comprehensive bibliography.

1574. Parker, Thomas A. "Navy Got It: Desert Storm's Wake-Up Call." *United States Naval Institute Proceedings* 120, no. 9 (September 1994): 32–36.

Asserts that naval aviation did not have plans for the kind of conflict represented by the Gulf War, but it has learned much and is now ready to participate more effectively in inter-service operations. After many years of defending its own ways of air warfare, the navy has accepted air force methods used in the Gulf War.

1575. Putney, Diane T. *Airpower Advantage: Planning the Gulf War Air Campaign, 1989–1991*. Washington, DC: Air Force History and Museums Air Operations Programs, U.S. Air Force, 2004.

Award-winning study that deals with every conceivable facet of planning for both Desert Shield and Desert Storm. This appears to be the definitive work on the subject. Much of the research and writing took place soon after the conclusion of Desert Storm, but years passed before the text could be declassified and authorized for issuance.

1576. Reynolds, Richard T. *Heart of the Storm: The Genesis of the Air Campaign Against Iraq*. Maxwell Air Force Base, AL: Air University Press, 1995.

Highly readable study of the air force leaders who crafted the doctrine applied in the Gulf War air campaign. Reynolds distributes a good deal of personal praise and personal blame in telling this story, which is based largely on oral history. This is a controversial but candid examination of strong differences of opinion within the air force about how to fight the rapidly approaching war. It is a companion piece to Edward C. Mann's less personal study (item 1572).

1577. Siegel, Adam B. "Scuds Against Al Jubayl?" *United States Naval Institute Proceedings* 128, no. 12 (December 2002): 34–36.

Warns that United States support facilities might be highly vulnerable during a second Gulf War. As an example of what could happen, Siegel describes an unsuccessful Iraqi cruise missile attack launched against the Al Jubayl pier in Saudi Arabia. The consequences of a hit would have been devastating. Missile systems are improving, which could make them much more dangerous than those used by Iraq in 1991.

1578. Vriesenga, Michael P. *From the Line in the Sand: Accounts of USAF Company Grade Officers in Support of Desert Shield/Desert Storm*. Maxwell Air Force Base, AL: Air University Press, 1994.

Consists of relatively long statements by air force officers at the rank of captain or first lieutenant with varied combat or support assignments during the Gulf War. The statements appear to be quite candid. They were selected from a number of contributions submitted voluntarily. The book also includes a large number of well-chosen illustrations.

1579. Winnefeld, James A., Preston Niblack, and Dana J. Johnson. *A League of Airmen: U.S. Air Power in the Gulf War*. Santa Monica, CA: RAND, 1994.

Thorough analysis that seeks to pinpoint to what degree the decisive victory in the Gulf War was owed to the effective application of air power by the United States. Clearly, air power was vital in this conflict, but future conflicts might be quite different. Therefore, the authors warn against either overestimating or underestimating the impact of this element in national power. The book is filled with fine maps and informative charts and diagrams that make the complex text easier to understand. The detailed list of abbreviations and acronyms fills an obvious need for users of the book.

G. Army land operations

1580. Bourque, Stephen A. *Jayhawk! The VII Corps in the Persian Gulf War*. Washington, DC: Center for Military History, 2002.

Fine example of an official military history. VII Corps provided much of the land force assembled for Desert Storm. This is the thoroughly documented and heavily illustrated account of the corps' experiences. The many attractive maps help make the movements on the battlefields quite clear. Bourque supplemented his archival and library research with a number of interviews with participants in the war. The bibliography is seemingly exhaustive.

1581. Brown, John S. "The Maturation of Operational Art: Operations Desert Shield and Desert Storm," pp. 439–81 in *Historical Perspectives of the Operation Art*, ed. Michael D. Krause and R. Cody Phillips. Washington, DC: Center of Military History, 2005.

To understand Brown's article, the reader needs first to study Bruce W. Menning's segment of the same book, entitled "Operational Art's Origins." Menning explains how the concept developed within the Red Army of the Soviet Union between the two world wars, in good part as a result of the increasing impact of technological developments on the conduct of war. The concept is difficult to explain briefly, but it involves planning and coordination well beyond fighting individual battles, and maintaining flexibility to change the configuration and approach of the forces engaged. Both Menning and Brown describe the influence of the operational art theories on United States military doctrine. Brown's article asserts that the limited number of United States casualties was a result of the "American mastery of the operational art" (p. 464).

1582. Caraccilo, Dominic J. *The Ready Brigade of the 82nd Airborne in Desert Storm*. Jefferson, NC: McFarland & Company, Inc., 1993.

Based on a number of sources, primarily then Captain Caraccilo's memories and the notes he kept, this is a vivid description first of the military experience in Saudi Arabia and later of the fighting during Desert Storm. In addition to being highly informative and extremely readable, the book includes a number of charts and other materials that further illustrate important points about Desert Shield and Desert Storm.

1583. Chapman, Craig S. "Gulf War: Nondeployed Roundouts." *Military Review* 72, no. 9 (September 1992): 20–35.

Highlights the fact that "round outs" (that is, reserve components of active army divisions) were hardly used in the Gulf War. The many advantages of the active army/round out concept were lost. Nevertheless, large numbers of army reservists were used.

1584. Sack, John. *Company C: The Real War in Iraq*. New York: Morrow, 1995.

Worthy successor to Sack's renowned *M* which recounted the training and deployment of an infantry company early in the Vietnam War. This book deals with Company C, 2nd Battalion, 34th Armor, United States Army as it left the United States, went into combat, and returned to its normal duty station. Sack filled the book with verbatim conversations that illustrate graphically the thoughts and emotions of the tankers.

1585. Scales, Robert H., Jr., and the Desert Storm Study Project. *Certain Victory: United States Army in the Gulf War*. Washington, DC: Office of the Chief of Staff, United States Army, 1993.

Scales led the Desert Storm Study Project in its effort to produce an extensive, yet readable account of the army's role in the Gulf War. The text of this well researched study is considerably enhanced with a careful selection of striking color photographs and by the inclusion of numerous colorful maps and charts.

1586. Schubert, Frank N. and Theresa L. Kraus, eds. *The Whirlwind War: The United States Army in Operations Desert Shield and Desert Storm*. Washington, DC: Center of Military History, 1995.

The editors state modestly and accurately that their book is not "definitive." Nevertheless, this is an impressive start, indeed, in describing the army's contribution to victory in the Gulf War. A concise chapter provides helpful historical background to the conflict. Army historians ensure that their works are well-illustrated and are provided with appropriate maps, and this volume is no exception. Important appendices discuss the "Patriot Air Defense System" and the equipment used by United States and Iraqi forces.

1587. Vernon, Alex, David Trybula, Greg Downey, Neal Creighton Jr., and Rob Holmes. *The Eyes of Orion: Five Tank Lieutenants in the Persian Gulf War.* Kent, OH: Kent State University Press, 1999.

Extremely candid and highly valuable accounts of the officers' experiences. Interestingly, by the time the book was published, only one of the officers was still on active duty. The readable text is enhanced by charts showing typical tank formations, numerous photographs, an extensive glossary, and a brief, annotated bibliography.

H. Marine operations

1588. Brown, Ronald J. *United States Marine Corps in the Persian Gulf, 1990–1991: Humanitarian Operations in Northern Iraq, 1991: With Marines in Operation Provide Comfort.* Washington, DC: History and Museums Division, Headquarters, U.S. Marine Corps, 1995.

Thoroughly documented and extensively detailed description of marine contributions to this mission. There are numerous valuable appendices, including a chronology, an explanation of acronyms and abbreviations, and a brief discussion of the Kurds.

1589. ———. *United States Marine Corps in the Persian Gulf, 1990–1991: With Marine Forces Afloat in Desert Shield and Desert Storm.* Washington, DC: History and Museums Division, Headquarters, U.S. Marine Corps, 1998.

Massive study of the marines who were assigned to ships to constitute a significant threat to Iraqi forces. Some amphibious operations were undertaken as part of the successful effort to draw some Iraqi troops away from the main battlefield. The study includes an extremely detailed chronology.

1590. Cureton, Charles H. *United States Marine Corps in the Persian Gulf, 1990–1991: With the 1st Marine Division in Desert Shield and Desert Storm.* Washington, DC: History and Museums Division, Headquarters, U.S. Marine Corps, 1993.

Another fine entry in the series. Unfortunately, the book seems to have been lost in the federal depository system. Unlike every other volume in the series, this one is not listed in the *Catalog of U.S. Government Publications.* Thus, there is no Superintendant of Documents number for Cureton's study. Few libraries seem to own copies.

1591. Lehrack, Otto J. *America's Battalion: Marines in the First Gulf War.* Tuscaloosa: University of Alabama Press, 2005.

Consists of extremely short, but highly candid, recollections by members of the 3rd Battalion, 3rd Marines, an organization the author wrote about in *No Shining Armor: The Marines at War in Vietnam.* Before proceeding to the marines' stories, Lehrack briefly outlines the experiences the force underwent and explains why most of the marines who gave statements were officers or NCOs. Much of the book deals with the Battle of Khafji, Saudi Arabia.

1592. Melson, Charles D., Evelyn A. Englander, and David A. Dawson. *U.S. Marines in the Persian Gulf, 1990–1991: Anthology and Annotated Bibliography.* Washington, DC: History and Museums Division, Headquarters, U.S. Marine Corps, 1992. Reprinted 1995.

Fine compilation of first-hand accounts by marine veterans of the Gulf War. There is also a rather detailed chronology and a bibliography of articles from the *Marine Corps Gazette* and the *United States Naval Institute Proceedings* about the conflict.

1593. Mroczkowski, Dennis P. *United States Marine Corps in the Persian Gulf, 1990–1991: With the 2D Marine Division in Desert Shield and Desert Storm.* Washington, DC: History and Museums Division, Headquarters, U.S. Marine Corps, 1993.

The author carefully warns readers of his book that it is only "intended to be a first effort to present the division's actions, operations, and contribution to victory" (v). Nevertheless, this is a solid contribution to the literature of the Gulf War. There are a full marine order of battle information, a fine chronology, an extensive section of references, and a useful explanation of acronyms and abbreviations.

1594. Quilter, Charles J., II. *United States Marine Corps in the Persian Gulf, 1990–1991: With the I Marine Expeditionary Force in Desert Shield and Desert Storm.* Washington, DC: History and Museums Division, Headquarters, U.S. Marine Corps, 1993.

General account of the marines ashore during the Gulf War. Although this is a "preliminary accounting" (iii), it is an impressive work fully grounded in primary materials and interviews with people directly involved in the marine effort. The text is almost evenly divided between Desert Shield and Desert Storm. Detailed statistical information about force strength and equipment is included.

1595. Quinn, John T., II. *United States Marine Corps in the Persian Gulf, 1990–1991: Marine Communications in Desert Shield and Desert Storm.* Washington, DC: History and Museums Division, Headquarters, U.S. Marine Corps, 1996.

Important study of an often neglected facet of military operations. This is a careful examination of all aspects of marine communications whether ashore or afloat during Desert Shield and Desert Storm. Essential appendices review Fleet Marine Communications on the eve

1596. Stearns, Leroy D. *United States Marine Corps in the Persian Gulf, 1990–1991: The 3d Marine Aircraft Wing in Desert Shield and Desert Storm*. Washington, DC: History and Museums Division, Headquarters, U.S. Marine Corps, 1999.

Authoritative study. The book is well-written and contains important ancillary material, including an extensive chronology and full order of battle information.

1597. Swofford, Anthony. *Jarhead: A Marine's Chronicles of the Gulf War and Other Battles*. New York: Scribner, 2003.

Somewhat rambling account by a member of a marine corps sniper team. Much of the book deals with Swofford's life before entering the marines, but, nevertheless, the author's memories of his military duty are well worth reading. At times, the book may remind the reader of Tim O'Brien's *The Things They Carried* or Stephen Wright's *Meditations in Green*.

1598. Zimmeck, Steven M. *United States Marine Corps in the Persian Gulf, 1990–1991: Combat Service Support in Desert Shield and Desert Storm*. Washington, DC: History and Museums Division, Headquarters, U.S. Marine Corps, 1999.

This large study of marine logistics in the Gulf War is filled with photographs, maps, and charts that enhance the text. There is also a detailed chronology. Among the other problems faced by the marines was the unexpected flow of surrendered Iraqi service personnel.

I. Naval operations

1599. Blaker, James R. "What Now, Navy?" *United States Naval Institute Proceedings* 118, no. 5 (May 1992): 58–62, 65.

Discusses six lessons for the navy from the Gulf War. These include the inevitability of joint operations and the increased usefulness of strategic air power. These principles have important implications for navy doctrine and structure.

1600. Marolda, Edward J. and Robert J. Schneller, Jr. *Shield and Sword: The United States Navy and the Persian Gulf War*. Washington, DC: Government Printing Office, 1998.

Massive, thoroughly documented, official study. The text is supplemented by numerous maps and illustrations, by an extensive bibliography that contains pithy remarks about many of the books and articles included, and by a large list of abbreviations used.

1601. Palmer, Michael A. *On Course to Desert Storm: The United States Navy and the Persian Gulf War*. Washington, DC: Government Printing Office, 1998.

Focuses largely on the United States naval presence after World War II. The book does not deal at all with Desert Shield or Desert Storm, but it is an important background publication with extensive notes, a large bibliography, and some interesting illustrations.

1602. Pokrant, Marvin. *Desert Shield at Sea: What the Navy Really Did*. Westport, CT: Greenwood Press, 1999.

Pokrant, Marvin. *Desert Storm at Sea: What the Navy Really Did*. Westport, CT: Greenwood Press, 1999.

Thoroughly professional analyses of United States naval operations. The skillful integration of text, maps, and illustrations make these volumes a model of military writing. The value of the set is further enhanced by biographical sketches of many leaders and by extensive bibliographies.

1603. Truver, Scott C. "Exploding the Mine Warfare Myth." *United States Naval Institute Proceedings* 120, no. 10 (October 1994): 36–43.

Defends the United States Navy against charges that its minesweeping capabilities are not as advanced as other countries' navies. There were problems in United States minesweeping operations during the Gulf War, but overall performance was thoroughly competent. Moreover, the navy learned much from the Gulf War, and it has developed an operational plan for future challenges.

J. Special military operations

1604. Carney, John T., Jr., and Benjamin F. Schemmer. *No Room for Error: The Covert Operations of America's Special Tactics Units from Iran to Afghanistan*. Chap. 14, pp. 224–44. New York: Ballantine Books, 2002.

Despite their success in Panama in 1989, Joint Special Operations Forces (JSOF) only won the right to participate in the Gulf War and its antecedent and subsequent operations by asserting their value to higher commanders, notably General Norman Schwarzkopf. About half this chapter deals with humanitarian aid to Iraqi Kurds through Operation Provide Comfort.

K. Special military studies

1605. Mauroni, Albert J. *Chemical-Biological Defense: U.S. Military Policies and Decisions in the Gulf War*. Westport, CT: Praeger, 1998.

One of the major concerns of Coalition planners before the Gulf War was the considerable possibility of friendly forces encountering chemical and biological weapons unleashed by the Iraqis. Mauroni discusses how this real possibility impacted planning and execution of the Coalition's liberation of Kuwait and further military moves. This is an essential book for understanding the war. It is rather technical, and it requires careful reading. There is an instructive glossary that most readers will find useful.

L. Logistics

1606. Pagonis, William G. with Jeffrey L. Cruikshank. *Moving Mountains: Lessons in Leadership and Logistics from the Gulf War.* Boston, MA: Harvard Business School Press, 1992.

Remarkable book that describes the unusually complicated logistics situation before, during, and after the Gulf War in a thoroughly readable fashion. Pagonis, who headed the logistics effort and who has gone on to a most successful career in the business world, makes logistics comprehensible and downright interesting to non-military readers. This is an indispensable part of the literature of the Gulf War.

1607. Pierce, Terry C. "Voodoo Logistics Sink Triphibious Warfare." *United States Naval Institute Proceedings* 122, no. 9 (September 1996): 74–77.

Interesting comparison between the battle of Guadalcanal and the Gulf War. From a logistical standpoint, the Guadalcanal invasion had serious flaws. A lack of logistical planning occurred in the Gulf War, too, but with less serious effects because Iraq was much weaker than Japan and because facilities existed for cargo handling in the Persian Gulf that were totally unavailable on Guadalcanal.

M. Civil affairs/civic action

1608. Nash, Douglas E. "Civil Affairs in the Gulf War: Administration of an Occupied Town." *Special Warfare* 7, no. 4 (October 1994): 18–27.

Emphasizes the many types of assignments given to United States Army civil affairs units in the Gulf War. Nash describes the deployment of a civil affairs company at As Salman in support of the 6th Light Armored Division of the French Army. This is a highly informational account of the many complexities of governing the town. One advantage was that As Salman had not been heavily damaged during the fight.

1609. Rudd, Gordon W. *Humanitarian Intervention: Assisting the Iraqi Kurds in Operation Provide Comfort, 1991.* Washington, DC: United States Army Center of Military History, 2004.

Thorough study by an army historian who was assigned to record the events of this operation. Rudd later wrote a Ph.D. dissertation at Duke University about the episode. This book is a condensation and revision of the dissertation. It may be the most extensive discussion of a humanitarian effort by military forces. An introductory chapter provides essential information about the Kurds and their history. Outstanding graphics, including numerous maps and charts, clarify the text.

1610. Whitters, Robert A. "Civil Affairs in the Gulf War." *United States Naval Institute Proceedings* 118 (November 1992): 107–8.

Discusses the experiences of the marine corps' 3rd and 4th Civil Affairs Groups (CAGs) in the Gulf War. Whitters notes that commanders "did not know how to employ ... [CAG] personnel effectively" (p. 107), and describes differences between the army's and the marine's views of civil affairs responsibilities.

N. Political aspects

1611. Collins, John M. "Options in the Middle East." *United States Naval Institute Proceedings* 116, no. 10 (October 1990): 119–22.

Interesting capsule comments about a number of events and possible events and the responses that might be made to them.

1612. Curriden, Lorrie Lynn. "The Gulf Between Congress and the President Over War Powers: An Analysis of the Persian Gulf War." M.A. thesis, University of Nevada, Las Vegas, 1995.

The Gulf War rekindled the perpetual debate over the relative powers of the executive and legislative branches in respect to the commitment of United States military forces. President George H.W. Bush justified his actions not only on his authority as commander-in-chief of the armed forces, but also on the various actions of the United Nations that he asserted gave him further authority. Congress emphasized its role in making such decisions under the terms of the War Powers Resolution.

1613. Mahnken, Thomas G. "A Squandered Opportunity? The Decision to End the Gulf War," pp. 121–48 in *The Gulf War of 1991 Reconsidered*, eds. Andrew J. Bacevich and Efraim Inbar. London: Frank Cass, 2003.

Answers the question posed in the title affirmatively. Mahnken contends that there were possibilities for a settlement in the Gulf that would have ensured a long-lasting period of peace without undertaking a complete occupation of Iraq. Much of the essay examined discussions within the Bush administration about how and when to end the war. Mahnken seems to be especially critical of General Norman Schwarzkopf.

1614. Menos, Dennis. *Arms Over Diplomacy: Reflections on the Persian Gulf War.* Westport, CT: Praeger, 1992.

Thoughtful examination of the origins of the Gulf War and the role that diplomacy could have played in the crisis, had the United States and Iraq been so inclined. Much of the relatively brief text consists of essays, some written during the war and some after the end of the fighting. There are a number of appendices that contain the texts of relevant documents, and there is a fairly detailed chronology. As Menos admits, a year or so after the Gulf War was "much too early to judge whether the fears and concerns of the war's opponents were truly justified" (p. 8). Interestingly, in view of later developments, Menos asserts that "Iraq has a worldwide terrorist apparatus ready to strike on signal from Baghdad" (p. 33).

1615. Rashid, Nasser Ibrahim and Esber Ibrahim Shaheen. *Saudi Arabia and the Gulf War.* Joplin, MO: International Institute of Technology, Inc., 1992.

Interesting Saudi Arabian perspective on the Gulf War that includes considerable invective against Saddam Hussein. The text is supplemented by detailed maps, a chronology, and a summary of pertinent United Nations resolutions.

O. Legal aspects

1616. Borch, Frederic L. *Judge Advocates in Combat: Army Lawyers in Military Operations from Vietnam to Haiti.* Chap. 4, "Army Lawyers in Desert Shield, 1990–91," pp. 121–64; Chap. 5, "Operation Desert Storm, 1991," pp. 165–97. Washington, DC: Government Printing Office, 2001.

Special regulations were put in place to facilitate the stationing of U.S. forces in Saudi Arabia prior to the Gulf War. Many judge advocates became involved in supply matters, and they rendered important contributions. Legal operations were complicated to a degree by the appearance of many reservists.

1617. ———. *Judge Advocates in Combat: Army Lawyers in Military Operations from Vietnam to Haiti.* Chap. 8, "Judge Advocates in Operations Other Than War, 1965–94 (Operation Provide Comfort)," pp. 281–87). Washington, DC: Government Printing Office, 2001.

During relief efforts for the Kurds in Iraq, judge advocates were faced with complex questions regarding rules of engagement and damage claims submitted by Kurds.

1618. Roberts, Adam. "Laws of War in the 1990–91 Gulf Conflict." *International Security* 18, no. 3 (Winter 1993–94): 131–81.

Begins with a review of major pertinent treaties that had numerous parties to the Gulf War as signatories. There was considerable uncertainty about Iraq's acceptance of major international standards of warfare. Roberts then reviews the legal issues that developed in relation to the war, giving significant attention to the legal facets of air operations.

1619. Van Creveld, Martin. "Persian Gulf Crisis of 1990–91 and the Future of Morally Constrained War." *Parameters* 22, no. 2 (Summer 1992): 21–40.

Noted scholar believes that the Gulf War was different from many earlier wars, "that much new ground has been broken" (p. 21). Van Creveld compares several "rules of war" to what happened in this conflict. He argues that "the traditional rules of war ... are being weakened and bastardized" (p. 39), and he supports efforts to stop the process.

P. The Gulf War and the media and the role of public opinion

1620. Bennett, W. Lance and David L. Platz, eds. *Taken by Storm: The Media, Public Opinion, and U.S. Foreign Policy in the Gulf War.* Chicago: University of Chicago Press, 1994.

Wide-ranging collection of essays by leading scholars in political science, sociology, journalism, and communication, such as Marvin Kalb. The thoroughness of *Taken by Storm* is suggested by the inclusion of an essay by Gladys Engel Lang and Kurt Lang subtitled "Media Coverage of Saddam's Iraq, 1979–90."

1621. Denton, Robert E., Jr., ed. *The Media and the Persian Gulf War.* Westport, CT: Praeger, 1993.

Stimulating collection of essays analyzing numerous aspects of the topic, including "Media Coverage of Women in the Gulf War" by Anne Johnston, and a discussion of United States films relating to the Middle East by Stephen Prince. Each essay includes a thorough bibliography, and, finally, there is an extensive bibliography for the entire work. Unfortunately, space concerns prevented listing and annotating all of these valuable essays.

1622. Greenberg, Bradley S. and Walter Gantz, eds. *Desert Storm and the Mass Media.* Cresskill, NJ: Hampton Press, Inc., 1993.

Large collection of intensively researched articles that generally employ sophisticated quantitative methods. Five entries are devoted to "Children's Response to the War." With few exceptions, the many contributors are professors of journalism or communication. Given the numerous topics explored in the articles and the wealth of data gathered and presented, Greenberg's "Summary and Commentary" is of particular importance.

1623. Keeble, Richard. *Secret State, Silent Press: New Militarism, the Gulf and the Modern Image of Warfare.* Luton, U.K.: University of Luton Press, 1997.

Controversial, but important, study from a British perspective. Keeble carefully examined the role of "myth" in the Gulf War. Arguing that there "is no massive conspiracy to con the public," he suggests that myths have developed from broad trends. This book draws from a great many sources, ranging from interviews with journalists to studies of numerous newspapers. The author condemns the war, which he regards as no "war" at all, but simply a series of "massacres."

1624. Licatovich, Bill. "Iraqi Propaganda Ploys." *Soldiers* 46, no. 1 (January 1991): 6–8.

Discusses the propaganda war of nerves waged by Iraq against the presence of United States forces in Saudi Arabia. Several anti-United States Iraqi cartoons are illustrated and translated.

1625. McCain, Thomas A. and Leonard Shyles, eds. *The 1,000 Hour War: Communication in the Gulf.* Westport, CT: Greenwood Press, 1994.

Broad approach to the role of the media in the conflict. Separate articles deal with C-SPAN and CNN. Although this is a major work, only one article is separately annotated, which discusses psychological warfare, which seems by its nature to be quite different from the other segments of the book.

1626. Mueller, John. *Policy and Opinion in the Gulf War.* Chicago: University of Chicago Press, 1994.

Indispensable analysis of the flow of opinion during the war. Mueller carefully describes the nature of public opinion research and his own methods. Approximately half of the book consists of the texts of questions asked by polling organizations and statistical breakdowns of the responses. There is also an excellent section of "references."

1627. Nohrstedt, Stig A. "Images of the UN in *Dagens Nyheter* and the *Washington Post* during the Gulf War," pp. 185–201 in *Journalism and the New World Order: Studying War and the Media*, eds Wilhelm Kempf and Heikki Luostarinen. Goeteborg: Nordicom Goeteborg University, Sweden, 2003.

Compares the treatment of three major events related to the Gulf War: the UN Security Council's vote on resolution 678, the authorization of military force by the United States Congress, and the effort to aid the Kurds in northern Iraq, in the newspapers noted. The purpose of the study was to ascertain "whether, and to what extent, the US perspective successively [sic] came to dominate the reporting of one of Sweden's leading newspapers" (p. 186). The authors carefully described the qualitative and quantitative measurements used in their analysis of news coverage. They found that the presentations of the two newspapers converged because of "a shift in the position of the Swedish paper's reporting toward that of the US paper" (p. 200).

1628. Parker, Jay M., and Jerold L. Hale. "Psychological Operations in the Gulf War: Analyzing Key Themes in Battlefield Leaflets," pp. 89–109 in *The 1,000 Hour War: Communication in the Gulf*, ed. Thomas A. McCain and Leonard Shyles. Westport, CT: Greenwood Press, 1994.

Goes beyond discussing the leaflets aimed at enemy combatants by analyzing United States psychological warfare generally, emphasizing the Gulf War effort and considering the effectiveness of those leaflets. Although the techniques employed were not uniformly successful and although some errors were made, overall, psychological operations appeared to have played a significant role in the campaign.

1629. Smith, Hedrick, ed. *The Media and the Gulf War*. Washington, DC: Seven Locks Press, 1992.

Valuable collection of essays, statements, and other materials that include various points of view about media involvement with the Gulf War. Many journalists express concerns about the restrictions and say that they felt hampered in their work. Many items are reprints, but, nevertheless, they are usefully gathered here. A detailed chronology of the war supplements the text.

1630. Swank, Eric. "The Ebbs and Flows of Gulf War Protests." *Journal of Political and Military Sociology* 25, no. 2 (Winter 1997): 211–29.

Used content analysis of newspaper accounts of anti-war meetings as a primary tool for assessing the development of the opposition movement. Swank carefully explains his research techniques, making this a useful contribution to the literature concerning content analysis. Swank concludes that protests occurred in numerous United States cities well before the beginning of hostilities.

1631. Taylor, Philip D. *Propaganda and Persuasion in the Gulf War*. 2nd ed. Manchester, U.K.: Manchester University Press, 1998.

Focuses on the ways in which Iraq and its opponents waged war through the media. Taylor's work is based heavily upon the media presentations recorded at the Institute of Communications Studies at the University of Leeds, which recorded a huge amount of material related to the Gulf War. The book emphasizes three episodes: the battles of Khafji and the Mutlah Gap and the air attack at Amiriya. Taylor asserts that although there were many journalists from a highly diverse selection of agencies, those journalists "were all essentially dependent upon the coalition military for their principal source of information about the progress of the war" (p. 268).

1632. Thomson, Alex. *Smokescreen: The Media, The Censors, The Gulf.* Tunbridge Wells, U.K.: Laburnham and Spellmount, Ltd., 1992.

Discusses the experiences of British and United States journalists who reported the Gulf War either within the "pool" or independently. Thomson describes a number of incidents in which journalists and the British and United States military clashed. The book is based on several hundred interviews with journalists and military censors and what Thomson calls "minders."

1633. Williams, Peter. "The Press and the Persian Gulf War." *Parameters* 21, no. 3 (Autumn 1991): 2–9.

The Assistant Secretary of Defense for Public Affairs seeks to rebut assertions that the media did badly in reporting the Gulf War and did not have an adequate opportunity to do so owing to government restrictions. Williams argues that a media pool was the only way for reporters to enter Saudi Arabia. He states that United States media policies during the Gulf conflict were based on World War II, Korean War, and Vietnam War precedents.

XXVIII

Somalia, 1992–94

United States forces operated in Somalia as part of a United Nations effort to provide humanitarian assistance to the people of a country that had been ravaged by civil war. Divisions within Somali society and uncertainty within the international community made the situation extremely complex and highly volatile.

A fully successful relief effort would probably have required an extensive, mandatory disarmament operation under United Nations auspices. Instead, the intervention forces sometimes negotiated with, sometimes assailed, one of the most important Somali leaders, Mohamed Farah Aideed (name spelled in various fashions in Western publications).

President George H.W. Bush sent United States forces to Somalia in 1992 near the end of his term of office, in response to a United Nations decision. President Bill Clinton inherited the Bush administration's commitment, and most of the involvement in Somalia occurred during the Clinton presidency.

A number of clashes occurred between United Nations and Somali forces. Finally, there was a major battle in Mogadishu in October 1993. United States forces incurred casualties, and President Clinton decided to withdraw most of the United States element of the international force.

Chronology

1960
July 1   After years as two European colonies and, later, as an Italian-administered trusteeship, Somalia received independence.

1969   Mohammed Siad Barre established a dictatorship in Somalia.

1977	Somalia attacked Ethiopia but was defeated. Barre continued to lead Somalia, despite increasing resistance to his rule.
1991	
September 27	Barre left Mogadishu, the capital of Somalia, after heavy fighting in the city. A number of contending factions threw the country into chaos. A massive drought severely worsened living conditions.
1992	
January 23	The United Nations Security Council proposed a cease-fire and asked United Nations members to provide assistance to the starving population of Somalia.
July 23	Initial United Nations group arrived in Somalia.
July 30	A Somalia Working Group was established by President H.W. Bush to deal with the problems of starvation and instability.
August 28	First United States humanitarian flight made to Belet Uen, Somalia. The use of military rather than civilian aircraft caused a controversy that persisted. Meanwhile, the United States became committed to helping establish "points of security" to facilitate food distribution.
November 25	On the advice of military advisers, President H.W. Bush decided to make a major commitment of United States forces to Somalia to undertake disarmament of warring groups and other measures in the face of the collapse of Somali society.
December 4	President Bush made known his decision to send forces to Somalia in what became known as Operation Restore Hope. During early December a number of other countries pledged support.
December 9	First United States forces, consisting of marine and naval personnel, arrived at Mogadishu. Eventually, a Unified Task Force for Somalia (UNITAF) was established. United Nations Operation in Somalia (UNOSOM) would gradually assume responsibility from UNITAF.
1993	
January 28	United States Marine Corps Lieutenant General Robert B. Johnson, commanding United States forces, reported that conflict in Somalia had decreased and that aid was being supplied satisfactorily. Further planning of the transition from UNITAF to UNOSOM (now UNOSOMII) was now considered feasible.
May 4	UNOSOMII replaced UNITAF. The security situation began deteriorating, partly because disarmament of rival groups had been far from complete.

June 6	United Nations Security Council resolution 837 issued in the wake of a serious attack on Pakistani forces in Somalia. It authorized vigorous action to ensure punishment of the attackers. During the next several months United Nations, including United States, forces tried to open roads and to reduce the military capability of some Somali factions.
August	Various attacks killed or wounded United States military personnel.
October 3	A United States raid in Mogadishu encountered significant Somali resistance, and heavy fighting occurred.
October 7	President Clinton announced that additional United States forces were to be sent to Somalia but that most of the United States commitment would end by March 31, 1994.
1994	
March 25	The final military contingent, consisting of 1,100 marines, except for guards protecting the United States Liaison Office, departed from Mogadishu.
September 19	Marine detachment of 55 left Somalia when the United States Liaison Office was ended.

A. General studies

1634. Brune, Lester H. *The United States and Post–Cold War Interventions: Bush and Clinton in Somalia, Haiti, and Bosnia, 1992–1998.* Chap. 2, "The Somalia Intervention," pp. 13–36. Claremont, CA: Regina Books, 1998.

Briefly sketches the historical development of Somalia and then examines the events that prompted intervention in the troubled country. Brune is highly critical of the performances of the United Nations and both the Bush and Clinton administrations. The author aptly describes the United Nations effort as "a quick police raid that provided no effective aid after the police left" (p. 23).

1635. Cassidy, Robert M. *Peacekeeping in the Abyss: British and American Peacekeeping Doctrine and Practice After the Cold War.* Chap. 5, "The American Military in Somalia—Into the Abyss," pp. 149–73. Westport, CT: Praeger, 2004.

Suggests that just as the outcome of the Vietnam War temporarily discouraged the United States from becoming involved in large-scale, open-ended conflicts, the Somalia experience discouraged the United States temporarily from becoming engaged in international humanitarian efforts under most circumstances. Cassidy skillfully analyzes the interplay of military and political events during the Somalia mission.

He identifies the critical problem in Somalia as the confusion of purely military and peacekeeping functions. Although the objectives were supposed to be humanitarian relief and a reduction in violence, the propensity of the United States armed services to use relatively heavy firepower helped change the situation. They came "to abandon the OOTW [operations other than war] principle of restraint, and thus legitimacy" as peacekeepers (p. 165).

1636. Connaughton, Richard. *Military Intervention and Peacekeeping: The Reality.* Chap. 6, "'If you liked Beirut, you will love Somalia'— Ambassador Smith Hempstone," pp. 111–37. Aldershot, U.K.: Ashgate Publishing Limited, 2001.

In contrast to Cassidy (item 1635), Connaughton stresses the political aspects of the Somalia mission, although he hardly ignores its military events. He discusses the major tensions and disagreements that existed between institutions and individuals, which made progress in Somalia so difficult. In contrast to the Italians and some others who advocated keeping to the original peacekeeping objective, some United States and United Nations leaders pursued broader aims that involved outright restructuring of Somalia's political life. Such ambitious plans naturally encountered considerable opposition from many in Somalia.

1637. Hendrickson, Ryan C. *The Clinton Wars: The Constitution, Congress, and War Powers.* Chap. 3, "Somalia," pp. 21–42. Nashville, TN: Vanderbilt University Press, 2002.

Begins with a capsule, but helpful, discussion of Somalia and its problems. Hendrickson then usefully summarizes and analyzes the reaction of Congress to the development of a United States policy toward the Somalia crisis. Although the Clinton administration's view of the War Powers Resolution and its requirements was not significantly different than earlier Democratic and Republican administrations, Clinton did keep Congress informed of his actions.

1638. Hippel, Karin von. *Democracy by Force: U.S. Military Intervention in the Post–Cold War World.* Chap. 3, "Disappointed and Defeated in Somalia," pp. 55–91. Cambridge: Cambridge University Press, 2000.

This is a highly straightforward account that effectively uses the available sources of information. Von Hippel is not afraid to assign blame for bad results in her detailed discussion, first of political developments in Somalia and second of United Nations and United States efforts to remedy the situation. Despite the unhappy result in Somalia, the endeavor was not without positive effects. Both the United Nations and the United States learned some significant lessons, which were fruitfully applied in Bosnia and Haiti. Especially in view of the complexities of the situation, von Hippel's lucid account is a major contribution to the literature about this episode.

1639. Huchthausen, Peter. *America's Splendid Little Wars: A Short History of U.S. Military Engagements: 1975–2000*. Chap. 12, "President Bush Responds to Starvation," pp. 161–69; chap. 13, "President Clinton Crosses the Mogadishu Line," pp. 170–82. New York: Viking, 2003.

Sees the Somalia intervention as simply a humanitarian effort, which thus differed from many other United States military commitments. Huchthausen provides a good summary of the events in Somalia that led to United Nations and United States actions. Although he refrains from offering outright judgments, Hutchhausen presents material that is implicitly critical of the Clinton administration. He also emphasizes the skill of the Somali opponents of the intervention in both the political and military spheres.

1640. Kirkpatrick, Jeane J. *Making War to Keep Peace*. Chap. 3, "Saving Somalia," pp. 59–111. New York: HarperCollins, 2007.

Highlights the initial struggle between President George Bush and UN Secretary-General Boutros-Ghali over the depth and scope of the United States' commitment to Operation Restore Hope. Bush wished to keep the operation within strict limits, but Boutros-Ghali saw the effort as a broad-gauged endeavor that would set the pattern for future UN interventions. The Clinton administration tended to take Boutros-Ghali's broad view of the situation, a stance that Kirkpatrick criticizes sharply.

1641. Klarevas, Louis J. "The United States Peace Operation in Somalia." *Public Opinion* 64, no. 4 (2000): 523–40.

Asserts that humanitarian aid efforts are becoming the typical United States military involvements abroad. Public opinion polling data shows that the public was firmly in favor of the United States commitment in Somalia at first, but views changed radically as United States forces began to sustain casualties. Although disillusionment with the Somalia involvement was high during the Clinton administration, H.W. Bush was largely blamed for the lack of a successful outcome. Much of the article consists of detailed polling data.

1642. Phillips, Jason. "International Norms, American Foreign Policy and African Famine: A Comparative Study of U.S. Humanitarian Policy in Nigeria/Biafra, Ethiopia, and Somalia." Ph.D. diss., Johns Hopkins University, 1998.

Stimulating study that examines the roots of United States involvement in famine relief. Phillips identifies "norm entrepreneurs" in the United

States that vary from members of the Congress to media figures who persuade the federal government to act against famines. This is an important contribution to the research about decision-making.

1643. Weiss, Thomas G. *Military–Civilian Interactions: Humanitarian Crises and the Responsibility to Protect.* 2nd ed. Chap. 4, "Somalia, 1992–95: The Death of Pollyannaish Humanitarianism?" pp. 55–70. Lanham, MD: Rowman & Littlefield Publishers, Inc., 2005.

Highly precise discussion of this complicated situation, which had been further aggravated by significant arms shipments to Somalia by many countries. Weiss reviews the mounting conflict in Somalia and then describes the various foreign efforts to aid the country. He pinpoints the numerous deficiencies in the assistance programs, including a fundamental lack of coordination between the intervening powers. This chapter is a fine introduction to the disheartening story of Somalia in the 1990s.

1644. Woodward, Peter. *US Foreign Policy and the Horn of Africa.* Chap. 4, "Intervention in Somalia," pp. 59–75. Aldershot, U.K.: Ashgate Publishing Ltd., 2005.

Incisive study that usefully puts the Somalia operation into the context of United States relations with the various states near the Horn of Africa. This is a skillful examination of the political development of the mission in Somalia. Military aspects are also, and necessarily, addressed appropriately. Woodward makes the point several times that both the United States and the United Nations were not unacquainted with the situation in Somalia, and, therefore, they should have been much better prepared for the problems that occurred.

B. Military studies: general

1645. Allard, C. Kenneth. *Somalia Operations: Lessons Learned.* Washington, DC: National Defense University Press, 1995.

Searching analysis of the lessons to be learned from Somalia. This is a notable contribution to the literatures of both the Somalia effort and peace operations in general. The "lessons learned" relate to such subjects as planning issues and deployment methods. There is a useful bibliography of relevant publications issued by the armed services and the National Defense University.

1646. Cooling, Norman L. "Operation Restore Hope in Somalia: A Tactical Action Turned Strategic Defeat." *Marine Corps Gazette* 85, no. 9 (September 2001): 92–106.

Thorough survey of military action during the Somalia operation. This is an impressive analysis of the shortcomings of the United

States/United Nations endeavor, especially the complicated command and control structure, problems in establishing reasonable rules of engagement, and limitations of the amount of intelligence available. Cooling is not afraid to name names. He is sharply critical of some decisions of major commanders and of President Bill Clinton.

1647. Garcia, Elroy. "Hoping for the Best, Expecting the Worst." *Soldiers* 49 (February 1994): 13–16.

Describes the increased United States military presence in Somalia in the wake of the fighting on October 3–4, 1993 and the anger United States personnel felt toward those they termed "bandits." Garcia emphasizes the usefulness of armored vehicles both as protection for United States personnel and as a deterrent against actions by insurgents.

1648. Hoffman, Frank G. "One Decade Later: Debacle in Somalia." *United States Naval Institute Proceedings* 130, no. 1 (January 2004): 66–71.

Summarizes a conference held at the McCormick Tribune Foundation in 2003 to review events in Somalia and to draw appropriate lessons from them. Hoffman describes the Somalia effort in some detail and with considerable clarity. There were some positive periods in Somalia, but there was confusion generated by the seemingly conflicting goals established at some points.

1649. Kendrick, William A. "Peacekeeping Operations in Somalia." *Infantry* 85, no. 3 (May–June 1995): 31–35.

Discusses both the advantages United States forces had in Somalia and the areas of improvement that the experience identified. More needed to be done in civil affairs, including the establishment of checkpoints. There were limitations in the effectiveness of light infantry because of the lack of armored forces. Kendrick also provides a general assessment of the relevance of several current army field manuals to Somalia.

1650. Miles, Donna. "Farewell to Somalia." *Soldiers* 49 (May 1994): 24–25.

Summarizes Operation Provide Hope and lists army personnel who were killed.

1651. Poole, Walter S. *The Effort to Save Somalia, August 1992–March 1994*. Washington, DC: Office of the Chairman of the Joint Chiefs of Staff Joint History Office, 2005.

Devoted entirely to the roles of the Joint Chiefs of Staff and its Joint Staff in regard to the operation. For an official document, this history is unusually outspoken in recording the misgivings of the Joint Chiefs of Staff and other governmental bodies about this humanitarian effort.

1652. *United States Forces, Somalia After Action Report and Historical Overview: The United States Army in Somalia, 1992–1994.* Washington, DC: Center of Military History, 2003.

The authoritative official history to date of army activities in Somalia. The extensive after action report is preceded by a fine essay written by Richard W. Stewart, a Center of Military History historian, that summarizes the Somalia experience. The extensive text is supplemented by excellent maps and imaginative graphics that convey complicated information in an easily understood fashion. The detailed tables of contents for the several sections are helpful, but they do not entirely replace a subject index. Non-military readers may be confused by the extensive use of acronyms. Unfortunately, there is no list of those acronyms and their meanings. Nonetheless, this is an essential source.

1653. Zaalberg, Thijs W. Brocades. *Soldiers and Civil Power: Supporting or Substituting Civil Authorities in Modern Peace Operations.* Chap. 6, "'Peacekeeping' in a Power Vacuum: The Reluctant American Occupation of Somalia," pp. 161–97. Amsterdam: Amsterdam University Press, 2006.

This is a thoughtful, highly informative, and heavily documented survey of Operation Restore Hope. Zaalberg describes the military aspects of the mission in considerable and lucid detail. He believes the United States needed to take a broader view of its responsibilities than it did, and he compares United States policies and actions unfavorably with those of the Australians, which he discusses in chapter 7 of this work.

C. Military studies: intelligence

1654. Brand, David L., Paul J. Bryson, and Alfredo Lopez, Jr. "Intelligence Support to the Logistician in Somalia." *Military Intelligence* 20, no. 4 (October/December 1994): 5–8.

Discusses the elements in providing intelligence to the United States Army Corps Support Command in Somalia. Intelligence elements had to assist this command in a combat environment where a variety of hostile Somali forces might be encountered. The authors conclude that more needs to be done to prepare military intelligence personnel assigned to support units and commands.

1655. Murphy, John R. "Memories of Somalia." *Marine Corps Gazette* 82, no. 4 (April 1998): 20–25.

Murphy served as an intelligence specialist in Somalia. He notes what he learned from his duty in the country, especially the need to avoid downplaying the enemy's abilities.

1656. Shelton, David L. "Intelligence Lessons Known and Revealed During Operation Restore Hope Somalia." *Marine Corps Gazette* 79, no. 2 (February 1995): 37–40.

General survey of intelligence operations in Somalia, which Shelton asserts were generally quite effective.

D. Military studies: logistics

1657. Hasenauer, Heike. "Beyond Beans and Bullets." *Soldiers* 49 (May 1994): 26–28.

Describes the work of the 507th Combat Support Group, which formed a "command and control logistical headquarters" during both the Gulf War and the Somalia operation. It provided a wide variety of services beyond supply.

1658. Jespersen, David M. "Coalition Logistics in Somalia." *Marine Corps Gazette* 78, no. 4 (April 1994): 32–34.

Provides a general review of logistics efforts during the Somalia operation.

E. Other military studies

1659. Butler, Keith D. "Captive in Somalia." *Soldiers* 49, no. 1 (January 1994): 22–23.

Although severely wounded, army warrant officer Michael Durant, a helicopter pilot, survived some days as a prisoner of Somali fighters. He benefitted greatly from having had Survival, Evasion, Resistance and Escape training in 1988. Durant was fearful of being killed, but he received some medical treatment, and the International Red Cross was permitted to visit him. He was eventually released.

1660. Casper, Lawrence E. "The Aviation Brigade as a Maneuver Headquarters." *Army* 45, no. 3 (March 1995): 20–23.

Significant criticism was leveled at the decision to have the 10th Mountain Division (Light Infantry)'s 10th Aviation Brigade to direct the "quick reaction force" in Somalia. The concept was not entirely new, but this was the first instance of such an assignment in a combat area. After the battle of Mogadishu still more troop units were added to the 10th Aviation Brigade's responsibilities, but it coped with this.

1661. Celeski, Joseph D. "A History of SF Operations in Somalia, 1992–95." *Special Warfare* 15, no. 2 (June 2002): 16–27.

Important survey of special operations forces that emphasizes the need for such forces to fully coordinate their activities. This is a chronological account that demonstrates the wide variety of tasks given to special operations forces personnel.

1662. Eichelberger, K.L. "Making the Navy's Case in Somalia." *United States Naval Institute Proceedings* 121, no. 5 (May 1995): 126–28.

Recommends a stronger navy commitment to joint force operations. More attention needs to be given to the selection of liaison officers and their placement in joint commands.

1663. Garcia, Elroy. "Violence in Somalia." *Soldiers* 48, no. 8 (August 1993): 6–7.

Describes air strikes and other attacks on installations, including a key radio station, maintained by followers of Mohamed Farah Aideed, who became a major opponent of the intervention forces. The attacks were a response to the ambush of Pakistani peacekeeping forces.

1664. Hornbeck, Donald. "Operation Shining Hope Owes Somalia." *United States Naval Institute Proceedings* 125, no. 12 (December 1999): 69–71.

Asserts that Operation Shining Hope, a relief effort for Kosovars, demonstrated that the United States armed services learned important lessons from Somalia. Hornbeck gives a rather detailed description of what was done to shelter and otherwise care for the Kosovars.

1665. "Marines Return to Somalia." *Marine Corps Gazette* 79, no. 2 (February 1995): 4–5.

Discusses the marine role in Operation United Shield in helping to effect the evacuation of United Nations personnel from Somalia.

1666. Stanton, Martin N. "Cordon and Search." *Infantry* 84, no. 6 (November/December 1994): 18–21.

Describes operations in support of humanitarian relief efforts at Marka, Somalia, on the coast south of Mogadishu. Stanton emphasizes the importance of mobility provided by helicopter transportation. He also analyzes cordon operations in detail.

1667. ———. "Lessons Learned from Counter-Bandit Operations." *Marine Corps Gazette* 78, no. 2 (February 1994): 30–32.

Uses the Somalia experience to emphasize the need to provide protection for humanitarian aid efforts with well-trained and led military forces.

F. The Battle of Mogadishu

1668. Barela, Timothy P. "'Bloody Sunday.'" *Airman* (February 1994): 2–7.

   Discusses the initial treatment and aerial evacuation to Germany of rangers wounded in the battle of Mogadishu. Barela notes that most of the medical personnel were members of the Army National Guard and Army Reserve on active duty.

1669. Carney, John T., Jr., and Benjamin F. Schemmer. *No Room for Error: The Covert Operations of America's Special Tactics Units from Iran to Afghanistan*. Chap. 15, pp. 245–60. New York: Ballantine Books, 2002.

   Describes the extensive training that the various types of special operations personnel involved in the Mogadishu battle had received and then discusses the fighting briefly.

1670. Ecklund, Marshall V. "Analysis of Operation Gothic Serpent: TF Ranger in Somalia." *Special Warfare* 16, no. 4 (May 2004): 38–47.

   Thoroughly describes an effort to apprehend several leaders in Mohamed Farah Aideed's band, which led to severe fighting in Mogadishu. After discussing the Somalia intervention generally, Ecklund provides a candid analysis of the United States effort on October 3, 1993.

1671. Ferry, Charles P. "Mogadishu, October 1993: A Company XO's Notes on Lessons Learned." *Infantry* 84, no. 6 (November/December 1984): 31–38.

   Emphasizes leadership, training, fire control, and tactics. Ferry discusses at length the equipment, ammunition, and weapons employed. This is a detailed account from the United States viewpoint of how fighting was conducted.

1672. Garcia, Elroy. "We Did Right That Night." *Soldiers* 49 (February 1994): 17–20.

   Points out that Task Force Ranger successfully captured a number of enemy Somali militiamen, although the assault is often viewed as a defeat for United States forces. Garcia discusses Somali tactics and other aspects of the fighting, largely in the words of participants.

1673. Glenn, Russell, ed. *Capital Preservation: Preparing for Urban Operations in the Twenty-First Century: Proceedings of the RAND Arroyo-TRADOC-MCWL-OSD Urban Operations Conference, March 22–23, 2000*. Santa Monica, CA: RAND, 2001.

   Essential study for understanding the fighting in Mogadishu. The book contains a number of long articles on the subject written for the most part by army and marine corps officers and several army noncommissioned officers. The detailed articles contain a number of useful maps and charts. Other combat situations covered during the conference were the battle of Grozny in Chechnya and peacekeeping efforts in East Timor and Bosnia–Herzegovina.

1674. Hollis, Mark A.B. "Platoon Under Fire: Mogadishu, October 1993." *Infantry* 88, no. 1 (January–April 1998): 27–34.

Hollis led a platoon that was isolated in Mogadishu during October 3–4, 1993. This is a valuable, detailed description of the fighting, supplemented by several excellent maps. Hollis provides a step-by-step account of the fight and offers a number of "lessons learned."

1675. Rhodes, Philip E. "No Time for Fear." *Airman* (May 1994): 22–31.

Describes the experiences of air force personnel operating on the ground during the battle of Mogadishu on October 3, 1993. The article includes a detailed diagram of the battle.

G. Legal aspects

1676. Borch, Frederic L. *Judge Advocates in Combat: Army Lawyers in Military Operations from Vietnam to Haiti.* Chap. 6, "Judge Advocates in Africa, 1992–94," pp. 199–228. Washington, DC: Government Printing Office, 2001.

Chaotic conditions in Somalia raised questions of applying international law and interpreting the various United Nations resolutions aimed at solving problems in Somalia. As in Panama, there were complex rules of engagement. Another challenging matter was the assessment of payments for damaged property.

1677. Dworken, Jonathan T. "Rules of Engagement: Lessons from Restore Hope." *Military Review* 74, no. 9 (September 1994): 26–34.

Examines the rules of engagement in Somalia within the context of rules of engagement as generally stipulated by the United States armed forces. The author finds that the rules applied in Somalia were effective in preventing many serious encounters with the Somalian population.

1678. Lorenz, F.M. "Confronting Thievery in Somalia." *Military Review* 74, no. 8 (August 1994): 46–55.

Begins with a brief description of conditions in Somalia, especially the problems of its youth. The deployment of nonlethal weapons was either disapproved or slow in being deployed. Lorenz recommends a number of steps to be taken in future operations resembling Somalia.

These include provision of nonlethal weapons and the deployment of adequate numbers of military police and civil affairs and psychological operations personnel.

H. The Somalia intervention and the media

1679. Mermin, Jonathan. "Television News and American Intervention in Somalia: The Myth of a Media-Driven Foreign Policy." *Political Science Quarterly* 112, no. 3 (Fall 1997): 385–403.

   Suggests that both the George H.W. Bush administration and the media were instrumental in bringing a demand for United States involvement in Somalia. Mermin asserts that the administration did not simply respond to pressure from the media and media-generated popular opinion.

1680. Robinson, Piers. "Operation Restore Hope and the Illusion of a News Media Driven Intervention." *Political Studies* 49, no. 5 (December 2001): 941–55.

   Denies that the media had a determining role in the United States decision to join the United Nations effort in Somalia. Robinson asserts that the media began to give significant attention to Somalia only after the United States government had decided to act in this situation.

XXIX

Bosnia, 1992–96

This brief summary the chronology that follows it only hint at the complex developments that occurred in the Bosnia area after the collapse of former Yugoslavia. For some time United States involvement in the protracted conflict in Bosnia and Herzegovina was limited primarily to air operations either to provide humanitarian assistance or to attack Serb forces or both. These efforts were undertaken by United States and other forces acting as NATO partners but under United Nations authority. Ironically, perhaps, there was only a large-scale commitment of United States ground forces after a peace agreement was achieved late in 1995. From maintaining a cautious stance based on the assumption that Europe could solve its own problems, the United States under President Bill Clinton became the primary mediator in the termination of the conflict and a major enforcer of peace in Bosnia.

Chronology

1992

February 29– A referendum indicated that Bosnian Catholics and Muslims
March 1   wanted to establish an independent state of Bosnia and Herzegovina, but the Serbian Orthodox Christians objected. A civil war developed, despite the rapid recognition of Bosnia and Herzegovina by many countries. By July the United Nations imposed an arms embargo on Serbia.

July 2   Operation Provide Promise began to provide humanitarian assistance in Bosnia and Herzegovina and to carry out other missions. Although other NATO countries participated in the effort, United States Air Force and United States Navy aircraft carried the bulk of the cargo. Other aspects of the

	operation included maintenance of a hospital in Zagreb and border surveillance at various points.
October 9	United Nations Security Council resolution 781 established what became Operation Sky Monitor, an effort to document flights in the air space over Bosnia and Herzegovina and thereby to discourage the use of aircraft in the civil war.

1993

March 31	United Nations Security Council resolution 816 demanded that unauthorized flights over Bosnia and Herzegovina cease and permitted United Nations member states to take action to prevent such flights.
April 12	Operation Deny Flight was initiated by NATO to maintain a "no-fly zone" across Bosnia–Herzegovina to carry out resolution 816 and thus hamper military use of air space.
April 16	The United Nations Security Council began establishing safety zones in Bosnia and Herzegovina. Despite such designations, which were made from time to time, Serb forces often attacked them.
June 4	The United Nations Security Council passed resolution 836, which permitted use of NATO aircraft to support UN ground troops when air support was requested.

1994

February 28	When Serbian aircraft attacked a Bosnian target, most of them were shot down by United States Air Force fighters operating as a part of a NATO contingent. Although the no-fly zone was quite effective in regard to fixed-wing aircraft, helicopter use by combatants in the Bosnia and Herzegovina conflict was often disguised in a fashion that prevented NATO apprehension.
April	Serbian forces threatened United Nations peacekeeping forces and the "safe areas" they had established in Bosnia. In response, NATO aircraft attacked Serbian positions, and the Serbs shot down several of them.
November 21	NATO aircraft began attacks on Serb positions near or in Croatia, after UN Security Council resolution 958 permitted such action.

1995

May 24	Beginning of an effort by Serbs to employ captured United Nations personnel as "shields" to prevent NATO air attacks.
August 30–September 14	Heavy NATO air attacks on positions held by Serbs in Bosnia finally forced the Serbs to accept the independence of Bosnia and Herzegovina.
September 26	A new state of Bosnia and Herzegovina established.
October 12	A ceasefire was agreed to for Bosnia and Herzegovina.

	November 1–21, Conference held at the Wright–Patterson Air Force Base near Dayton, Ohio, brought about an agreement to end the Bosnian War. The agreement is known by different names, such as the Dayton Agreement or the Dayton Accords. Bosnia and Herzegovina was finally established, but the agreement did not end all of the problems associated with the dissolution of Yugoslavia.
December 14	Dayton Treaty formally signed in Paris.
December 20	A NATO contingent known as the IFOR or "Implementation Force," including United States military elements, assumed responsibility for supervising the implementation of the Dayton Accords, thereby replacing an earlier United Nations military force.

1996

January 9	Operation Provide Promise, largely a NATO humanitarian assistance airlift, ended. The last phase of the mission was observation of the border area between Serbia and Macedonia, a task which was soon transferred to United States Army Europe.
December 12	A new entity, the Stabilization Force (SFOR) established under UN and NATO auspices to continue monitoring the situation in Bosnia. United States forces remained in Bosnia until 2004.

A. General studies

1681. Brune, Lester H. *The United States and Post–Cold War Interventions: Bush and Clinton in Somalia, Haiti, and Bosnia, 1992–1998*. Chap. 4, "The Disintegration of Yugoslavia," pp. 65–79; chap. 5, "The U.S. and Bosnia," pp. 81–95; chap. 6, "Clinton's Bosnian Policies: The 1995 Accords," pp. 97–116; chap. 7, "Implementing the Dayton Accords," pp. 117–54. Claremont, CA: Regina Books, 1998.

Lucid treatment of a highly complicated sequence of events. Brune carefully puts the Bosnia and Herzegovina events into the context of profound political changes in the Balkans as the result of the breakup of Yugoslavia. Several fine maps enhance the narrative. Brune also provides an interesting chronology that interweaves events in Somalia, Haiti, and Bosnia, helping readers to understand how United States decision-makers were confronted with concurrent crises.

1682. Collins, John M. "Military Options in Bosnia." *United States Naval Institute Proceedings* 121, no. 8 (August 1995), pp. 37–39.

Offers several strategies for the United States to deal with the conflict, including enforcement of a settlement, which might be the best solution but would be difficult to achieve, and altering the United States stance toward the arms embargo.

1683. Hendrickson, Ryan C. *The Clinton Wars: The Constitution, Congress, and War Powers.* Chap. 4, "Bosnia," pp. 68–98. Nashville, TN: Vanderbilt University Press, 2002.

Reviews the strong opposition in Congress to a United States involvement in the problems of what had been Yugoslavia, but, as in other cases, the executive branch eventually took and maintained a determined stance on the question of intervention. Most of the chapter focuses on congressional reactions to presidential moves and the limited efforts of Congress to restrict further the president's prerogatives in respect to the projection of military force abroad.

1684. Henriksen, Thomas H. *American Power After the Berlin Wall.* Chap. 5. "Bosnia: War and Intervention," pp. 79–100. New York: Palgrave Macmillan, 2007.

Usefully explores European attitudes to the Bosnian crisis and the efforts European powers made to deal with it. Henriksen sees sharp differences between European and United States perspectives on the nature of the conflict. He provides an interesting view of how and why the Clinton administration switched from a relatively passive position to take a leadership role in the effort to stop what amounted to a civil war.

1685. Hippel, Karin von. *Democracy by Force: US Military Intervention in the Post–Cold War World.* Chap. 5, "UNPROFOR, IFOR, and SFOR: Can Peace Be FORced on Bosnia?", pp. 127–67. Cambridge: Cambridge University Press, 2000.

Precise review of events in Bosnia. Von Hippel observes that, as in some other international interventions, attempting to combine mediation efforts and the outright enforcement of United Nations decisions against one party in a conflict leads to confusion and worse. She describes a number of peace proposals in some detail and then considers prospects for lasting peace in Bosnia. The several maps in this chapter help to illustrate the ethnic and religious complexities that made reaching a settlement so difficult.

1686. Huchthausen, Peter. *America's Splendid Little Wars: A Short History of U.S. Military Engagements: 1975–2000.* Chap. 14, "Intervention in Bosnia," pp. 185–211. New York: Viking, 2003.

Reviews concisely and clearly the conditions and events in the former Yugoslavia that gave rise to the Bosnian crisis. Huchthausen also examines the complex relations between NATO and the United Nations that hindered the development of effective responses to Serbian actions. He describes more briefly the disagreements that arose among United States decision-makers.

1687. ———. "Back to the Balkans." *United States Naval Institute Proceedings* 119, no. 6 (June 1993), pp. 43–48.

Reviews the post–World War II history of Yugoslavia, emphasizing the ethnic tensions that lingered during Tito's rule before exploding. Huchthausen also examines Yugoslav defense planning.

1688. ———. "Revisiting the Balkan Caldron." *United States Naval Institute Proceedings* 120, no. 5 (May 1994), pp. 73–76.

Predicts that the Bosnia situation will improve and describes the configuration of political forces in some detail. The author makes useful references to the Yugoslav situation during World War II.

1689. Kaldor, Mary. *New and Old Wars: Organized Violence in a Global Era.* Chap. 3, "Bosnia-Herzegovina: A Case Study of a New War," pp. 31–68. Stanford, CA: Stanford University Press, 1999.

Warns that "new wars" derive from "identity politics," which involve "the claim to power on the basis of labels—in so far as there are ideas about political or social change, they tend to relate to an idealized nostalgic representation of the past" (p. 7). In the chapter on Bosnia and Herzegovina, Kaldor describes in some detail the end of Yugoslavia and then gives a skillful description of how political and military elements were blended during the war and briefly of what occurred after the Dayton Agreement. This book merits careful study by anyone interested in conflicts that are currently military or could become military.

1690. Kirkpatrick, Jeane J. *Making War to Keep Peace.* Chap. 4, "The Balkan Wars: Making War to Keep the Peace," pp. 155–238. New York: HarperCollins, 2007.

Surveys at length the dissolution of Yugoslavia and the rising nationalism of Serbia. Kirkpatrick notes that the Bush administration, well-served by key personnel with direct knowledge of the Balkans, disapproved of the destruction of Yugoslavia and discouraged the various independence movements. This stance slowly changed with the increase in violence. Kirkpatrick was an early advocate of assistance to the Bosnians and minces no words in criticizing the Bush administration's and the United Nations' hesitation and much of the Clinton administration's approach to the complex situation.

1691. Linder, James C. "Myths of the Balkans." *United States Naval Institute Proceedings* 120, no. 5 (May 1994), pp. 77–78.

Seeks to deflate a number of political and military assumptions about the Balkan conflict, including the view that a peace accord will end armed conflict.

1692. Weiss, Thomas G. *Military-Civilian Interactions: Humanitarian Crises the Responsibility to Protect.* 2nd ed. Chap. 5, "Bosnia, 1992–95: Convoluted Charity," pp. 71–93. Lanham, MD: Rowman & Littlefield Publishers, Inc., 2005.

In contrast to many other treatments of events in Bosnia, Weiss focuses on operations of the various United Nations ground forces employed in the area. Weiss also discusses developments in Slovenia and Croatia, thereby providing the reader with a broad perspective on problems arising from the dissolution of Yugoslavia. Having this kind of information is essential for an understanding of the situation as air attacks became the favored method for reaching a peace settlement.

B. Air operations

1693. Beale, Michael. *Bombs Over Bosnia: The Role of Airpower in Bosnia–Herzegovina*. Maxwell Air Force Base, AL: Air University Press, 1997.

This "thesis presented to the faculty of the School of Advanced Airpower Studies, Maxwell Air Force Base, Alabama, for Completion of Graduation Requirements, Academic Year, 1995–96," is one of the major sources about the military side of the Bosnia involvement. Beale discusses both Deny Flight and Deliberate Force.

1694. Caravella, Frank J. "Combined Air Operations in Bosnia." *Military Review* 77, no. 4 (July–August 1997), pp. 87–91.

Discusses the development of the Bosnia Common Operational Picture (BCOP) to guide army and air force coordination.

1695. Owen, Robert C., ed. *Deliberate Force: A Case Study in Effective Air Campaigning: Final Report of the Air University Balkans Air Campaign Study*. Maxwell Air Force Base, AL: Air University Press, 2000. http://purl.access.gpo.gov/GPO/LP20446

Essential source for anyone interested in the precisely executed aerial campaign. This is a massive collection of studies by air force officers, except for one about Yugoslavia written by a civilian professor at the Air University, Karl Mueller. The many subjects discussed range from planning to the aircraft employed to the tactics used. There are many informative tables and helpful illustrations. The volume is thoroughly indexed, moreover.

C. Other military studies

1696. Adams, Thomas K. "Psychological Operations in Bosnia." *Military Review* 78, no. 6 (December/January–February 1998/99), pp. 35–37.

Stresses the importance of psychological warfare in special operations and predicts that it will assume more significance in the future.

Broadcasting from aircraft proved useful in an area low in information channels. Adams examines countermeasures against Serb broadcasting and describes more traditional methods, such as air-dropped pamphlets.

1697. Burton, John T. "War Crimes During Operations Other Than War: Military Doctrine and Law 50 Years After Nuremberg and Beyond." *Military Law Review* 149 (Summer 1995), pp. 199–206.

Discusses legal background of the Bosnia operation, which includes chapter 7 of the United Nations Charter, and the peace accord. Burton emphasizes that the operation is multi-national, a NATO effort, and that this is a type of operation that involves maintaining peace with military force and preventing the commission of war crimes.

1698. Shanahan, Stephen W. and Garry J. Beavers. "Information Operations in Bosnia." *Military Review* 77, no. 6 (November–December 1997), pp. 53–62.

Reviews planning for the use of "information" in the coming deployment of the 1st Infantry Division (Mechanized). Information was identified immediately as an important component of the peacekeeping effort. The authors include a thoroughly detailed set of lessons learned.

D. The peace settlement

1699. Chollet, Derek. *The Road to the Dayton Accords: A Study of American Statecraft*. New York: Palgrave Macmillan, 2005.

The book is the outgrowth of a State Department document prepared by Chollet. It is based on numerous interviews with participants and on a considerable number of government records, quite a few of which remained classified at the time the book was published. A portion of the volume consists of a day-to-day examination of negotiations at Dayton. Chollet is enthusiastic about what the Clinton administration achieved, and he believes that the Dayton Accords considerably increased Clinton's stature.

1700. Daalder, Ivo H. *Getting to Dayton: The Making of America's Bosnia Policy*: Washington, DC: Brookings Institution Press, 2000.

Highly readable, but detailed, account of the negotiations. This is an important study of the Clinton administration's long-term reluctance to pursue a leading role in dealing with the continuing Bosnia and Herzegovina crisis and of the administration's major change of course. Daalder's research included some interviews, but much of the material comes from contemporary press accounts. Fine maps and a rather long chronology supplement the narrative.

XXX

Haiti, 1991–96

The overthrow of President Jean-Bertrand Aristide of Haiti by his country's military led to an extended debate internationally and, in particular, to discussions in the United States about the appropriate response to the military coup. A sharp battle and continued frustrations during the Somalia intervention had generated caution among political and military leaders in the United States about similar incursions, thereby delaying action.

Eventually, for a variety of reasons, President Bill Clinton did decide to employ the United States armed forces to dislodge the Haitian military junta and to restore Aristide to office. A skillful diplomatic effort headed by retired General Colin Powell persuaded Haitian military leaders to submit peacefully to the United States landings in their country, changing what might have been an invasion into what was largely an assistance program. An earlier attempt to disembark a small peacekeeping force had ended in a rapid and embarrassing retreat for the United States.

United States forces were joined by contingents from other countries, and the rather brief occupation of Haiti elapsed quietly for the most part. Orderly elections were held, and a new government was established. Unfortunately, Haiti continued to experience serious economic and political problems, and, eventually, Aristide again lost the presidency, which he had resumed after another Haitian leader held the office.

Chronology

1991

February 7   Jean-Bertrand Aristide assumed the presidency of Haiti after having been elected to the post in 1990.

September 29	Aristide was overthrown by the army after he came into conflict with the Haitian legislative body. He left the country to begin a campaign to return to the presidency.
1993	
June	The United Nations imposed an embargo on Haiti to express its displeasure about the ousting of Aristide, and the United States Navy executed the trade blockade and cared for the large number of Haitians who tried to flee their country amid worsening economic conditions.
July 3	After a long struggle with the Haitian military, which had been put under considerable pressure through economic sanctions against Haiti, Aristide and the military chiefs signed the "Governors Island agreement" in New York City. Among its provisions was a stipulation that Aristide would become president again and would return to Haiti on October 30.
October 12	The United States Navy landing ship *Harlan County* was ordered to leave Haitian waters after its arrival at Port-au-Prince had been greeted by a hostile mob. The *Harlan County* carried a small part of United States military personnel who were intended to assist the Haitian government.
1994	
July 31	The United Nations gave the United States the authority to eliminate military rule in Haiti.
September 19	A large force of United States military and naval service personnel landed in Haiti to restore order and to ensure that Aristide assumed the Haitian presidency once again. Resistance to the landings was anticipated, but Colin Powell, former armed forces chief of staff, negotiated an uncontested descent on the island nation. A number of United Nations member countries contributed military personnel to help constitute a Multinational Force (MFN). The intervention in Haiti was termed Operation Uphold Democracy.
September 29	The United Nations decided to remove its economic sanctions with the proviso that Aristide resume the presidency by October 15.
October 15	Aristide came back to Haiti to resume his role as president of the republic.
1995	
March 31	Operation Uphold Democracy ended as a United Nations Mission in Haiti (UNMIH) was installed in Haiti. Some United States service personnel remained in Haiti for another year as part of Operation New Horizons.

June 25 The first free elections held in Haiti for years were begun.
December 17 A new presidential election took place.

1996
February 7 Inauguration of Rene Preval as the new president of Haiti.
February 29 The last United States forces left Haiti.

A. General studies

1701. Adams, Thomas K. "Intervention in Haiti: Lessons Relearned." *Military Review* 76, no. 5 (September–October 1996): 45–56.

    Interesting comparative study between the United States occupation of Haiti between 1915 and 1934 and the multi-national intervention in Haiti from 1994 to 1996. Adams asserts that the fundamental deficiency in both episodes was the unwillingness of the occupation authorities to seek what Adams terms "basic social change" (p. 50). This failure in part resulted from the fact that those authorities did not really understand Haitian society.

1702. Brune, Lester H. *The United States and Post–Cold War Interventions: Bush and Clinton in Somalia, Haiti, and Bosnia, 1992–1998.* Chap. 3, "U.S. Intervention in Haiti," pp. 37–64. Claremont, CA: Regina Books, 1998.

    Begins with an examination of the various positions that United States governmental agencies assumed in regard to Haiti after the ouster of the country's president, Jean-Bertrand Aristide. Much of the chapter deals with the political evolution of Haiti and the effect of attempted immigration by individual Haitians to the United States on the development of United States policy. Bureaucratic infighting continued in Washington, and Clinton and the Congress clashed over the president's authority to intervene in Haiti. Brune follows the story of Haiti into the late 1990s.

1703. Cox, Ronald. "Private Interests and U.S. Foreign Policy in Haiti and the Caribbean Basin," pp. 187–207 in *Contested Social Orders and International Politics*, ed. David Skidmore. Nashville, TN: Vanderbilt University Press, 1997.

    Economic interpretation of U.S. relations with Haiti, emphasizing the interplay of U.S. economic and political aims. Important study for its discussion of the tangled relationship between the United States and Aristide.

1704. Girard, Philippe R. *Clinton in Haiti: The 1994 U.S. Invasion of Haiti.* New York: Palgrave Macmillan, 2004.

Thoroughly researched and highly astute analysis of Haiti and the Clinton administration's approach to its problems. Girard analyzes in detail the political developments in the United States that led to the invasion, including extensive lobbying by Aristide and his supporters and Clinton's concern about the effects of several setbacks to United States policies abroad. The inability to land forces in Port-au-Prince from the U.S.S. *Harlan County* became something of a symbol of weakness, which Clinton eliminated by taking a resolute decision to occupy Haiti. This is one of the few studies of the invasion to utilize Haitian as well as United States sources to any extent.

1705. Hendrickson, Ryan C. *The Clinton Wars: The Constitution, Congress, and War Powers.* Chap. 3, "Haiti," pp. 43–67. Nashville, TN: Vanderbilt University Press, 2002.

Describes the intense consideration of the situation in Haiti by Congress, and the factions within Congress that had definite attitudes toward intervention in Haiti, such as the Congressional Black Caucus. At times, there were efforts made to limit Clinton's options and to compel more consultation with Congress. Ultimately, however, as so often in the past, Congress was willing to concede the president's power to act, while reserving the right to criticize the actions taken at a point in the future. Hendrickson also discusses Clinton's efforts to gain support for his policies regarding Haiti from other Caribbean countries.

1706. Herspring, Dale R. *The Pentagon and the Presidency: Civil-Military Relations from FDR to George W. Bush.* Chap. 12, "The Military and William Clinton," pp. 351–55. Lawrence: University Press of Kansas, 2005.

Clinton moved from a policy based on using economic sanctions to military measures. Despite an initial setback when a small United States force had to retreat in the face of threats by some armed Haitians, the occupation of Haiti some time later was accomplished without fighting. Herspring describes discussions of United States options briefly.

1707. Hippel, Karin von. *Democracy by Force: US Military Intervention in the Post–Cold War World.* Chap. 4, "Heartened in Haiti," pp. 92–126. Cambridge: Cambridge University Press, 2000.

Begins with a highly useful exploration of the concept of "democratisation" (British spelling), as developed by several United States presidential administrations. This is a sometimes witty and sometimes scathing discussion. Von Hippel criticizes the Clinton administration policy toward Haiti before the intervention, but after a thorough review of the intervention itself she asserts that the proper moves were made and that there is hope for genuine improvements in the government and economy of the country.

1708. Kirkpatrick, Jeane J. *Making War to Keep Peace.* Chap. 3, "Haitians' Right to Democracy?", pp. 113–54. New York: HarperCollins, 2007.

Views the United States involvement with Haiti as a failure of both the Bush and Clinton administrations. There seemed little prospect that Haiti could develop into a democracy any time soon. Restoring President Jean-Bertrand Aristide to power seemed to Kirkpatrick to be a highly unwise goal, however. Negotiations over the intervention, the effect of intervention, and the 2000 elections in Haiti are discussed in considerable detail.

1709. Pezzullo, Ralph. *Plunging into Haiti: Clinton, Aristide, and the Defeat of Diplomacy.* Jackson: University Press of Mississippi, 2006.

Important contribution to the literature by a well-known author and the son of Lawrence Pezzullo, a career State Department official for many years, who was Clinton's special envoy to Haiti for a time. The book is almost entirely the product of interviews with five people: Lawrence Pezzullo, a prominent Haitian leader, and representatives of the State Department and the United Nations. The book should not be read in isolation, but it does highlight some problems involved in the complicated relations of the United States with Haiti.

B. Military studies: general

1710. Ballard, John R. *Upholding Democracy: The United States Military Campaign in Haiti, 1994–1997.* Westport, CT: Praeger, 1998.

Valuable study that is clearly the most important examination of Operation Uphold Democracy from a military perspective. Ballard does not ignore the political side of the story, however, and he devotes several chapters to political developments in Haiti. His research included a number of interviews with key United States military figures and the use of many specialized reports. There is an extensive and much needed glossary of acronyms and abbreviations.

1711. Baumann, Robert E. "Operation Uphold Democracy—Power Under Control." *Military Review* 77, no. 4 (July–August 1997): 13–21.

Argues that in "operations other than war," goals are likely to be unclear, thereby making assessments of progress difficult. Moreover, military forces are attuned to warfare, thereby making adjustments to operations other than war quite challenging. This is an important article, and not all the points can be summarized in an annotation.

1712. Fields, Damon, Bill Pope, and Patrick Sharon. "Adventures in Hispaniola." *United States Naval Institute Proceedings* 128, no. 9 (September 2002): 60–64.

Emphasizes the intervention in Haiti in 1994 but usefully compares it to the intervention in the Dominican Republic in 1965. This is a brief, but fine, analysis of events in 1994. The three authors, representing the army, air force, and marines, seek lessons from the two interventions that might help improve future joint operations.

1713. Fishel, John T. "Operation Uphold Democracy: Old Principles, New Realities." *Military Review* 77, no. 4 (July–August 1997): 22–30.

Assesses the various segments of the operation in terms of United States military doctrine. This precise analysis is an important element of the literature about the intervention in Haiti. Although its focus is on military moves, the article gives appropriate attention to political developments.

C. Special military studies

1714. Borch, Frederic L. *Judge Advocates in Combat: Army Lawyers in Military Operations from Vietnam to Haiti.* Chap. 7, "Judge Advocates in Haiti, 1994–96," pp. 229–65, 292–96. Washington, DC: Government Printing Office, 2001.

Rules of engagement changed once it was clear that landings would not be opposed. Judge advocates were heavily involved with procurement issues and damage claims. Captured Haitians were considered prisoners of war, but dealing with violence between Haitians presented significant problems. Subsequent issues were created by attempted mass emigrations from Haiti.

1715. Brown, Stephen D. "PSYOP in Operation Uphold Democracy." *Military Review* 76, no. 5 (September–October 1996): 57–64, 67–73.

Psychological operations were utilized well before the beginning of the intervention. Brown briefly reviews events in Haiti and then discusses psychological operations in relation to United States Army doctrine that was current at the time of the Haiti occupation. He also describes in detail the changes in the content of messages to the Haitians as the situation in Haiti changed.

1716. Carney, John T., Jr., and Benjamin F. Schemmer. *No Room for Error: The Covert Operations of America's Special Tactics Units from Iran to Afghanistan.* Chap. 16, pp. 261–70. New York: Ballantine Books, 2002.

Noting that "[s]pecial operations forces had primary responsibility for the invasion" (p. 261), the authors describe some of the civic action responsibilities that the United States undertook briefly.

1717. Quinn, Michael A., and Douglas W. Daniel. "Civil Affairs in Haiti." *Military Review* 78, no. 4 (July–August 1998): 3–10.

Describes the assignment of several United States Army civil affairs units as part of the Multinational Force in Haiti. Based on what occurred in Haiti, Quinn argues vigorously that civil affairs units should be deployed as units, not as individuals, and that the commanding officer of the civil affairs units should be part of overall command staff.

1718. Riehm, J.A. "The USS Harlan County Affair." *Military Review* 77, no. 4 (July–August 1997): 31–36.

Exceptionally well written account of an encounter that could have become a serious incident. This is a detailed account of the assemblage of the force embarked on the *Harlan County* and its hostile reception in Haiti. The article is based largely on a number of interviews with United States participants in the incident.

1719. Shaw, Robert C. "Integrating Conventional and Special Operations Forces." *Military Review* 77, no. 4 (July–August 1997): 37–41.

Planners assumed that the second attempt to intervene in Haiti after the *Harlan County* incident would be contested, perhaps heavily, but as the situation developed, the occupation seemed less likely to involve significant combat. These changes in mission resulted in tensions within the United States contingent between conventional and special operations forces personnel.

1720. Shelton, H. Hugh, and Timothy D. Vine. "Winning the Information War in Haiti." *Military Review* 75, no. 6 (November–December 1995): 3–9.

Discusses in considerable detail the armed forces policy toward the media during the Haiti operation. The authors emphasize the power of the media to project images, which may have significant political impacts. This ability requires the armed forces to work with journalists and to ensure that the latter have a full understanding of the conditions that face the former. Shelton headed the Haiti intervention, and this article is an important command perspective.

XXXI

Kosovo, 1999–

The Kosovo crisis involved the United States first in a vigorous air campaign and then in a long-term commitment of ground forces to a peacekeeping mission. The Dayton Accords that brought peace to Bosnia did not recognize the aspirations of many Kosovars, most of whom are Albanians, for self-rule. Indeed, Serbia, of which the Kosovo region was a part, was bent on reducing the autonomy the Kosovars had enjoyed. After a period of claims and counterclaims, continued fighting in Kosovo, and the issuance of several threats to the Serbs by NATO, an intensive NATO air campaign was waged against Serbia, which brought an agreement for a new Kosovar state with an international peacekeeping force to protect it from further Serb attacks. The new arrangements did not end tensions, and the international force remains in Kosovo, even after its declaration of independence in 2008.

Chronology

1991	Albanian Kosovars voted in favor of a Republic of Kosovo, but their goal was reached only years later.
1992 December	The United States informed Serbia that it could expect United States military action should Kosovo be attacked. Further warnings went to Serbia after Bill Clinton became president of the United States.
1996 February	Militancy in Kosovo increased to the point at which a Kosovo Liberation Army (KLA) was formed, which unleashed a guerrilla campaign against the Serbian government.

1998

March 23	United Nations Security Council resolution 1160 condemned Serbian actions in Kosovo and initiated economic sanctions and an arms embargo against Serbia.
April 23	Almost unanimously, Serbs indicated in a referendum that they would not accept foreign mediation in the Kosovo dispute.
June 1	Ibrahim Rugova, the unofficial president of Kosovo, formally requested action by the United Nations and NATO.
August	Further military action by Serb forces against the Kosovars.
September 23	United Nations Security Council resolution 1199 ordered an end to fighting and the removal of Serb forces from Kosovo.
October 16	Serbia agreed to a plan for having foreign observers supervise a ceasefire between Serbian and Kosovar combatants.
October 27	Extensive removal of Serb forces from Kosovo.
December	Further fighting occurred in Kosovo.

1999

February 6	Formal talks concerning the Kosovo problem began in Rambouillet, France, although Serbia did not participate in them.
March 15	Rambouillet discussions resumed in Paris after delays.
March 18	Rambouillet Accords signed, establishing a quasi-independent province of Kosovo within Yugoslavia under United Nations protection.
March 20	Serb forces started new attacks in Kosovo.
March 24	NATO air attacks began against the Serbs.
April 3	NATO missiles struck Belgrade, the Serbian capital.
April 9	Owing to continued air attacks, Boris Yeltsin, the Russian president, warned that there could be a major war as the result of the worsening Balkan situation.
April 14	Russia moved away from its previous militant position concerning the Kosovo crisis.
April 21	Further air attacks on Belgrade.
April 30	Additional air attacks on Belgrade.
May 7	Chinese embassy in Belgrade bombed with casualties within the embassy.
June 10	United Nations Security Council resolution 1244 authorized the establishment of an international regime for Kosovo.
June 11	NATO air campaign ended.
June 12	A Kosovo Force (KFOR) moved into Kosovo under the authority of Security Council resolution 1244.
June 20	Formal end of the NATO air campaign, as Serbian forces were withdrawn from Kosovo.
September 20	Kosovo Liberation Army completed its disarmament at the request of NATO.

2005   August, KFOR reorganized from brigades into task forces to achieve more efficiency and flexibility. The force has sustained a number of casualties since 1999, including some fatalities, but major fighting has not occurred.

2008
February 17   Kosovo declared itself an independent state. A number of countries recognized Kosovo as a sovereign nation, but many states still object to its action or remain undecided. Foreign troops, including members of the United States armed services, remain in Kosovo, at least for the time being.

A. General studies and political analyses

1721. Bieber, Florian, and Zidas Daskalovski, eds. *Understanding the War in Kosovo.* London: Frank Cass, 2003.

    More a series of studies of the political facets of the Kosovo issue than a group of military analyses, this is an important effort to present a number of points of view. Written largely by junior faculty members and Ph.D. candidates, the essays delve into some topics that have rarely been discussed in NATO countries. Their work hardly leads to a consensus about the problems related to Kosovo, but the essays promote, as the title of the book suggests, a greater "understanding" of the extremely complex situation.

1722. Clark, Wesley K. *Waging Modern War: Bosnia, Kosovo, and the Future of Combat.* New York: Public Affairs, 2001.

    Important command perspective on Kosovo. Clark led NATO forces during the Kosovo war. He discusses at considerable length the conflicting views of major players about the conduct of political and military moves. This is a candid look at decision-making from Clark's point of view. Using the experience of the Bosnia and Kosovo wars, Clark emphasizes the continuing importance of air operations in future military encounters.

1723. Hendrickson, Ryan C. *The Clinton Wars: The Constitution, Congress, and War Powers.* Chap. 6, "Kosovo," pp. 117–37. Nashville, TN: Vanderbilt University Press, 2002.

    Reviews the Clinton administration's actions in respect to Kosovo and then turns to the vigorous debates within Congress about Clinton's actions and about the relevance of the War Powers Resolution to the conflict. As in other cases, despite the fact that there was a good deal of discussion, Congress effectively accepted the authority of the president to act in military matters, largely unimpeded by congressional concerns.

1724. Kirkpatrick, Jeane J. *Making War to Keep Peace.* Chap. 5, "Kosovo," pp. 239–69. New York: HarperCollins, 2007.

Reluctantly gives the Clinton administration credit for the settlement of the Kosovo problem. Kirkpatrick details the Serbian political and military attacks on the people of Kosovo that continued with little interference for some time, and the complicated negotiations that were required to forge a decisive NATO response to this crisis.

1725. Weller, Marc. *Peace Lost: The Failure of Conflict Prevention in Kosovo.* Leiden, Netherlands: Martinus Nijhoff Publishers, 2008.

This impressive example of social science research applies rigorous methods to examining four phases of the crisis and the ways in which the intensification of the conflict could have been avoided during each phase. The book includes innovative graphs and tables, although, regrettably, the former are difficult to read because of the extremely small print. There is also a long and enlightening chronology.

1726. Weymouth, Tony, and Stanley Henig, eds. *The Kosovo Crisis: The Last American War in Europe?* London: Pearson Education, 2001.

Searching look at the ways in which a number of nations viewed the crisis, and the actions that they and the United Nations, NATO, and the European Union took in the face of it. Although only one chapter deals directly with the United States and another with the air campaign, this is an important source for understanding the complicated politics of the Kosovo war. Overall, the tone of the book is pessimistic about future developments in the Balkans.

B. Military studies: general

1727. Connaughton, Richard. *Military Intervention and Peacekeeping: The Reality.* Chap. 8, "Kosovo—'Only a bunch of bad options'—Bill Clinton," pp. 191–239. Aldershot, U.K.: Ashgate Publishing Ltd., 2001.

Skillful analysis of the political and military elements of the crisis and the subsequent war. Connaughton does a fine job of explaining the complex NATO approach to Kosovo and its relations with Russia. He also gives a good deal of attention to what he views as the exaggerated belief within NATO circles that air attacks are a highly effective tool for achieving political goals. Connaughton charges that information about the impact of air attacks was deliberately manipulated to provide better "results" than were actually achieved.

1728. Huchthausen, Peter. *America's Splendid Little Wars: A Short History of U.S. Military Engagements: 1975–2000.* Chap. 15, "Intervention in Kosovo," pp. 212–18. New York: Viking, 2000.

Describes how the Dayton Accords that brought peace to Bosnia allowed Serbia to turn its full attention to the Albanian Muslims in Kosovo in an effort to expel them. Despite NATO air attacks the Serbians achieved considerable success in their "ethnic cleansing" campaign, but finally air attacks on Serbia itself led to the introduction of peacekeeping forces.

1729. Nardulli, Bruce, Walter Perry, Bruce Pirnie, John Gordon, and John McGinn. *Disjointed War: Military in Kosovo, 1999*. Santa Monica, CA: RAND, 2002.

Although this study gives a good deal of attention to the air campaign, its focus is on ground operations, including the important role played by Task Force Hawk. This was a force in being that consisted of army helicopters and various army formations. Its primary usefulness was the image it presented of being the advance party of a possible ground offensive into Serbia. The prospect of such an offensive presumably concerned Serbia a good deal and thereby contributed to the eventual peace settlement.

C. The air campaign

1730. Cordesman, Anthony H. *The Lessons and Non-Lessons of the Air and Missile Campaign in Kosovo*. Westport, CT: Praeger, 2001.

Authoritative and exhaustive study by a leading defense analyst. This book closely examines all phases of the air war from the origins of the Kosovo conflict and the decision to attack Serbia through the establishment of an international peacekeeping mission. Cordesman discusses weapons, tactics, and numerous other aspects of the war, including aerial refueling of aircraft. The author pinpoints deficiencies revealed. Although he focuses on the military aspects of the war, he by no means ignores the political implications of military decisions.

1731. Haave, Christopher E. and Phil M. Haun. *A-10s Over Kosovo: The Victory of Airpower Over a Field Army as Told by Those Airmen Who Fought in Operation Allied Force*. Maxwell Air Force Base, AL: Air University Press, 2003.

Important collection of statements by pilots and others involved in the employment of A-10s in the Kosovo campaign. A-10s participated in a variety of ways, including aid with search and rescue missions, and this book describes all of them. There is much information about tactics used and the role of intelligence in the air effort.

1732. Henriksen, Dag. *NATO's Gamble: Combining Diplomacy and Airpower in the Kosovo Crisis 1998–1999*. Annapolis, MD: Naval Institute Press, 2007.

Although his primary concern is with NATO's use of air power as both a threat to back diplomatic demands and as a weapon once its demands were rejected by Serbia, Henriksen also provides ample information about and analysis of diplomatic efforts to end the Kosovo crisis short of fighting. Henriksen objects to what he views as NATO's lack of a strategy and its continuing problems in agreeing upon such a strategy, despite the eventual success of the air campaign. This well-researched study is based not only on a variety of published sources, but also on a number of interviews the author conducted with important military and political figures.

1733. Lambeth, Benjamin S. *NATO's Air War for Kosovo: A Strategic and Operational Assessment*. Santa Monica, CA: RAND, 2001.

Important survey of the air campaign. Lambeth provides rich detail about operations and includes a considerable amount of statistical information in a number of easily understood tables and graphs. There is an extensive bibliography. Although the effort was a clear success, Lambeth analyzes the numerous operational problems, including the attack on the Chinese Embassy. Despite the necessary emphasis on combat, the book includes a chapter about the Serbian decision to submit to NATO.

D. The international force in Kosovo

1734. Barnett, Gary G. "MI Tactical HUMINT Team Operations in Kosovo." *Military Intelligence* 27, no. 1 (January–September 2001): 20–22.

Examines "force protection teams" of active duty and reserve personnel who had counterintelligence and language capabilities. There is much detail on the lessons learned, including additional types of equipment that were needed but not available.

1735. Bigge, Lauren. "Mission of Hope." *National Guard* 60, no. 6 (June 2006): 32–35.

Briefly describes the activities of Michigan Army National Guard aviation elements in Kosovo. The guard personnel Bigge interviewed noted the persistence of tension between Serbs and Albanians, but they believed it was decreasing.

1736. Chamberlain, Robert L., and Ralph Kluna. "Long-range Surveillance Operations in Kosovo—Complementing Existing Capability." *Military Intelligence* 27, no. 1 (January–September 2001): 47–52.

Discusses the Long-range Surveillance Company of the 205th MI Brigade, which performed a unique role in intelligence gathering. The authors include much on training.

1737. Culp, Robert A. "Ground Surveillance Operations: The Nightstalkers of Vitina—Countermortar Operations in Kosovo." *Military Intelligence* 27, no. 1 (January–September 2001): 23–26.

Describes 101st MI Battalion support for operations by Task Force 1–77 Armor against random mortar firing into Serb towns.

1738. ———. "Sustained Company Operations: Lessons from the GS MI Company in Kosovo." *Military Intelligence* 27, no. 1 (January–September 2001): 53–55.

Emphasizes the development of "standing operating procedures" in such areas as "weapons and ammunition safety" as the 101st MI Battalion moved from garrison life in Germany to active duty in Kosovo. Personnel from services other than the army and some civilians were added to the battalion.

1739. Redmon, David. "Electronic Warfare Operations in Kosovo." *Military Intelligence* 27, no. 1 (January–September 2001): 29–32.

Chronicles operations of D Company, 101st MI Battalion. Operating primarily in squads, members of the company performed a variety of missions, including use of SIGINT (Signal Intelligence).

1740. Rovergno, John S. "Kosovo: A Year of Intelligence Operations." *Military Intelligence* 27, no. 1 (January–September 2001): 8–13.

Surveys military intelligence operations in Kosovo from the perspective of the commander of the 101st MI Battalion. Lack of personnel was a major problem, but, eventually, elements of other battalions arrived to reinforce the 101st. Kosovo operations utilized experience from Bosnia–Herzegovina.

1741. Sanchez, Ricardo S., with Donald T. Phillips. *Wiser in Battle: A Soldier's Story.* Chap. 7, "Kosovo and Coalition Warfare," pp. 117–33. New York: HarperCollins, 2008.

Sanchez, who came to command United States ground troops assigned to Kosovo, discusses the problems of the coalition effort and severely criticizes both the United Nations and the United States for making insufficient preparations for the intervention. This is a brief, but important, critique of the operation.

1742. Schaefer, Robert W. "The 10th SF Group Keeps Kosovo Stable." *Special Warfare* 15, no. 2 (June 2002): 52–55.

After briefly outlining the geographical area covered, Schaefer emphasizes responsibilities of small "liaison teams" and stresses the importance of 10th SF Group's language abilities.

1743. Seiple, Chris. "The Lessons of Kosovo." *Marine Corps Gazette* 84, no. 6 (June 2000): 39–41.

For Seiple, the "lessons of Kosovo" relate to the marines' future missions and configuration. Seiple imaginatively portrays some possible structures of military and non-military forces tasked to deal with crises. The future of military action seems to be expeditionary forces. The marines have an advantage in this environment, given their notable expeditionary experience, but they must be ready to continue staking out their own role when all the armed services will be concerned with expeditionary duties.

1744. Steele, Dennis. "Remember Kosovo? The U.S. Army Effort That Began Our Involvement." *Army* 58, no. 4 (April 2008): 28–30, 31–32, 34, 36–38, 40, 42–43.

Informative review of the Kosovo war and the coming of United States forces (Task Force Falcon) into Kosovo. Steele's emphasis is on the early days of the peacekeeping assignment, but he notes that now most of the United States contingent consists of members of the National Guard. Steele also describes the recent renewed attention that Kosovo has received because of the country's declaration of independence.

1745. Wood, Donald K., and Joan B. Mercier. "Building the ACE in Kosovo." *Military Intelligence* 27, no. 1 (January–September 2001): 33–36.

Discusses the ACE (Analysis and Control Element) of the 101st MI Battalion, which was responsible for collection, management, and dissemination of intelligence. The initial emphasis was on HUMINT (Human Intelligence), but other sources were employed. The article gives much emphasis to lessons learned.

1746. Zaalberg, Thijs W. Brocades. *Soldiers and Civil Power: Supporting or Substituting Civil Authorities in Modern Peace Operations.* Part IV, "Kosovo: Military Government by Default," pp. 287–413. Amsterdam: Amsterdam University Press, 2006.

Five chapters consider various aspects of the Kosovo peacekeeping effort. The topics discussed include establishment of a police organization and the treatment of members of the Serb minority in Kosovo. Although the initial international force was strong enough to maintain order, it suffered from a lack of prior planning, and it really needed a different configuration than it had. The force was not ready to assume military government responsibilities, yet it found itself confronted with a variety of demanding assignments.

XXXII

Doctrines and general studies in the age of the War on Terrorism

The 9/11 attack and the consequent War on Terrorism, especially the invasions of Afghanistan and Iraq, have produced another flood of writing about counterinsurgency. Use of the broad descriptive phrase "low-intensity conflict" seems to have receded, although some types of operations being discussed currently are probably better described by that phrase than by the word "counterinsurgency." In any event, the amount of relevant literature produced during the past few years is large, and the citations below represent no more than a sampling of some important works.

A. Counterinsurgency doctrine

1747. United States Army. *Field Manual No. 3–07. Stability Operations.* Washington, DC: United States Department of the Army, 2008.

The purpose of this volume is to provide authoritative guidance in assisting friendly governments to improve their internal security during and after a period of conflict. This publication necessarily benefitted considerably from the continuing United States experiences in Afghanistan and Iraq, although the foreword to the manual states that "America's future abroad is unlikely to resemble Afghanistan or Iraq."

1748. United States Army/United States Marine Corps. *Field Manual No. 3–24/Marine Corps Warfighting Publication no. 3–33.5. Counterinsurgency.* Washington, DC: United States Department of the Army/Headquarters, United States Marine Corps, 2006.

Much discussed and highly important publication that arose in response to the continuing challenge of insurgency in Iraq. The army developed a provisional manual that was issued in 2004, but there needed to be a further effort. Lieutenant General David Petraeus, as newly appointed head of the Combined Arms Center, led the project. This is a remarkable accomplishment, not least because it marked the first time that the army and the marines had collaborated to produce such a guide.

B. Counterinsurgency doctrine: critiques

1749. Alderson, Alexander. "US COIN Doctrine and Practice: An Ally's Perspective." *Parameters* 37, no. 4 (Winter 2007): 33–45.

    An experienced British officer acknowledges the difficulties inherent in waging a counterinsurgency but finds some signs of progress in the Middle East. Although Field Manual 3–24 (item 1748) deals with the political facets of counterinsurgency only to a limited degree, its publication is a notable event. According to Alderson, the manual "provides the intellectual foundation for success" (p. 44), and its principles are being applied usefully in Afghanistan and Iraq.

1750. Armstrong, Charles L. "Fresh Counterinsurgency Doctrine (Field Manual 3–24/Marine Corps Warfighting Publication 3–33.5)." *Marine Corps Gazette* 91, no. 3 (March 2007): 49–50.

    Briefly outlines some of the criticisms made of Field Manual 3–24. Armstrong believes it makes many good points, but there are some deficiencies. Not enough attention is given to learning and respecting the local culture and language, for example. Nevertheless, with some improvements, "it should serve well for generations" (p. 50).

1751. Beebe, Kenneth. "The Air Force's Missing Doctrine: How the US Air Force Ignores Counterinsurgency." *Air and Space Power Journal* 20, no. 1 (Spring 2006): 27–34.

    Argues that the "Air Force has no workable doctrine" for counterinsurgency, despite the fact that counterinsurgency and similar types of unconventional conflicts are of increasing importance. There are doctrinal statements related to counterinsurgency, but not enough effort has been made to focus on this challenge. Many air force and other military personnel still do not distinguish sufficiently between high- and low-intensity conflict, thereby obstructing the development of appropriate counterinsurgency doctrine.

1752. Celeski, Joseph D. "Strategic Aspects of Counterinsurgency." *Military Review* 86, no. 2 (March–April 2006): 35–41.

Given the aftermath of Vietnam, the United States armed services were slow to consider the possibility of renewed counterinsurgency responsibilities after the end of the Cold War. Celeski points to the Joint Special Operations University Report 05–2, "Operationalizing COIN" as a key doctrinal statement. Imperatives in fighting insurgents, as well as terrorists, include close cooperation with other countries and working with and learning from law enforcement agencies. Celeski has much to say about the need to understand the role of religion in present-day insurgencies and the process for developing future counterinsurgency leaders in the United States.

1753. Cohen, Eliot, Conrad Crane, Jan Horvath, and John Nagl. "Principles, Imperatives, and Paradoxes of Counterinsurgency." *Military Review* 86, no. 2 (March–April 2006): 49–53.

Notes the considerable experience that the United States military services have had with counterinsurgency and asserts that "a successful counterinsurgency requires an adaptive force led by agile leaders." When this article was written, the authors were working on Field Manual 3–24, the completion of which was a major step in developing contemporary counterinsurgency doctrine. Thus, this is an important contribution because it shows counterinsurgency specialists as their thinking developed. Their conclusion is that counterinsurgents "achieve the most meaningful successes by gaining support and legitimizing the host government, not by killing insurgents" (p. 52), thereby foreshadowing the principles of Field Manual 3–24.

1754. Crane, Conrad C. "Minting COIN: Principles and Imperatives for Combating Insurgency." *Air and Space Power Journal* 21, no. 4 (Winter 2007): 57–62.

A major architect of Field Manual 3–24, Crane explains that this publication is based not only on historical examples but also the latest encounters with insurgency in Afghanistan and Iraq. Crane identifies a number of factors that must be understood when undertaking a counterinsurgency. Among those factors is the necessity of tailoring the use of force precisely to meet the need for it. Crane briefly defends the manual from air force criticisms for its alleged overemphasis on land operations.

1755. Downs, William Brian. "Unconventional Airpower." *Air and Space Power Journal* 19, no. 1 (Spring 2005): 20–25.

In Downs' view, the air force must develop doctrine suited to fight terrorism and wage counterinsurgency. His article is an effort to provide such doctrine, which includes the development of aircraft planned specifically for counterterrorism and counterinsurgency. Downs stipulates that "we must employ Air Force assets *selectively* to avoid creating more enemies" and then proceeds to discuss "force structure" and tactics. He says much about the Thrush Vigilante, an aircraft suitable for counterinsurgency and quotes several advocates of the Vigilante.

1756. Dunlap, Charles J., Jr. "Air-Minded Considerations for Joint Counterinsurgency Doctrine." *Air and Space Power Journal* 21, no. 4 (Winter 2007): 63–74.

Criticizes Field Manual 3–24 for being too tied to ground warfare and, thereby, for slighting the advantages that air supremacy, including airlift and air drop capacity, offers to counterinsurgents. Dunlap is careful not to overemphasize the significance of air assets, but he does insist upon fully integrating air power into the writing of counterinsurgency doctrine. He describes the use of air power in Afghanistan and argues that complete "impotency in the face of superior weaponry and the denial of a meaningful death crush war-fighting instincts."

1757. Gentile, Gian P. "Our COIN Doctrine Removes the Enemy from the Essence of War." *Armed Forces Journal* 145, no. 6 (January 2008): 39.

Gentile points out that AirLand Battle doctrine, which was intended to meet Cold War needs, was discussed extensively in professional journals, but emerging counterinsurgency doctrine has encountered a much smaller response. He worries that using the new Field Manual 3–24 as a cookbook for dealing with counterinsurgency may generate too much confidence and may lead to a dangerous failure to keep the enemy and his strategy in focus.

1758. Kramer, Jacob M. "The Two Sides of COIN: Applying FM 3–24 to the Brigade and Below Counterinsurgency Fight." *Infantry* 96, no. 4 (July–August 2007): 25–31.

Discusses efforts at the National Training Center to prepare soldiers and units to carry out "non-lethal operations," because combat lessons have presumably already been learned in earlier training. Field Manual 3–24 is an appropriate guide for training in "non-lethal operations." It teaches the principle that "the purpose of counterinsurgency is to legitimatize a nation's government" rather than overemphasizing combat. Kramer goes on to write "[n]on-lethal operations are DECISIVE in counterinsurgency."

1759. Metz, Steven, and Raymond Millen. "Insurgency and Counterinsurgency in the 21st Century: Reconceptualizing Threat and Response." *Special Warfare* 17, no. 3 (February 2005): 6–21.

Analyzes the complex nature of insurgencies, which the authors divide into several categories, such as "liberation insurgencies." The authors usefully remind readers that "building an effective insurgency is difficult" (p. 8), a factor that aids counterinsurgents who react promptly and appropriately. They provide helpful summaries of and commentaries on several recent statements of army counterinsurgency doctrine and stress the need to assist foreign governments to fight insurgencies rather than try to fight them largely with United States forces. This thoughtful piece clearly recognizes the obstacles facing outsiders who involve themselves in waging counterinsurgencies.

1760. Orellana, Manuel A., Jr. "American Counterinsurgency Doctrine and El Salvador: Review Essay." *Military Review* 85, no. 6 (November–December 2005): 89–90.

Important review of Benjamin C. Schwarz's *American Counterinsurgency Doctrine and El Salvador*, which Orellana hopes will provide lessons to apply in Iraq. El Salvador and Iraq seem quite similar in Orellana's view. Despite the inability of the United States to achieve its goals in El Salvador, pinpointing the reasons for its failure may help avoid similar errors in Iraq.

1761. Peters, Ralph. "Dishonest Doctrine: A Selective Use of History Taints the COIN Manual." *Armed Forces Journal* 145, no. 5 (December 2007): 42–44.

Peters terms Field Manual 3-24 "a politically correct document for a politically correct age," primarily because it downplays what he calls "deadly force." He objects to the view that counterinsurgency is primarily political, not military, asserting that "[w]ithout the will to establish and maintain security for the population, nothing else works." This is a biting critique of the manual and the way Peters believes it was written. Peters suggests the manual was prepared largely by academics who ignored examples of successful counterinsurgencies, such as suppression of the Mau Mau in Kenya, that did not fit their theories.

1762. ———. "Progress and Peril: New Counterinsurgency Manual Cheats on the History Exam." *Armed Forces Journal* 144, no. 7 (February 2007): 34–37.

Praises Field Manual 3-24 for the improvements made over an earlier draft, but, nevertheless, Peters views it as a "stopgap," because counterinsurgency doctrine requires further refinement. Peters asserts that the United States armed forces have encountered realities in their post–Cold War military involvements that have effectively scrapped many assumptions about counterinsurgency. He makes a number of specific criticisms of the new manual. This is a hard-nosed critique that requires careful study.

1763. Rogers, Chris. "More Soup, please: COIN Manual Provides Guidance for Modern-Day Tactical Commanders." *Armed Forces Journal* 145, no. 6 (January 2008): 36–38.

Praises Field Manual 3–24 for providing material that commanders can use effectively in their encounters with insurgency. Rogers believes that Gentile (item 1757) has given too little credit to commanders. Rogers, moreover, takes a broad view of the concept of "tactics" in counterinsurgency to include actions well beyond those associated with combat, asserting that "we have come to realize that fighting encompasses [a] ... greater range of options available to the tactical commander."

1764. Sewall, Sarah. "Modernizing U.S. Counterinsurgency Practice: Rethinking Risk and Developing a National Strategy." *Military Review* 86, no. 5 (September–October 2006): 103–8.

A counterinsurgency expert traces the revision of counterinsurgency doctrine in the United States and warns that problems continue to confront this effort. She fully understands and clearly demonstrates the complexities of counterinsurgency. The military aversion to counterinsurgency is a persistent problem after decades during which the armed forces prepared primarily for high-intensity conflict. A totally military approach is inadequate. One of Sewall's insights is that there is the high risk of casualties for friendly forces in a counterinsurgency, whereas the armed forces have been trying to minimize casualties.

C. Small wars: general studies

1765. Kirkpatrick, Jeane J. *Making War to Keep Peace*. New York: Harper Collins, 2007.

This is a highly detailed account of United States interventions from the Gulf War through the beginning of the invasions of Afghanistan and Iraq. The Clinton administration, in particular, receives a good deal of criticism, although the Bush administration is not spared in this study. Most chapters in the book are listed and annotated in the appropriate sections of this bibliography. Notwithstanding her remarks about United States leaders and policies, Kirkpatrick may well have regarded former UN Secretary-General Boutros Boutros-Ghali as a symbol of the concern she expressed in the title of the book. In regard to Somalia, she argued that "Boutros-Ghali even tried to absorb war into the category of peace operations" (p. 71).

D. Counterinsurgency: general studies

1766. Ackerman, Robert K., and Beverley P. Mowery. "Innovative Approaches Key to Warfighting, Military Posture." *Signal* 61, no. 8 (April 2007): 82–92.

Some of the article deals with the new army/marine Field Manual 3–24. The authors emphasize that both gathering needed information about insurgents and building positive popular attitudes toward the government are essential in waging counterinsurgency. A number of other topics were examined at a major conference in January–February 2007 at San Diego, including narcotic interdiction and counterinsurgency especially in Iraq.

1767. Baker, Jim. "Systems Thinking and Counterinsurgencies." *Parameters* 36, no. 4 (Winter 2006): 26–43.

Carefully analyzes the obstacles to developing an effective counterinsurgency strategy, such as the difficulty in establishing appropriate measures of progress, and briefly describes the nature of systems theory. Baker refers to "the core dilemma for the counterinsurgent—walking the thin line between too forceful and too limited military actions related to security." This is an excellent introduction to the problems of counterinsurgency, which uses graphics imaginatively and effectively.

1768. Carafano, James Jay. *Private Sector, Public Wars; Contractors in Combat—Afghanistan, Iraq, and Future Conflicts.* Westport, CT: Praeger Security International, 2008.

Seeks to debunk the view that the "military-industrial complex" in the United States has been a major factor in the Iraq War and offers a vigorous defense of the roles private contractors play in modern United States military operations.

1769. Chamberlain, Robert M. "Lies, Damned Lies and Counterinsurgency: Not All Insurgencies Have Been Protracted Affairs." *Armed Forces Journal* 145, no. 10 (May 2008): 41–43.

Severely criticizes studies that Chamberlain believes take unduly pessimistic views of prospects for winning counterinsurgencies, attributing most of these assessments to the work of the Dupuy Institute, which makes quantitative studies of military conflict. Chamberlain believes such analyses are distorted because they focus on only a few insurgencies. In reality, he argues, starting and winning an insurgency are difficult tasks. Thus, it "is simply untrue that counterinsurgencies are inherently long-term struggles."

1770. Cropsey, Seth. "Janus and the God of Jointness." *Armed Forces Journal* 143, no. 11 (June 2006): 18–20+.

Argues that the Goldwater-Nichols Defense Reorganization Act of 1986, which fostered further integration of the United States armed forces, was aimed not at coping with unconventional warfare but with high-intensity conflict, which was the major concern during the

Cold War years. The United States must retain high-intensity conflict capability, but it must also be ready to meet the very real threats of terrorism and insurgency.

1771. Daly, Terence J. "Classic Counterinsurgency: The Key to Victory Against Today's Insurgents." *Marine Corps Gazette* 90, no. 12 (December 2006): 53–57.

Regards "classic counterinsurgency doctrine" as stemming from the British success in Malaya and the French failure in Indochina. Daly rightly observes that both "security" and "development" are vital to the local population, and they must be pursued concurrently.

1772. Erwin, Sandra I. "Know the Enemy—Mathematical Models: The Latest Weapons Against Urban Insurgencies." *National Defense* 92, no. 649 (December 2007): 46+.

Carefully written article discusses the many uses of increasingly sophisticated computer simulations, which are valuable despite the objections of their critics. Simulations are frequently difficult to construct, but they can be important aids for decision-makers, provided that they are always regarded as aids, not devices that dictate particular approaches.

1773. ——. "Marines Probing New Ways to Fight Future Insurgencies." *National Defense* 90, no. 623 (October 2005): 14–15.

The marines are developing new ways to react to future unconventional wars, which will be tested in coming war games. Al Qaeda differs significantly from previous insurgencies encountered by the corps, and new doctrine must reflect those differences. A key finding of the marines is that it is more useful to view Al Qaeda as an insurgency rather than as a terrorist organization.

1774. Grdovic, Mark. "Understanding Counterinsurgency." *Special Warfare* 17, no. 2 (December 2004): 2–5.

Compares the worldwide Al Qaeda offensive with the Communist use of insurgency during the Cold War. Grdovic also makes an interesting comparison between a counterinsurgency theater of operations and a courtroom with the populace as the jury, which determines the legitimacy of one side or the other. This brief article is filled with insights, such as the finding that during a "counterinsurgency, perception can be more important than reality" (p. 5).

1775. Green, Dan. "Counterinsurgency Diplomacy: Political Advisors at the Operational and Tactical Levels." *Military Review* 87, no. 3 (May–June 2007): 24–30.

Emphasizing the interlocking of the military and political factors in a counterinsurgency, Green criticizes the view that only force was necessary for success in Afghanistan and Iraq. He usefully describes the increasing attention the United States is giving to counterinsurgency, including greater funding and the assignment of many more people to that sphere. Although much has been done to merge political and military responses to unconventional warfare, the State Department role needs to be expanded, even to the extent that some of its personnel are fully integrated into Special Forces teams.

1776. Grotto, Andrew. "Needing NATO: The Alliance Could Crumble if the U.S. Becomes the COIN Expert." *Armed Forces Journal* 145, no. 4 (November 2007): 12+.

Afghanistan is a testing ground for the long-term viability of NATO. One of the questions about NATO's future relates to the increasing commitment of the United States to developing its unconventional warfare capability. If it reaches a high degree of capability, it may no longer need the personnel and the experience provided by the European members of NATO, and, therefore, its support for NATO may be undermined.

1777. Grubbs, Lee K., and Michael J. Forsyth. "Is There a Deep Fight in a Counterinsurgency?" *Military Review* 85, no. 4 (July–August 2005): 28–31.

Innovative treatment of the "depth" concept in military operations which is generally seen as "'the extension of operations in time, space, and resources.'" (p. 28, quoting from army Field Manual 3–0, Operations). "Depth" refers to rear areas of enemy forces, which provide support to front line fighters. In an insurgency, a "rear area" can be mental, as well as geographical. Part of the counterinsurgent's task is to prevent successful insurgent adaptation to counterinsurgent initiatives and to their own losses.

1778. Hammes, Thomas X. "The Message is the Insurgency: Strategic Communications in the Society at War." *Marine Corps Gazette* 91, no. 11 (November 2007): 18, 20, 22, 24, 26, 28, 30.

Stresses the importance of information and communications campaigns in a counterinsurgency environment. Hammes cites North Vietnam's successful communications program when fighting the United States. The United States must build its programs for current struggles, while keeping in mind that they must be flexible and that their messages may vary at the local level. The situations the United States now faces are quite different from fighting Communist insurgencies. Today, "most insurgencies will involve aspects of a multisided civil war" (p. 22).

1779. Hayden, H. Thomas. "Measuring Success in Counterinsurgency." *Marine Corps Gazette* 89, no. 11 (November 2005): 50–51.

Examines the Hamlet Evaluation System and the Accelerated Pacification Program used in Vietnam and then describes how an accelerated pacification program for Iraq might be structured. As a basis, it "must stress the principle of community spirit" (p. 51).

1780. Hilburn, Matt. "Rethinking the Enemy: Joint Army/Marine Corps Counterinsurgency Center Aims to 'Change the Mindset' of Military Strategy." *Sea Power* 50, no. 4 (April 2007): 26–28, 30.

The Counterinsurgency Center at Fort Leavenworth opened in July 2006. Drawing on the Afghanistan and Iraq experiences, the center is attempting to focus the attention of the armed forces on counterinsurgency and to develop their capability in this area. The message for the armed forces is flexibility in dealing with insurgencies, which may take various forms and have various origins.

1781. Hillen, John. "Developing a National Counterinsurgency Capability for the War on Terror." *Military Review* 87, no. 1 (January–February 2007): 13–15.

Views the campaign against terrorism as a conflict with Muslim extremists who aim at establishing theocracies in many countries. Hillen, then Assistant Secretary of State for Political-Military Affairs, compares the insurgencies that these extremists are pursuing to a cancer, which presents a challenge that leaves the United States relatively unprepared. The emphasis in counterinsurgency must be on non-military factors, and, therefore, civilian agencies must be deeply involved in the United States practice of counterinsurgency.

1782. Hoffman, F.G. "Best Practices in Counterinsurgencies: Compressing the Learning Curve." *Marine Corps Gazette* 91, no. 10 (October 2007): 42–48.

Examines eight insurgencies since World War II to extract the "best practices" in terms of what the authors calls "campaign components," such as "integrated intelligence" and "resources control." This is a sophisticated effort in qualitative evaluation.

1783. ———. "Mind Maneuvers (the Psychological Element of Counterinsurgency Warfare)." *Armed Forces Journal* 144, no. 9 (April 2007): 28–32.

Technology has given insurgents important new capabilities for disseminating their views. This is a vital force-multiplier for insurgents given that "[i]deas and grievances are the seeds of most irregular wars." Hoffman discusses the communication technologies that are increasingly being employed. He usefully relates the new world of communications to traditional theories about waging counterinsurgency and then proposes ways to use technology to fight insurgents.

1784. Killebrew, Robert E. "Winning Wars." *Army* 55, no. 4 (April 2005): 25, 28, 32.

Observes that United States decision-makers did not anticipate the protracted guerrilla resistance that followed the conventional campaign in Iraq. Killebrew asserts that the war in Iraq is being won and asks what lessons can be learned from the army's experiences in Iraq. He notes that technology can materially assist insurgents, as well as conventional military forces. One major element in a successful counterinsurgency is the understanding that only host nation forces can finally defeat the insurgents.

1785. Lynn, John A. "Patterns of Insurgency and Counterinsurgency." *Military Review* 85, no. 4 (July–August 2005): 22–27.

Echoes United State Marine Corps General Anthony Zinn's plea to the armed forces that they recognize counterinsurgency as their major responsibility for some time to come. This is a rather detailed survey of insurgencies during the second half of the twentieth century, which is intended to identify lessons for this century. Lynn emphasizes the importance of providing physical security to the population of an area and gaining information about the insurgents.

1786. McFate, Montgomery. "Anthropology and Counterinsurgency: The Strange Story of Their Curious Relationship." *Military Review* 85, no. 2 (March–April 2005): 24–38.

Many military leaders and civilian analysts are calling for greater "cultural awareness" within the United States armed forces as a result of experiences in Afghanistan and Iraq. McFate argues that "[s]uccessful counterinsurgency depends on attaining a holistic total understanding of local culture" (p. 37). An inability to pick up cultural cues can be disastrous or at the least can catch the military utterly unprepared for many contingencies. Anthropology as a field useful for the military fell out of favor dramatically after the Vietnam War, however. McFate usefully reviews British and United States experiences in utilizing anthropology in war.

1787. McFate, Montgomery, and Andrea V. Jackson. "The Object Beyond War: Counterinsurgency and the Four Tools of Political Competition." *Military Review* 86, no. 1 (January–February 2006): 13–26.

Like many students of counterinsurgency, the authors emphasize that insurgencies are efforts to put forth governmental alternatives to ruling authorities. McFate and Jackson identify the measures that can be taken either to sustain an insurgency or to suppress one as "coercive force, economic incentive or disincentive, legitimating ideology and traditional authority" (p. 13). What follows are skillful analyses of each of these factors. This fine study is rich in the use of historical examples to reinforce the points made.

1788. Mackinlay, John, and Alison Al-Baddawy. *Rethinking Counterinsurgency.* Santa Monica, CA: RAND Corporation, 2008.

Begins with an extremely brief discussion of insurgency and counterinsurgency after World War II and then proceeds to an analysis of insurgencies in Muslim societies. The authors argue that to be successful counterinsurgents need to focus on the environment in which an insurgency is generated rather than concerning themselves exclusively with the insurgency.

1789. Mahnken, Thomas G. "5. The American Way of War in the Twenty-first Century." pp. 73–84, in *Democracies and Small Wars*, ed. Efraim Inbar. London: Frank Cass, 2003.

In this excellent review of United States military traditions, Mahnken distinguishes between the "strategic level," "the operational level," and the "tactical level." This approach provides a precise way to view those traditions, which is more useful than discussing a global "American way of war." Mahnken makes an important contribution to the literature by showing to what extent attitudes and policies developed to meet the armed forces of major powers for high-intensity conflicts can apply to small wars.

1790. Nagl, John A., and Paul L. Yingling. "New Rules for New Enemies." *Armed Forces Journal* 144, no. 3 (October 2006): 25–26+.

Notes the success of the United States in recent conventional conflicts, such as Panama in 1989, and the challenges it now faces from insurgencies, which is the kind of warfare the United States is likely to confront in the future. The United States armed forces are changing to meet insurgencies, as they must. Nagl proposes a number of innovations that would facilitate this transformation. Reducing the hierarchical nature of the army is just one of these innovations.

1791. Palazzo, Louis J. "Can We Learn From the Past? There Is No Single Answer to Waging a Successful Counterinsurgency." *Marine Corps Gazette* 92, no. 4 (April 2008): 49–53.

Argues strongly that the question posed in the title should be answered affirmatively. Palazzo summarizes several classic studies of counterinsurgency and then applies some key lessons from those studies to Iraq. The most important step for the United States in Iraq is to regain the initiative from the insurgents.

1792. Tiron, Roxana. "Irregular Warfare: Counterinsurgency in Iraq Provides Template for Fighting Terrorism." *National Defense* 89, no. 617 (April 2005): 40–41.

As the title indicates, at least some United States military leaders believe that counterinsurgency lessons from the Iraq experience can be applied generally in the Global War on Terrorism (GWOT). The United States defense establishment is giving much more emphasis to political and economic approaches than in the past, and a number of new programs are being considered or implemented by both military and civilian components of the United States government, including the State Department.

1793. Wendt, Eric P. "Strategic Counterinsurgency Modeling." *Special Warfare* 18, no. 2 (September 2005): 2–13.

Basing his observations in part on his extensive experience in the Philippines and in Iraq, Wendt shows how employing counterinsurgency models developed by the Department of Defense Analysis of the Naval Postgraduate School can provide the kind of understanding of counterinsurgency that may bring success. Wendt presents a number of his own views and usefully distinguishes between the roles of special operations and conventional forces in counterinsurgency.

E. Counterinsurgency: past lessons

1794. Cassidy, Robert M. "Back to the Street Without Joy: Counterinsurgency Lessons from Vietnam and Other Small Wars." *Parameters* 34, no. 2 (Summer 2004): 73–83.

Reviews the army's retreat from counterinsurgency after Vietnam and its neglect of counterinsurgency lessons from that conflict. Cassidy attempts to revive interest in counterinsurgency generally and to draw some lessons from Vietnam. He highlights and praises many of the marine and Special Forces experiences but does not ignore shortcomings in their efforts. Cassidy has quite a bit to say about the marines' *Small Wars Manual, 1940* and about the Philippine–American War. Overall, he is optimistic about the army's ability to meet twenty-first-century challenges.

1795. ———. "Winning the War of the Flea: Lessons from Guerrilla Warfare." *Military Review* 84, no. 5 (September–October 2004): 41–46.

Sees the global War on Terrorism as "a global counterinsurgency." The army must finally learn lessons from counterinsurgency victories to meet the latest threats. Much of the article is a historical and analytical survey of United States encounters with insurgents from the post–Civil War West through Vietnam, with much emphasis on Vietnam. Cassidy notes the many successes in Vietnam, but he also describes the failures.

1796. Deady, Timothy K. "Lessons from a Successful Counterinsurgency: The Philippines, 1899–1902." *Parameters* 35, no. 1 (Spring 2005): 53–68.

Suggests that the Iraq War has many similarities to the Philippine–American War. This is an informed summary of the Philippine–American War that includes many relevant facts and figures. Deady emphasizes the importance of the United States election of 1900 in the context of the conflict. Filipino nationalists hoped to sicken United States voters to further fighting by continuing to inflict meaningful casualties on United States forces, but William McKinley's reelection as president significantly damaged the prospects of a nationalist victory.

1797. Eno, Russell A. "Mekong Delta 1968: Counterinsurgency Then and Now." *Infantry* 96, no. 4 (July–August 2007): 36–38.

Describes Eno's experiences as a "District Assistant Advisor" in the aftermath of the Tet Offensive. Despite differences between Vietnam, on the one hand, and Afghanistan and Iraq, on the other hand, Eno nevertheless sees many similarities between these insurgency situations. He stresses the need to learn as much of the local language as possible, to use translators effectively, and to develop an effective public information program.

1798. Kilcullen, David. "Counter-Insurgency Redux." *Survival* 48, no. 4 (Winter 2006): 111–30.

Asserts that present-day insurgencies differ considerably from those of the past. Nevertheless, Kilcullen believes that much can be learned from older writings about counterinsurgency. Using material from his dissertation, Kilcullen attempts to pinpoint the unique characteristics of the newer insurgencies, in order to encourage counterinsurgents to adapt their strategies to these elements. Insurgent campaigns' use of the media is relatively new, for example. Such efforts greatly intensify the political nature of insurgencies/counterinsurgencies. Current insurgencies are often transnational, moreover, which forces counterinsurgents to think in terms of regions rather than countries. Kilcullen recognizes that this article is very much an essay, subject to correction, but it is a thought-provoking contribution.

1799. Leslie, Mark S. "Integrating Cultural Sensitivity into Combat Operations." *Armor* 116, no. 1 (January–February 2007): 35–38.

Analyzes the influence of the cultural patterns of an operational area on the conduct of counterinsurgency. Leslie maintains that "[c]ultural awareness is a combat multiplier [that is, an aid] in a counterinsurgency environment." Extensive, practical training in cultural awareness is, therefore, essential for soldiers and units to be deployed. Leslie goes on to give examples of how this awareness facilitates United States operations in Afghanistan and Iraq, and to warn that United States fire in whatever form it takes must be carefully controlled to do what it must but not to impact civilians and their property.

1800. Linn, Brian. "The Philippines: Nationbuilding and Pacification." *Military Review* 85, no. 2 (March–April 2006): 46–54.

Presents a highly realistic view of United States attitudes in the early years of United States rule in the Philippines. Imperialism or colonialism was a far more accurate description than the modern term "nationbuilding." Nevertheless, a number of United States programs, such as those in public health, did benefit many Filipinos. One of the keys to United States success was the decentralization of authority that permitted local commanders to adopt policies suited to local conditions. Linn gives considerable attention to the policies and activities of the Philippine government headed by Aguinaldo.

1801. Long, Austin. *Doctrine of Eternal Recurrence: The U.S. Military and Counterinsurgency Doctrine, 1960–1970 and 2003–2006*. Santa Monica, CA: RAND Corporation, 2008.

This is a brief, but fascinating, review of the development of counterinsurgency doctrine in the United States armed forces. The coverage is broader than indicated in the title. Long stresses that military services focusing on high-intensity conflict have great difficulty in adapting to the really more complicated environment of an insurgency. Much of the book is devoted to the Vietnam experience, but the importance of Long's analysis goes far beyond that conflict.

1802. ———. *On "Other War": Lessons from Five Decades of RAND Counterinsurgency Research*. Santa Monica, CA: RAND Corporation, 2006.

The RAND Corporation has studied insurgency and counterinsurgency for years. What Long does is to draw some overall conclusions or recommendations from a massive amount of research. This is a highly significant effort which states many of the problems of effective counterinsurgency in plain language. In addition to a section of "References," the book contains an annotated bibliography of a number of RAND reports.

1803. Metz, Steven. "New Challenges and Old Concepts: Understanding 21st Century Insurgency." *Parameters* 37, no. 4 (Winter 2007): 20–32.

This highly provocative study begins by noting the variations in the army's interest in counterinsurgency from Vietnam to the present. When an insurgency challenge emerges, so does concern with counterinsurgency. When there is no immediate threat, counterinsurgency is left to the Special Forces community. Today, Metz asserts, the United States is devoting too much attention to such conflicts as Vietnam and Algeria and is not giving sufficient attention to recent confrontations, such as Somalia and Bosnia. Metz stresses the economic facets of current insurgencies, moreover, even arguing that "contemporary insurgencies emulate corporations in a hyper, competitive (violent) market [which] shapes their operational methods" (p. 25).

1804. Moyar, Mark. "The Phoenix Program and Contemporary Warfare." *Joint Force Quarterly*, no. 47 (October 2007): 155–59.

The author of *Phoenix and the Birds of Prey* relates the well-known assault on the Vietcong infrastructure in South Vietnam to current operations in Afghanistan and Iraq. Moyar makes numerous astute comparisons between insurgencies in these three countries, emphasizes the importance of sharing intelligence between civilian and military authorities engaged in counterinsurgency, and offers other lessons from the Vietnam experience.

1805. Rabasa, Angel, Lesley Anne Warner, Peter Chalk, Ivan Khilko, and Paraag Shukla. *Money in the Bank: Lessons Learned from Past Counterinsurgency (COIN) Operations*. Santa Monica, CA: RAND Corporation, 2007.

Discusses six insurgencies, some of which ended years ago and some of which are continuing today, with a view to drawing specific lessons about the "tactics, techniques, and procedures" that may be useful in the twenty-first century. The analyses presented are relatively brief, but this book is a convenient introduction to the study of counterinsurgency.

1806. Sepp, Kalev I. "Best Practices in Counterinsurgency." *Military Review* 85, no. 3 (May–June 2005): 8–12.

Examines a large number of insurgencies and counterinsurgencies from the Philippine–American War through Afghanistan and Iraq to detect the "best practices." Successful counterinsurgents need to be concerned with "human rights," "law enforcement," "population control," the "political process" (psychological warfare), "counterinsurgent warfare," "securing borders," and a (centralized) "executive authority," and should avoid over-militarizing operations.

1807. Steele, Dennis. "The Army Magazine Hooah Guide to COIN (Counterinsurgency)." *Army* 57, no. 7 (July 2007): 43–66.

Summarizes Field Manual 3-24 and observes that although current counterinsurgency is heavily focused on Afghanistan and Iraq it is really a worldwide effort, the "first global counterinsurgency campaign." Steele emphasizes the value of studying previous counterinsurgencies for the contemporary counterinsurgent. He focuses on the Roman Empire, the British experience in the American Revolution, the Civil War, and the Philippine–American War.

1808. Tomes, Robert R. "Relearning Counterinsurgency Warfare." *Parameters* 34, no. 1 (Spring 2004): 16–28.

Points to a period in the 1990s when counterinsurgency theory and practice was virtually ignored by the United States armed forces. Tomes stresses learning from several older, but important, analyses of counterinsurgency. These books are David Galula's *Counterinsurgency Warfare: Theory and Practice* (1964); Roger Trinquier's *Modern Warfare: A French View of Counterinsurgency* (1961); and Frank Kitson's *Low Intensity Operations: Subversion, Insurgency and Peacekeeping*. They offer many useful lessons, such as the need for highly centralized control in counterinsurgency. The United States seems to be using some techniques recommended by Trinquier, such as a high degree of attention to intelligence gathering and direct attacks on the insurgent infrastructure. Population control is a key factor in distinguishing insurgents from ordinary citizens.

1809. Tovo, Ken. "From the Ashes of the Phoenix: Lessons for Contemporary Counterinsurgency Operations." *Special Warfare* 20, no. 1 (January–February 2007): 6–15.

Notes the army's virtual abandonment of counterinsurgency planning after Vietnam when it concentrated almost exclusively on conventional warfare. Although many people emphasize differences between Vietnam and current counterinsurgencies, there are lessons to be learned from the Vietnam experience. Tovo reviews the development of Civil Operations and Revolutionary Development and its Operation Phoenix that was intended to assail the Vietcong infrastructure. Despite significant abuses, distortions, and mistakes, Operation Phoenix helped to some degree, but it could not alter the course of the war. There is a need to strike at the contemporary terrorist infrastructure, but the campaign should avoid the problems of Operation Phoenix.

1810. Ward, Robert, Jr. "Oil Spot: Spreading Security to Counter Insurgency." *Special Warfare* 20, no. 2 (March–April 2007): 8–17.

Asks how the Iraqi government can suppress insurgency, if the United States armed forces are unable to do so. Ward reviews insurgency and counterinsurgency historically, emphasizing the United States defeat in Vietnam, and expresses approval for the French colonial "oil spot" strategy that involves developing firmly held bases, usually in cities or larger towns, that permit the gradual extension of what Ward terms "government base areas" (p. 11).

F. Counterinsurgency: tactics

1811. Brown, Robert B. "The Agile-Leader Mind-Set: Leveraging the Power of Modularity in Iraq." *Military Review* 87, no. 4 (July–August 2007): 32–44.

The ability to process information into intelligence rapidly and to react flexibly to enemy moves are the keys to effective operations in Iraq. Appropriate and realistic training is necessary for developing these capabilities. The increased information available on the counterinsurgency battlefield requires the delegation of decision-making to lower-level commanders. Much of the article centers on the "modular brigade combat team" that was introduced into the United States Army recently.

1812. Celeski, Joseph D. "Attacking Insurgent Space: Sanctuary Denial and Border Interdiction." *Military Review* 86, no. 6 (November–December 2006): 51–57.

Much progress has been made in Afghanistan against Al Qaeda forces, but they still have sanctuaries in Pakistan and, thus, the opportunity to regroup and begin new attacks in Afghanistan. The sanctuary problem is not the same as in Iraq and is more serious. Celeski offers numerous suggestions about assaulting insurgent sanctuaries. Contrary to many students of counterinsurgency, he welcomes insurgent development of sanctuaries, which can become counterinsurgent targets, especially as insurgents enter or exit their sanctuaries. This is an important study of a surprisingly neglected topic.

1813. Church, Iain J. "Assuring Mobility in a COIN Environment." *Engineer* 37, no. 1 (January–March 2007): 21–23.

Emphasizes that "protecting the populace rather than the COIN [counterinsurgency] force and ... maintaining legitimacy" (p. 21) are the keys to winning a counterinsurgency. Church examines efforts of the United States Army Engineer School to enhance army capability for dealing with explosive devices. Those abilities will help protect civilians and keep roads open. The courses offered by the school are described in some detail.

1814. Cone, Robert W. "The Changing National Training Center." *Military Review* 86, no. 3 (May–June 2006): 70–79.

The National Training Center, which was established in 1981, helped prepare the army for the Gulf War and Afghanistan, but it is now revamping its services to meet current insurgency threats. The author describes modifications in training, in large part the result of communication with United States forces in combat. Training at the center is not lock-step; it is fashioned to meet the specific needs of specific units. A highly impressive array of realistic training areas and innovative techniques are part of the center.

1815. D'Aria, Dorian. "CEHC (Counter Explosive Hazards Center) Support to COIN Operations." *Engineer* 37, no. 1 (January–March 2007): 16–20.

According to D'Aria, United States forces have to adapt constantly to the changing nature of their opponents. He stresses the unique demands that counterinsurgency imposes on the United States. Like many other analysts, D'Aria emphasizes the need to provide security to the civilian population during a counterinsurgency. An important part of maintaining security is the detection and disposal of explosive devices. This is an impressively detailed account of the army's effort to cope with this problem, in part through the training center of the Counter Explosives Hazards Center.

1816. Demarest, Geoffrey B., and Lester W. Grau. "Maginot Line or Fort Apache? Using Forts to Shape the Counterinsurgency Battlefield." *Military Review* 85, no. 6 (November–December 2005): 35–40.

Notes the negative view taken of fortifications by many military historians and defense analysts. The authors, however, argue that forts can play a valuable role in counterinsurgency by blocking or at least obstructing insurgent lines of communication. One effective current example is the use of numerous police stations by Colombia. Pakistan, too, is employing forts properly. Demarest and Grau question the United States pattern of fortifications in Iraq. They provide extensive lists of the desirable characteristics of fortifications.

1817. Gellman, Brian. "Assessing Stability During Counterinsurgency (COIN) or Stability and Support Operations Through Patrol Debriefs." *Military Intelligence* 31, no. 1 (January–March 2005): 8–12.

Describes a situation in Baghdad when such an assessment had to be made under extreme time pressure in 2003. Those responsible for measuring stability responded to the assignment successfully. Stability, which is critical in defeating insurgents, must be subject to quantitative analysis. Gellman offers a number of definitions of factors related to stability and demonstrates relatively easy ways to study them on a continuing basis.

1818. Kyser, Giles, Matt Keegan, and Samuel A. Musa. "Applying Law Enforcement Technology to Counterinsurgency Operations." *Joint Force Quarterly*, no. 46 (2007): 32–38.

The authors see many similarities between counterinsurgency and law enforcement efforts in the United States that are aimed at large-scale criminal activities, such as gangs. Knowledge of local areas is one way to suppress both crime and insurgency, but there are obstacles to acquiring such knowledge. "Biometric technology" can be applied to such situations by specifically identifying individuals and can be a powerful tool when used in conjunction with computer networks.

1819. Pappalardo, Joe. "Training Center Tries to Keep Focus on Future Fights." *National Defense* 90, no. 620 (July 2005): 41–42.

The National Training Center at Fort Irwin, California, must provide training for current low-intensity conflicts, but it cannot ignore the possibility of larger-scale military challenges in the future. Thus, while training troops for imminent deployments in Afghanistan or Iraq, it must also demonstrate the capability of preparing soldiers for high-intensity or conventional warfare.

1820. Wilson, Gregory. "Anatomy of a Successful COIN Operation: OEF Philippines and the Indirect Approach." *Military Review* 86, no. 6 (November–December 2006): 2–12.

Observes that the United States has difficulty adjusting to conflicts in which its opponents are guerrillas. Wilson argues that successful counterinsurgencies depend ultimately on the proper deployment of host nation armed forces and police, and asserts that United States operations in Afghanistan and Iraq have failed to succeed because of the burdensome United States military presence. Most of the article discusses United States–Filipino operations in the Muslim areas of the southern Philippines. Those operations, which began in 2002, have involved considerable civic action to benefit the local population.

G. Counterinsurgency: logistics

1821. Vlasak, Marian E. "The Paradox of Logistics in Insurgencies and Counterinsurgencies." *Military Review* 87, no. 1 (January–February 2007): 86–95.

The logistics of both insurgencies and counterinsurgencies have largely been ignored. Vlasak poses a number of highly pertinent questions about these matters and then examines insurgencies and counterinsurgencies in the post–World War II years, emphasizing the Chinese Civil War and the Vietnam conflict. The author notes that today insurgents often have access to materiel that reduces differences between the two sides in an insurgency. This is a sophisticated treatment of a critical, but much under-analyzed facet of unconventional warfare.

H. Intelligence and counterinsurgency

1822. Erwin, Sandra I. "Army Trying to Get Better Grasp on War Zone Intelligence." *National Defense* 89, no. 611 (October 2004): 18.

Army training is being changed to emphasize the importance of gathering information that can be processed as intelligence. The new approach is that "every soldier is a sensor." Involving additional personnel at the battalion level is an example of what can be done to enhance United States forces' intelligence capability. Another advance is the use of computers while troops are on patrol, which moves information to analysts quickly. There are still many problems, which often stem from insufficient "cultural awareness."

1823. Hart, Douglas, and Steven Simon. "Thinking Straight and Talking Straight: Problems of Intelligence Analysis." *Survival* 48, no. 1 (Spring 2006): 35–59.

Examines current obstacles to recruiting and training appropriate intelligence personnel and coordinating their efforts to understand "jihadism" in order to counter terrorism. The authors range usefully over a number of issues, including the lack of people with competency in languages relevant to dealing with terrorism. Greater use of technology may, to a degree, help reduce the significance of the problems mentioned.

1824. "Intelligent Design: COIN (Counterinsurgency) Operations Require Soldiers to Put a New Emphasis on Intelligence Collection and Analysis." *Special Warfare* 19, no. 3 (May–June 2006): 22–29. Reprinted in the *Military Review* 86, no. 5 (September–October 2006): 30–37.

Part of the task of learning how to wage counterinsurgency is developing an emphasis on data collection and analysis that is different from the patterns inherited from the Cold War. Among other changes is a greater need for "human intelligence." This article is largely devoted to describing the various phases of intelligence gathering. This is a useful reminder of the difficulty in shifting gears from the Cold War to another "counterinsurgency era," to use Blaufarb's term (item 994).

1825. Teamey, Kyle. "Effects-Based Targeting at the Bridge." *Military Intelligence* 31, no. 3 (July–September 2005): 50–54.

Proper targeting based on applicable and highly current intelligence is vital in a counterinsurgency environment. In turn, a military intelligence effort to support a counterinsurgency campaign is much more complex than in a conventional war situation. Teamey describes the process of target acquisition and the typical personnel configuration of a "target cell," that is, a group of armed forces members assigned to determining targets.

1826. Teamey, Kyle, and Jonathan Sweet. "Organizing Intelligence for Counterinsurgency." *Military Review* 86, no. 5 (September–October 2006): 24–29.

Asserts that traditional military intelligence organization and methods are unsuitable for counterinsurgency. The authors identify six key elements in counterinsurgency intelligence, such as the principle that "intelligence flows from the bottom up in counterinsurgency."

Thus, more intelligence personnel need to move from headquarters into combat forces. This article provides detailed guidance for organizing intelligence efforts and channeling the flow of intelligence.

1827. Williams, Vernon J. "Air Combat Intelligence: Tailoring Wing Intelligence for Counterinsurgency Operations." *Marine Corps Gazette* 90, no. 1 (January 2006): 34, 36.

Using his experience in Iraq, Williams underlines the problems of gathering intelligence about insurgents. The answer for the 2nd Marine Air Wing was to develop a centralized intelligence facility to support both its aerial and land operations, the latter including convoy protection.

I. Aviation and counterinsurgency

1828. Bellflower, John W. "The Indirect Approach (Thinking Continues to be Grounded in a Theory of Strategic Attack That Fails to Maximize Air Power's Utility in Small Wars)." *Armed Forces Journal* 144, no. 6 (January 2007): 12–16, 37.

Warns that aerial bombardment is only one possible application of air power in small wars and is one that can sometimes be counterproductive. Airlift, including the provision of humanitarian relief, intelligence collection, and psychological campaigns are other ways for the United States to exploit its air supremacy in counterinsurgency.

1829. Brown, Jason M. "To Bomb or Not to Bomb? Counterinsurgency, Airpower, and Dynamic Targeting." *Air and Space Power Journal* 21, no. 4 (Winter 2007): 75–85.

Refers to the many uses of air assets in counterinsurgency but warns that they should usually be coordinated with land operations. Brown also cautions that insurgencies vary considerably and that the response to them, including the use of air power, must be adapted to those different movements and the transformations that they undergo. This important study includes a detailed analysis of armed forces doctrines for targeting and proposes ways to improve the process.

1830. Comer, Richard. "An Irregular Challenge (Irregular Warfare)." *Armed Forces Journal* 145, no. 7 (February 2008): 12–15.

Argues that the air force needs to develop a more vigorous commitment to irregular warfare. The air force has contended that irregular warfare is simply high-intensity warfare reduced in scale, and, therefore, its high-end weapons and systems can readily be applied to lower levels of conflict intensity. There are signs of change in the air force position as shown by modifications in its budgetary requests, however. Comer usefully outlines the many ways in which the air force can contribute to successful outcomes as the United States assists its allies in developing their internal defense capabilities.

1831. Dunlap, Charles J., Jr. "America's Asymmetric Advantage." *Armed Forces Journal* 144, no. 2 (September 2006): 20–22+.

Briefly reviews and dismisses the current use of past practices for suppressing insurgencies, such as the use of a military and police presence that has numbers of personnel far in excess of the insurgents and the use of civic action projects to benefit the local population. Dunlap considers these and other methods as totally inapplicable to Afghanistan and Iraq because of the nature of the insurgencies and other factors. Air power seems to be the answer. Dunlap views opponents of the application of the full range of air power as "neo-Luddites."

1832. Macgregor, Douglas. "Washington's War." *Armed Forces Journal* 145, no. 3 (October 2007): 16+.

Extremely pessimistic view of the Iraq War and the "surge" in military personnel. An entirely new strategy for the employment of the military is needed in the present environment. Macgregor insists that the United States should avoid undertaking military occupations, which would almost certainly be counterproductive. Instead, he favors an air-maritime strategy which suits United States requirements and capabilities.

1833. Maitre, Benjamin R. "The Paradox of Irregular Airpower." *Air and Space Power Journal* 21, no. 4 (Winter 2007): 36–41.

Notes the difficulty of identifying air targets in irregular warfare and in avoiding impacts on civilians not participating in an insurgency. Maitre usefully defines several terms related to irregular warfare and then discusses its unique demands on air force personnel. The employment of air power in irregular warfare is necessarily quite different from its use in a strategic role as part of a high-intensity conflict.

1834. Owen, Robert C., and Karl P. Mueller. *Airlift Capabilities for Future U.S. Counterinsurgency Operations.* Santa Monica, CA: RAND Corporation, 2007.

The authors examined several questions related to the ability of the United States to respond promptly and effectively to insurgency challenges through the use of airlift. They found that for the most part present air force resources can be used to accomplish counterinsurgency airlift assignments, but some missions might require the development of a specialized airlift segment to support them. There is a brief, but highly useful, bibliography of current works. This study includes little information about the historical development of airlift for counterinsurgencies.

1835. Peck, Allen G. "Airpower's Crucial Role in Irregular Warfare." *Air and Space Power Journal* 21, no. 2 (Summer 2007): 10–15.

Suggests that unconventional warfare strategies and tactics used by enemies of the United States may be viewed as responses to United States domination of the air. Peck is careful not to ignore the role of ground forces in counterinsurgency, but he asserts the importance of air power, including space technology. Aerial weapons and systems are highly adaptable, and Peck cites a number of examples of this adaptability.

1836. Read, Robyn. "Irregular Warfare and the US Air Force: The Way Ahead." *Air and Space Power Journal* 21, no. 4 (Winter 2007): 42–52.

Discusses some of the findings of an Air Force Symposium on Counterinsurgency held in April 2007. The air force can perform well in battling an insurgency but loses what it has learned soon after the campaign is over. The air force cannot win a counterinsurgency on its own but must "partner" with other United States services and United States civilian agencies, and, most important, with indigenous forces. "Solutions in IW [Irregular warfare]/COIN [Counterinsurgency] will be based on [the] unique local circumstances of each conflict and result largely from political rather than military initiatives."

1837. Samuel, David F. "Making Better Use of Aviation Assets." *Marine Corps Gazette* 89, no. 11 (November 2005): 16–17.

There needs to be better communication and coordination between marine ground and air elements in counterinsurgency and stability environments. The capabilities of fixed-wing aircraft and helicopters differ considerably and the differences between them should govern their employment.

1838. Slife, Jim. "Shootdown Solution (Flying High May be the Best Protection Against Shootdowns)." *Armed Forces Journal* 144, no. 11 (June 2007): 16–20+.

Helicopter pilots in Iraq need to forget the old lesson of combat flying that lower altitudes are safer than staying high. Fighter doctrine began moving toward higher altitudes in the 1970s, and now helicopter pilots need to be more aware of natural hazards in Iraq and enemy weapons, such as surface-to-air missiles.

J. Counterinsurgency and civic action

1839. Andrade, Dale, and James H. Wilbanks. "CORDS (Civil Operations and Revolutionary Development Support)/Phoenix: Counterinsurgency

Lessons From Vietnam for the Future." *Military Review* 86, no. 2 (March–April 2006): 9–23.

Stresses that lessons can be learned from the counterinsurgency effort in Vietnam. The authors review the phases of the Vietnam conflict, concentrating on the period from 1966 onward. They describe CORDS organization and activity in some detail and outline and defend the Phoenix attack on Vietcong infrastructure. There is much to praise in the CORDS program, especially the combination of the military and civilian elements of counterinsurgency, and Andrade and Wilbanks argue that the Provincial Reconstruction Teams in Afghanistan and Iraq could benefit from knowing more about the CORDS experience.

1840. Brown, Kendall. "The Role of Air Force Civil Engineers in Counterinsurgency Operations." *Air and Space Power Journal* 22, no. 2 (Summer 2008): 44–50.

Traces the history of civil engineering in the air force from World War I onward, noting that the army provided engineering support for the air force for years even after the latter became an independent service branch in 1947. In 1965, the air force did begin to develop its own engineering capability. In addition to constructing and maintaining air force facilities, the civil engineer components can perform civic action projects, such as building hospitals and schools, for local civilians.

1841. McIntosh, Scott E. "Building a Second-Half Team: Securing Cultural Expertise for the Battlespace." *Air and Space Power Journal* 21, no. 1 (Spring 2007): 61–70.

Despite the overwhelming military power of the United States, it often lacks the ability to deal with opponents who use unconventional warfare. In such situations, safeguarding and servicing the civilian population is crucial. McIntosh uses a number of the military science journal articles to bolster his points and suggests several ways for the United States armed forces to enhance their ability to interact with foreign cultures.

1842. Ryan, Mick. "The Military and Reconstruction Operations." *Parameters* 37, no. 4 (Winter 2004): 58–70.

An Australian officer discusses the increasing commitment of Western armed forces engaged in counterinsurgency to what has often been termed "civic action," which is now more often called "humanitarian assistance." The Provincial Reconstruction Teams deployed in Afghanistan are important, but they need improved intelligence and security capabilities. Indeed, intelligence is critical for effective reconstruction. Civilian organizations, which might be highly useful, cannot function in combat environments. Ryan also recommends using local labor for many reconstruction projects, a practice that has many benefits for the host country.

K. Counterinsurgency: legal aspects

1843. Dungan, C. Peter. "Fighting Lawfare at the Special Operations Task Force Level." *Special Warfare* 21, no. 2 (March–April 2008): 9–15.

Law suits have become a weapon that is frequently used against United States unconventional warfare campaigns. The law suit is essentially an element of enemy information and psychological warfare operations. Law suits by some prisoners held at Guantanamo Bay, Cuba, are examples of this response to United States anti-terrorist measures. Dungan uses the experiences of the 1st Battalion, 3rd Special Forces Group in Afghanistan in 2006–7 to teach tactical commanders how to avoid facing legal entanglements that obstruct their performance of their duties. Legal attacks on United States forces generally involve alleged "detainee abuse." Responding to legal thrusts can often require skillful use of the media.

1844. Perry, David L. "Why Hearts and Minds Matter: Humanity, Even in Counterinsurgency, Is Not Obsolete." *Armed Forces Journal* 144, no. 2 (September 2006): 40+.

Although some members of the United States armed forces may wish to apply extremely harsh measures against insurgents, because of their ruthless tactics, Perry insists that restraint is needed even in the face of great provocation. He recalls the Mylai massacre in 1968 and cautions against repetitions of it, warning that "no soldier is authorized to be judge, jury and executioner of suspect insurgent sympathizers." Commanders must exemplify and enforce "chivalry and humanity."

1845. Teamey, Kyle B. "Arresting Insurgency." *Joint Force Quarterly*, no. 47 (October 2007): 117–22.

The apprehension and detention of insurgents is a vital component of successful counterinsurgency. Arresting the wrong people is, of course, highly counterproductive, and Teamey discusses the reasons for such errors. Releasing insurgents can also be a serious mistake in many instances. Teamey identifies a number of other problems in the apprehension and internment process and offers many principles for maintaining an effective system.

L. Counterinsurgency forces: general

1846. Book, Elizabeth G. "Role of Special Ops Evolves Over Time." *National Defense* 86, no. 579 (February 2002): 18–19.

Briefly reviews the history of United States special operations forces, emphasizing recent developments, such as the establishment of the United States Special Operations Command. Book quotes several United States military leaders who cite the importance of special operations troops in asymmetrical warfare, and she notes the small amount of financing and personnel devoted to special operations by this country.

1847. Worley, Robert. "A Small Wars Service." *Joint Forces Quarterly*, no. 44 (January 2007): 28–33.

Proposes the establishment of a Special Operations Corps to fight small wars. Worley notes that some steps have already been taken in this direction, notably the formation of the United States Special Operations Command. Now, the marines should be moved into this command where they would largely shape the new service. Worley usefully describes the various existing components of the command. The thought that the marines should make small wars their responsibility is not altogether new, but this seems to be the most detailed proposal.

M. Counterinsurgency forces: army and air force

1848. Foresman, Henry J., Jr. "Culture Battle: Selective Use of History Should Not be Used to Justify the Status Quo." *Armed Forces Journal* 145, no. 1 (August 2007): 38–41.

Asks whether the traditional structure of the army, including divisions and corps, is appropriate in an era of unconventional warfare. Indeed, through much of its history, the army has been committed to unconventional conflict rather than major wars. Foresman believes that the army is overly rigid in its thinking and is not sufficiently open to innovation, but he states that the army is making progress toward adapting to present conditions. To him, Field Manuals 3–24 and 3–0 are "revolutionary."

N. Counterinsurgency forces: marines and navy

1849. Cooper, David E. "An Organizational Model for Marines Fighting an Insurgency." *Marine Corps Gazette* 89, no. 6 (June 2005): 48–50.

Discusses the need to work with non-military agencies in order to combat an insurgency successfully. Sometimes, it is necessary for the marines to assume non-military duties themselves. A better solution is to have the marines coordinate both military and non-military operations. Cooper says much about both Vietnam and Iraq.

1850. Kennedy, Harold. "Shift to Special Operations Will Not 'Gut' the Marine Corps, General Says." *National Defense* 90, no. 628 (March 2006): 144+.

Discusses the corps' new Marine Special Operations Command, which consists, so far, of two battalions designed for special duties, such as raiding, target identification, and rescue missions. Most of the marines in these battalions come from a reconnaissance background, which concerns some commanders who fear they are losing highly skilled people. Some other marines are also critical of the new formations.

1851. Murphy, Martin N. "Blue Berets." *Armed Forces Journal* 144, no. 9 (April 2007): 20–22+.

The navy must adapt further to meet the threats represented by opponents waging unconventional warfare against the United States. Appropriate moves are being made, such as the introduction of Littoral Command Ships, but more needs to be done. The navy can certainly contribute to fighting guerrillas by imposing blockades, but it must also develop applied political and cultural knowledge.

XXXIII

Afghanistan, 2001–

In a dramatic move after the 9/11 attack, the United States and a number of its NATO allies and other countries attacked Afghanistan on the ground that it had sheltered and encouraged the elements that had planned and carried out the 9/11 attack and other hostile actions. In addition to being a demanding combat challenge, Operation Enduring Freedom was a daunting logistical effort, given Afghanistan's landlocked position.

Coalition forces allied themselves with the many dissident elements in Afghanistan and undertook successful offensives and special operations. Fierce resistance by part of the population and the geography of Afghanistan continues to make stabilizing the country extremely difficult.

For a time, Afghanistan was the major theater of the global War on Terrorism, but that changed with the invasion of Iraq. That country became the primary concern of the United States and its allies for years. More recently, an upsurge in violence in Afghanistan and the partial stabilization of Iraq have demanded more human and other resources for the allied mission in Afghanistan.

A factor that has further complicated the situation in Afghanistan is the spread of violence to Pakistan and the use of Pakistan by Afghan guerrilla forces as a sanctuary. Coalition air attacks both in Afghanistan and in Pakistan continue to create controversy. Often they are perceived as necessary to support ground troops, but frequently there are non-combatants killed, which results in protests. War weariness in coalition countries, moreover, seems to be increasing. At the same time that more troops appear to be needed, popular support for the war is dropping.

Chronology

1978
April 27   Leftists seized control in Kabul. Within a month opponents of the new regime began military training in Pakistan.

April 30	A Revolutionary Council formally took control of Afghanistan.
December 5	A treaty with the U.S.S.R. brought Afghanistan into the Eastern bloc.

1979
January 28	Anti-government military operations began in some parts of Afghanistan. Numerous battles and revolts by dissident military forces occurred over a long period.

1980
January	The Carter administration began assisting anti-government elements in Afghanistan. Other countries, notably Egypt, also provided aid for anti-government forces. The United Nations objected strongly and repeatedly to Soviet involvement in Afghanistan.

1988
April 14	After years of fighting and complicated negotiations the Soviets agreed to withdraw from Afghanistan. Soviet troops began leaving the country on August 8, and their presence ended by February 14, 1989.

1996
September 27	After years of fighting between various factions, Afghanistan was largely won by the Taliban, when its forces captured the capital of Kabul. That victory did not end combat within Afghanistan, however.

1998
August 20	United States cruise missiles fired against suspected Al Qaeda positions in Afghanistan in retaliation for attacks on United States embassies in Tanzania and Kenya.

1999
November 14	The United Nations applied economic and communications sanctions to Afghanistan. As time passed, both the United Nations and the United States demanded that Afghanistan agree to the extradition of Osama bin Laden for what were deemed terrorist activities.

2000
November 7	The United Nations refused to recognize the Taliban regime and would not give Afghanistan's seat in that body to the Taliban.

2001
September 11	Attack on the World Trade Center brought the Afghan crisis to a head.

October 7	Operation Enduring Freedom, the invasion of Afghanistan, begun by the United States.
November 13	Northern Alliance troops, allied with the United States, captured Kabul.
December 6	Kandahar, the last major Afghan city held by the Taliban, surrendered.
December 19	International Security Assistance Force formed under United Nations auspices to work for stability in Afghanistan.

2002

March 1	Operation Anaconda, a major offensive against the Taliban, was begun.
April 17	During Operation Anaconda, four Canadian soldiers were killed by United States Air Force warplanes. Errors in target identification on the ground continued to occur, creating tensions with the Afghans and Pakistanis.

2003

August 11	NATO assumed responsibility for the International Security Assistance Force in Kabul.

2004

January 26	New constitution adopted for Afghanistan.
October 9	Elections in Afghanistan gave Hamid Karzai the presidency. He had headed the Afghan government since December 11, 2001.

2005

September 15	Elections for parliament and the provinces held.

2006

May 15	Beginning of Operation Mountain Thrust, which resulted in heavy fighting for some time.
October	NATO took command of all forces in Afghanistan.

2008

April	NATO reaffirmed its support for the Afghanistan operation.
August	Numerous civilians killed by air attacks in Herat Province.
November	Taliban stated it will not participate in negotiations to end fighting until all foreign troops have left Afghanistan.

2009

March	Additional United States troops dispatched to Afghanistan owing to the deterioration of security in that country.
June 15	General Stanley McChrystal assumed command of United States forces in Afghanistan.

July 2 Operation Strike of the Sword, a combined attack primarily by United States marines but also by some Afghan soldiers, initiated in the Helmand River Valley in southern Afghanistan.
August 20 In national elections, Karzai was said to have won the presidency again, but both international authorities and many Afghans disputed the validity of balloting and vote counts.

A. Encyclopedia

1852. Clements, Frank A. *Conflict in Afghanistan: A Historical Encyclopedia.* Santa Barbara, CA: ABC-CLIO, 2003.

Impressive reference work for a complex area of the world. The title should have been "Conflicts in Afghanistan," because much of the content relates to wars and other problems before the invasion of 2001. Nevertheless, this volume is especially helpful for the current war. A hugely detailed chronology and a large bibliography add to the value of the work. The book continues to be important, although, of course, much has happened in the country since August 2003, when Clements ended his research.

B. General studies

1853. Donini, Antonio, Norah Niland, and Karin Wermester, eds. *Nation-Building Unraveled? Aid, Peace and Justice in Afghanistan.* Bloomfield, CT: Kumarian Press, 2004.

Major collection of essays. Several essays are of special importance to those concerned with the military side. Kate Clark's discussion of "The Struggle for Hearts and Minds: The Military, Aid, and the Media" demonstrates the problems of collecting and transmitting accurate information about Afghanistan, for example. All the contributors have had considerable experience in Afghanistan, and their varied backgrounds further enhance the value of the book.

1854. Donnelly, Thomas. "The Islamabad Dilemma: Success in Afghanistan Depends on Success in Pakistan." *Armed Forces Journal* 144, no. 4 (November 2006): 49–51.

While acknowledging Pakistan's considerable assistance to the coalition during the invasion and occupation of Afghanistan, Donnelly cautions that Pakistan is really oriented toward a possible confrontation with India. Moreover, Pakistan has its own insurgency and its own agenda with respect to Afghanistan. The United States must not ignore the complexity of Pakistan's political situation.

1855. Dorronsoro, Gilles. *Revolution Unending Afghanistan: 1979 to the Present*. Trans. from the French by John King. New York: Columbia University Press, 2005.

Thorough analysis of political developments in Afghanistan that provides a necessary background to the final chapter about the invasion of 2001 and what the author calls "the return of fragmentation." Dorronsoro carefully describes the profound changes that have occurred in Afghanistan, and he states that "there is no indication that a rapid pacification of the country can be expected" (p. 354). There is a brief, but important, segment about the relationship of Pakistan to the situation in Afghanistan.

1856. Dreyer, Vincent M. "Retooling the Nationbuilding Strategy in Afghanistan." *Joint Force Quarterly*, no. 43 (October 2006): 34–39.

Nationbuilding became a prime commitment of the Bush administration in Afghanistan once the conventional war had been won. This is an excellent overview of United States strategy in Afghanistan and its practical application. The failures of the strategy are described in detail. Lack of adequate funding and deficiencies in cooperation and coordination between NATO partners are among the problems. There is a great deal of conflicting advice being given about solving the various problems, moreover.

1857. Goodson, Larry P. *Afghanistan's Endless War: State Failure, Regional Politics, and the Rise of the Taliban*. Seattle: University of Washington Press, 2001.

The fact that Goodson's study was published before the 9/11 attack and the Western invasion of Afghanistan in a sense makes the volume all the more useful. Goodson's fine analysis of Afghanistan's problems is thus uncomplicated by the changes brought by those events. The focus is on the political evolution of Afghanistan, although economic and military factors are not omitted. Goodson emphasizes what he views as the shortcomings of the Taliban regime which has "had virtually no positive presence in the wider world" (p. 80) and which seems to lack any "ideas about how to confront Afghanistan's myriad development problems" (p. 115).

1858. Harrison, Marshal Tony. "Coalitions in Crisis: Without a Plan for Getting Out, Nations are Unlikely to Pitch In." *Armed Forces Journal* 144, no. 5 (December 2006): 33–35+.

British Army Chief of Staff Richard Dannatt was highly critical of a British commitment to Iraq. Dannatt believes that the situations in Iraq and Afghanistan are quite different, asserting that the outlook in Afghanistan is promising, but prospects for a favorable outcome in Iraq are fading.

1859. Hayes, Geoffrey, and Mark Sedra, eds. *Afghanistan: Transition Under Threat*. Waterloo, Ontario: Wilfrid Laurier University Press, 2008.

Although written primarily for Canadians, this collection of essays is highly useful for readers in the United States because of its broad perspective and its relative currency. The authors discuss political, economic, and security elements in the current Afghan situation. A good deal of attention is given to Pakistan and to the border between Afghanistan and Pakistan.

1860. Kolhatkar, Sonali, and James Ingalls. *Bleeding Afghanistan: Washington, Warlords, and the Propaganda of Silence*. New York: Seven Stories Press, 2006.

Extremely critical account of the occupation of Afghanistan that is best described as an anti-war text. The authors have a good deal to say about the many Afghan warlords who became allies of the United States against the Taliban and whom they view as close to the Taliban in many of their policies. They also see significant continuity between the policies toward Afghanistan of the Clinton and George W. Bush administrations.

1861. Mills, Greg, and Terence McNamee. "Challenges and Choices [in Afghanistan]." *Armed Forces Journal* 144, no. 4 (November 2006): 16–21.

Although definite progress has been made in Afghanistan, internal security challenges are mounting, at the same time that many Afghans are dissatisfied with government ineffectiveness and weakness. This is a thoughtful analysis of the problems facing Afghanistan and the coalition and the ways in which those problems are being tackled. The authors recommend development of an exit strategy from Afghanistan.

1862. Rashid, Ahmed. *Descent into Chaos: The United States and the Failure of Nation Building in Pakistan, Afghanistan, and Central Asia*. New York: Viking Penguin, 2008.

Assails the failure of the United States to pursue nationbuilding in Afghanistan and assesses United States policies more generally in the area mentioned in the book's subtitle. Assistance in developing of economies in the countries of this region is essential to counter Muslim extremists. This is an important analysis of complex international and internal relationships by a highly knowledgeable journalist. Outstanding maps and a helpful glossary supplement the text.

1863. Rotberg, Robert I., ed. *Building a New Afghanistan*. Cambridge, MA: World Peace Foundation, 2007.

The many essays make a highly significant contribution to the literature. There is a strong emphasis on economic development, not only in Afghanistan itself but also the whole region. Nevertheless, other aspects of the reconstructing of the country are hardly ignored. Hekmat Karzai and Ali A. Jalali, for example, provide fine chapters about the security challenges in Afghanistan. Overall, the book is an optimistic statement about the possible future of Afghanistan, notwithstanding the many problems, including the role of opium in the economy, which the contributors analyze skillfully.

C. General military studies

1864. Ackerman, Robert K. "Afghanistan is Only the Tip of the Network-Centric Iceberg." *Signal* 56, no. 8 (April 2002): 45–47.

   Electronic devices are contributing significantly to United States successes in Afghanistan, in part because opposition elements lack, by far, the capabilities of United States forces. Communication between sensing systems and firepower is a key element in these developments. Information technology will surely be of increasing importance as a part of United States defense measures, moreover.

1865. "Afghanistan Imagery Reveals Snapshot of Future Challenges." *Signal* 56, no. 6 (February 2002): 21–23.

   The National Imagery and Mapping Agency faced major challenges after 9/11. It had to develop new approaches to data collection and processing and had to work with commercial firms to produce the information needed to support coalition forces in Afghanistan. Information must be available quickly, moreover, which also increases the agency's dependence on commercial sources.

1866. Bailey, Timothy J. "Why Not Afghanistan? This Mission is Still to be Accomplished." *Marine Corps Gazette* 91, no. 8 (August 2007): 14–17.

   Closely examines renewed Taliban activity in Afghanistan. Bailey correctly predicts an increase in the United States commitment in Afghanistan, and he strongly advocates a larger marine presence in the country. He also gives considerable attention to Taliban operations in Pakistan and on the Afghanistan–Pakistan border.

1867. Bolduc, Donald C., and Mike Erwin. "The Anatomy of an Insurgency: An Enemy Organizational Analysis." *Special Warfare* 20, no. 4 (July–August 2007): 14–17.

   Insightful study of the Afghan insurgent organization, which is highly valuable because many authors have dismissed the insurgents as an entirely ad hoc, even chaotic, body of opponents. The authors pinpoint the roles of the various elements within the insurgent forces.

The insurgents have mastered the skills that they need to survive and to make themselves felt, including the extensive use of technology and the effective employment of public relations techniques.

1868. Brookes, Peter. "Flashpoint: Peril in Pakistan: Mountainous Border with Afghanistan Provides Haven for the Taliban." *Armed Forces Journal* 144, no. 11 (June 2007): 12–14.

The Pakistan government's settlement with Taliban sympathizers has given anti-coalition forces in Afghanistan, both Taliban and Al Qaeda, a virtual sanctuary in Pakistan. Brookes discusses the turbulent situation in Pakistan and describes the involvement of some Pakistanis in planning terrorist attacks in Great Britain and elsewhere.

1869. Burgess, Mark. "The Afghan Campaign One Year On." *Defense Monitor* 31, no. 8 (September 2002): 1–3.

Notes the concerns expressed about the perils of an Afghan invasion, based on historical experiences, but assesses the present situation (2002) rather favorably. Air power, use of special operations forces, and alliances with anti-Taliban elements in Afghanistan brought swift defeats to the Taliban. Burgess stresses the essential role of small numbers of special operations personnel in bringing successes. He takes a rather critical view of our Afghan allies.

1870. Coburn, Matthew D. "It Takes a Village to Counter an Insurgency." *Special Warfare* 20, no. 4 (July–August 2007): 8–13.

Examines the application of the "clear, hold and build" strategy of counterinsurgency in Afghanistan. Coburn discusses the Special Forces contribution to building the Afghan National Army, which is a key element in defeating the insurgents, and focuses on the army's deployment at the local level. Paramilitary police and militia are needed to supplement and to work with the army. Obviously, Coburn has the Vietnam experience very much in mind, and at one point he refers to the Civilian Irregular Defense Groups used there.

1871. Collins, Joseph J. "Planning Lessons from Afghanistan and Iraq." *Joint Force Quarterly*, no. 41 (April 2006): 10–14.

Briefly compares planning for these two wars. Despite limited planning, the Afghan operation has gone relatively well. At the same time what Collins views as the extensively planned Iraq invasion has been less successful. Collins praises the Provincial Reconstruction Teams in Afghanistan. He identifies a number of errors in Iraq and suggests that in an operation like Iraq, relatively rapid progress is needed to deflect media criticism. Participation by the State Department and the United States Agency for International Development in such an operation is essential from the beginning.

1872. Corbin, Marcus. "Operation Enduring Freedom and Military Transformation." *Defense Monitor* 31, no. 8 (September 2002): 4–5+.

The United States armed forces are moving away from a Cold War frame of reference to deal with less militarily sophisticated opponents, but they are still emphasizing technological advances to stay ahead of challenges. Possible lessons from the Afghanistan war are being analyzed, but care is being taken to avoid reading too much into experiences from that conflict. The article is based heavily on Secretary of Defense Rumsfeld's 2002 annual report.

1873. Croot, Edward C. "Digging Deeper: Cultural Understanding Requires ARSOF (Army Special Operations Forces) to Go Beyond Surface Understanding." *Special Warfare* 20, no. 2 (March–April 2007): 26–29.

Understanding Islam was crucial for the 3rd Battalion, 3rd Special Forces Group during its assignment to Afghanistan in 2003. Treatment of dead Muslims was a case in point. Croot describes several encounters with Afghans to illustrate the importance of respecting their culture.

1874. Darley, William M. "Strategic Imperative: The Necessity for Values Operations as Opposed to Information Operations in Iraq and Afghanistan." *Air and Space Power Journal* 21, no. 1 (Spring 2007): 33–41.

Argues that discussions of "culture" have been insufficient in relation to recent wars. "Culture is not merely one dimension of these conflicts; it is the battlefield." Darley says candidly that the United States goal must be to adapt some foreign cultures to the model it represents. He reviews ways in which culture has been imposed on conquered groups in the past. The imperative for United States policy in Afghanistan and Iraq must be to eliminate the political power and influence of Muslim clergy.

1875. Eikenberry, Karl W. "The Afghanistan Campaign: Progress and Challenges." *Army* 56, no. 10 (October 2006): 113–14+.

Overview of the Afghanistan situation by the coalition commander. Despite considerable progress toward stability, significant fighting is still occurring. Civic action is an important part of the coalition effort, as is the strengthening of Afghan governmental institutions.

1876. Friedman, Norman. "More Lessons from Afghanistan." *United States Naval Institute Proceedings* 129, no. 2 (February 2003): 4, 6.

Highly critical account of the planning and execution of Operation Anaconda. Friedman attributes problems to inter-service rivalries, especially to friction between the army and the air force. The intention was to use extensive surveillance and then to make pinpoint air and land attacks. The army, in particular, was at fault for depending too much on its own aviation and not utilizing air force support effectively.

1877. Griffin, Christopher. "A Working Plan: Hope Isn't the Only Strategy for Afghanistan." *Armed Forces Journal* 144, no. 9 (April 2007): 34–38.

Griffin concisely reviews events in Afghanistan from 2001 onward, emphasizing the political aspects. He predicts significant fighting with the arrival of spring 2007 and says that current United States military and diplomatic leaders may need to fashion a response to Taliban attacks by learning from their own predecessors, who developed an effective military-diplomatic team.

1878. Hawkins, William R. "What Not to Learn from Afghanistan." *Parameters* 32, no. 2 (Summer 2002): 24–32.

Warns emphatically that air power is not the whole answer in waging war, despite its critical role in Afghanistan. Hawkins argues, too, against bracketing Afghanistan with the interventions in the Balkans in the 1990s. He surveys United States military history briefly and asserts that the invasion of Afghanistan is a "punitive expedition."

1879. Hernandez, Prisco R. "Developing Cultural Understanding in Stability Operations: A Three-Step Process." *Field Artillery* (January–February 2007): 5–10.

Basing his article, in part, on army doctrine, Hernandez emphasizes that United States military personnel must understand the cultures of the countries where they operate and offers a plan for developing "cultural understanding," which is something more than "cultural awareness" but less than the knowledge and empathy required to make someone a "regional or cultural expert" (p. 6).

1880. Jalali, Ali Ahmad. "Afghanistan: The Anatomy of an Ongoing Conflict." *Parameters* 31, no. 1 (Spring 2001): 85–98.

Sophisticated analysis of the traditional role of the Afghan armed forces, including the many militias and the increasing militarization of Afghan society as a result of the Soviet invasion. Jalali shows how the Taliban gained power and how it lost a good deal of its backing. The article is especially useful for its discussion of the militias and for its emphasis on the role of pro-Taliban Pakistani fighters.

1881. Kenyon, Henry S. "Effects-Based Approach Reshapes Strategic Landscape." *Signal* 60, no. 10 (June 2006): 75–76+.

Describes the testing of an effects-based approach through a Multinational 4 (MNE4) analysis related to operations in Afghanistan. The "effects-based approach" amounts to what could be termed a total, interlocked, strategy, seemingly similar to a systems approach.

Numerous NATO partners participated in the tests that integrated purely military and pacification efforts.

1882. Kosnik, Mark E. "The Military Response to Terrorism." *Naval War College Review* 53, no. 2 (Spring 2000): 13–39.

Analyzes three United States responses: the air attacks against Libya, 1986; the missile firings against Iraq, 1993; and the missile attacks against Afghanistan and Sudan, 1998. The background and results of each response are described in detail. Kosnik notes that military answers are especially fruitful when they involve attacks on state actors rather than loosely organized terrorist groups.

1883. McElroy, Robert H. "Afghanistan: Fire Support for Operation Anaconda: Interview [with] Major General Franklin L. Hagenbeck." *Field Artillery* (September–October 2002): 5–9.

Hagenbeck led ground troops in Operation Anaconda against Al Qaeda elements. In the interview he discussed the challenging terrain of Afghanistan and the highly changeable climatic conditions. Among the most useful weapons the United States has employed are mortars and Apache helicopters.

1884. O'Hara, Patrick M. "The Carrot and the Stick: Can It Still Work?" *Special Warfare* 20, no. 5 (September–October 2007): 30–35.

Questions the effectiveness of counterinsurgency offensives being undertaken in Afghanistan and Iraq. O'Hara views the suppression of the Communist insurgency in the Philippines as a model counterinsurgency operation and draws lessons from that experience for present day application. The Philippine government used positive and civic action measures while at the same time insurgents were threatened with criminal penalties for their actions. O'Hara promotes similar approaches in Afghanistan and Iraq.

1885. Sepp, Kalev I. "The Campaign in Transition: From Conventional to Unconventional War." *Special Warfare* 15, no. 3 (September 2002): 24–26.

Describes the unexpectedly rapid success of NATO and the Northern Alliance at the beginning of the Afghan war, and, as the title suggests, examines the conversion of conventional fighting into a protracted counterinsurgency effort. Thus, special operations forces continued to have extensive responsibilities in Afghanistan.

1886. Steele, Dennis. "Afghan Scenes." *Army* 52, no. 7 (July 2002): 40–41.

Various photographs show Afghan fighters against the Taliban, artifacts from the war-torn county, including a Soviet rifle, and other aspects of life in Afghanistan.

1887. Ullman, Harlan. "Unity of Command." *United States Naval Institute Proceedings* 133, no. 7 (July 2007): 8.

Argues that the chain of command for operations in Afghanistan and Iraq is uncertain. Ullman considers the prospects of success for LTG Douglas Lute, who is slated to coordinate the two wars. He carefully examines the many obstacles facing Lute. The assignment given to Lute might better have been assigned to the national security adviser.

1888. Vego, Milan N. "What Can We Learn from Enduring Freedom?" *United States Naval Institute Proceedings* 128, no. 7 (July 2002): 28–33.

Asserts that air strikes launched from aircraft carriers were decisive efforts in Afghanistan. Vego celebrates the successes achieved in the invasion but warns that trying to learn from the campaign is not easy. There is a danger of learned "lessons" that might lead United States commanders along false paths. "Too much emphasis on technology is unsound because the human element of warfare is being dismissed as irrelevant to our modern age" (p. 33). There is still much to be done in Afghanistan. Many Al Qaeda leaders have evaded NATO forces, although their elimination was one of the principle objectives of the invasion.

1889. "'We Will Not Relent': The Army's Role in America's War on Terrorism." *Army Logistician* 34, no. 4 (July–August 2002): 44–45.

Pictorial essay shows United States soldiers from airborne and other forces in Afghanistan.

1890. Wells, Christopher B. "Breaking the Afghan Insurgency." *Special Warfare* 20, no. 5 (September–October 2007): 20–29.

Advocates approaching the campaign in Afghanistan by analyzing the insurgents and their structure and by using a "task-force" stance that incorporates an array of military and non-military personnel and resources. Wells decries the tactic of "attack and holding" areas temporarily, which he believes cannot bring lasting pacification. He describes the successful efforts of Task Force 31 (1st Battalion, 3rd Special Forces Group) during its five tours in Afghanistan. This is an impressive analysis that clearly identifies errors made in the war.

1891. Williamson, Joel E., and Jennifer D.P. Moroney. "Security Pays Off: A Lesson from the Afghan War: Reprint from the DFI Government Practice, Inc." *DISAM Journal* 24, no. 3 (Spring 2002): 79–82.

The United States laid the groundwork for Operation Enduring Freedom by building relationships with Uzbekistan, Kazakhstan, and Kyrgyzstan well before 9/11. The author advocates continuing and expanding "security co-operation" with other nations across the globe to help meet future unanticipated demands on the United States armed forces.

D. Air operations

1892. "Air Force Communicators Move Faster Lighter." *Signal* 57, no. 1 (September 2002): 47–50.

Remarkable strides have been made in reducing the size of communications equipment, which makes it much more adaptable to ground warfare, including special operations. This is a rather technical article about technology that is vital in Afghanistan.

1893. Burgess, Richard R. "Beneath the Surface: A Navy Flight Squadron Conducts a Geological Survey of Afghanistan." *Sea Power* 49, no. 11 (November 2006): 20–22.

A survey employing conventional techniques would have required a number of years. Working with Canadian personnel and with the United States Geological Survey, the navy's Scientific Development Squadron One used sensing devices to provide information to Afghanistan's Ministry of Mines and Industry. Surface techniques and aircraft are examined in considerable detail.

1894. Carlson, Marnee. "Reserve Aircrew Honored for Heroic Mission." *Officer* 78, no. 7 (September 2002): 23.

An MC–130 aircraft, ordinarily used to move troops and equipment, was pressed into service to refuel four helicopters in enemy territory, thereby preventing the loss of personnel and highly expensive aircraft. The assistance mission was not without problems, but all the crews and their helicopters returned.

1895. Friedman, Norman. "Dealing with Friendly Fire." *United States Naval Institute Proceedings* 129, no. 3 (March 2003): 4, 6.

Focuses on the disturbing fact that coalition forces in Afghanistan are suffering casualties almost entirely from friendly fire. Friedman judiciously reviews a tragic incident in which Air National Guard aircraft bombed some Canadian soldiers. He emphasizes the possibility that pilot fatigue can explain the error, but he also recommends giving pilots better information about possible targets to prevent such incidents in future.

1896. Friscolanti, Michael. *Friendly Fire: The Untold Story of the U.S. Bombing That Killed Four Canadian Soldiers in Afghanistan.* Mississauga, Ontario: John Wiley & Sons Canada, Ltd., 2005.

Exhaustive analysis of the United States Air Force attack, the controversy, the subsequent courts-martial, and the aftermath. Friscolanti interviewed numerous people who had a part in the incident or later events, and he offers what appears to be a remarkably objective view of the incident. The importance of the book goes far beyond the tragic deaths of the four soldiers, because it provides an illuminating view of the character of much of the air warfare in Afghanistan.

1897. Gourley, Scott R. "US Employs a Broad Array of Equipment." *Armada International* 26, no. 3 (June–July 2002): 10+.

The article emphasizes the ways in which air support is being given to ground forces in Afghanistan, using many kinds of equipment and technologies. Unmanned aerial vehicles are of considerable assistance in guiding gunship aircraft to targets quickly, so that the gunships have less exposure to ground-to-air attacks. Other weapons are discussed, including a modified rifle developed by the marines.

1898. Grant, Rebecca. "An Air War Like No Other." *Air Force Magazine* 85, no. 11 (November 2002): 30–37.

Emphasizes adaptability in the employment of air power during the initial stages of the invasion of Afghanistan. In most cases aircraft were given target assignments only after they were airborne. This is a brief but highly informative description of air and land operations against Al Qaeda and the Taliban.

1899. ———. "The War Nobody Expected." *Air Force Magazine* 85, no. 4 (April 2002): 34–40.

Credits the air force and navy aviation with facilitating the Northern Alliance's offensive in Afghanistan through a limited application of air power, including both attacks and support operations. There were problems and some failures, partly as a result of delays in the authorization of some attacks, but overall the response of air assets was rapid and pinpointed.

1900. Grier, Peter. "The Combination That Worked." *Air Force Magazine* 85, no. 4 (April 2002): 30–32.

General Tommy Franks emphasized the importance of achieving coalition goals in Afghanistan. He praised precision in the application of air assets, which was much more efficient and effective in Afghanistan than during the Gulf War.

1901. Jansen, John M., Nicholas Dienna, Todd Bufkin II, David I. Oclander, Thomas Tomasso, and James B. Di Sisler. "JCAS (Joint Close Air Support) in Afghanistan: Fixing the Tower of Babel." *Field Artillery* (March–April 2003): 22–30.

Asserts that joint close air support during Operation Anaconda revealed serious deficiencies. The authors describe an anti-mortar mission during which a pilot received confusing directions from multiple controllers, making for a dangerous situation. This is a complex article filled with acronyms, but the problems are readily apparent. The authors offer numerous proposals for improving close air support.

1902. Lambeth, Benjamin S. *Air Power Against Terror: America's Conduct of Operation Enduring Freedom.* Santa Monica, CA: RAND, 2005.

Massive study of the use of air assets in Afghanistan by a leading scholar in the field of air power and its employment. In addition to presenting an authoritative analysis of air fighting early in the Afghanistan invasion, Lambeth discusses the formation of an alliance to enter that country and the use of aircraft to provide humanitarian assistance. The book includes what may be a comprehensive bibliography up to the time the manuscript was completed.

1903. Newman, Richard J. "Masters of Invisibility." *Air Force Magazine* 85, no. 6 (June 2002): 36–41.

Reviews the wide range of air force special operations in Afghanistan. These include directing air strikes, with air force personnel often being attached to army special forces detachments; refueling helicopters; evacuating wounded personnel from all services; and airlifting supplies.

1904. "Operation Enduring Freedom Highlights." *Airman* 46, no. 9 (September 2002): 14.

Statistical information about the air force commitment to operations in Afghanistan.

1905. Schanz, Marc V. "Watch on Afghanistan: Airmen at Bagram Air Base Help Deliver Justice to the Bad Guys in the Mountains." *Air Force Magazine* 90, no. 9 (September 2007): 46–48+.

Briefly describes the restoration of the Bagram air field and concentrates on its current operations, which go far beyond air support. Bagram is a kind of base camp for varied activities. Providing supplies and equipment by air to outposts is highly challenging because of the mountainous terrain. Weather damage makes constant aircraft maintenance necessary.

1906. Schroder, James A. "'Have Tools, Will Travel': Company D, 109th Aviation Battalion." *Special Warfare* 15, no. 3 (September 2002): 22–3.

Emphasizes the importance of Army Special Operations Aviation to the campaign against extremists in Afghanistan. After moving to active duty, the Army National Guard company discussed in the article was separated into "four contact teams" that were assigned to three battalions and a training company of the 160th Special

Operations Aviation Regiment to provide "intermediate maintenance" for helicopters.

1907. Seydel, Cane A. "Business as Usual." *Airman* 46, no. 9 (September 2002): 42–48.

Describes life in Afghanistan for some air force special operations personnel. Some of them related their experiences in building a facility for aircraft landings and takeoffs. Others talked about their search and rescue duties. Another segment of the article addresses locating Taliban forces with modern technology.

1908. Steele, Dennis. "Ghost Town of the al Qaeda." *Army* 52, no. 7 (July 2002): 49.

United States air attacks completely destroyed a major Al Qaeda training facility at Tannac Puhl, Afghanistan, early in the Afghan invasion. The article is heavily illustrated.

1909. "Striking Back (Air and Ground Elements Press the War in Afghanistan)." *Air Force Magazine* 85, no. 1 (January 2002): 32–39.

Pictorial treatment of operations in Afghanistan and some of the air force weapons used in the conflict. The Predator unmanned aerial vehicle is given prominence.

1910. Sweetman, Bill. "USAF Predators Come of Age in Iraq and Afghanistan as Reaper Waits in the Wings." *Jane's International Defence Review* 39, no. 12 (December 2006): 52–54+.

Describes the accomplishments of the Predator unmanned aerial vehicle, which has assumed an important combat role in Afghanistan and Iraq. The MQ–9A Reaper (once termed the Predator B) is an even more powerful and versatile weapons system.

E. Ground combat

1911. Adams, Terrence A. "Current Operations Section of the Modular Brigade FEC (Fires and Effects Cell) in Afghanistan." *Field Artillery* (July–August 2006): 44–45.

The "fire and effects cell" not only directs traditional firepower but also oversees information, psychological, and "civic-military" operations. Adams focuses on such a cell within the 10th Mountain Division. Although his cell functioned well, Adams candidly points to several areas that need improvement.

1912. Bentley, Christopher F. "Afghanistan: Joint and Coalition Fire Support in Operation Anaconda." *Field Artillery* (September–October 2002): 10–14.

Much changes when artillery fire is delivered in a coalition environment. Targeting is a key factor, including the precise definition of "restricted-fire areas" and "no-fire areas." Bentley includes a worthwhile discussion of the role of air support in conventional warfare environments.

1913. Betancourt, Alberto. "Coalition Team Clears Land Mines." *Soldiers* 57, no. 5 (May 2002): 8–9.

Millions of land mines threaten life in Afghanistan. A force of Norwegian soldiers and members of the 101st Airborne Division have been clearing mines near the Kandahar airfield.

1914. Blatt, Darin J., Scott. T. McGleish, and Peter G. Fischer. "Taming the Tagab Valley: Planning and Executing a Full–Spectrum Operation in the Afghan Theater." *Special Warfare* 20, no. 4 (July–August 2007): 18–27.

After a successful offensive by coalition special operations forces in 2005 cleared the Tagab Valley of insurgents, the valley was re-infiltrated by hostile forces who regained control of the area. A new special operations effort was made in 2006 that included elements from the United Arab Emirates and many Afghan soldiers and police. Psychological operations and civic action efforts were skillfully interwoven with military measures.

1915. Carr, Darrell. "In Sync: EBA (Effects-Based Approach) Synchronizes Operations for the SF Headquarters." *Special Warfare* 19, no. 4 (July–August 2006): 8–15.

The 2nd Battalion, 1st Special Forces Group used the effects-based approach (EBA) to counterinsurgency in its area of responsibility in Afghanistan. This approach requires precise coordination of activities. Carr carefully explains the EBA. He states that effective counterinsurgency requires closely controlling the level of violence. Fine graphics enhance this article. The frequent use of acronyms necessitates careful reading.

1916. Conner, Carol. "Signal Command Provides Commercial Communications in Afghanistan." *Army* 53, no. 4 (April 2003): 45–47.

Describes the use of civilian communications capabilities in the Bagram area of Afghanistan to support United States and other military forces. Systems were and are needed for both "secure" and "non secure" uses.

1917. Davis, Anthony. "Afghan Security Deteriorates as Taliban Regroup." *Jane's Intelligence Review* 15, no. 5 (May 2003): 10–15.

Discusses Taliban leaders and their threats to expel NATO forces from Afghanistan. Rocket attacks are increasing, and the Taliban is showing more confidence in making attacks. Even international aid workers are not immune from Taliban assaults. An important response is to move members of anti-government militiamen into the Afghan National Army or job training.

1918. Exun, Andrew. *This Man's Army: A Soldier's Story from the Front Lines of the War on Terrorism.* New York: Gotham Books, 2004.

After being commissioned, Exun received airborne and ranger training, which he describes in detail. After the ordeal of ranger school, in particular, it is easy to realize that the author was prepared for combat. This is an excellent description of small unit, specifically platoon-level, operations in Afghanistan. Exun's observation that "eleven-year-olds we came across might be combatants just as dedicated to killing us as their fathers were" (p. 132) underlines the tragedy of insurgency and counterinsurgency.

1919. Geibel, Adam. "Operation Anaconda, Shah-i-Khot Valley, Afghanistan, 2–10 March 2002." *Military Review* 82, no. 3 (May–June 2002): 72–77.

Thorough survey of Operation Anaconda, the largest offensive in the early part of the Afghanistan conflict. Coalition leaders were aware of the hostile build-up in a cave complex but did not intercept the gathering forces in order to bring about a large battle to eliminate as many Taliban and Al Qaeda troops as quickly as possible. This is a valuable, well-documented chronological treatment.

1920. Krym, Jaime. "A Typical Day in Afghanistan." *Army* 57, no. 4 (April 2007): 53–54+.

Story of a security patrol carried out by National Guard combat engineers. A village chief had to be prevented from diverting a load of gravel from its intended use. On a more positive note, later on the engineers distributed school supplies to Afghan children. Other episodes occurred, including the correction of inattentive soldiers by Sergeant Krym.

1921. *Long Hard Road: NCO Experiences in Afghanistan and Iraq.* Fort Bliss, TX: U.S. Army Sergeants Major Academy, 2007.

Important, partly because of the range of NCO experiences presented. The NCOs who described their challenges came from the active and reserve forces and had duties in combat units and support activities. Their candor should make this a helpful guide for subsequent generations of NCOs. Although there is a useful "Glossary," not all the acronyms used are explained.

1922. Mackey, Chris, and Greg Miller. *The Interrogators: Inside the Secret War Against Al Qaeda.* New York: Little, Brown and Company, 2004.

Riveting account by an army reservist and a journalist of prisoner interrogation in Afghanistan. The authors describe the training given to interrogators both in the United States and the United Kingdom. Of particular importance are the chapters dealing with encounters between interrogators and prisoners. Mackey and Miller contrast the approaches used in Afghanistan with those employed at Abu Ghraib in Iraq. Their contempt for the Abu Ghraib interrogators and their methods is unquestionable.

1923. MacPherson, Malcolm. *Roberts Ridge: A Story of Courage and Sacrifice on Takur Ghar Mountain, Afghanistan.* New York: Delacorte Press, 2005.

Extremely detailed narrative of an intense engagement between United States military personnel and guerrilla forces at Takur Ghar on March 3–4, 2002. The military effort planned was a part of a much larger move, Operation Anaconda. MacPherson's book is based largely on interviews with a considerable number of participants in the battle. The author had to deal with the understandable concerns at various levels about discussing this episode because of the possibility that information about United States tactics might assist the enemy. Nevertheless, MacPherson managed to put together a coherent and informative report.

1924. Maloney, Sean S. *Enduring the Freedom: A Rogue Historian in Afghanistan.* Washington, DC: Potomac Books, Inc., 2005.

Canadian military historian provides a graphic and readable view of ground operations in Afghanistan as experienced by members of the coalition forces in 2003, emphasizing United States participation. Individual chapters deal with coalition efforts in Kabul, Bagram, and Kandahar. Although much of the material in the book comes from Maloney's direct observations, there are extensive notes and a long bibliography. The author's viewpoint about the outcome of the war seems to be cautiously optimistic.

1925. Marck, David. "EOD (Explosive Ordnance Disposal) in Afghanistan." *Soldiers* 57, no. 6 (June 2002): 32–33.

In conjunction with some Canadian soldiers, the army's 710th Ordnance Company is clearing explosive devices that are the heritage of years of war in Afghanistan. Priorities for dealing with ordnance are essential because of the great quantity of munitions in the country.

1926. Mitchell, Joshua D. "Afghanistan: Firing Artillery Accurately with Air Force Met Support." *Field Artillery* (January–February 2003): 38–41.

The extreme conditions in Afghanistan make meteorological information essential for accurate firing. Artillery deployed to Afghanistan as part of the Task Force Panther of the 82nd Airborne Division lacked organic meteorological support; so it worked out alternative methods with the cooperation of the Bagram air field staff. This is a highly detailed account of how the problems were overcome using computer software.

1927. Morgan, Matthew J. "Civilians on the CTC (Combat Training Center) Battlefield—Threat, Opportunity, or Distraction?" *Military Intelligence* 29, no. 1 (January–March 2003): 49–51.

Examines the varied roles of "civilians" (role-playing members of the army) at the Joint Readiness Training Center at Fort Polk, Louisiana, and their real-life counterparts. Civilians may be insurgents, of course, but they can also provide information that can be transformed into highly useful intelligence. Much of the article deals with the informational handling of civilians.

1928. Prochniak, Scott E., and Dennis W. Yates. "Counterfire in Afghanistan." *Field Artillery* (September–October 2002): 15–18.

Discusses counterfire employed to protect the Kandahar airfield in support of the 3rd Brigade Combat Team, 101st Airborne Division. Given the proximity of large numbers of Afghan civilians, precision in firing is essential. Much of the article deals with the role of radar in these operations.

1929. Redmore, James. "A Cycle of Victory: Manning System Contributes to 'Spartan' Success in Afghanistan." *Infantry* 96, no. 2 (March–April 2007): 8–11.

Discusses the advantages of employing a "modular brigade combat team," using the 3rd Brigade Combat Team of the 10th Mountain Division as an example. Redmore describes the formation of the brigade from existing elements of the 10th Division. Considerable training and cross-training were needed to develop the skills of new soldiers.

1930. Schroder, James A. "Ambush at 80 Knots: Company B, 3/160 SOAR (Special Operations Aviation Regiment)." *Special Warfare* 15, no. 3 (September 2002): 39–41.

Describes an attack planned on some Al Qaeda troops, using army helicopters and SEAL assault personnel. Helicopters were drawn from several companies owing to earlier losses. This is a graphic depiction of a rapid attack by army and navy elements. The character of the terrain had a significant impact on the events that occurred.

1931. Steele, Dennis. "A U.S. Army Line Battalion in the War on Terrorism: The Soldiers." *Army* 52, no. 6 (June 2002): 42–46+.

Brief statements from members of the 10th Mountain Division show how NCOs and lower ranking enlisted men felt about their duties.

Soldiers departing on missions are relatively quiet, but they are generally exuberant when returning. Their relief of coming through unharmed is obvious.

1932. ———. "A U.S. Army Line Battalion in the War on Terrorism: The Valleys." *Army* 52, no. 6 (June 2002): 32–36+.

Describes patrol activity by troops of the 10th Mountain Division near Sur Bagham Khwar, Afghanistan, near the border with Pakistan. The primary objective was locating munitions, and several dumps were found. Steele observed that although the area had been cleared for the time being, Al Qaeda elements were likely to re-infiltrate from Pakistan.

1933. Tiron, Roxana. "Fear of Fratricide Prevailed in Afghanistan, Troops Say." *National Defense* 87, no. 584 (July 2002): 26.

Various technologies are increasingly being used to prevent casualties from "friendly fire" in Afghanistan. There have been considerable improvements in this area since the Gulf War, although real concerns continue about the problem and about limitations in the technologies designed to prevent "friendly fire" incidents.

1934. "The U.S. Army in Afghanistan." *Army* 57, no. 3 (March 2007): 78+.

Pictorial study shows the highly rugged terrain of the country and some of the equipment used there.

F. Special military operations

1935. Kiper, Richard. "'We Support to the Utmost': The 528th Special Operations Support Battalion." *Special Warfare* 15, no. 3 (September 2002): 13–15.

The battalion was tasked with a considerable variety of duties at Kanabad, Uzbekistan. No other support troops would be working with battalion personnel. Despite a number of problems the battalion built Camp Stronghold Freedom in a short time before moving to Fort Bragg.

1936. Sepp, Kalev I. "Change of Mission: ODA (Operational Detachment A) 394." *Special Warfare* 15, no. 3 (September 2002): 27–29.

A Special Forces team experienced a dizzying series of changes in orders. Its members had come to Afghanistan as a reconnaissance force, but they were trained for close combat and later as a "quick-reaction force" as part of a diverse NATO contingent, before returning to a reconnaissance role. Further changes occurred before the team finally went into combat.

1937. ——. "Meeting the 'G-Chief': ODA (Operational Detachment A) 595." *Special Warfare* 15, no. 3 (September 2002): 10–12.

A special forces detachment commander was tasked for work with Abdul Rasid Dostum, an anti-Taliban Uzbek with Soviet military experience, who led a considerable force. The Special Forces team coordinated combat air operations to support Dostum's men.

1938. ——. "Uprising at Qala-i Jangi: The Staff of the 3/5th SF (Special Forces) Group." *Special Warfare* 15, no. 3 (September 2002): 16–18.

A surrender of Al Qaeda troops near Mazar e Sharif early in the Afghan invasion led to an insurrection by prisoners, beginning with the detonation of hidden grenades. NATO forces responded. Several days of heavy fighting followed, during which most of the prisoners were killed.

G. Intelligence

1939. Moores, Drew. "The 101st Airborne Division (Air Assault) Deployable Intelligence Support Element (DISE) in Operation Enduring Freedom." *Military Intelligence* 28, no. 4 (October–December 2002): 38–40.

Asserts that the "Afghanistan campaign ... [is] the first conflict in which intelligence was the primary U.S. weapon" (p. 38). Much of the article deals with the configuration of intelligence elements attached to the 101st Airborne Division. Those intelligence elements participated fully in Operation Anaconda against the Taliban. Civilian contract personnel played an important role in the intelligence-gathering effort.

H. Military advisory activities

1940. Gamble, Krishna M. "ANA (Afghan National Army) Aviators Partner with U.S. Counterparts." *Army Aviation* 56 nos. 8–9 (August–September 2007): 46.

There are Afghan pilots with considerable experience, but United States Army orientation and training will strengthen their performance. Afghans have noted with interest and appreciation the United States Army concern with safety, both of crews and their aircraft.

1941. Jalali, Ali Ahmad. "Rebuilding Afghanistan's National Army." *Parameters* 32, no. 3 (Autumn 2002): 72–86.

Predicts that developing a new Afghan army will be a long process. Jalali usefully reviews the army's role in Afghanistan historically, noting that it has always been merely one of various armed bodies in the country. Stresses between the loyalty a modern army demands and ethnic and tribal affiliations remain a major challenge in building a reliable armed force.

1942. Kipps, Richard L. "An Army for Afghanistan: The 1st Battalion, 3rd SF (Special Forces) Group, and the Afghan Army." *Special Warfare* 15, no. 3 (September 2002): 42–43.

The imperative of building a new Afghan National Army was obvious to all decision-makers, and the Kabul Military Academy was cleaned and upgraded to receive Afghan recruits. A special effort was made to develop a group of NCOs from early recruits. All recruits needed thorough orientation to understand their role in the new Afghanistan.

1943. Portman, Charles. "Training an Afghan Army." *Soldiers* 57, no. 8 (August 2002): 4–7.

Pictorial story shows Afghan recruits through their entry into the army and into their training. Portman notes that recently, United States forces have had experience in training foreign troops in several countries before the Afghanistan war.

1944. Prawdzik, Christopher. "Scaling New Heights (Vermont National Guard's Mountain Warfare School)." *National Guard* 57, no. 5 (June 2003): 26–27.

The Vermont National Guard sent instructors from its Mountain Warfare Center to develop a similar training facility at Kabul, Afghanistan. The author reviews the history of the center and describes its courses in some detail. The low literacy of the Afghan trainees will be one of a number of problems faced by the instructors.

1945. Townsend, Kellard N., Jonathan E. Marion, Joseph W. Boler, and Madison M. Carney. "Hoosier Redlegs Train Afghan Kandaks: Sempre Gumbi." *Field Artillery* (May–June 2005): 32–34.

Personnel from three Indiana National Guard battalions helped train and advise Afghan gunners in 2004, while also maintaining liaison with United States forces. These artillerymen also acted as paymasters and purchased supplies. Like many authors writing about unconventional warfare, they stress the need for adaptability in a constantly changing environment. They also offer a detailed outline of the kinds of training needed by United States forces that may be deploying.

I. Logistics

1946. Betancourt, Alberto. "Kandahar's Supply Hub." *Soldiers* 57, no. 5 (May 2002): 4–7.

Kandahar is an extremely busy air facility. Work crews perform 24 hours a day in shifts, but there is still a great need for more personnel and for equipment, such as forklifts. The article focuses on the 626th Forward Support Battalion in the 101st Airborne Division.

1947. Clintron, David. "MTMC (Military Traffic Management Command) Surface Shipments Sustain Troops in Afghanistan." *Army Logistician* 34, no. 5 (September–October 2002): 26–28.

Largely a pictorial treatment of army transportation in Afghanistan, the article is valuable because it shows graphically some of the obstacles and rough conditions that confront private vehicles operating under army contracts.

1948. Hall, Clark. "Host Nation Trucking in Afghanistan: Good for Afghanistan, Good for the US." *Defense Transportation Journal* 63, no. 1 (February 2007): 17–19.

The National Defense Transportation Association helped recruit Hall to aid in improving transportation in Afghanistan. Hall describes the Afghan trucking firms that support United States military efforts, the dangerous environment in which they operate, and the many problems facing them on a day-to-day basis. The author recommends that safety programs be developed to help protect drivers.

1949. Harper, Sue, and Gregory Jones. "Supporting Afghanistan from Europe: U.S. Army Europe's 21st Theater Support Command." *Army* 52, no. 8 (August 2002): 41–42.

The 21st Theater Support Command (TSC) is providing great quantities of supplies to the United States Army in Afghanistan and to Afghans who need food and other items. Both active army and reserve organizations and companies are components of the 21st TSC.

1950. Steele, Dennis. "Afghanistan SITREP." *Army* 52, no. 6 (June 2002): 52.

Describes United States and British efforts to locate weapons and munitions in Afghanistan and the composition of United States Army forces in the country. Much of the article, however, deals with the United States Army Security Assistance Command, which provides various types of support and supplies, including the sale of some items, to foreign armed forces.

1951. ———. "Unconventional Logistics." *Army* 52, no. 11 (November 2002): 58–59.

The 200th Material Management Center (MMC), 21st Theater Support Command, supplied special operations personnel, including CIA members, with whatever they needed during the initial phase of the invasion of Afghanistan. Needed items included saddles and feed for horses. Many items had to be purchased quickly from civilian sources in Germany.

1952. Tripp, Robert S., David George, Mahyar A. Amouzegar, and C. Robert Roll. *Supporting Air and Space Expeditionary Forces: Lessons From Operation Enduring Freedom.* Santa Monica, CA: RAND, 2004.

Masterful study of United States Air Force airlift accomplishments in connection with the invasion of Afghanistan. The authors make important comparisons between air force efforts in Yugoslavia and Afghanistan and show how the Yugoslav experience helped improve the response to the presidential call for action in Afghanistan. There are attractive and highly informative graphics. The authors wisely included a glossary of acronyms used; without it, the text could not be understood.

J. Civil affairs/civic action

1953. Avallone, Paul. "A Morning with an FST (Forward Surgical Team) in Rural Afghanistan." *Army* 57, no. 6 (June 2007): 30–32+.

Examines the contributions of the 758th Forward Surgical Team (FST) of a combat surgical hospital (CSH). Much of the story deals with an Afghan girl, who was severely wounded by a land mine and who received extensive treatment from United States medical personnel.

1954. Briscoe, C.H. "Coalition Humanitarian Liaison Cells and PSYOP (Psychological Operations) Teams in Afghanistan." *Special Warfare* 15, no. 3 (September 2002): 36–38.

Describes the work of two "humanitarian liaison cells" of the 489th Civil Affairs Battalion, an army reserve organization. They sought to identify local needs by communicating with village elders and then worked to satisfy them by making contracts for supplies and services. One team helped maintain and expand a radio station.

1955. Desjarlais, Orville F., Jr. "On the Road to Restoration: Bagram Provincial Reconstruction Team Helps Build Bridges, Roads, and Schools." *Airman* 50, no. 4 (Fall 2006): 30–35.

The team discussed is one of thirteen such groups in Afghanistan, seven of them air force teams. They fulfill a variety of functions that differ widely, depending upon local conditions. Road building and medical services are among their common activities.

1956. Espinoza, Robert. "Army Engineers Pursue Humanitarian Tack Along Afghan Road Project." *Army* 57, no. 2 (February 2007): 45–46+.

During a large-scale road construction project in Afghanistan, the 37th Engineer Battalion employed many methods for helping local

Afghans to accept its necessarily intrusive stay in the area. A number of civic action projects were undertaken, even including the provision of school books and other materials.

1957. Keenan, Sean. "The Doctor is in: Task Force 31 Uses Host-Nation Medical Care to Support Its COIN Efforts." *Special Warfare* 20, no. 3 (May–June 2007): 8–11.

Describes the provision of emergency medical treatment to Afghans as a counterinsurgency tactic. The area of Afghanistan where these services are provided lacks significant medical facilities and personnel.

1958. Kennedy, Harold. "Pentagon Broadens Duties for Its Civil Affairs Teams." *National Defense* 87, no. 591 (February 2003): 36–39.

Small numbers of civil affairs personnel, mostly army reservists, are working in Afghanistan in teams under the Special Operations Command. A considerable variety of projects are being undertaken in such fields as education, medical services, and irrigation. These efforts build goodwill and provide jobs for Afghans, thereby contributing to economic recovery.

1959. Philpott, Tom. "Rebuilding a Nation: U.S. Navy Sailors Head a Half-Dozen PRT (Provincial Reconstruction Teams) Teams Providing Hope for the Future in Afghanistan." *Sea Power* 50, no. 4 (April 2007): 45–46+.

A number of the Provincial Reconstruction Teams (PRT) employed for civic action in Afghanistan are led by navy personnel and others by the air force or by personnel from other countries. This story focuses on one PRT positioned near the Afghanistan–Pakistan border. It included soldiers, sailors, civil service employees, and civilians.

1960. Plata, Holly. "The Other Afghan Campaign." *Soldiers* 57, no. 7 (July 2002): 14–15.

United States soldiers are assisting Afghans in the wake of a serious earthquake by providing supplies of various kinds, including food, water, and clothing. Medical services are also being offered.

1961. Schanz, Marc V. "The Medical Middlemen (the 455th Aeromedical Evacuation)." *Air Force Magazine* 90, no. 10 (October 2007): 38–40.

The 455th Aeromedical Evacuation Flight, made up largely of air force reserve and Air National Guard volunteers, takes care of service personnel and Afghan civilians who need advanced medical attention in Germany or in the United States. Preparing patients for transportation is a complex task, especially in view of the possible effects of long flights on the patients.

1962. Sepp, Kalev I. "'Deminimus Activities' at the Bagram Clinic: CA (Civil Affairs) Team A-41." *Special Warfare* 15, no. 3 (September 2002): 44–45.

Describes the members of a tiny civil affairs team whose members were interviewed on "Sixty Minutes Two." A dramatic incident occurred when a young Afghan man needed medical assistance. The Afghan was helped quickly, and the entire episode was recorded.

1963. "A Sigh of Relief." *Airman* 45, no. 12 (December 2001): 8–9.

C-17 aircraft were deployed to airdrop thousands of meals to Afghans daily. Flights were conducted from Ramstein Air Force Base, Germany. Both active and reserve air force personnel worked in this humanitarian effort.

K. Psychological operations

1964. Kiper, Richard L. "'Of Vital Importance': The 4th PSYOP (Psychological Operations) Group." *Special Warfare* 15, no. 3 (September 2002): 19–21.

Examines planning by the 4th Psychological Operations Group for Afghanistan and the deployment of the 8th Psychological Operations Battalion to that country and their activities. Real successes were achieved, including a significant role in inducing the surrender of Konduz.

1965. ——. "To Educate and to Motivate: The 345th PSYOP (Psychological Operations) Company." *Special Warfare* 15, no. 3 (September 2002): 32–33.

Owing to the limited number of active-duty psychological operations battalions, reservists had to be called for active duty. The 345th worked with the 3rd Special Forces Group and later with the 19th Special Forces Group. Kiper notes that psychological operations in Afghanistan were especially challenging, in good part because of the small percentage of the population that is literate.

L. Engineering operations

1966. Cooke, Leonard W.W. "A Deployment to Remember: The Navy's Seabees in Afghanistan." *Sea Power* 45, no. 10 (October 2002): 55–57.

Focuses on elements of Naval Mobile Construction Battalion 133, which was directed to provide engineer support for Task Force 58 in Afghanistan. Airfield repair was the major responsibility for the

Seabees, but they performed many other duties, including digging wells, and maintained security for their own work parties.

1967. Dobie, Paul W., and Vicki Flack. "A Project of Patience." *Engineer* 35, no. 4 (October–December 2005): 28–29.

Describes the construction of an engineer camp at Bagram, Afghanistan, by United States, Polish, and Korean troops and private contract personnel. The buildings were Sprung Instant Structures made of aluminum. Various problems prevented the completion of the project for some time.

1968. Hutson, Charles E. "NGIC (National Ground Intelligence Center) Uses Web-Based Visualization Technology to Inform Soldiers of Minefield Locations in Afghanistan." *Military Intelligence* 28, no. 4 (October–December 2002): 28–30.

The National Ground Intelligence Center was tasked to gather information about landmines in Afghanistan before the invasion of that country. GIS makes possible the production of maps that are far more flexible than older types. This is an extremely detailed description of the mapping project.

1969. McNulty, Dennis J. "Repairing Runways and Clearing Mines in Afghanistan." *Engineer* 32 (July 2002): 8–14.

The experiences of army engineers in Afghanistan may contribute to enhancing Army Corps of Engineers doctrine. Airfields are critical for operations in Afghanistan in view of deficiencies in roadways. McNulty describes runway repair and maintenance in considerable detail. He devotes even more attention to minefield clearance. This is an essential article for understanding these problems.

1970. Runge, Jared E. "Team Hellraiser: The Soldiers of a Headquarters Company." *Army* 57, no. 6 (June 2007): 47–49.

Examines the contributions of Headquarters Company, 37th Engineer Battalion at Forward Operating Base Sharana in eastern Afghanistan. Advanced maintenance of all types of all types of equipment is a vital service of the company. Other personnel are members of a security platoon that undertakes local patrols, guards the airstrip, and supervises entries and exits at the base.

XXXIV

Iraq War, 2003–

In contrast to the Afghanistan intervention, the Iraq War became highly controversial. The principal reasons cited by the George W. Bush administration for the invasion, that Iraq was developing weapons of mass destruction and that it was cooperating with Middle Eastern extremists who were responsible for the 9/11 attack, have been discredited. The next assertions were that the Iraqi people had to be freed from a dictator and that a democratic Iraq would be a stabilizing factor in the Middle East; these have been maintained, but they remain controversial.

The initial conventional offensive into Iraq was highly successful, but as has been the case with a good many wars and campaigns, including the Philippine–American War of 1899–c.1902, a guerrilla phase evolved after the end of large-scale fighting, which was considerably more difficult to confront than the conventional resistance that lasted only briefly.

The complex religious, ethnic, and political divisions in Iraq grew especially acute at times, creating a strong possibility of outright civil war. Eventually, the level of violence stabilized. After it was reduced, however, violence sometimes reappeared at a relatively high level.

After protracted negotiations the new Iraq government and the occupying forces agreed that the latter would depart completely from Iraq by the end of 2011. In the meantime, foreign troops were removed from urban areas to become a kind of reserve force to meet new challenges. Those troops were active in fighting guerrillas in rural parts of Iraq.

Chronology

1991
April 3   United Nations Security Council resolution 687 required Iraq to permit United Nations inspection of nuclear installations in the country.

July 25	Despite a United States warning of possible air attacks, Iraq did not comply with a demand that it reveal information about its nuclear efforts.
August 26	A "no-fly zone" in Iraq was established by the United States, and numerous incidents followed over a long period of time.

1997

November 13	United Nations inspectors temporarily withdrawn from Iraq in response to Iraqi reluctance to comply with the supervision system.

1998

October 31	President Bill Clinton signed the Iraq Liberation Act of 1998, which stated that the United States favored the end of Saddam Hussein's rule in Iraq.
December 17–20	United States air attacks on Iraq were mounted because Iraq had refused to allow the continuation of the inspection system.

2001

February 16	British and United States forces undertook air attacks on Iraqi air defense installations, and such attacks continued for some time.
September 15	Bush administration leaders secretly discussed the possibility of invading Iraq in the wake of the 9/11 attacks on the United States.

2002

January 29	In his State of the Union address, President George W. Bush included Iraq in what he termed the "axis of evil," which included North Korea and Iran. Bush added that pre-emptive attacks might be made against those countries.
September 12	President Bush addressed the United Nations General Assembly on the subject of Iraq. Although he asked for United Nations action, Bush declared that the United States would act in any event.
October 11	Congress passed resolutions supporting Bush's strong stance in respect to Iraq.
November 8	United Nations Security Council passed resolution 1441 that once again demanded Iraqi cooperation with arms inspections. The resolution emphasized the possibility of what it termed "serious consequences" in the face of continued Iraqi obstruction. Weapons inspections were then resumed on November 27.

2003

January 28	In his State of the Union address President Bush argued that Iraq was developing "weapons of mass destruction" and must be stopped.
February 3	Secretary of State Colin Powell addressed the United Nations Security Council and presented what was said to be evidence that Iraq was, indeed, producing "weapons of mass destruction."
March 17	President Bush demanded that Saddam Hussein and other leaders depart from Iraq almost immediately.
March 19	Heavy air attacks on Iraq.
March 20	Various coalition land forces, including special operations units, entered Iraq at many points.
March 21	Gains of up to 100 miles were registered, as the war progressed rapidly.
May 1	President Bush announced the end of organized resistance in Iraq.
May 6	L. Paul Bremer selected to lead the occupation of Iraq. A few weeks later Bremer ordered the dissolution of Iraqi military forces, a move that continued to be controversial.
December 13	Former Iraqi dictator Saddam Hussein captured by United States forces.

2004

March 31	The first battle of Fallujah was ignited when a United States convoy was attacked in that city, and several private security guards were killed and then dragged through the streets.
April 28	The abuse of Iraqis held at Abu Ghraib prison in Baghdad by United States personnel was publicly revealed.
June 28	The occupation regime in Iraq was converted into an interim Iraqi governing structure, pending national elections.
September 14	Much of the funding slated for reconstruction efforts was transferred to security expenditures.
November 7	Second battle of Fallujah began.

2005

January 30	National elections established a new government for Iraq that was also an interim structure.
May 11	The United States approved another aid package for Iraq, which, again, consisted largely of expenditures for security.
October 15	The Iraqi National Assembly approved a new constitution.
October 19	Saddam Hussein stood trial.
December 15	Another round of parliamentary elections was held in order to establish a new government to replace the interim structure.

2006
August 20   President Bush reiterated that United States must remain in Iraq as part of the global War on Terrorism.
December 30   Saddam Hussein executed.

2007
January 10   The further commitment of United States forces to Iraq was officially announced. This was the beginning of what was termed a "surge" in troop strength.

2008
January 12   Lower-ranking members of Saddam Hussein's ruling party allowed to enter government service. This was an important step, because part of the instability in Iraq was attributed to the widespread dismissal of former government workers.
July 16   Strengthening of United States forces in Iraq ("the surge") completed.
November 27   Signing of a Status of Forces Agreement that forecast the gradual drawdown of United States forces from Iraq until their departure by the end of 2010.

2009
January 31   National elections held in a relatively peaceful atmosphere.
February 27   Agreement reached that forecast the removal of United States forces from a combat stance by August 31, 2010, prior to their full withdrawal at the end of 2010.
June 30   United States troops left major urban areas in Iraq.

A. General studies

1971. Allawi, Ali A. *The Occupation of Iraq: Winning the War, Losing the Peace.* New Haven, CT: Yale University Press, 2007.

Important primary source. Allawi was, for a time, prime minister of Iraq. This is a detailed and well-documented account that emphasizes events after the fall of Saddam Hussein. Allawi is hardly a neutral observer of developments in Iraq, as demonstrated by his favorable view of Ahmad Chalabi, a highly controversial figure, but his book is a sophisticated study that deserves careful reading.

1972. Carlisle, Rodney P. *Iraq War.* Updated ed. New York: Facts On File, Inc., 2007.

Excellent introduction to the conflict. The Library of Congress describes the volume as "juvenile literature" about Iraq and the Iraq War; nevertheless, this highly concise, but thorough, survey is valuable for anyone seeking information about the topics. Carlisle provides a good deal of historical background information that is essential to an understanding of issues and events. The illustrations and the numerous maps are exceptionally good. The text is complemented by a well-organized index.

1973. Cockburn, Patrick. *The Occupation: War and Resistance in Iraq*. London: Verso, 2006.

Interprets the war as an attempt by the United States to establish a "protectorate" over Iraq rather than to free it from tyranny. Cockburn asserts, moreover, that domestic political objectives in the United States have played a disastrous part in the Bush administration's moves in Iraq. This is an important study of the impact of the invasion on the Iraqi population. In its emphasis on groups within Iraq, it can be contrasted with Anderson, *The Fall of Baghdad* (item 2060), which provides more stories of individuals.

1974. Diamond, Larry. *Squandered Victory: The American Occupation and the Bungled Effort to Bring Democracy to Iraq*. New York: Times Books/Henry Holt, 2005.

Detailed analysis by a noted political scientist and adviser in Iraq of the numerous failures of the United States in carrying out its objectives. Giving overall control of the situation to the Defense Department is said to have been one of the key errors, but there were many other poor decisions and much evidence of a lack of preparation. Diamond believes that despite the many serious obstacles the United States could have achieved its goals.

1975. Doppler, Ryan J., ed. *Iraq Revisited*. Hauppauge, NY: Nova Science Publishers, Inc., 2003.

Largely a collection of signed articles that discuss topics related to the early stages of the occupation of Iraq, such as Iraq's "military confrontations with the United States" and "Al Qaeda after the Iraq conflict." Jeremy M. Sharp's "Iraq: A Compilation of Legislation Enacted and Resolutions Adopted by Congress, 1990–2003," is a useful feature. There is an intriguing CIA study entitled "Iraq's Weapons of Mass Destruction Programs." There is little historical perspective, as compared with Carlisle (item 1972). This book has been dated by developments during the past few years, moreover, but, nevertheless, it is an interesting compilation of materials with an adequate index. It is best used with the Jeffries collection (item 2000).

1976. Engel, Richard. *A Fist in the Hornet's Nest: On the Ground in Baghdad Before, During and After the War*. New York: Hyperion, 2004.

Engel, a United States correspondent, remained in Baghdad during the invasion in 2003, and he presents a gripping account of the last days of the Saddam regime and the fighting before the protracted insurgency. The author praises some features of United States policy, such as granting Iraqis considerable freedom of expression, even when their observations are critical of the invasion and occupation.

1977. ——. *War Journal: My Five Years in Iraq*. New York: Simon & Schuster, 2008.

Consists largely of the author's experiences and his observations. There are no extensive interviews, and the book is undocumented. Not the first book to read to understand more about Iraq, but it is important for the "feel" it provides of the Iraq situation. This is a sometimes grisly account.

1978. Fawn, Rick, and Raymond Hinnebusch, eds. *The Iraq War: Causes and Consequences*. Boulder, CO: Lynne Rienner Publishers, 2006.

Vital for its broad perspective of the conflict. The essays discuss the stakes that a number of countries had and still have in Iraq and the ways they were involved in the war. The numerous contributors to the book are largely senior scholars with significant publication records. Despite the sophistication of the essays, their brevity and careful organization make this a relatively fast read, but the book is a valuable source for understanding the meaning of the war.

1979. Ferguson, Charles H. *No End in Sight: Iraq's Descent into Chaos*. New York: Public Affairs, 2008.

Detailed transcripts of interviews with named interviewees about all aspects of the Iraq occupation. The book is an outgrowth and updating of Ferguson's film *No End in Sight*. In addition to discussing other topics, many interviewees refer to the very haphazard planning for the war.

1980. Galbraith, Peter W. *The End of Iraq: How American Incompetence Created a War Without End*. New York: Simon & Schuster, 2006.

Analyzes United States policy toward Iraq during the earlier Reagan and G.H.W. Bush administrations, thereby providing an indispensable perspective. Galbraith includes a detailed plan for United States disengagement. He believes Iraq has been an artificial political entity and asserts that "holding ... Iraq together by force ... has been destabilizing" (p. 206).

1981. Glantz, Aaron. *How America Lost Iraq*. New York: Jeremy P. Tarcher/Penguin, 2005.

Detailed description of the impact on the Iraqis of continued fighting, by a persistent opponent of the war. Glantz has much to say, too, about the United States mistakes that generated renewed fighting.

An important feature of the book is its discussion of the role of Sayyed Muqtada al-Sadr.

1982. Hendrickson, David C., and Robert W. Tucker. "Revisions in Need of Revising: What Went Wrong in the Iraq War." *Survival* 47, no. 2 (Summer 2005): 7–32.

Asserts that although many of the criticisms made of the Bush administration's handling of the war are valid, many of the problems were inherent in waging war in that country rather than specific failures in United States planning and United States actions early in the occupation. This is an important study.

1983. Hughes, Solomon. *War on Terror, Inc. Corporate Profiteering from the Politics of Fear.* London: Verso, 2007.

Critical discussion of the use of mercenaries and other theoretically non-military individuals in recent and current United States and British conflicts. Hughes emphasizes the Iraq War but includes information about Bosnia and Afghanistan. His conclusion is that corporate militaries have done great damage to coalition efforts to pacify Iraq.

B. The decision for war

1984. Alfonsi, Christian. *Circle in the Sand: Why We Went Back to Iraq.* New York: Doubleday, 2006.

Examines the Iraq War in terms of "what really happened in the first Bush White House after the 1991 Gulf War" (p. 1). Alfonsi argues that George W. Bush and his advisers became obsessed with the possibility that Saddam might take actions that would jeopardize Bush's 2004 re-election bid. Alfonsi also usefully discusses the much debated relationships between the administration and those people described as "neoconservatives."

1985. Bamford, James. *A Pretext for War: 9/11, Iraq, and the Abuse of America's Intelligence Agencies.* New York: Doubleday, 2004.

Charges that the Bush administration manipulated intelligence to further its goals. Bamford argues emphatically that 9/11 occurred because of United States support for Israel, and that Iraq was not involved in the attack.

1986. Blix, Hans. *Disarming Iraq.* New York: Pantheon Books, 2004.

Key statement by the man who led the United Nations effort to investigate Iraqi weapons of mass destruction. Blix believes that whatever weapons of this sort Iraq possessed were destroyed at some point and asserts that the "UN and the world ... succeeded in disarming Iraq without knowing it" (p. 259).

1987. Boyer, Douglas A. "Inverse Engagement: Lessons from U.S.–Iraq Relations, 1982–90." *Parameters* 33, no. 2 (Summer 2003): 51–65.

Sees "engagement" as involving economic pressures rather than military moves. Boyer traces increasing economic ties between the United States and Iraq. He asserts that much more effective use could have been made of economic factors to influence Iraq, but United States interest groups seem to have prevented their employment.

1988. Byman, Daniel L., and Matthew C. Waxman. *Confronting Iraq: U.S. Policy and the Use of Force Since the Gulf War.* Santa Monica, CA: RAND, 2000.

Succinct, but expert, analysis of the conflict. The authors discuss the role of coercion in international affairs and the goals of both Iraq and the United States. Chapter 5 is a careful review of the steps that have been taken to lead Iraq to change its behavior. Byman and Waxman propose that, if necessary, the United States foster a smaller group of nations that will be united on stronger measures toward Iraq.

1989. Carstens, Roger D. "Iraq—Let's Look Before Leaping." *United States Naval Institute Proceedings* 128, no. 9 (September 2003): 65.

Asserts that the United States must explain its objectives very carefully in Iraq if that country is to be attacked. A major question is how stability will be established after a second Gulf War. A clear declaration of principles will garner support from many nations.

1990. Daalder, Ivo H., and James M. Lindsay. *America Unbound: The Bush Revolution in Foreign Policy.* Washington, DC: Brookings Institution Press, 2003.

In sharp contrast to the interpretation of the Iraq War offered by Holmes (item 1997), the authors contend that the Bush administration had been committed to a much more assertive, if not unilateral, stance in world affairs before 9/11. The attack gave Bush an opportunity to demonstrate that the United States could operate largely, although not completely, without support and assistance from allies.

1991. Drogin, Bob. *Curveball: Spies, Lies, and the Con Man Who Caused a War.* New York: Random House, 2007.

Attributes the theory that Iraq was developing "weapons of mass destruction" to a single, unreliable Iraqi defector. Describes the mobile "germ warfare laboratories" that were in reality seed containers. Drogin includes a detailed chronology of the steps in this intelligence operation.

1992. Everest, Larry. *Oil, Power and Empire: Iraq and the U.S. Global Agenda*. Monroe, ME: Common Courage Press, 2004.

Thoroughgoing attack on the United States for waging what Everest calls "an unjust war" against Iraq. As the title indicates, Everest believes the determination of the United States to control Middle Eastern oil was the reason for the invasion of Iraq, although the decision to attack was merely part of United States plans for a world empire. He begins the book by looking at British imperialism in the oil countries of the Middle East and continues by discussing United States–Iraqi relations over several decades before the Gulf War. A long appendix is titled "Dissecting U.S. Pretexts for War."

1993. Friedman, Norman. "United States Must Deal with Iraq." *United States Naval Institute Proceedings* 128, no.11 (November 2002): 4, 6.

Warns that even if Al Qaeda is scattered, the United States will face enemies who are hostile to "modernization" (p. 4). Friedman believes that despite his defeat in the Gulf War, Saddam Hussein enjoys considerable prestige among Muslims. This is a concise, but thoughtful, examination of the problem of Iraq. Ultimately, despite the serious problems associated with an attack, Friedman believes that the United States may have to go to war with Iraq.

1994. Gardner, Lloyd C. *The Long Road to Baghdad: A History of U.S. Foreign Policy from the 1970s to the Present*. New York: New Press, 2008.

Witty, incisive, but thorough, account of the highly complex maneuvers that brought the United States to invade Iraq. Gardner successfully traces the origins of the invasion to the Vietnam era and the theorizing of Walt Rostow, and then follows developments through several administrations that had surprisingly similar goals not only in regard to the Middle East but also in respect to strengthening or even reclaiming the powers of the president in security affairs. "What will stand out one day is not George W. Bush's uniqueness but the continuum from the Carter doctrine to 'shock and awe' in 2003" (p. 63).

1995. Graham-Brown, Sarah. *Sanctioning Saddam: The Politics of Intervention in Iraq*. London: I.B. Tauris, 1999.

Massively documented study that is essential to understanding much of what occurred before the invasion of Iraq. Graham-Brown's impressive text is enhanced by outstanding maps and an important appendix: "The Use and Threat of Force Against Iraq, July 1991–February 1999." She expresses the view that in its actions with regard to Iraq, the "international community" continues to be much less than that. National political interests consistently trump larger concerns.

1996. Hayes, Stephen F. *Cheney: The Untold Story of America's Most Powerful and Controversial Vice President.* New York: HarperCollins, 2007.

Detailed and highly supportive discussion of Cheney's role in the Iraq War. Much of the material about Iraq centers on intelligence assessments before the U.S. invasion and the subsequent controversies over those estimates.

1997. Holmes, Stephen. *The Matador's Cape: America's Reckless Response to Terror.* New York: Cambridge University Press, 2007.

Asserts that the United States reacted to the 9/11 attack by becoming panic-stricken and that the Bush administration was duped into attacking Iraq by Middle Eastern extremists who thus manipulated United States behavior. Holmes is severely critical of the administration for giving primary responsibility for the War on Terrorism to the Defense Department, rather than the State Department.

1998. Isikoff, Michael, and David Corn. *Hubris: The Inside Story of Spin, Scandal, and the Selling of the Iraq War.* New York: Crown Publishers, 2006.

Discusses in detail the handling of intelligence related to Iraq and the development of the Bush administration's case against that country. Much of the book is based on what the authors term "anonymous sources," although they also conducted many interviews with persons who are identified.

1999. Jackson, Robert J., and Philip Towle. *Temptations of Power: The United States in Global Politics after 9/11.* Houndmills: Palgrave Macmillan, 2006.

Argues that the sanctions and no-fly zone stances toward Iraq were declining before 9/11, which gave the Bush administration an opportunity to pursue its objective of fighting Iraq, even before ending combat operations in Afghanistan. The larger goals of the administration in attacking Iraq were to complete the unfinished business of the Gulf War, to increase the presence of the United States in the Middle East by developing further military bases, and to expand the power of the president within the political system.

2000. Jeffries, Leon M. *Iraq: Issues, Historical Background, Bibliography.* Hauppauge, NY: Nova Science Publishers, Inc. 2003.

Interesting compilation of signed articles that was published on the eve of the invasion of Iraq. A number of important topics are discussed, such as Saddam Hussein's opponents among the Iraqis and the "Oil-for-Food" program. There is a long historical essay by Helen Chapin Metz that fills the gap in the Doppler volume (item 1975). The maps are poor, and the bibliography is undistinguished. Otherwise, it is a useful source that is best used in connection with the Doppler collection.

2001. Kirkpatrick, Jeane J. *Making War to Keep Peace.* Chap. 6, "Conclusion: Afghanistan and Iraq," pp. 271–305. New York: HarperCollins, 2007.

Voices the conviction that, although fostering the extension of democracy should be an important United States policy goal, efforts in support of democracy are unlikely to be successful unless there are already "institutions that are receptive and willing to support democracy" (p. 279). Despite her misgivings, Kirkpatrick publicly supported the Bush administration's policy against Iraq in 2003 on the assumption that it was part of the earlier Gulf War. Much of the chapter attempts to refute UN Secretary-General Kofi Annan's charge that the invasion of Iraq was "illegal."

2002. Record, Jeffrey. "The Bush Doctrine and War with Iraq." *Parameters* 33, no. 1 (Spring 2003): 4–21.

Discusses components of the Bush Doctrine, as embodied in the *National Security Strategy* (2002), which moved away from Cold War perspectives. This is a detailed critique of the Bush administration's case against Iraq.

2003. ——. "Why the Bush Administration Invaded Iraq: Making Strategy after 9/11." *Strategic Studies Quarterly* 2, no. 2 (Summer 2008): 63–92.

Points to the welter of explanations and accusations surrounding the United States decision to attack Iraq. Record points to the "neoconservatives" within the Bush administration and other decision-makers and counselors who insisted on the invasion in order to give the world an example of United States power. The specific goals of the invasion were cloudy and contentious, however. Record believes that a probing investigation of the Iraq War is needed to prevent similar disasters in the future.

2004. Ritchie, Nick, and Paul Rogers. *The Political Road to War with Iraq: Bush, 9/11 and the Drive to Overthrow Saddam.* London: Routledge, 2007.

Important study that calls the Iraq War "all but inevitable after the 9/11 attacks." The authors provide a valuable service by precisely comparing the Clinton and Bush administrations' policies toward Iraq. They emphasize the continuing importance of the Gulf region to the United States, and they predict that "a full withdrawal from Iraq would only happen in the event of a quite extraordinary change in the political environment in the United States" (p. 171).

2005. Roston, Aram. *The Man Who Pushed America to War: The Extraordinary Life, Adventures, and Obsessions of Ahmad Chalabi.* New York: Nation Books, 2008.

At one time the deputy premier of Iraq after the invasion, Chalabi had previously headed the Iraqi National Congress. Roston describes the complicated relationships between that organization, the State Department, the CIA, and the Defense Intelligence Agency. To many in Washington, Chalabi seemed to be an ideal leader in the struggle to unseat Saddam, but he never was able to build a mass following for himself in Iraq.

2006. Russell, Richard L. "War and the Iraq Dilemma: Facing Harsh Realities." *Parameters* 32, no. 3 (Autumn 2002): 46–61.

Praises Bush's response to terrorism directed against the United States and criticizes other NATO members, except Britain, for failing to follow Bush's lead with respect to Iraq. Russell's assertions are based heavily on what he perceives as weapons of mass destruction held by Iraq.

C. General military studies

2007. Bensahel, Nora, Heather S. Gregg, Keith Crane, Olga Oliker, and Richard R. Brennan. *After Saddam: Prewar Planning and the Occupation of Iraq.* Santa Monica, CA: Rand Corporation, 2008.

Close study, made at the request of the United States Army, of the planning done for the invasion of Iraq and the post-invasion period. The analysis ends on June 28, 2004, when the Coalition Provisional Authority was terminated. All aspects of the challenges that invading forces would encounter, including governmental, economic, and humanitarian relief, are considered. The authors have not blinked in discussing shortcomings in the planning process.

2008. Boyne, Walter J. *Operation Iraqi Freedom: What Went Right, What Went Wrong, and Why.* New York: Tom Doherty Associates, 2003.

Deals specifically with events between March 19, and May 1, 2003, that is, the invasion itself. Boyne praises military decision-makers and other members of the services for what he views as a superbly executed operation and downplays the significance of the absence of weapons of mass destruction in Iraq. The appendices, especially the descriptions of weapons systems employed, make this a highly useful study.

2009. Clark, Wesley K. *Winning Modern Wars: Iraq, Terrorism, and the American Empire.* New York: Public Affairs, 2003.

Early, perceptive account of the war. Clark asserts that the United States put most of its efforts into Iraq, although international terrorism was the real problem. This assertion has been repeated in many later analyses of the United States stance toward terrorism and toward the Middle East. Clark gives the armed forces credit for an outstanding performance during the initial attack but also points to costs to the armed forces, warning that the United States Army is "not an Army of empire," meaning a long-term occupation force (p. 168).

2010. Conway, James T. "'Farther and Faster' in Iraq." *United States Naval Institute Proceedings* 131, no. 1 (January 2005): 48–52.

States that Operation Iraqi Freedom went through four stages, ending with "a reconstruction phase" (p. 48), which may well be the most difficult segment. Counterinsurgency is now the focus after a highly successful conventional campaign. Progress in developing effective Iraqi forces is slow, but there have been improvements.

2011. Cordesman, Anthony H. *The Iraq War: Strategy, Tactics, and Military Lessons.* Westport, CT: Praeger, 2003.

Magisterial account of the war. This is strictly a military study. Relatively little is said of the political and economic background of the conflict, but the objective of this study was military analysis. Given the extreme detail of the book, it may not be the best volume to consult for an introduction to the subject, but anyone seriously interested in the waging of the war needs to read this book.

2012. Erwin, Sandra I. "Iraq Lessons Pervade Army War Games." *National Defense* 89, no. 619 (June 2005): 12–13.

Although the army is wary of committing itself to "fighting the last war," Iraq experiences are clearly influencing the war games conducted. Erwin describes war game "United Quest" in considerable detail. United States access to oil supplies was a key concern in the game.

2013. Gordon, Michael R., and Bernard E. Trainor. *Cobra II: The Inside Story of the Invasion and Occupation of Iraq.* New York: Pantheon Books, 2006.

Perhaps the most detailed account of the war to date, the book is based heavily on interviews with participants at all levels. The text is enhanced by a fine section of maps and an extensive appendix that contains a chronology and numerous documents.

2014. Murray, Williamson, and Robert H. Scales, Jr. *The Iraq War: A Military History.* Cambridge, MA: The Belknap Press of Harvard University Press, 2003.

Thorough, but lightly documented, description of the invasion in 2003. Political facets of the conflict are discussed extensively from the perspective of the Bush administration. The text is enhanced by numerous maps and photographs and by a long section about the weapons used in the war.

2015. Perret, Geoffrey. *Commander in Chief: How Truman, Johnson, and Bush Turned a Presidential Power into a Threat to America's Future.* Chap. 28, "Armed Missionaries," pp. 364–74; chap. 29, "Iraq Syndrome," pp. 375–89. New York: Farrar, Straus and Giroux, 2007.

Scathing critique of the performance of these three post-World War II presidents as commanders in chief with respect to Korea, Vietnam, and Iraq. Perret is a well-known student of military affairs, and his book is especially valuable as a comparative study. Nevertheless, his readable volume might have been improved by moderating the many highly personal attacks on major actors in each administration. Perret has an entirely pessimistic view of the future of the United States, evidently believing that China is now the emerging superpower.

2016. Record, Jeffrey. *Beating Goliath: Why Insurgencies Win.* Chap. 3, "The Iraqi Insurgency: Vietnam Perspectives," pp. 67–102. Washington, DC: Potomac Books, 2007.

Penetrating comparative study of the Iraq and Vietnam conflicts, that states precisely the differences and the similarities between them. Among the differences was the fact that in Vietnam, the United States was trying to maintain an existing state, but in Iraq the state had to be rebuilt. Record scores the United States armed forces for ignoring the possibility after Vietnam that the nation could be involved in new counterinsurgency campaigns.

2017. Ricks, Thomas E. *Fiasco: The American Military Adventure in Iraq.* New York: Penguin Press, 2006.

Detailed study of the entire Iraq operation from the invasion through several years of insurgency. Ricks emphasizes what he views as the errors of the United States that ignited the insurgency. At the beginning of his "Notes" section, the author discusses the huge amount of information, including formal documents, interviews, and emails used to produce the book, observing that such a wealth of materials would have been unavailable, especially so quickly, in earlier years.

2018. Sanchez, Ricardo S., with Donald T. Phillips. *Wiser in Battle: A Soldier's Story.* New York: HarperCollins, 2008.

Sanchez commanded United States and Coalition forces in Iraq for some time, and he was blamed for the Abu Ghraib abuses. Despite its defensive tone, Sanchez's book is valuable for its detailed discussion of many of the problems of the Iraq conflict.

2019. West, F.J. Bing. "American Military Performance in Iraq." *Military Review* 86, no. 5 (September–October 2006): 2–7.

This is a brief, but important, statement of the many errors and misunderstandings associated with the Iraq occupation. There is so much material that the article is difficult to summarize properly. Among other points, West asserts that the insurgency has been a Sunni undertaking, hostile to coalition forces, Kurds, and Shi'ites. He argues, too, that Iraq was in utter disarray, a failed state. According to West, "[r]esponsible Iraqi leadership was the commodity in least supply in post-Saddam Iraq."

2020. Zais, Mitchell M. "U.S. Strategy in Iraq." *Military Review* 87, no. 2 (March/April 2007): 105–8.

Highly critical of the implicit strategy of the United States in Iraq. Inadequate funding and substantial shortfalls in personnel are among the factors that severely constrict the war effort in Iraq. Top United States war leaders are knowledgeable about military technology but not about the situation in the Middle East.

2021. Zeman, P.M. "Goat-grab Diplomacy in Iraq." *United States Naval Institute Proceedings* 131, no. 11 (November 2005): 20–25.

Argues that the United States Marine Corps' *Small Wars Manual, 1940* is directly pertinent to the Iraq conflict. Zeman also highlights the importance of working with Iraqi sheiks in the restoration of stability.

D. Air operations

2022. Schanz, Marc V. "A Complex and Changing Air War: The Top Airman in Southwest Asia Discusses Air Operations Over Iraq and Afghanistan." *Air Force Magazine* 89, no. 1 (January 2006): 68–69.

LTG Walter E. Buchanan III discusses the use of air power against terrorism, including conventional and non-crewed aircraft and the application of aviation assets in urban environments. Buchanan candidly relates the many problems of using aviation in Iraq and Afghanistan.

2023. Speredelozzi, John M. "Special Operations Support: One Aviator's Perspective." *United States Naval Institute Proceedings* 131, no. 6 (June 2005): 77–79.

Describes a "composite" force of marines, naval personnel, and members of special operations units from other coalition countries. Both helicopters and fixed-wing aircraft are useful in its duties. The author believes the navy can expand its role in anti-terrorism efforts by supporting land forces.

E. Ground combat

2024. Alfieri, Paul, and Don McKeon. "Stryker Suitability Challenges in a Complex Threat Environment." *Defense Acquisition Review Journal* 15, no. 1 (April 2008): 49–63.

Assesses the suitability of the Stryker system for its tasks. Although the Stryker is working well in Iraq, it was originally intended for a very different environment, that of low-intensity conflict during peacekeeping operations, rather than intense fighting. Iraq is definitely taking a toll on these vehicles, which require considerable maintenance. This is a highly technical article that stresses the need to detect deficiencies early to avoid acquiring systems that may not meet United States military needs.

2025. Angeles, Ren. "Examining the SBCT (Stryker Brigade Combat Team) Concept and Insurgency in Mosul, Iraq." *Infantry* 94, no. 3 (May–June 2005): 35–39.

Extremely detailed examination of Stryker capabilities, focusing on their various configurations in the Iraq War. Angeles emphasizes the Stryker is a new concept. Much needs to be learned in order to exploit its already considerable usefulness even further.

2026. Atkinson, Rick. *In the Company of Soldiers: A Chronicle of Combat.* New York: Henry Holt and Company, 2004.

Distinguished war correspondent and military historian's view of the invasion and occupation from the vantage point of the headquarters of the 101st Airborne Division, then commanded by Major General David H. Petraeus. Essential reading for Atkinson's portrayal of Petraeus. Although the book is largely the product of Atkinson's experiences and interviews, the author cites a number of other sources.

2027. Batiste, John R.S., and Paul R. Daniels. "The Fight for Samarra: Full-Spectrum Operations in Modern Warfare." *Military Review* 85, no. 3 (May–June 2005): 13–21.

Discusses an attack on Samarra (Operation Baton Rouge) by the 1st Infantry Division and Iraqi troops in October 2004 to oust antigovernment forces. The battle had considerable political significance, because it was fought before the general election in Iraq. The authors emphasize the integration of military, political, and economic action. The article includes much on the role of Iraqi police in counterinsurgency.

2028. Briscoe, Charles H., Kenneth Finlayson, and Robert W. Jones, Jr. *All Roads Lead to Baghdad: Army Special Operations in Iraq.* Fort Bragg, NC: USASOC History Office, 2006.

Major source. The scope of the book extends well beyond the role of army special operations forces, as important as it was. Much is said about the significance of Iraq in United States strategy and the composition of Iraqi forces just before the beginning of combat operations in 2003. A great deal of material is presented from the viewpoint of participants. The whole text is supplemented by maps and other graphics.

2029. Brown, Todd S. *Battleground Iraq: Journal of a Company Commander.* Washington, DC: Government Printing Office, 2007.

Straightforward narrative by an officer who served in Iraq in the 3rd Infantry Division during 2003 and 2004. Episodic and sometimes slow reading because of the heavy use of acronyms, which are explained, in part, by a glossary, Brown's account is, nevertheless, a highly graphic account of the author's experiences and, more broadly, of the nature of fighting in Iraq.

2030. Burgess, Richard R. "Remote Control: Ordnance Disposal Robots Are Being Deployed Against Deadly Concealed Explosives in Iraq, Afghanistan." *Sea Power* 48, no. 10 (October 2005): 10–12.

Discusses robots used to disarm explosives in Iraq and Afghanistan. The navy is the lead agency in developing such devices. The "Bombot" is a new device that is not particularly costly compared with other systems. Various other devices are described in detail.

2031. Buzzell, Colby. *My War: Killing Time in Iraq.* New York: G.P. Putnam's Sons, 2005.

Given the attention created by his generally well-received blog about the Iraq War, Buzzell was hardly a typical enlisted infantryman, but his highly skillful assembly of letters, news reports, and other items, such as the "rules of engagement" distributed to United States service personnel, provides a genuine feel of the fighting.

2032. Elkhamri, Mounir. "Dealing with the Iraqi Populace: An Arab-American Soldier's Perspective." *Military Review* 87, no. 1 (January/February 2007): 110–13.

Recommends much increased familiarization with Middle Eastern culture and languages for United States personnel heading for Iraq. The needs of locals must be understood, and such understanding cannot be achieved in a vacuum. Much of the article deals with the equitable distribution of vital resources and the provision of employment to Iraqis.

2033. Erwin, Sandra I. "'Protective Cover': Army to Expand Array of Armored Vehicles in Iraq." *National Defense* 90, no. 623 (October 2005): 42+.

Conditions in Iraq have shown a need for vehicles with larger troop capacity and weapons capability. Erwin manages to describe a number of vehicles in a short article, both those that are already being employed and those that are in the planning stage. Developing the most effective vehicles appears to be a complex process.

2034. Fabey, Michael. "U.S. EA-6B Jets Jam Cell Phones in Iraq, Block IEDs." *Defense News* 20, no. 42 (October 31, 2005): 1+.

Describes the communications jammers mounted in EA-6B Prowler aircraft. These jammers showed impressive results as electronic warfare weapons, although there have been instances in which friendly communications have been jammed.

2035. Friedman, Norman. "Have We Found a Weapon to Fight IEDs (Improvised Explosive Devices) in Iraq?" *United States Naval Institute Proceedings* 131, no. 1 (January 2005): 4, 6.

Discusses the Naval Surface Warfare's promising new device to deal with IEDs. The Marine Corps Warfighting Laboratory also has a counter-weapon that may make it possible to defeat IEDs from unmanned aerial vehicles. Electrical jamming techniques and the use of microwaves are other possibilities for fighting these devices.

2036. Goldenberg, Richard. "Another Chapter: New York's Rainbow Division Adds to its Illustrious History by Maintaining Relative Calm in Volatile Sunni Triangle." *National Guard* 59, no. 9 (September 2005): 30–32.

Very briefly outlines the history of the 42nd Infantry Division (Rainbow Division) but focuses on its role as part of Task Force Liberty in Iraq. Goldenberg stresses the fact that United States forces are holding the line until Iraqi troops and police can finally defeat the insurgency.

2037. Grant, Greg. "Naval Gatling Gun Given Anti-Mortar Role: U.S. Army Eyes Defense Against Iraq Insurgents." *Defense News* 20, no. 24 (June 13, 2005): 26.

In order to reply effectively to the use of mortars and rockets by Iraqi insurgents, the army has borrowed a weapons system from the navy, the Counter-Rocket Artillery Mortar System, which operates with radar and both a multiple-barrel gun and a crewless aerial vehicle. Significant modifications had to be made to adapt the system to urban combat.

2038. ——. "U.S. Army Casts Wide Net for IED (Improvised Explosive Devices) Defenses." *Defense News* 20, no. 25 (June 20, 2005): 1+.

Emphasizes the importance of improvised explosive devices to insurgents in Iraq and the adaptability of the insurgents in employing these relatively simple weapons. IED incidents are concentrated at present largely in Baghdad. The availability of explosives in Iraq is a factor that aids the insurgents. Meanwhile, the army is hard put to cope with IEDs.

2039. Hickey, Christopher. "Principles and Priorities in Training for Iraq." *Military Review* 87, no. 2 (March/April 2007): 22–32.

Describes the seemingly overwhelming challenges faced by the United States military in Iraq and then itemizes the major elements in providing appropriate training. The key is to develop soldiers who can exercise independent judgment.

2040. Hilburn, Matt. "'Generation of Warriors': Iraq Operations Make Battle-Savvy Marines, but Some Core Competencies Have Faded: Interview." *Sea Power* 51, no. 5 (May 2008): 32–34.

Interview with Assistant Marine Corps Commandant General Robert Magnus. The general reviews the prospects of marines remaining in Iraq and analyzes the ways marines have gained or lost from the Iraq experience. Recently, the corps' primary commitment has been to counterinsurgency, not amphibious operations. Amphibious competency must be maintained, however.

2041. Jean, Grace V. "Stryker Brigades Training for Upcoming Deployment." *National Defense* 90, no. 634 (October 2005): 36–37.

Discusses training at Fort Lewis, Washington, using video telephone conferencing technology. The Mission Support Training Facility (MSTF) was able to employ information from the first Stryker brigade, which was in Iraq. Other uses of technology are described more briefly.

2042. Knights, Michael. "Battle for Iraq Lies in the South." *Jane's Intelligence Review* 17, no. 6 (June 2005): 12–15.

Asserts that Iraq's future will be decided in the Shi'ite area in the south, not the north, which is receiving the most attention because of the many clashes there. Knights focuses on the Supreme Council for Islamic Revolution (SCIRI) and its strategies and predicts that Shi'a elements will probably gain more influence in Iraq's army and police. This is an important analysis of a highly complex situation.

2043. McFate, Montgomery. "Iraq: The Social Context of IEDs." *Military Review* 85, no. 3 (May–June 2005): 37–40.

Discusses an alternative method for dealing with IEDs by focusing on the bombers rather than the technology of the bombs. McFate reports that the use of IEDs stems largely from the effective development of the weapons by Saddam Hussein's military intelligence service.

2044. McKenna, Ted. "No Silver Bullet for IEDs." *Journal of Electronic Defense* 28, no. 12 (December 2005): 18–20.

Iraq War, 2003–

Asserts that the United States has yet to develop effective countermeasures against IEDs, which have caused the predominant number of Coalition casualties. A number of responses have been developed to deal with the problem, and some have real promise.

2045. ——. "Shutting the Bomb Factory: Technology So Far Has Proven No Match for Relatively Low-Tech Bombs in Iraq and Afghanistan." *Journal of Electronic Defense* 28, no. 9 (September 2005): 14–17.

IEDs are, by far, the most effective weapon used against Coalition forces. McKenna discusses the problem of stopping the manufacture of bombs, particularly in view of the masses of munitions in Iraq, which are difficult to guard properly, and the adaptability of the bombers. There are various ways to counteract IEDs, but better methods are needed to eliminate the threat.

2046. Miller, James L., and Michael A. Harp. "Counterstrike at the NTC (National Training Center): Reversing Negative Trends." *Field Artillery* (September–October 2005): 26–28.

Discusses the lessons from preparing units for combat assignments. This is a seemingly exhaustive study that emphasizes the importance of proactive higher commanders who will maintain a "combined arms" approach to fighting, because artillery alone is insufficient. Another key element is effective intelligence-gathering, including the proper use of radar.

2047. Moore, Robin. *Hunting Down Saddam: The Inside Story of the Search and Capture.* New York: St. Martin's Press, 2004.

Highly detailed and thoroughly documented account that is enhanced by maps, numerous illustrations, and an extensive glossary.

2048. Pappalardo, Joe. "Greetings from Insurgent City: Guardsmen Act as Foes to Aid Iraq-Bound Soldiers." *National Defense* 90, no. 620 (July 2005): 36–40.

Reviews a training session for elements of the 28th Infantry Division (Pennsylvania National Guard) at "Insurgent City," a component of the National Training Center at Fort Irwin, California. Valuable in-depth account of how urban combat training is carried out. Role playing by Iraqi–Americans helps deepen the experience in these exercises.

2049. Pengelley, Rupert. "Curbing the Roadside Bomber." *Jane's International Defence Review* 39, no. 1 (January 2006): 39–40+.

The highly successful use of IEDs fired electronically by insurgents in Afghanistan and Iraq has spurred many manufacturers to try to develop countermeasures. Such efforts are receiving publicity, despite the desirability of keeping the technology as discreet as possible. A number of systems are considered in the article.

2050. Petraeus, David H. "Transcript: General Petraeus on the Way Ahead in Iraq." *Military Review* 87, no. 2 (March/April 2007): 2–4.

Examines the chaos in Iraq after the fall of Saddam. Petraeus stresses the need to enhance personal security for Iraqis and to develop further non-military approaches to the problems of Iraq.

2051. Reynolds, Nicholas E. *U.S. Marines in Iraq, 2003: Basrah, Baghdad and Beyond*. Washington, DC: History Division, U.S. Marine Corps, 2007.

Detailed description of marine operations during the high-intensity phase of the Iraq war. One chapter deals with United States planning for the conflict. Other chapters deal with various topics, including the role of the marines in Kurdistan. There is interesting material about cooperation between the marines and the British contingent in Iraq, moreover. The book is well illustrated, and it contains useful maps.

2052. Risberg, Robert H. "'Down and Dirty': Training an FA Bn (Field Artillery Battalion) to be a TF (Task Force) in Iraq." *Field Artillery* (November–December 2005): 35.

Describes the responsibilities of a field artillery battalion within a task force that includes infantry and engineer troops and that performs a wide spectrum of duties. Much of the article discusses the kinds of training needed for Iraq. Risberg cautions that rules of engagement orientation are essential before troops reach the combat zone.

2053. Roth, Jason, Bill Heard, Richard C. Sperando, and Stephen P. Boyle. "Overhead Protection: The U.S. Army Corps of Engineers Has Developed a New Method of Providing Wide-Area Protection of U.S. Base Camps in Iraq." *Military Engineer* 99, no. 647 (May–June 2007): 53–54.

Structures in United States base camps in Iraq often do not offer much protection during enemy attacks. The Engineer Research and Development Center of the Army Corps of Engineers pioneered efforts to bring out new construction methods and the use of innovative materials to help mitigate the effects of attacks.

2054. Steele, Dennis. "Commanders in Iraq: Some Lessons Learned." *Army* 55, no. 6 (June 2005): 24–26, 28, 30.

Brief observations by four battalion commanders in Iraq. The reports do not follow a uniform format, but they contain insightful remarks on a number of topics. Frequent themes are the ways of dealing with Iraqis and the understanding shown by families of United States service members when troops are killed.

2055. ——. "The 278th RCT (Regimental Combat Team) in Iraq: A 'Volunteer' Heritage and a Critical Mission." *Army* 55, no. 6 (June 2005): 38–41, 44, 46.

Stresses the close relations between members of this Tennessee National Guard force. Even the vast majority of replacements come from Tennessee. The RCT commander believes that the Iraqi elections have radically changed the situation for the better.

2056. Tiron, Roxana. "Unconventional Weapons Can Help U.S. Troops Fight Insurgents in Iraq." *National Defense* 89, no. 610 (September 2004): 38–39.

Discusses efforts to provide new and more sophisticated non-lethal weapons to deal with insurgent tactics, such as using "human shields." A wide variety of such weapons exists, but modifications are being developed in response to problems reported by field troops, such as the short range of some weapons.

2057. Turay, Ismail, Jr. "Honing Search Skills in Iraq." *Soldiers* 60, no. 5 (May 2005): 12–13.

Relates a search coordinated by the 1st Infantry Division and Pennsylvania National Guard troops at Siniyah, Iraq. Little information about insurgents emerged and no arms were found, but there were no civilian casualties, either.

2058. Van Sickle, Jeffrey B. "Stability Operations in Northern Iraq: Task Force Altun Kupri." *Infantry* 94, no. 1 (January–February 2005): 25–29.

Reviews various lessons learned about effective stability operations. United States forces have had to avoid being drawn into ethnic conflicts, to work with the Muhtar, a kind of neighborhood notable, and to maintain an effective information program to offset hostile propaganda directed at Coalition forces.

2059. Vines, John R. "The XVIII Airborne Corps on the Ground in Iraq." *Military Review* 86, no. 5 (September–October 2006): 38–46.

Summarizes XVIII Airborne Corps experiences and lessons learned in Iraq during 2005 and into January 2006. Vines notes the successful elections in 2005, but he says that the obstacles to building viable Iraqi security forces have been underestimated. On the positive side, a successful effort has been made to reconfigure United States vehicles to counter IED attacks. The article is wide-ranging and candid in stating problems in many operational spheres.

F. Battles: Baghdad

2060. Anderson, Jon Lee. *The Fall of Baghdad*. New York: Penguin Press, 2004.

Relates the stories of Iraqis who had varied experiences before and during the war, beginning in 2003. Anderson thus provides perspectives that are hardly touched upon in most other books about Iraq.

2061. Steele, Dennis. "Haifa Street: Purple Heart Boulevard Redux." *Army* 55, no. 9 (September 2005): 22–24+.

Haifa Street was one of the most dangerous parts of Baghdad at one time. Massive deployment of United States troops changed the situation, and, gradually, Iraqi forces assumed more responsibility for the area. An Iraqi officer describes tactics for maintaining control. The security situation is now vastly different from what it was earlier.

G. Battles: Fallujah

2062. Ballard, John R. *Fighting for Fallujah: A New Dawn for Iraq*. Westport, CT: Praeger Security International, 2006.

Highly detailed and thoroughly documented account. The text is enhanced by an extensive bibliography and a large glossary.

2063. Gentry, Keil R. "RCT-1 Fires in the Battle of Fallujah." *Field Artillery* (November–December 2005): 26–29.

Argues that "basic fire support principles" (p. 26) are applicable to all types of warfare, including counterinsurgency. Gentry notes that a Regimental Combat Team headquarters had to coordinate counter-fire rather than the usual battalion or regimental headquarters. He presents a seemingly exhaustive description of counterfire procedures.

2064. Keiler, Jonathan F. "Who Won the Battle of Fallujah?" *United States Naval Institute Proceedings* 131, no. 1 (January 2005): 57–61.

Describes the United States' opponents in Fallujah as an "indifferently armed rabble" and speculates about the consequences of encounters with Syrian or Palestinian elements. At Fallujah, after some initial fighting, a drive was called off because of the desire to minimize Iraqi casualties. The Coalition attack in November 2004 was much more powerful than the one in the spring, but restrained Israeli-style tactics were employed.

2065. Thompson, Neal. "A Marine Drops in, Turns on Naval Institute Warfare Conference." *United States Naval Institute Proceedings* 131, no. 11 (November 2005): 26–27.

Briefly describes the Battle of Fallujah, Iraq, in 2004.

H. Insurgency and the counterinsurgency campaign

2066. Alywin-Foster, Nigel R.F. "Changing the Army for Counterinsurgency Operations: Originally Published in the 'Seaford House Papers.'" *Military Review* 85, no. 6 (November–December 2005): 2–15.

A significant opportunity to stabilize Iraq was lost immediately after the successful invasion in 2003, owing to inadequate planning for a possible insurgency phase. Alywin-Foster, a British general who served with the United States Army in Iraq, praises members of the armed forces but criticizes their lack of cultural awareness, which is close to "institutional racism" (p. 3). Balancing between a supportive, understanding stance toward the civil population and waging war effectively is difficult, but, unfortunately, United States actions have tended to emphasize the latter to the detriment of the former. There has also been too much reliance on intelligence-gathering through technology and not enough on human intelligence.

2067. Daly, Terence J. "How to Win in Iraq." *United States Naval Institute Proceedings* 131, no. 12 (December 2005): 14–19.

Argues that traditional counterinsurgency methods are needed in Iraq, specifically the restoration of law and order and the implementation of population controls through using such measures as a precise personal identification system. Daly stresses that an effective counterinsurgency effort requires much time.

2068. Donnelly, Thomas. "The (Iraqi) Political Battles Ahead." *Armed Forces Journal* 143, no. 6 (January 2006): 18–20.

Analyzes counterinsurgency strategy in Iraq, which Donnelly summarizes as "clear, hold, build." This involves large-scale operations in urban areas, and it is Donnelly's view that the "hold" element has been the least successful element, because of the considerable demands made on limited personnel. Donnelly notes, too, the limited support Bush's policies are receiving in the United States.

2069. ———. "A Split on Strategy?" *Armed Forces Journal* 143, no. 5 (December 2005): 30–31.

Discusses an intercepted message between major Al Qaeda leaders concerning strategy and tactics. Bidding for support among Muslims, Ayman al-Awahiri argued for focusing on the elimination of United States forces in Iraq. This is a detailed analysis of the message and of the strains within Al Qaeda.

2070. Gibson, Chris. "Battlefield Victories and Strategic Success: The Path Forward in Iraq." *Military Review* 86, no. 5 (September/October 2006): 47–59.

Discusses in detail counterinsurgency as waged in Iraq. Gibson stresses that "security" for local inhabitants is an all-essential element in defeating the insurgents. Improving their security increases the flow of information from them to the military.

2071. Grant, Greg. "Iraqi Insurgents Find Ways to Bounce Back." *Defense News* 21, no. 8 (February 20, 2006): 25.

Killing its major leaders has little long-term effect on the insurgency. The insurgents seemingly have no problem in replacing personnel, in good part because of the high unemployment in Iraq.

2072. Hollis, Patricia Slayden, "The 1st Cav in Baghdad-Counterinsurgency EBO (Effects-Based Operations) in Dense Urban Terrain: Interview (with) Major General Peter W. Chiarelli." *Field Artillery* (September–October 2005): 3–8.

Chiarelli, commander of the 1st Cavalry Division, notes that many members of Iraqi security forces were absent in the early days of his experience but also asserts that proper training brought Iraqis to a high standard. Chiarelli also describes the various types of opposition elements (not really groups, in his view) that Coalition forces have encountered. This is a thoughtful discussion of the requirements for successful counterinsurgency that includes much information about civic action and economic development.

2073. Johnson, Lee. "Build Trust Through Communication." *United States Naval Institute Proceedings* 131, no. 1 (January 2005): 54–56.

Begins by describing insurgent attacks on Iraq's oil platforms in April 2004 and the United States response. In the aftermath, Johnson found an eager response to his questions from Iraqis when he spoke Arabic. He further noted Iraqi concern about an abrupt departure of Coalition forces that would leave many Iraqis with an uncertain future.

2074. Kagan, Frederick W. "Measuring Success." *Armed Forces Journal* 143, no. 6 (January 2006): 20–24.

Discounts many of the numerical measures used to assess progress toward winning in Iraq. Using the number of Iraqi troops can be misleading, for example, because of the difficulty of evaluating real degrees of training. What is important is the extent to which the insurgents are losing the sympathy of the Sunnis. Kagan makes interesting proposals for measuring Sunni views.

2075. Knights, Michael, and Zack Snyder. "The Role Played by Funding in the Iraq Insurgency." *Jane's Intelligence Review* 17, no. 8 (August 2005): 8–15.

Emphasizes the need to reduce the movement of foreign money into Iraq that maintains the insurgent offensive. The economic situation in Iraq remains grim, thus creating severe unemployment. Some of the unemployed will work for those elements with money who oppose

Coalition goals in the country. Placement of an IED is worth US $100–150, for example. Important article for understanding a major facet of the insurgency.

2076. Mann, Morgan. "The Power Equation: Using Tribal Politics in Counterinsurgency." *Military Review* 87, no. 3 (May/June 2007): 104–8.

Describes successful efforts to direct local rulers, who were tied to Saddam because of the patronage he had provided to them, away from support for the insurgency. Mann stresses the significance of tribes in rural areas of Iraq. The United States military must understand this point in order to wage effective counterinsurgency.

2077. Peters, Ralph. "No Silver Bullets: Fighting the Insurgency in Iraq." *Armed Forces Journal* 143, no. 6 (January 2006): 36–39.

Criticizes the United States tendency to try to over-simplify complex problems, such as waging a counterinsurgency campaign. Peters recommends flexibility and adaptability and condemns seizing on a single apparent countermeasure. Highly critical of the Bush administration for not sending enough troops and for playing politics with the war, and of the media in the United States.

2078. Pray, Jeremiah. "Kinetic Targeting in Iraq at the Battalion Task Force Level: From Target to Detainee." *Infantry* 94, no. 4 (July–August 2005): 30–33.

Interprets the insurgency as driven largely by economic problems, such as unemployment and the lack of public services, rather than ideological commitments, at least in the Kikrit area. There is much in this article about the formation and organization of insurgent groups, the role of "human intelligence" in countering insurgent threats, and the conduct of anti-insurgent raids.

2079. Pruett, Jesse P. "Good Cops, Bad Cops, Carrots and Sticks." *Special Warfare* 18, no. 2 (September 2005): 18–21.

In the War on Terrorism, military operations are the "stick" and civil affairs and civic action efforts are the "carrot." Applying these tactics demands flexibility and precise knowledge of the environment. Pruett points out that combat troops and civil affairs teams may not be fully aware of each other's work, but the civilian population will learn the distinctions quickly.

2080. Voorhies, David. "Stability Operations: The Legacy of the Search and Attack." *Infantry* 94, no. 3 (May–June 2005): 27–34.

Argues that current army doctrine gives appropriate guidance to those who are waging counterinsurgency in Iraq, provided that commanders study it. The article usefully summarizes doctrine regarding "stability operations" that appears in several army field manuals and other publications, and identifies and discusses some key articles pertinent to the subject. Voorhies carefully relates doctrine to conditions in Iraq, making this an especially valuable piece.

I. Naval operations

2081. Axe, David. "Being a Seabee Ain't Easy." *United States Naval Institute Proceedings* 132, no. 4 (April 2006): 66–67.

Opens with a description of a mortar attack on Seabees at Ar Ramadi, Iraq. Attacks are only one of the problems facing the Seabees. Those include maintenance, supply, and guarding truck convoys.

2082. ———. "On Its Own: The Iraqi Navy in 2005." *United States Naval Institute Proceedings* 131, no. 8 (August 2005): 34–36.

Describes capabilities and operations of the new Iraqi Navy. Key points on Iraq's short coastline are the Kohr al Amaya and Mina al Bakr oil platforms. In two years the Iraqi Coast Guard moved to become the Iraqi Navy. More personnel and ships are on the way.

2083. Jean, Grace V. "Into Brown Waters: Navy Riverine Force to Report for Iraq Duty in 2007." *National Defense* 90, no. 626 (January 2006): 34–36.

The navy plans to send three "riverine squadrons" to Iraq in 2007 to replace marine personnel. Various types of vessels may be needed for patrol and for transport duties. The article quotes many navy and marine leaders directly concerned with the riverine program.

2084. Johnson, J. Lee. "The Expeditionary Sailor in the War on Terror." *Military Review* 86, no. 2 (September/October 2006): 96–99.

After very briefly discussing United States Navy and Royal Navy experiences in land operations, Johnson examines some ways the United States Navy has contributed to the Iraq conflict and suggests ways for enhancing its role.

2085. Klamper, Amy. "River War: Navy Riverines are in Demand in Iraq to Deny Insurgents' use of Rivers as Transport Routes, Avenues of Escape." *Sea Power* 49, no. 2 (February 2006): 10–12.

Discusses the development of riverine forces by the United States Navy and their project role in Iraq. Very limited riverine assets exist, which means considerable development and commitment of resources are needed. The article notes differences between riverine operations in Vietnam and Iraq.

2086. Tiron, Roxana. "Defense of Iraqi Oil Depots Tests Mettle of U.S. Sailors." *National Defense* 89, no. 619 (June 2005): 40–42.

Examines security measures carried out by United States naval personnel to protect the flow of oil from Iraq. Oil platforms ABOT and KAAOT are the main points that must be guarded. Iraqi forces are gradually assuming responsibility for protecting the sites. Coalition warships are also an important part of the security plan. Tiron also explains procedures for boarding foreign ships in the area as part of the effort to guard the oil platforms.

2087. Wary, Robert O., Jr. "Baghdad, Help Wanted." *United States Naval Institute Proceedings* 131, no. 1 (January 2005): 45–47.

Charges that the navy is not providing an adequate number of personnel to serve in staff roles in Iraq and suggests that the navy ask for at least 300 volunteers, perhaps all or at least many of them from the navy reserve. Wary describes the various structures that need additional people.

J. Intelligence

2088. Ackerman, Robert K. "Army Intelligence Incorporates Iraq Lessons." *Signal* 59, no. 2 (October 2004): 16–18.

Presents a highly positive view of military intelligence lessons learned. There has been a pattern of rapid and continuous adaptation. This is a detailed account of the blending of people and technology in intelligence operations.

2089. Erwin, Sandra I. "Information Technology: Army to Deploy Web Based Intelligence Network." *National Defense* 90, no. 623 (October 2005): 46+.

There has been no centralized source of information about individual insurgents and suspected insurgents. Now, a Joint Intelligence Operations Capability is being constructed. There is much data available, but it is not being disseminated properly. Some are skeptical about the program, however.

K. Prisons and prisoners

2090. Danner, Mark. *Torture and Truth: America, Abu Ghraib, and the War on Terror.* New York: New York Review Books, 2004.

This is an extensive analysis of the role of torture in the global War on Terrorism. Of great importance are the appendices, which offer a wide variety of documents. Many of those documents issued from United States political and military leaders. There are also a number of reports by investigating bodies, including the International Red Cross.

2091. Dunlap, Charles J. "Learning from Abu Ghraib: The Joint Commander and Force Discipline." *United States Naval Institute Proceedings* 131, no. 9 (September 2005): 34–38.

Stresses the fact that the Abu Ghraib scandal made the United States effort in Iraq much more difficult. Joint force commanders must ensure that actions by their men and women do not give our opponents opportunities to wage what Dunlap calls "lawwar" through the media to discredit the entire United States commitment. Dunlap warns that discipline is an issue that extends beyond the service branches composing a joint force. Thus, discipline must be a major concern for the joint force commander.

2092. Greenberg, Karen J., and Joshua L. Dratel. *The Torture Papers: the Road to Abu Ghraib.* Cambridge: Cambridge University Press, 2005.

This is a larger collection of documents than Danner's volume (item 2090), but it lacks the long analysis and some other features of that book. This collection provides much information on how administration officials and advisers arrived at their decision to treat captives outside the protection of the Geneva Conventions.

2093. Houlgate, Kelly P. "Detention Center Conviction Highlights Supervisory Deficiencies." *United States Naval Institute Proceedings* 131, no. 2 (February 2005): 2.

Discusses the court-martial of a marine officer held responsible for the mistreatment of prisoners held at a camp in Iraq. This is a rather detailed analysis of the case. Ultimately, Houlgate argues that the "men of the detention center simply were unqualified for their tasks."

2094. Mestrovic, S.G. *The Trials of Abu Ghraib: An Expert Witness Account of Shame and Honor.* Boulder, CO: Paradigm Publishers, 2007.

Strongly asserts that the real culprits in the scandal were high-ranking military and civilian authorities rather than the perpetrators of the abuses at the prison. The author reviews several of the major trials in which he participated as a witness.

2095. Trapp, Brian. "A Better Place for Detainees." *Soldiers* 60, no. 5 (May 2005): 14–17.

Describes extensive efforts to improve a prisoner facility called Camp Bucca, which was needed for holding some suspected insurgents. It was more secure than Abu Ghraib, which was a frequent target for attacks. The article includes much detail about the processing of detainees.

L. Military advisory activities

2096. Davidson, Brian. "New Iraqi Airmen Stage First Operational Airlift Mission." *Mobility Forum* 14, no. 5 (September–October 2005): 10–13.

The first Iraqi airlift mission was relatively simple, moving equipment from Baghdad to Basrah for the Ministry of Electricity, but it was executed with little participation by United States personnel. Moreover, it was an early operation staged from the new Al Muthana Air Base.

2097. Erwin, Sandra I. "Complex Realities Lie Behind U.S. Rush to Train Iraqi Army." *National Defense* 90, no. 626 (January 2006): 13.

Discusses the difficulties in the development of an Iraqi force that can replace United States personnel. The Bush administration's *National Strategy for Victory in Iraq* offers many ways to measure Iraqi military effectiveness but does not present any time-frame. Erwin quotes a number of officers and defense analysts regarding challenges in building Iraqi forces.

2098. Martin, John R. "Training Indigenous Security Forces at the Upper End of the Counterinsurgency Spectrum." *Military Review* 86, no. 6 (November/December 2006): 58–64.

Usefully pinpoints two different types of insurgents and their tactics. Thus, efforts to train Iraqi forces had to be kept flexible as the security environment changed. Includes a very helpful discussion of training for police. This is an excellent presentation of the problems faced in training indigenous forces.

2099. Scales, Robert H. "The Willpower Quotient, SITREP Iraq." *United States Naval Institute Proceedings* 131, no. 10 (October 2005): 21, 23.

Asserts that Iraqi security forces are expanding rapidly and are improving their capability. Scales usefully discusses the importance of the Iraqi police in fighting the insurgency.

2100. Steele, Dennis. "The CSM: Training and Empowering NCOs." *Army* 55, no. 9 (September 2005): 29–30+.

United States military personnel and selected Iraqis have had to build a training process for developing NCOs for the Iraqi Army, which, under Saddam Hussein, did not have significant roles for its NCOs. Steele uses Command Sergeant Major Hassan Abdul Fadum Radi as a model of what Iraqi NCOs should be.

2101. ———. "Gutting It Out." *Army* 55, no. 9 (September 2005): 34–36.

Praises Iraqi soldiers as they pursue their training. In the case of an Iraqi company he studied, United States advisers had to provide what amounted to basic training. Morale remains high despite extremely poor living conditions and dangers when Iraqi soldiers are out with passes.

2102. Sullivan, Mike. "From the Ashes: Rebuilding the Iraqi Army." *Armor* 114, no. 4 (July–August 2005): 44–47.

Discusses Sullivan's experiences in helping train an Iraqi infantry battalion. A 10-man team consisting mainly of NCOs went from Germany to Kirkush, Iraq, in 2004 to train Iraqi officers and NCOs, who would then train lower-ranking enlisted men. Supply shortages were among the many problems encountered.

2103. Trowbridge, Gordon. "Building an Iraqi Army." *Armed Forces Journal* 143, no. 6 (January 2006): 26–28.

Candid picture of the strengths and weaknesses of the new Iraqi Army. Trowbridge emphasizes the absence of a viable supply system in the Iraqi force and the pilfering by soldiers that results from that lack. Many United States officers have been highly critical of the Iraqi Ministry of Defense for its failure to provide logistical support for its field forces.

2104. Vines, John R. "The XVIII Airborne Corps on the Ground in Iraq." *Military Review* 86, no. 5 (September/October 2006): 38–46.

Examines efforts to build Iraqi security forces, which undertook more and more responsibility for security in the XVIII Airborne Corps area. Vines highlights the many deficiencies of Iraqi forces that were inherited from the old regime. Much of the article deals with improvements in the gathering and processing of information.

M. Logistics

2105. Fick, Charles W., Jr. "Army Field Support Brigades Move Out in Europe and Iraq." *Army* 55, no. 12 (December 2005): 23–24+.

These brigades are new force structures designed to provide high levels of logistical support. They are attempting to be proactive by preparing for demands well in advance of these demands arising. Every effort is made to assess requirements by communication with combat forces in Iraq. The article briefly notes the increasing use of contractors.

2106. Harding, Steve. "Movement Masters." *Soldiers* 60, no. 4 (April 2005): 8–17.

Focuses on the 7th Transportation Group, consisting of both active army and reserve personnel, that supports forces in Iraq using all forms of transportation. In 2005, over half the group was in Iraq and Kuwait. The article describes the flow of supplies from the United States through many hands to their destinations.

2107. Pappalardo, Joe. "A Single Day Changed Supply Strategy in Iraq." *National Defense* 90, no. 620 (August 2005): 60–62.

That day was April 7, 2004, when insurgents made numerous attacks on United States supply routes, targeting bridges in particular. The Iraq War has become a major challenge to United States logistical doctrine as Iraq's highways have become battlegrounds. More expensive, decentralized supply seemed to be the answer to insurgent tactics. The adaptability of the insurgents has to be matched by the United States.

N. Civil affairs/civic action

2108. Chandrasekaran, Rajiv. *Imperial Life in the Emerald City: Inside Iraq's Green Zone*. New York: Alfred A. Knopf, 2006.

Chandrasekaran, a *Washington Post* editor, who was in Iraq between 2002 and 2004, describes the recruitment of the United States staff that administered the occupation and their activities once in Iraq. There is much detail about bureaucratic conflicts, including Department of Defense antagonism toward the Department of State, and about the attitudes and policies of U.S. ambassador Lewis Paul Bremer III.

2109. Davids, Douglas J. "Shaping the Conditions for a Post Conflict Environment," pp. 103–25 in *Warfare in the Age of Non-State Actors: Implications for the US Army, The Proceedings of the Combat Studies Institute 2007 Military History Symposium*, ed. Kendall D. Gott and Michael G. Brooks. Fort Leavenworth, KS: Combined Arms Center, Combat Studies Institute Press, 2007.

Extremely critical of administrator Lewis Paul Bremer's handling of the Iraq situation, especially the demobilization of the Iraqi Army and the purging of the Iraqi government of Ba'ath party members. Davids skillfully employs historical precedents in his critique of Bremer's policies. In addition to being an important discussion of the early period in Iraq from the perspective of earlier United States experiences with post-conflict situations, Davids' article includes insightful comments about recent army doctrine.

2110. Drumsta, Raymond. "A Friendly Patrol." *Soldiers* 60, no. 7 (July 2005): 12–13.

Reviews the work of the 411th Civil Affairs Battalion, which was tasked with a wide variety of assignments, such as supporting public health services, aiding in the establishment of local elections, and working on school repairs. Emphasis is being given to helping Iraqis manage their own affairs.

2111. Gordon, Jack. "New Beginnings in Iraq." *Soldiers* 60, no. 4 (April 2005): 18–22.

Focuses on a public-safety team led by a captain from the 416th Civil Affairs Battalion, an army reserve organization, in the Mosul area working in full cooperation with Iraqi authorities. A number of problems were encountered, including fire-fighting and prison administration.

2112. Gould, Nani, and Fauwaz Hanbali. "Army Engineers Develop Model to Support Iraqi Water Management." *Engineer* 34, no. 4 (October–December 2004): 10–11+.

A civilian employee of the United States Army Corps of Engineers and a consultant to the army discuss water management in the large marsh areas of Iraq. Under Saddam Hussein water management was far behind the practices used in many countries. The authors discuss use of computer technology in developing a new approach, which is gaining favorable attention well beyond Iraq.

2113. ———. "Army Engineers Help Restore Iraqi Wetlands." *Engineer* 34, no. 4 (October–December 2004): 9.

Programs undertaken in marsh areas of Iraq by Saddam Hussein's regime made refugees of many living in these parts of the country. Significant drainage and other problems had to be overcome by the United States Army Corps of Engineers and the Agency for International Development to assist Iraqis to return to their homes. Reservoirs are essential for the effectiveness of the new system of water control.

2114. Kimmons, Sean. "Soldiers Help Preserve Archeological Sites." *Engineer* 34, no. 4 (October–December 2004): 4.

Archeological finds are being made by United States forces in the course of their operations in Iraq. When they are located, military personnel from civil affairs units who have appropriate backgrounds ensure that the sites are properly preserved and that Iraqi specialists are involved in these efforts.

2115. Lacey, James G. "Sense and Nonsense (the Iraq Economy in War Time)." *United States Naval Institute Proceedings* 131, no. 6 (June 2005): 32–36.

Stresses the importance of economic factors in stabilizing Iraq. Lacey derides the approaches of many economists and seeks to state the basic facts, as he sees them, of the economic situation in Iraq. To Lacey, "economic growth equals victory" over the insurgents (p. 33). He discusses the underground economy and recommends that Iraq default on all its debts and that property rights be firmly established.

2116. Mateo, Lek. "Making a Difference in Iraq." *Soldiers* 60, no. 5 (May 2005): 8–11.

Experiences of an artillery captain who moved to civil affairs duties to work on such projects as building schools. Given his Sudanese origin and his knowledge of Arabic, the officer was able to communicate and work effectively with Iraqis. Mateo notes more briefly an NCO's duties in maintaining road security in Iraq.

2117. Miller, T. Christian. *Blood Money: Wasted Billions, Lost Lives, and Corporate Greed in Iraq.* New York: Little, Brown and Company, 2006.

Scathing attack on what the author views as the Bush administration's colossal mishandling of reconstruction. Corruption is said to have overshadowed any positive steps toward rebuilding the country.

2118. Runnels, Al, and Wesley Miller. "Bankrolling the Rebuilding of Iraq." *Armed Forces Comptroller* 50, no. 3 (Summer 2005): 34–35.

Electronic fund transfers were not generally feasible in wartime Iraq. A considerable amount of adaptability was required for successful financial operations, especially during the currency reform that was implemented.

O. Information programs/psychological operations

2119. Freeburg, John, and Jess T. Todd. "The 101st Airborne Division in Iraq: Televising Freedom." *Military Review* 84, no. 6 (November–December 2004): 39–41.

Describes efforts of the 318th Tactical PSYOP [psychological operations] Company (Reserve) and elements of the 101st Airborne Division to deal with the challenges of providing information to the Iraqis. Although posters and handbills were used to communicate with the public, newspapers and television were the key media used. Much was done to rebuild the Iraqi media, but care had to be taken to avoid the perception that they were simply tools of the United States.

2120. Grier, Peter. "The Military Meets Madison Avenue: A New RAND Study Argues the Pentagon Could Use Some Salesmanship." *Air Force Magazine* 90, no. 11 (November 2007): 58–60.

Notes the obstacles to employing United States advertising techniques for winning local support for the counterinsurgency in Iraq, but usefully summarizes a RAND report recommending such an approach. The report proposes "branding" the army and marines as part of the effort to persuade many Iraqis to change their behavior in the direction of supporting United States goals. United States personnel, moreover, must live up to the "brand" in day-to-day encounters with the Iraqis.

P. The media and the war

2121. Conway, James T. "Tell It to the Troops, SITREP Iraq." *United States Naval Institute Proceedings* 131, no. 10 (October 2005): 20, 22.

Deplores the pessimistic view the media often take toward the struggle in Iraq and asserts that participants are much better able to assess the situation than the media. Conway points to continued successes, such as the Iraqi elections, and asserts that United States forces are protecting their homeland from terrorism by fighting in Iraq. Presents a highly optimistic view of the Iraq situation. Foreign troop morale is high. A key factor is public opinion in the United States.

2122. Mitchell, Greg. *So Wrong for So Long: How the Press, the Pundits—and the President—Failed on Iraq*. London: Union Square Press, 2008.

Despite the title, the book is largely an indictment of the media for the caution they have shown in reporting events in Iraq. Mitchell, editor of *Editor and Publisher*, seemingly directs as much criticism to "liberal" sources, such as *The New York Times*, as to the Bush administration. The book is enriched by the frequent references to media coverage of developments in Iraq in other countries.

2123. Stephenson, Scott. "The Embeds' War: Review Essay." *Military Review* 85, no. 3 (May–June 2005): 96–97.

Credits the use of embedded correspondents with the reduction of distance between the military and the media.

Q. Terminating the war

2124. Arnove, Anthony. *Iraq: The Logic of Withdrawal*. New York: New Press, 2006.

Socialist perspective on the invasion of Iraq and the subsequent occupation. Arnove argues that the desire for an easily controlled source of oil caused the Iraq War and that the behavior of United States personnel inside and outside Iraq has been based on highly racist assumptions. Ultimately, he blames the theory of the "exceptionalism" of the United States that is often used to exempt this country from international standards of conduct.

2125. De Atkine, Norvell B. "Iraq Withdrawal—A Tragedy in Slow Motion." *United States Naval Institute Proceedings* 132, no. 3 (March 2006): 12–15.

Advises President Bush not to authorize premature withdrawal of forces from Iraq. If troops are pulled out, Iraq will gradually deteriorate. De Atkine probes the complicated religious and ethnic composition of Iraq and the difficulty of reaching stable conditions. A sudden withdrawal may well discredit democracy in the Middle East.

2126. Fettweis, Christopher J. *Losing Hurts Twice as Bad: The Four Stages to Moving Beyond Iraq.* New York: W.W. Norton & Company, 2008.

Controversial and highly stimulating discussion of how the United States may adjust to what Fettweis views as a certain defeat in Iraq. Offering many comparisons between Iraq and earlier conflicts, especially the United States experience with Vietnam, Fettweis recommends the healthiest and most constructive ways for dealing with the loss of another war by the United States. This is a daring work. Fettweis argues strongly that there would be hardly any ill effects if the United States withdrew from Iraq unilaterally and almost immediately.

2127. Gold, Philip. "To Guard an Era: American Purpose After Iraq." *United States Naval Institute Proceedings* 131, no. 9 (September 2005): 18–22.

Tries to unravel the loose thinking about the military challenges facing the United States by offering something like George F. Kennan's article about a "containment" policy regarding the Soviet Union. Gold believes that there is a worldwide divide between people who embrace the twenty-first century, those who reject it, and those who want to be admitted to it.

2128. Hinote, Clint. "The Drawdown Asymmetry: Why Ground Forces Will Depart Iraq But Air Force Will Stay." *Strategic Studies Quarterly* 2, no. 2 (Summer 2008): 31–62.

Argues that discussions of United States forces in Iraq and their size have overemphasized land components and that air assets must remain even as army and marine elements are drawn down. Limited numbers of personnel available in both those service branches are putting pressure on the army and marines to reduce their presence. Some Coalition forces are necessary, however, to maintain stability in Iraq, and air elements must support them through both airlift and combat capability.

2129. Korb, Lawrence, and Brian Katulis. "Strategic Redeployment." *Armed Forces Journal* 143, no. 6 (January 2006): 30–33.

Thoroughly pessimistic article about the ability of the United States to keep large numbers of troops in Iraq with the All-Volunteer Force. The authors strongly recommend taking United States troops out of all urban locations and starting redeployment to reduce troop levels drastically by the end of 2007. The authors argue that diplomacy must play a greater role in United States strategy for the Middle East.

2130. Matthews, William. "Exit Strategy in Iraqi Elections: Congress Sees Opportunity to Withdraw Troops." *Armed Forces Journal* 143, no. 5 (December 2005): 10–12.

A successful democratic election in Iraq will assist the Bush administration, according to members of Congress who favor the relatively rapid withdrawal of United States forces. Matthews discusses at length statements by Rep. Walter A. Jones, a Republican from North Carolina, and other members of Congress about the effect of elections in Iraq.

2131. Robinson, Linda. *Tell Me How This Ends: General David Petraeus and the Search for a Way Out of Iraq.* New York: Public Affairs, 2008.

Highly readable, journalistic account of what may be the latter stages of the Iraq War and Petraeus' approach to the conflict. The book includes some material about the development of a new statement of doctrine concerning the United States approach to counterinsurgency, Field Manual 3–24.

Author index

This index includes names of authors, co-authors, editors, and compilers. Institutions, associations, and other "corporate authors" are not listed.

A

Ackerman, Carl W. 827
Ackerman, Robert K. 1766, 1864, 2088
Adams, Cyrus C. 572
Adams, Raymond E. 1062
Adams, Terrence A. 1911
Adams, Thomas K. 639, 1696, 1701
Adkin, Mark 1455
Agnew, James B. 298
Aguirre, Cesar Reyes 621
Agwani, M.S. 1162
Ahern, Neal J. 1063
Ainsworth, Troy 585
Al-Aiban, Bandar Mohammed 1163
Al-Baddawy, Alison 1788
Albertson, Ralph 798, 815
Alderson, Alexander 1749
Alfieri, Paul 2024
Alfonsi, Christian 1984
Aliano, Richard A. 1054
Alin, Erika G. 1164
Allard, C. Kenneth 1645.
Allawi, Ali A. 1971
Allen, Inez V. 549
Allman, T.D. 1445
Altieri, Jayson A. 163
Alywin-Foster, Nigel R.F. 2066
Amory, John Forth 1055
Amouzegar, Mahyar A. 1952
Ancker, W.M. 741
Anderson, Charles R. 929
Anderson, Eddie 1083
Anderson, Jon Lee 2060
Anderson, William T. 1383
Andrade, Dale 1839
Andrews, William F. 1563
Angeles, Ren 2025
Aplington, Henry II 1153, 1156
Archer, Jules 526
Armstrong, Charles L. 1277–78, 1750
Arnove, Anthony 2124
Ashby, Timothy 1447
Ashton, N.J. 1187
Asprey, Robert R. 1064
Astor, Gerald 988
Atkin, Ronald 527
Atkinson, Rick 2026
Auletta, Anthony J. 1117
Avallone, Paul 1953
Axe, David 2081–82
Axelrod, Alan 71, 72

B

Bacevich, Andrew J. 249, 936, 1534–35, 1613
Bacon, Eugene Howard 778
Bahnsen, Peter 1279
Bailey, Timothy J. 1866
Bain, David Haward 199
Bainbridge, Mrs. William E. 314
Baker, C.S. 663, 930
Baker, Caleb 1492
Baker, Jim 1767
Balch, Emily Greene 640
Baldridge, Elward F. 1196
Ballard, John R. 1710, 2062
Bamford, James 1985
Banister, W.B. 353
Barber, Willard F. 1118
Barela, Timothy P. 1668
Barnard, Roy S. 7
Barnes, Rudolph C., Jr. 1280–81, 1391
Barnet, Richard J. 989, 1165, 1206, 1448
Barnett, Gary G. 1734
Barrett, Drew J., Jr. 73
Barrett, Raymond J. 990, 1380
Barrow, Clay 1084
Barry, Richard 261
Bashore, Boyd T. 1085
Bass, Jeff D. 1207
Bateman, C.C. 262–63, 264
Batiste, John R.S. 2027
Baughman, C.C. 758
Baumann, Robert E. 1711
Baumer, William H. 86, 304, 533, 795
Baylen, Joseph L. 845–46
Beale, Michael 1693
Beasdale, Victor F. 888
Beaumont, Roger A. 991
Beavers, Garry J. 1698
Beck, Robert J. 1449–50
Beebe, John E. 960
Beebe, Kenneth 1751
Beede, Benjamin R. 74
Beisner, Robert L. 47, 235
Belfrage, Cedric 860
Bell, J. Bowyer 48, 75, 1405
Bell, Raymond A. 1473
Bellamy, Christopher 1536
Bellflower, John W. 1828
Bellinger, John B., Jr. 1119
Belton, Fred 666, 703, 704
Bennett, Ralph Kinney 1451

Bennett, W. Lance 1620
Bensahel, Nora 2007
Bentley, Christopher F. 1912
Bentz, Harold F., Jr. 1120
Berger, Paul A. 1242–44, 1246
Berle, Adolf A. 788, 1209
Betancourt, Alberto 1913, 1946
Betts, Richard K. 76
Bevan, James 327, 334
Beverage, Harold G. 1282
Bevilacqua, A.C. 1386
Bickel, Keith B. 117
Bieber, Florian 1721
Bigge, Lauren 1735
Biggs, Chester M., Jr. 335, 403
Bin, Alberto 1537
Birnbaumer, F.F. 483
Birtle, Andrew J. 118, 200, 973
Bishku, Michael B. 1188
Bjelajac, Slavko N. 992
Black, Edwin F. 993
Blackwell, James A. Jr. 1547
Blair, Arthur H. 1532
Blair, Edison T. 1065, 1111
Blaker, James R. 1599
Blanchard, George S. 1066
Blant, Stephen 1353
Blassingame, John W. 664
Blatt, Darin J. 1914
Blaufarb, Donald S. 994
Blechman, Barry, 77, 1505
Bliese, Paul D. 1515
Blix, Hans 1986
Bloomfield, Lincoln P. 995
Blumenson, Martin 579
Bodron, Margaret M. 1178
Boehnke, E.C. 362
Bolduc, Donald C. 1867
Boler, Joseph W. 1945
Bolger, Daniel P. 1458
Bolling, Landrum R. 1271
Bond, Peter A. 1283
Book, Elizabeth G. 1846
Booker, Simeon 1485
Bookout, Jack Eugene 462
Boot, Max 78, 164, 299, 356, 528, 603, 641, 793, 847
Booth, Waller B. 961
Borch, Frederic L. 1247, 1459, 1498, 1616–17, 1676, 1714
Borg, Dorothy 372
Bourque, Stephen A. 1580
Bowman, Stephen Lee 974
Boyden, Hayne D. 753–54
Boyer, Douglas A. 1987
Boyle, Stephen P. 2053
Boyne, Walter J. 2008
Braddy, Haldeen 529
Brady, John D. 712
Brady, Robert G. 1392

Braestrup, Peter 1203
Brainard, Edwin H. 905–6
Braisted, William Reynolds 300, 373, 839
Brand, David L. 1654
Brands, H.W. 996
Brask, James J. 975
Bratton, Joseph K. 997
Brennan, Richard R. 2007
Bresnahan, Roger G. 236
Bride, Frank L. 682
Briggs, Clarence E. III 1499
Briscoe, C.H. 1954, 2028
Broderick, J.M. 865
Brodhead, Frank 1217
Brookes, Peter 1868
Brooks, David C. 866
Brown, Campbell H. 143
Brown, Jason M. 1829
Brown, John Clifford 201
Brown, John S. 1581
Brown, Kendall 1840
Brown, Meredith 202
Brown, Richard M. 1252
Brown, Robert B. 1811
Brown, Ronald J. 1588–89
Brown, Seyom 1208
Brown, Stephen D. 1715
Brown, Todd S. 2029
Brownell, Atherton 265–66
Brune, Lester H. 1506, 1634, 1681, 1702
Bruscino, Thomas A., Jr. 490
Bruss, Donald J. 1393
Bruton, James K. 1284
Bryson, Paul J. 1654
Bryson, Thomas A. 1197
Bubur, K.F. 681
Buchanan, William J. 998–99
Buckley, Kevin 1486
Buckner, David N. 404, 678, 742
Buell, Raymond Leslie 937, 953
Bufkin, Todd II 1901
Bullard, Robert L. 119, 267–68, 451
Bum-Joon, Lee Park 51
Bundt, Thomas S. 165
Burdett, Thomas F. 166
Burgess, Mark 1869
Burgess, Richard R. 1893, 2030
Burgess, Susan Renee 79
Burk, James 1507
Burrowes, Reynold A. 1436
Burt, Richard 49
Burton, F. 405
Burton, John T. 1697
Buskirk, Lawrence E. 1121
Butler, Keith D. 1659
Butler, Smedley D. 324, 337, 406–9, 463–64, 604, 642, 667
Buzzell, Colby 2031
Byler, Charles 250
Byman, Daniel L. 1988

C

Cabell, De R.C. 328
Cable, James 80
Cable, Larry E. 976
Cahill, Howard F.K. 363
Calder, Bruce J. 728
Caldwell, Ruth A. 1109
Calhoun, Christopher S. 1198
Calhoun, Frederick S. 81, 530, 605, 643, 729, 779
Callcott, Wilfrid Hardy 441, 693, 730, 772
Campbell, H. Denny 141
Campbell, Kenneth J. 1508
Canan, James W. 1460
Caraccilo, Dominic J. 1582
Carafano, James Jay 1768
Caravella, Frank J. 1694
Cardillo, John A. 163
Carey, John 1209
Carlisle, Rodney P. 1972
Carlson, Evans F. 357, 411–12, 921
Carlson, Marnee 1894
Carmony, Donald F. 212
Carney, John T., Jr. 1260, 1461, 1500, 1604, 1669, 1716
Carney, Madison M. 1945
Carns, Edwin H.J. 1394
Carr, Darrell 1915
Carstens, Roger D. 1989
Carter, C.B. 922
Carter, William Harding 301
Case, Frank L. 583, 587
Casmus, David M. 1463
Casper, Lawrence E. 1285, 1660
Cassidy, Robert M. 977, 1509, 1635, 1794–95
Castle, William R., Jr. 82
Castleman, James T. 519
Caterini, Dino J. 167
Celeski, Joseph D. 1661, 1752, 1812
Chalk, Peter 1805
Chamberlain, Robert L. 1736
Chamberlain, Robert M. 1769
Chamberlain, Weston P. 501
Chandrasekaran, Rajiv 2108
Chapman, Charles E. 442
Chapman, Craig S. 1583
Chapman, Gregory 203
Chaput, Donald 204, 269
Chern, Kenneth S. 1147
Chester, Eric Thomas 1210
Chew, Peter T. 1248
Chollet, Derek 1699
Chomsky, Noam 1555
Church, George J. 1423
Church, Iain J. 1813
Churley, Robert A. 1135
Clagett, John H. 777
Clancy, Tom 1564
Clark, George B. 338, 414, 867, 1134
Clark, Wesley K. 1722, 2009

Clay, James P. 396
Clements, Frank A. 1852
Clendenen, Clarence C. 491, 531
Cliff, Andrew D. 232
Clingham, James H. 1249
Clintron, David 1947
Coats, George Yarrington 145
Coble, Donald W. 1086
Coburn, Matthew D. 1870
Cochrane, James A. 1034
Cockburn, Patrick 1973
Codevilla, Angelo M. 1286
Coerver, Don M. 492–93
Coffey, R.B. 694
Coffman, Edward M. 364
Cohen, Eliot A. 1565, 1753
Coker, William S. 614
Colbourn, Colin 1136
Colby, Elbridge 83, 588
Cole, Bernard D. 358
Cole, C.E. 931
Cole, Ronald H. 1261, 1462 1491
Coletta, Paolo Enrico 50
Collins, James L. 270, 1463
Collins, John M. 84. 1611, 1682
Collins, Joseph J. 1871
Comer, Richard 1830
Condit, D.M 51
Cone, Robert W. 1814
Conley, Michael C. 1000
Connaughton, Richard 1510, 1636, 1727
Conner, Carol 1916
Conrad, Charles 705
Conway, James T. 2010, 2121
Cooke, Leonard W.W. 1966
Cooling, Norman L. 1646
Cooper, David E. 1849
Corbin, Marcus. 1872
Corcoran, Edward Francis 978
Cordesman, Anthony H. 1730, 2011
Corn, David 1998
Cornebise, Alfred E. 366
Cosmas, Graham A. 743
Costello, Harry J. 799
Costello, Thomas J. 1287
Cotten, Lyman A. 624
Cox, Isaac Joslin 848
Cox, Ronald 1703
Coyle, Randolph 713
Craig, Edward A. 557
Craige, John H. 695
Cramer, Stuart W. 532
Crandall, Russell 1001
Crane, Conrad C. 1753–54
Crane, Keith 2007
Crawford, James Grant 168
Crawley, Eduardo 849
Creighton, Neal, Jr. 1587
Crockett, Cary I. 271
Croot, Edward C. 1873

Cropsey, Seth 1770
Crossland, Roger L. 1369
Cruikshank, Jeffrey L. 1606
Cudahy, John 800
Culp, Robert A. 1737–38
Cumberland, Charles C. 502
Cureton, Charles H. 1590
Currey, Cecil B. 1288
Curriden, Lorrie Lynn 1612
Curry, Earl H. 764
Curtis, Donald John, Jr. 817
Cusick, John J. 1464

D

D'Aria, Dorian 1815
Daalder, Ivo H. 1700, 1990
Daggett, Aaron S. 302
Dallam, Samuel F. 574
Daly, Terence J. 1771, 2067
Daniel, Douglas W. 1717
Daniels, John A. 868
Daniels, Paul R. 2027
Danis, Jan Shelton 2
Danner, Mark 2090
Dare, James A. 1240
Darley, William M. 1874
Darling, Roger 1289
Darragh, Shaun M. 1362
Daskalovski, Zidas 1721
Davids, Douglas J. 2109
Davidson, Brian 2096
Davidson, Scott 1479
Davidson, W.C. 303
Davis, Anthony 1917
Davis, Donald E. 789
Davis, H.P. 644
Davis, Henry C. 142
Davis, Oscar K. 272, 329–30
Davis, Richard G. 1566–67
Davis, Robert B. 1387
Dawson, David A. 1592
De Camp, William T. 1480
Deady, Timothy K. 169, 1796
Dean, David J. 1290–91, 1301, 1334, 1340, 1344, 1363, 1368, 1370, 1375, 1389, 1393
DeAtkline, Norvell B. 1538, 2125
Decker, David A 1395
Decker, William T. 962
Deitchman, Seymour J. 1056
Del Valle, Pedro A. 129
Delaney, Robert Finley 1002
Delgadillo, Roberto Carlos 850
Demarest, Geoffrey B. 1816
Denig, Robert L. 923, 932
Denny, Harold Norman 851
Denton, Robert E., Jr. 1621
Desjarlais, Orville F., Jr. 1955
Diamond, Larry 1974
Dienna, Nicholas 1901
Dickman, Joseph T. 331

Diokno, Maria Serena I. 237
Di Sisler, James B. 1901
Divine, Robert A. 1539
Dix, Billy 1363
Dobbs, Charles M. 1137
Dobie, Paul W. 1967
Dodd, Thomas J. 938
Dodds, Harold W. 939
Dodge, R.A. 716
Dodson, Charles A. 1067
Donini, Antonio 1853
Donnell, Guy Renfro 606
Donnelly, Thomas 1492, 1854, 2068–69
Donovan, Consorcia Lavadia 170
Doppler, Ryan J. 1975
Dorronsoro, Gilles 1855
Dougherty, Kevin J. 1517
Douglas, Paul H. 645
Downey, Edward F. 963
Downey, Greg 1587
Downie, Richard Duncan 85
Downs, William Brian 1755
Draper, Theodore 1211
Dratel, Joshua L. 2092
Dreyer, June 1292
Dreyer, Vincent M. 1856
Drinnon, Richard 1003
Drogin, Bob 1991
Drumsta, Raymond 2110
Dubois, Thomas R. 1556
Duffy, John J. 1122
Duncan, G.H. 706
Dungan, C. Peter 1843
Dunlap, Charles J. 1756, 1831, 2091
Dunn, Peter M. 1437
Dunn, Robert 575
Dunscomb, Paul E. 818
Dupuy, R. Ernest, 86, 304, 533, 794–95
Dworken, Jonathan T. 1677

E

Eagleton, Thomas F. 87
Eales, Anne B. 2
Easterbrook, Ernest F. 1087
Ecklund, Marshall V. 1670
Edson, Merritt A. 890
Edwards, T. C. 1088
Edwards, Warrick Ridgely, III 534
Eichelberger, K.L. 1662
Eikenberry, Karl W. 1875
Eisenhower, John S.D. 494, 607
Ekbladh, David 819
Elarth, Harold Hanne 146
Elkhamri, Mounir 2032
Ellis, E.H. 130
Ellis, E.M. 374
Ellis, L. Ethan 717, 841
Ellison, Herbert J. 1435
Ellsworth, Harry Allanson 88
Elser, Frank B. 535

Engel, Richard 1976–77
Englander, Evelyn A. 1592
Engle, Eloise 707
Eno, Russell A. 1797
Erwin, Mike 1867
Erwin, Sandra I. 1772–73, 1822, 2012, 2033, 2089, 2097
Escalante, Rene R. 238
Esherick, Joseph 305
Espinoza, Robert 1956
Estes, George H. 503
Etzold, Thomas H. 1370
Evans, David L., III 1100
Evans, Ellwood W. 561
Evans, Ernest 1293
Evans, Frank E. 124, 646
Everest, Larry 1992
Exun, Andrew 1918

F

Fabey, Michael 2034
Fabian, Robert 1568
Fairchild, David Ross 979
Fastabend, David 1518
Faust, Karl Irving 171
Fawn, Rick 1978
Fegan, J.C. 725
Feis, Herbert 1148
Fellowes, Edward A. 755
Felten, Peter Gerhard 1212–13
Fenn, Courtenay Hughes 339
Ferguson, Charles H. 1979
Ferry, Charles P. 1671
Fettweis, Christopher J. 2126
Feuer, A.B 147
Fic, Victor M. 780, 828
Fick, Charles W., Jr. 2105
Fields, Damon 1712
Filene, Peter G. 790
Filiberti, Edward J. 205
Finlayson, Kenneth 2028
Finney, Charles G. 367
Fischer, Peter G. 1914
Fishel, John T. 1713
Fisher, Albert L. 1101
Fitzgibbon, Russell H. 443
Flack, Vicki 1967
Flanagan, Edward M., Jr. 1493
Flavin, Mark J. 1464
Fleming, Lawrence J. 563
Fleming, Thomas J. 273
Fletcher, Marvin 52
Flint, Roy K. 120, 1094
Flores, Susan J. 1396
Foglesong, David S. 781
Foner, Philip S. 438
Forbes, W. Cameron 274
Foresman, Henry J., Jr. 1848
Forster, Merlin H. 608
Forsyth, Michael J. 1777

Forsyth, William W. 326
Foster, Gregory D. 1294
Fox, Frank W. J. 465, 498
Fox, Terrance Maurice 1272
Francia, Luis H. 192
Francona, Rick 1557
Frank, Benis M. 1138, 1406
Frazier, Charles E., Jr. 852
Freeburg, John 2119
Freeman, Joseph 101
Friedman, Norman 1540, 1876, 1895, 1993, 2035
Friscolanti, Michael 1896
Fritz, David L. 206, 240
Fromm, Joseph 1424
Fuller, Stephen N. 743
Fulton, John S. 1295

G

Gaddo, Randy 1407
Gaddy, G. Dale 653
Gaillard, Regina 1397
Galbraith, Peter W. 1980
Gale, Esson M. 375
Gamble, Krishna M. 1940
Ganley, Eugene E. 207
Gann, Lewis H. 1004
Gantz, Walter 1622
Garcia, Elroy 1647, 1663, 1672
Garcia, Rogelio 954
Gardiner, J.B.W. 796
Gardner, K.N. 376
Gardner, Lloyd C. 1994
Gates, John Morgan 172–75, 241, 1296
Gatewood, Willard B. 246
Geary, James W. 153
Geibel, Adam 1919
Geis, Robert K. 51
Gellman, Brian 1817
Gendzier, Irene L. 1166
Gentile, Gian P. 1757
Gentry, John A. 1511
Gentry, Keil R. 2063
George, David 1952
Ghareeb, Edmund 1545
Gibean, Victor Hugo, Jr. 773–74
Gibson, Chris 2070
Giffin, Frederick C. 829
Gifford, Robert K. 1515
Gilboa, Eytan 1487
Gilderhus, Mark T. 536
Gill, Burgo D. 121
Gillert, Gustav J. 1005
Gillett, Mary C. 231
Ginsburgh, Robert N. 176
Girard, Philippe R. 1704
Girling, John L.S. 1297
Glantz, Aaron 1981
Gleijeses, Piero 1215
Glenn, Russell 1673

Glick, Edward Bernard 1123
Gobat, Michel 466
Goede, Peter 1234
Gold, Philip 2127
Goldberg, Harold J. 782
Goldenberg, Richard 2036
Goldhurst, Richard 783
Goldoni, John E. 1089
Goldwert, Marvin 756, 924
Goodman, Allan E. 1298
Goodson, Larry P. 1857
Goodwin, Paul B. Jr. 1440
Gordon, Dennis 804
Gordon, Jack 2111
Gordon, John 1729
Gordon, Michael R. 1541, 2013
Gosnell, P. Wayne 1398–99
Gould, Nani 2112–13
Gourley, Scott R. 1897
Gowing, Peter Gordon 275
Graham-Brown, Sarah 1995
Granstaff, Bill 1262
Grant, Greg 2037–38, 2071
Grant, Rebecca 1898–99
Grant, Thomas Alexander 980
Grau, Lester W. 1816
Graves, William S 836
Gray, Colin S. 1057
Gray, David W. 1179
Gray, John A. 144, 668–69, 870
Grdovic, Mark 1774
Greathouse, G.H. 670
Green, Dan 1775
Greenberg, Bradley S. 1622
Greenberg, David Robert 1006
Greenberg, Karen J. 2092
Greenberg, Lawrence M. 1250
Greene, Joseph I. 271
Greene, Wallace M. 415
Greenwood, John 53
Greer, Virginia L. 484
Gregg, Heather S. 2007
Gregson, Wallace C. 1299
Grieb, Kenneth J. 615, 718, 765
Grier, Peter 1900, 2120
Griffin, Christopher 1877
Grinter, Lawrence E. 1300, 1353
Grotto, Andrew 1776
Grover, David H. 392
Grubbs, Lee K. 1777
Gruening, Ernest H. 647, 719–20, 731
Gulliver, Louis J. 397
Guo, Xixiao 1139–40
Gurtov, Melvin 1007
Guttieri, Karen 1008

H

Haave, Christopher E. 1731
Habib, Douglas F. 732
Hadd, H.A. 1189, 1204

Haffner, L.E. 1009
Hagedorn, Hermann 251
Hahn, Peter L. 1190
Hale, Jerold L. 1628
Haley, P. Edward 499, 537, 616, 1216
Hall, Clark 1948
Hall, David K. 77
Hall, David Locke 1263, 1408, 1481
Hall, Linda B. 492–93
Hall, Steven R. 853
Hallenbeck, Ralph A. 1409
Halliday, E.M. 805
Hallion, Richard P. 1569
Hamby, Larry B. 1301
Hamerman, Frederick 84
Hamilton, Louis McLane 131
Hamlett, Barksdale 1068
Hammel, Eric M. 1410
Hammes, Thomas X. 1778
Hanbali, Fauwaz 2112–13
Hanna, Matthew Elting 452
Hanneken, Herman H. 925
Harding, Steve 2106
Hardy, Earl B. 770, 940
Hargreaves, Reginald 340
Harned, Glenn M. 1519
Harp, Michael A. 2046
Harper, Gene 1465
Harper, Gilbert S. 1466
Harper, Sue 1949
Harrington, Samuel N. 132
Harris, Albert E. 964
Harris, Charles H. III 504
Harris, Harold D. 891
Harris, Larry A. 521
Harris, Richard Boyd 1010
Harrison, Benjamin 467
Harrison, Marshal Tony 1858
Hart, Douglas 1823
Hartness, William M. 1011
Hartzog, William W. 1504
Hasenauer, Heike 1657
Haun, Phil M. 1731
Hawkins, Tammy 1428
Hawkins, William R. 1878
Hayden, H. Thomas 1779
Hayes, Geoffrey 1859
Hayes, Harold B., III 830
Hayes, Stephen F. 1996
Head, William 1542
Healy, David 696
Heard, Bill 2053
Heilbrunn, Otto 1012–13
Heinl, Nancy Gordon 649
Heinl, Robert Debs, Jr. 341–42, 468, 649,
    671, 744, 871, 965
Hendrickson, David C. 1982
Hendrickson, Ryan C. 1512, 1637, 1683,
    1705, 1723
Henig, Stanley 1726

Hennessy, Juliette A. 580
Henriksen, Dag 1732
Henriksen, Thomas H. 1684
Heritage, Gordon W. 907
Herman, Edward S. 1217
Hernandez, Prisco R. 1879
Herspring, Dale R. 1411, 1467, 1558, 1706
Hess, Gary R. 1014
Hessler, William H. 1015
Hibschman, Harry J. 816
Hickey, Christopher 2039
Hicks, Melinda M. 496
Higham, Robin 54, 616, 1216
Hilburn, Matt 1780, 2040
Hill, Howard C. 444
Hill, Richard 1537
Hill, Roscoe R. 485
Hill, Walter N. 672
Hillen, John 1781
Hinckley, Ted C. 617
Hines, Charles A. 1364
Hines, Richard K. 208
Hinnebusch, Raymond 1978
Hinote, Clint 2128
Hippel, Karin von 1638, 1685, 1707
Hirsch, Ralph 133
Hirschfeld, Burt 306
Hitchcock, Norman E. 1302
Hitt, Parker 276
Hittle, James D. 1154
Hobbs, Horace P. 277
Hodges, Donald C. 872–73
Hoe, Susanna 315
Hoffman, Frank G. 1648, 1782–83
Hoffman, Jon T. 650, 874–75
Hoganson, Kristin L. 247
Holbrook, Francis X. 343
Holli, Melvin G. 209
Holliday, Sam C. 1016
Hollis, Mark A.B. 1674
Hollis, Patrecia Slayden 2072
Holly, Alonzo Potter Burgess 721
Holmes, David R. 1400
Holmes, Maurice G. 876
Holmes, Rob 1587
Holmes, Stephen 1997
Hooker, Richard D., Jr. 1453
Hopper, James 505, 522
Hornbeck, Donald 1664
Horner, Chuck 1564
Horvath, Jan 1753
Houchins, Lee Stretton 1141
Houlgate, Kelly P. 2093
House, Art 1468
Houston, Donald E. 590
Howe, Jerome W. 538
Howell, Charles F. 947
Howell, Glenn 377–81
Howland, H.S. 279

Howland, Harold J. 445
Huchthausen, Peter 1264, 1412, 1469, 1494, 1543, 1639, 1686–88, 1728
Huebner, Michael F. 1394
Hughes, Solomon 1983
Hughes, Wayne P. 1482
Huidekoper, Frederic Louis 307
Hunt, Geoffrey R. 210
Hunt, John B. 1303, 1520
Hunt, Michael H. 355
Hunter, Horace L. 1304, 1521
Hurley, David W. 1305
Hurley, Vic 148, 280
Hurst, James W. 523, 539
Hutson, Charles E. 1968
Hyatt, Robert A. 998–99

I

Ifedi, John-Patrick Afamefuna 1265
Inbar, Efraim 1534–35
Ingalls, James 1860
Inglis, William 453
Innes, John O. 1365
Isaacs, Alvin C. 1107
Isikoff, Michael 1998
Ivey, Robert A. 1306

J

Jablon, Howard 432
Jackson, Andrea V. 1787
Jackson, Chester V. 813
Jackson, Robert J. 1999
Jacobs, Walter Darnell 1017–18
Jalali, Ali Ahmad 1863, 1880, 1941
Jamieson, Perry D. 1570
Jansen, John M. 1901
Jaunal, Jack W. 344
Jean, Grace V. 2041, 2083
Jeffries, Leon M. 2000
Jenks, Jeremiah W. 281
Jenks, Leland Hamilton 446
Jennings, Kenneth A. 908
Jespersen, David M. 1658
Joes, Anthony James 89–90
Johns, Christina Jacqueline 1489
Johns, Lewis E. 802
Johnson, Benjamin Heber 495
Johnson, Bess Sellers 212
Johnson, Dana J. 1579
Johnson, Felix L. 382
Johnson, G.A. 134
Johnson, J. Lee 2084
Johnson, J. Reuben 211, 966
Johnson, Lee 2073
Johnson, P. Ward 1489
Johnson, Robert Bruce 540
Johnson, William R. 212
Johnson, Wray Ross 981
Johnston, Harry 447
Johnston, Lucius F. 417

Johnstone, John H. 55
Jones, Archer 1537
Jones, Gregory 1949
Jones, J.K 1199
Jones, Richard A. 1022
Jones, Robert Keith 1266
Jones, Robert W., Jr. 2028
Jornacion, George William 252
Juarez, Joseph Robert 766
Just, Ward 1307

K

Kafkalas, Peter N. 1366
Kagan, Frederick W. 1544, 2074
Kalb, Marvin 1620
Kaldor, Mary 1689
Kamman, William 854
Kane, William Everett 91, 1218
Kaplan, Stephen S. 77
Karabell, Zachary 1019
Karzai, Hekmat 1863
Katulis, Brian 2129
Katz, Friedrich 541
Keaney, Thomas A. 1565
Kecskemeti, Paul 1020
Keeble, Richard 1623
Keegan, Matt 1818
Keenan, Jerry 156
Keenan, Sean 1957
Keeney, C. Belmont 496
Keiler, Jonathan F. 2064
Keith, LeeAnna Yarbrough 1219
Kelley, Hubert, Jr. 1199
Kells, H.R. 213
Kelly, George A. 1021
Kelsey, Carl 651, 733
Kelso, Quinten Allen 1220
Kemp, Geoffrey 49, 53, 1413
Kempf, Wilhelm 1627
Kendrick, William A. 1649
Kengla, W.A. 418
Kennan, George F. 791
Kennedy, Harold 1850, 1958
Kennedy, William V. 1571
Kent, Irvin M. 1022, 1109
Kenyon, Henry S. 1881
Keown-Boyd, Henry 308
Kestenbaum, Justin Louis 506
Keyser, Ralph S. 135, 926
Khadduri, Majid 1545
Kniko, Ivan 1805
Kievit, James 1525
Kilcullen, David 1798
Kilgore, Joe E. 1523
Killebrew, Robert E. 1784
Killen, Linda 784
Kilmartin, Robert C. 746
Kimmons, Sean 2114
Kinard, William H. 1069
Kindall, Sylvian G. 837

King, Edward L. 1023
Kingaman, Jerome W. 1353
Kiper, Richard 1935, 1964–65
Kipps, Richard L. 1942
Kirkpatrick, Jeane 1308, 1559, 1640, 1690, 1708, 1724, 1765, 2001
Klamper, Amy 2085
Klare, Michael T. 1309–10
Klarevas, Louis J. 1641
Klein, Edwin 1151–52
Kleinman, Forrest K. 1070
Kluna, Ralph 1736
Knight, Melvin M. 734
Knights, Michael 2042, 2075
Knopp, Anthony 609
Koburger, Charles W., Jr. 1024–25, 1191
Koch, Noel C. 1359
Koginos, Manny T. 393
Kolb, Richard K. 92, 282
Kolhatkar, Sonali 1860
Kondracke, Morton 1414
Korb, Lawrence 2129
Korbani, Agnes G. 1167
Kosnik, Mark E. 1882
Kramer, Jacob M. 1758
Kramer, Paul A. 177
Krasin, Chaiyo 1124
Kraus, Theresa L. 1586
Krause, Michael D. 1581
Krym, Jaime 1920
Kubik, Chris 1544
Kyser, Giles 1818

L

Lacey, James G. 2115
Ladd, Jonathan F. 1026
Lambeth, Benjamin S. 1733, 1902
Lambeth, Carl L. 1311
Lambrakis, George Basil 1168
Lane, Jack C. 56, 253
Lane, Rufus H. 759
Lang, Daniel 1090
Lang, Gladys Engel 1620
Lang, Kurt 1620
Langley, Lester D. 93, 448, 469, 626, 948
Lansdale, Edward G. 1125
Larson, Arthur D. 57
Lasswell, A. Bryan 419
Leacacos, John P. 1126
Leach, C.R., 1102
Lederer, W.J. 398
Lehman, Christopher M. 1312
Lehman, John 1267
Lehrack, Otto J. 1591
Leiss, Amelia C. 995
Lejeune, John A. 892–93
LeMay, Curtis E. 1110
Leonard, Henry 325
Leschorn, Helen 735

Leslie, Mark S. 1799
Leventhal, Robert M. 420–21
Lewis, Gordon K. 1439
Lewis, Peter 213
Ley, Wes 747
Leyden, Andrew 1546
Licatovich, Bill 1624
Linder, James C. 1691
Lindsay, Franklin A. 1027
Lindsay, James M. 1990
Lindsey, J.R. 316
Lininger, Clarence 584
Link, Arthur S. 94, 542, 618
Linn, Brian McAllister 178–81, 214–15, 1800
Linn, Thomas C. 1371
Linsert, E.E. 136, 941
Lister, Florence C. 543
Lister, Robert H. 543
Little, Douglas 1192
Livingston, Hoyt R. 1127
Livingston, Michael John 855
Lockmiller, David A. 449, 460
Logan, Rayford W. 652
London, Jack 627
Long, Austin 1801–2
Long, William F. 1028–30, 1313
Longoria, Alfred, Jr. 1314
Lopez, Alfredo Jr. 1654
Lord, Rebecca Ann 760
Lorenz, F.M. 1678
Lou, Dennis Wingsou 507
Lowenthal, Abraham F. 1221
Lowry, Mrs. E.K. 317
Luckett, Judith A. 831
Lybrand, William A. 59
Lynn, John A. 1785
Lyon, Harold C., Jr. 1103

M

Macaulay, Neill 879–81
MacCloskey, Monro 1032
MacDonald, Scott B. 1440
Macgregor, Douglas 1832
Machette, Robert B. 2
MacGleish, Scott T. 1914
Mackey, Chris 1922
Mackinlay, John 1788
MacKinnon, Michael G. 1514
MacMichael, David Charles 736
MacPherson, Malcolm 1923
Maddox, Robert J. 820
Maechling, Charles 1317
Mahnken, Thomas G. 1613, 1789
Mahoney, Tom 524
Maitre, Benjamin R. 1833
Major, H.C. 909
Maloney, Sean S. 1924
Mann, Edward C. 1572
Mann, Morgan 2076
Manney, Henry N. 125

Manning, Clarence A. 821
Manwaring, Max G. 1268, 1318
Marble, William Sanders 582
Marck, David 1925
Marion, Jonathan E. 1945
Marolda, Edward J. 1600
Marsh, John O., Jr. 1319
Martin, C.H. 567
Martin, John Barlow 1224
Martin, John R. 2098
Martino, Joseph P. 1097
Marvin, George 576, 592
Mason, Gregory 593
Mason, Herbert Molloy, Jr. 544
Massing, Michael 1454
Mateo, Lek 2116
Matthews, William 2130
Mattox, Henry E. 1533
Mauroni, Albert J. 1605
May, Glenn Anthony 183–84, 216–17
Mazarr, Michael J. 1547
McCain, Thomas A. 1625, 1628
McCallus, Joseph 201
McClellan, Edwin N. 424–25, 470, 748, 840, 877–78, 894, 942
McClintock, Robert 1193
McCombie, Ryan J. 1513
McCormick, Medill 628–29
McCracken, Alan R. 399
McCrocklin, James H. 688
McCuen, John J. 1031
McDevitt, John Richard 471
McDougall, Walter A. 1434
McDowell, John D. 1349
McElroy, Robert H. 1883
McEnroe, Sean 182
McFate, Montgomery 1786–87, 2043
McGee, Vernon E. 126, 910
McGinn, John 1729
McGlasson, W.D. 1071, 1091
McGleish, Scott T. 1914
McGwire, Michael 1372
McHugh, James M. 426
McInnis, Charles W. 1316
McIntosh, Scott E. 1841
McKenna, Ted 2044–45
McKenzie, Scott W. 895
McKeon, Don 2024
McMillen, Fred E. 630, 708
McNamee, Terence 1861
McNulty, Dennis J. 1969
McPherson, Alan 1222–23
McPherson, Karen A. 1294
Mead, James M. 1417
Meade, Harry H. 802
Meernik, James David 1033
Mehl, S.P. 1384
Meixsel, Richard B. 154
Melanson, Richard A. 1443
Mellman, Harry George 95

Melshen, Paul 1373
Melson, Charles D. 1592
Melton, Carol Willcox 822
Menning, Bruce W. 1581
Menos, Dennis 1614
Meo, Leila M.T. 1169
Mercier, Joan B. 1745
Meriwether, Walter Scott 472
Merk, Frederick 239
Mermin, Jonathan 1051, 1679
Mestrovic, S.G. 2094
Metcalf, Clyde H. 96, 345, 454, 749
Metz, Helen Chapin 2000
Metz, Steven 1320, 1524–25, 1759, 1803
Meyer, Leo J. 773
Meyer, Michael C. 167, 619
Mickolus, Edward F. 58
Miles, Donna 1650
Millard, George A. 568
Millen, Raymond 1759
Miller, Charles J. 750
Miller, Greg 1922
Miller, Hope 59
Miller, J. Michael 346–48
Miller, James L. 2046
Miller, Roger G. 581
Miller, Roger J. 1034
Miller, Stuart Creighton 185–86
Miller, T. Christian 2117
Miller, Walter L., Jr. 1104
Miller, Wesley 2118
Millett, Allan R. 283, 427, 450, 455, 509, 674
Millett, Richard 97, 653, 751, 927
Mills, Greg 1861
Millspaugh, Arthur C. 654
Mitchell, Greg 2122
Mitchell, Joshua D. 1926
Mojares, Resil B. 218
Molloy, Ivan 1321
Montague, Ludwell 655
Montross, Lynn 911
Moore, Joel R. 801–2
Moore, John Norton 1322
Moore, Lynn 1155
Moore, Robin 2047
Moores, Drew 1939
Morales, Waltraud Q. 1269
Moran, Brian Patrick 761
Moran, John B. 60
Morelli, Donald R. 1323
Morey, Lewis S. 558
Morgan, Matthew J. 1927
Moroney, Jennifer D.P. 1891
Morton, Louis 359
Mosher, Norman G. 1081
Moskos, Charles C., Jr. 1251
Mott, Charles P. 1324
Moulis, Wallace J. 1252
Mowery, Beverley P. 1766
Moyar, Mark 1804

Mrazek, James E. 1035
Mroczkowski, Dennis P. 1593
Mrozek, Donald J. 54
Mueller, John. 1548, 1626
Mueller, Karl P., 1834
Mughisuddin, Margaret 51
Mulcahey, Mark 496
Mulcahy, Francis P. 912
Mulrooney, Virginia Frances 242
Munro, Dana G. 98, 99, 439, 473–74, 486, 697, 698, 722, 737, 767, 842, 944
Murphy, John R. 1655
Murphy, Martin N. 1851
Murphy, Robert 1170
Murray, Robert Hammond 254
Murray, Williamson 2014
Musa, Samuel A. 1818
Musicant, Ivan 100, 475, 656, 738, 856, 1235
Myers, John T. 318
Myers, Robert H. 709

N

Nagl, John A. 1753, 1790
Nalty, Bernard C. 857
Nardulli, Bruce 1729
Nash, Douglas E. 1608
Navarro-Génie, Marco Aurelio 858
Nearing, Scott 101
Nelson, Andrew G. 1081
Nelson, W.H. 1560
Neufeld, Gabrielle M. 428
Neville, Edwin L., Jr. 349
Newell, Clayton R. 1529
Newman, Richard J. 1903
Neymeyer, Robert James 739
Niblack, Preston 1579
Niland, Norah 1853
Ninkovich, Frank 187
Noble, Dennis L. 360
Nohrstedt, Stig A. 1627
Norman, Lloyd 1058, 1108
Nye, Joseph, Jr. 1549

O

O'Brien, William V. 1385
O'Connor, Richard 255, 545, 823
O'Hara, Patrick M. 1884
O'Shea, C.J. 1102
Ochosa, Orlino A. 219
Oclander, David I. 1901
Odom, William E. 1526
Ofcansky, Thomas P. 1353
Offutt, Milton 103
Oliker, Olga 2007
Olinger, Mark A. 1201
Olsen, John Andreas 1573
Olson, William J. 1325–27
Oppenheimer, Harold L. 967
Orellana, Manuel A., Jr. 1760
Orgill, Andrew 1528

Osborn, George K. 1328
Oseth, John M. 1388
Osgood, Robert Endicott 1329
O'Shaughnessy, Hugh 1441
Otsuki, Hiromi 1273
Ottaway, David 1330
Ovendale, Ritchie 1194
Owen, Norman G. 220
Owen, Robert C. 1695, 1834
Ozaki, John 1112

P

Paddock, Alfred H., Jr. 968
Pagano, Dom Albert 933
Page, George 1093
Pagonis, William G. 1606
Palazzo, Louis J. 1791
Palen, Don G. 1374
Palmer, Bruce, Jr. 1094, 1236, 1253,-1254
Palmer, Frederick 256, 332, 631–33
Palmer, Michael A. 1601
Palmer, Thomas A. 1157
Pamphile, Leon D. 657
Pappalardo, Joe 1819, 2048, 2107
Paret, Peter 1036, 1105
Park, Bum-Joon Lee 51
Parker, James 596
Parker, Jay M. 1628
Parker, Thomas A. 1574
Parks, W. Hays 1381
Paschall, Rod 1331
Paszek, Lawrence J. 3
Patton, George E. 579
Paulet, Anne 188
Payen, Cecile E. 319
Payne, Anthony 1442
Payne, Robert Bruce 221
Peard, Roger W. 882, 934
Pearson, Ivan 1195
Pease, Frank Chester 284
Peck, Allen G. 1835
Peers, W.R. 1072
Pence, Harvey J. 1113
Pengelley, Rupert 2049
Pepper, Suzanne 1150
Pérez, Louis A., Jr. 61, 440, 456, 775
Perkins, Dexter 104
Perkins, Stuart L. 1360
Perkins, Whitney T. 105, 477, 699, 768, 949, 1225
Perret, Geoffrey 2015
Perry, David L. 1844
Perry, Hamilton Darby 394
Perry, Walter 1729
Peters, Ralph 1761–62, 2077
Pethick, W.N. 320
Petit, Michael 1419
Petraeus, David H. 2050
Pezzullo, Ralph 1709
Pfaff, Roy 383

Phillips, Clinton A. 945
Phillips, Donald T. 1741, 2018
Phillips, Jason 1642
Philpott, Tom 1959
Pickering, Abner 510
Pierce, Philip N. 1185
Pierce, Terry C. 1607
Pineau, Roger 384
Pirnie, Bruce 1729
Pintak, Larry 1420
Planier, George S. 137
Plante, Trevor K. 350
Plata, Holly 1960
Platz, David L. 1620
Plummer, Brenda Gayle 665
Pokrant, Marvin 1602
Polete, Harry 1158, 1160
Pomeroy, William J. 189
Poole, Walter S. 1651
Pope, Bill 1712
Porter, John A. 571
Portman, Charles 1943
Posner, Walter H. 658
Potter, Michael C. 1472
Powers, Robert D., Jr. 1114
Pratt, Andrew N. 1375
Prawdzik, Christopher 1944
Pray, Jeremiah 2078
Preston, Diana 309
Prochniak, Scott E. 1928
Procter, David E. 1037
Proctor, Clarence B. 883
Pruett, Jesse P. 2079
Purkiss, Charles T. 1389
Pustay, John S. 1038
Putney, Diane T. 1575

Q

Quandt, William B. 77, 1171
Qubain, Fahim I. 1172
Quester, George H. 1332
Quilter, Charles J. II 1594
Quinn, John T. II 1595
Quinn, Michael A. 1717
Quirk, Robert E. 610

R

Rabasa, Angel 1805
Rabe, Stephen G. 1226
Rabh, John E. 1074
Radin, Robert M. 1473
Ramsey, Robert D. III 222–23
Rashid, Ahmed 1862
Rashid, Nasser Ibrahim 1615
Raymond, William Montgomery, Jr. 1039
Read, Robyn 1836
Reade, Philip 285
Reason, Barbara 51
Record, Jeffrey 1040, 1550, 2002–3, 2016
Redmon, David 1739

Redmore, James 1929
Reed, John Scott 224
Reese, Howard C. 1376
Reilly, Margaret Inglehart 225
Reisinger, H.C. 771, 928
Renda, Mary A. 700
Renshon, Stanley A. 1561
Rentflow, Frank Hunt 351, 898
Reveley, W. Taylor III 106
Reynolds, Nicholas E. 1501, 2051
Reynolds, Richard T. 1576
Rhodes, Benjamin D. 806
Rhodes, Charles D. 333
Rhodes, Philip E. 1675
Rhyne, Russell F. 1098
Rice, Michael David 478
Richardson, W.P. 803
Ricks, Thomas E. 2017
Rid, Thomas 1274
Riehm, J.A. 1718
Risbert, Robert H. 2052
Ritchie, Nick 2004
Rittgers, Courtney M. 1333
Roberts, Adam 1618
Roberts, Richard C. 598
Roberts, Stephen S. 402
Robertson, O. Zeller, Jr. 511
Robinson, Linda 2131
Robinson, Mary Ann 62
Robinson, Michael C. 248
Robinson, Piers 1680
Rogers, Chris 1763
Rogers, John J. 666, 689
Rogers, Paul 2004
Roll, Robert 1952
Ronning, C. Neale 1118
Roosevelt, E.L. 127
Roper, William L. 914
Rose, Donald Gregory 982
Rosen, Leora N. 1515
Ross, Stanley R. 500
Rosson, William B. 1073
Roston, Aram 2005
Rostow, Walt W. 1128, 1173
Rotberg, Robert I. 1863
Roth, Andrew 1234
Roth, Jason 2053
Roth, Margaret 1492
Roth, Russell 149
Rothwell, Richard D. 430
Roush, Maurice D. 1129
Rovergno, John S. 1740
Rowell, Ross E. 915–17
Rudd, Gordon W. 1609
Rudolph, Jack W. 286
Runge, Jared E 1970
Runnels, Al 2118
Russell, J. Thomas 4
Russell, John H. 701
Russell, Richard L. 2006

Ryan, Mick 1842
Rylander, R. Lynn 1377
Rystad, Goeran 243

S

Saborsky, Alan Ned 1341
Sack, John 1584
Sadler, Louis R. 504
Salisbury, Nathan F. 512
Salisbury, Richard V. 950
Samuel, David F. 1837
Sanchez, Ricardo S. 1741, 2018
Sanders, Ralph 1041
Sandos, James A. 497, 525
Sandstrom, Harald M. 1440
Santelli, James S. 428
Sardo, Americo A. 1074
Sarkesian, Sam C. 1334–37
Saund, Dalip 1099
Saunders, Richard M. 1180
Scales, Robert H., Jr. 1338, 1585, 2014, 2099
Schaefer, Robert W. 1742
Schafer, Charles H. 1474
Schaffer, Ronald 122
Schanz, Marc V. 1905, 1961, 2022
Schemmer, Benjamin F. 1260, 1461, 1475, 1500, 1604, 1669, 1716
Schirmer, Daniel. 190, 244
Schlaak, Thomas M. 1339.
Schlesinger, Arthur M., Jr. 107
Schlesinger, Robert J. 1042
Schmid, Clarence H. 1115
Schmidt, Hans 352, 431, 479, 611, 659–60
Schneller, Robert J. Jr. 1600
Schoenhals, Kai P. 1443
Schoonmaker, Herbert G. 1237
Schreadley, Richard Lee 661
Schreier, Konrad F. Jr. 269
Schroder, James A. 1906, 1930
Schroeder, Michael J. 859, 918, 951
Schubert, Frank N. 1586
Schubert, Richard H. 935
Schuon, Karl 1159
Schwartz, Richard A. 1530
Schwarz, Benjamin 1760
Schwarzkopf, Olaf 513
Scott, Hugh Lenox 546
Scott, Joseph L. 226
Scranton, Margaret E. 1490
Scully William L. 1337, 1352
Seabury, Paul 1434
Seaman, Louis Livingston 354
Sedra, Mark 1859
Seevers, James P. 84
Seigle, John W. 1074
Seim, H.B. 1082
Seiple, Chris 1743
Selser, Gregorio 860
Sepp, Kalev I. 1806, 1885, 1936–38, 1962
Sewall, Sarah 1764

Sexton, William Thaddeus 191
Seydel, Cane A. 1907
Shafer, D. Michael 983
Shaheen, Esber Ibrahim 1615
Shallcross, H.C. 1367
Shanahan, Stephen W. 1698
Shannon, Magdaline W. 662
Shapiro, Sumner 785
Sharf, Frederic Alan 310
Sharon, Patrick 1712
Sharp, Jeremy M. 1975
Shaw, Angela Velasco 192
Shaw, Henry I., Jr. 1138, 1143
Shaw, Robert C. 1719
Shaw, W.B. 577
Sheehan, J.M. 400–401
Sheehan, Kevin Patrick 984
Shelton, David L. 1656
Shelton, H. Hugh 1720
Sherry, Michael S. 1562
Shoup, David M. 432
Shrader, Charles R. 63
Shulimson, Jack 1200
Shulka, Paraay 1805
Shultz, Richard H. 1340–41, 1502
Shy, John W. 1036, 1105
Shyles, Leonard 1625, 1628
Siegel, Adam B. 1577
Sights, Albert P., Jr. 1205
Silbey, David J. 193
Simmons, Edwin H. 108
Simon, Steven 1823
Sivachev, Nikolai V. 786
Slater, Jerome N. 77, 1227–28
Slayden, James L. 620
Slife, Jim 1838
Sloan, Stephen 1342
Slover, Robert H. 1130
Slusser, Robert M. 77
Smallman-Raynor, Matthew 232
Smith, C.C. 287
Smith, D.H. 194
Smith, Emily Kaighn 715
Smith, Gibson Bell 824
Smith, Harry A. 514
Smith, Hedrick 1629
Smith, Herbert E. 838
Smith, Laun C. 1131
Smith, Lynn D. 1181
Smith, Myron J., Jr. 64–65
Smith, Oliver P. 128
Smith, Richard W. 150
Smith, Robert Freeman 515, 776
Smith, Roger K. 1549
Smith, Roy C., Jr. 361, 385
Smith, William E. 1425
Smyrl, Marc E. 1421
Smythe, Donald 257, 288–89, 547
Snider, Don M. 1547
Snyder, H.W. 679

Snyder, Zack 2075
Somin, Ilya 787
Sommers, Richard J. 5
Sonoski, Patrick J. 1275
Spark, Michael 1075
Spector, Robert M. 723–24
Sperardo, Richard C. 2053
Speredelozzi, John M. 2023
Spiller, Roger J. 1182
Spore, John B. 1058
Stack, Cecil 1444
Stahley, Carl E. 676
Stanton, Louise 1043
Stanton, Martin N. 1666–67
Stearns, Leroy D. 1596
Steel, Ronald 1044
Steele, Dennis 1744, 1807, 1886, 1908, 1931–32, 1950–51, 2054–55, 2061, 2100–101
Steeves, W.D., Jr. 1174
Stenberg, Richard K. 600
Stephenson, Scott 2123
Stern, Ellen P. 1045, 1300
Stevenson, Adlai E. 1229
Steward, H. Douglas. 1106
Stewart, John F. 1390
Stewart, Richard W. 1652
Stilwell, Richard G. 1343
Stimson, Henry L. 843
Stout, Joseph A., Jr. 548
Stowe, William M. III 163
Strakhovsky, Leonid I. 792, 807
Stratton, Ray E. 1368
Strobridge, William F. 227
Sturdevant, E.W. 109
Sullivan, Brian R. 1527
Sullivan, Mike 2102
Summers, Harry G. 1344–45
Sutliff, R.C. 386
Sutton, Ottie K. 66
Sutton, Paul 1442
Swain, Richard 1346
Swank, Eric 1630
Swanson, H.J. 395
Sweet, Jonathan 1826
Sweetman, Bill 1910
Sweetman, Jack 612
Sweezy, Paul 195
Swofford, Anthony 1597
Symanski, Michael W. 1347
Szafranski, Richard 1270
Szulc, Tad 1230

T

Tackenberg, William D. 602
Tallent, Robert W. 433
Taney, Dudley 710
Tang, Samuel Kong 291
Tannenbaum, Karen 212
Tarbuck, Ray D. 956

Taussig, J.K. 322
Taylor, James A. 1401
Taylor, John R.M. 159
Taylor, Philip D. 1631
Taylor, Richard H. 1348–49
Taylor, William J. Jr. 1328
Teamey, Kyle 1825–26, 1845
Thomas, Lowell 228, 337, 409, 464
Thomas, Robert S. 549
Thomason, John W. 110
Thompson, Dale L. 1476
Thompson, Denis William 1231
Thompson, Henry O. 229
Thompson, L.E. 677
Thompson, Loren B. 1350
Thompson, Neal 2065
Thompson, Wayne Wray 258
Thomson, Alex 1632
Thorndike, Tony 1442
Thorpe, George C. 762
Thrall, A. Trevor 1052
Thrasher, Thomas E. 680
Thurman, Maxwell B. 1504
Tierney, John Joseph, Jr. 861
Tilford, Earl H., Jr. 1542
Tillema, Herbert K. 1046, 1175, 1232
Tiron, Roxana 1792, 1933, 2056, 2086
Tiwathia, Vijay 1477
Tocchet, Gary J. 1351
Todd, Jess T. 2119
Tolley, Kemp 387–88, 814
Tomasso, Thomas 1901
Tomblin, Barbara B. 233
Tomes, Robert R. 1808
Tompkins, Frank 550
Tompkins, R. McC. 1245.
Toulmin, Harry Aubrey, Jr. 551
Tovo, Ken 1809
Towle, Philip 1999
Townsend, Henry Schuler 292
Townsend, Kellard N. 1945
Trager, Frank N. 1352
Trainor Bernard E. 1541, 2013
Trani, Eugene P. 789
Trapp, Brian 2095
Trask, David F. 67
Trask, Roger R. 67
Tripp, Robert S. 1952
Trowbridge, Gordon 2103
Truver, Scott C. 1603
Trybula, David 1587
Tsou, Tang 1161
Tsouras, Peter G. 1495
Tuchman, Barbara W. 369
Tucker, Phil 1246
Tucker, Robert W. 1982
Turay, Ismail, Jr. 2057
Turner, Frederick C., Jr. 1256
Turner, Thomas C. 920
Twichell, Heath, Jr. 151, 552

U

Ullman, Harlan 1887
Unterberger, Betty Miller 834–35
Utley, Harold H. 138–39
Utley, Robert Marshall 517

V

Valenta, Jiri 1435
Valentine, Lincoln G. 482
Van Creveld, Martin 1619
Vandiver, Frank E. 259, 553
Van Hassell, Agonstino 1422
Van Haute, Edward B. 1361
Van Sickle, Jeffrey B. 2058
Vandiver, Frank E. 259, 553
Vego, Milan N. 1888
Venzon, Anne Cipriano 155, 324, 407, 463, 604, 642
Vernon, Alex 1587
Villere, Paul, Jr. 389
Vincent, Ted 952
Vine, Timothy D. 1720
Vines, John R. 2059, 2104
Vlasak, Marian E. 1821
Voorhies, David 2080
Vriesenga, Michael P. 1578

W

Waddell, Ricky Lynn 1047
Wade, Gary H. 1202
Wade, S.S. 1186
Waghelstein, John David 112
Waite, Carleton Frederick 311
Walraven, J.G. 886
Walter, Richard J. 1059
Walterhouse, Harry F. 1132–33
Ward, Compton E. 1048
Ward, Robert, Jr. 1810
Ware, Lewis B. 1353
Warner, Lesley Anne 1805
Wary, Robert O., Jr. 2087
Wasielewski, G. 1378
Waters, Fleming A. 436
Watson, Bruce W. 1437, 1495, 1553
Watson, Francis M. 1127
Waxman, Matthew C. 1988
Weathers, Bynum E. 1353
Weed, Sylvia Lee 114
Weigley, Russell F. 1060
Weinhold, H.W. 489
Weinrod, W. Bruce 1354
Weiss, Thomas G. 1516, 1643, 1692
Weissheimer, J.W. 566
Welch, Richard E., Jr. 245
Weller, Jac 1049
Weller, Marc 1725
Welles, Sumner 740
Wellman, Leslie H. 763
Wells, Christopher B. 1890

Welsh, William E. 518
Welsome, Eileen 554
Wendt, Eric P. 1793
Wermester, Karin 1853
West, F.J. Bing 2019
West, Richard 1233
Westall, Virginia Cooper 825
Western, Jon W. 1053
Westphal, D'Val J. 1276
Weymouth, Tony 1726
Wharton, Wallace S. 390
Wheeler, W.R. 370
Whisnant, David E. 863
Whitaker, Mark 1426
White, John Albert 826
White, John Roberts 293
White, Thomas D. 140
White, Wolfred K. 1116
Whitters, Robert A. 1610
Wieczorek, George A. 459
Wilbanks, James H. 1839
Wilkins, Frederick 972
Williams, Charles F. 757
Williams, Dion 864, 887
Williams, John Hoyt 196
Williams, Laurin L. 371
Williams, Peter 1633
Williams, Robert Hugh 437
Williams, S.M. 559
Williams, Vernon J. 1827
Williams, Vernon Leon 234
Williamson, Joel E. 1891
Wilmeth, J.D. 902
Wilson, Gregory 1820
Wilson, James Henry 312
Windsor, Philip 77
Winkler, Melvin L. 1078
Winnefeld, James A. 1579
Winslow, Cameron McR., Jr. 391
Winslow, Richard Lawrence 1050
Wirkus, Faustin 710
Wise, Frederick May 711
Wise, H.D. 230
Wittes, Tamara Cofman 1505
Wolf, John Berchmans 1177

Wolff, Leon 197, 555
Wood, Bryce. 957
Wood, Donald K. 1745
Wood, Eric Fisher 260
Wood, John C. 902
Wood, John S. 1061
Wood, Samuel S. 1355
Woods, Randall B. 1234
Woodward, Mrs. M.S. 321
Woodward, Peter 1644
Woolard, James Richard 152
Worcester, Dean C. 613
Worley, Robert 1847
Wormuth, Francis D. 115
Wright, Kathleen M. 1515
Wurtsbaugh, Daniel W. 323

X

Xiang, Lanxin 313

Y

Yang, Zhiguo 1146
Yarborough, William P. 1079, 1356
Yates, Dennis W. 1928
Yates, Lawrence A. 1238, 1357, 1496–97
Yetiv, Steve A. 1554
Yingling, Paul L. 1790
Yokovlev, Nikolai N. 786
York, Dorothea 812
Young, Gene 1241
Young, Karl 560
Young, Kenneth Ray 198
Young, Oran R. 116

Z

Zaalberg, Thijs W. Brocades 1653, 1746
Zais, Mitchell M. 1358, 2020
Zajac, Wayne D. 1284
Zeman, P.M. 2021
Zimmeck, Steven M. 1598
Zinser, Larry R. 1382
Zoll, Atwell 1483
Zurat, Michael L. 1379

Subject index

Almost all of the entries in the bibliography are noted, sometimes more than once. The armed services of the United States are listed under their full titles titles, "United States Air Force," for example, not "air force." The references under a given military service are general; they do not include references to a particular small war or counterinsurgency campaign.

A

Afghanistan general 188, 1265, 1776, 1780, 1786, 1797, 1799, 1804, 1812, 1831, 1843, 1853–63; encyclopedia 1852; aerial operations 1892–1910; civic action 1953–57, 1959–63; civil affairs 1958; engineering operations 1966–70; intelligence 1939; logistics 1946–52; military advisory activities 1940–45; military operations, 1864–91, 1911–34; psychological operations 1964–65; special military operations 1935–38
African-Americans 177,182, 246, 248, 952, 1484–85; bibliography 153
Afro-Cubans 447
Aguinaldo y Fama, Emilio 184, 199
Algeria 983, 1803
Allen, Henry T. 141, 281, 552
American Revolution 1034, 1807
Angola 1265
Anti-imperialism 191, 235–45
Artillery and small wars/counterinsurgency 133, 1102
Aviation and small wars/counterinsurgency 141

B

Baldwin, Frank B. 272
Batraville, Benoit 669
Bayan 272
Belgium 731
Bell, J. Franklin 205, 216, 222
Berlin 1059
Biafra 1642
Bertol, Louis N. 666, 676
Bibliographies 47–70
Bliss, Tasker Howard 250, 252, 256, 258, 275
Bosch, Juan 1216
Bosnia general 981, 1509, 1512, 1518, 1638, 1673, 1681–92; aerial operations 1693–95; causes 1681, 1685–86; chronology 1681; legal aspects 1697; media1276; military operations 1518, 1682, 1696–98; peace 1685, 1691,1699–1700
Boxer Uprising, 1898–1901 general 294–313; cavalry operations 326, 328, 333; documents 294–97; marine operations 321, 327, 334–52; medical aspects 353–54; military government 355; naval operations 300, 303; relief of the legations in Beijing 298, 327–33; Seymour expedition, 322–23; siege of the legations in Beijing 314–21, 339–40, 342, 344, 346–47; Tianjin battle 324–25
Brooklyn (U.S. warship) 840
Buchanan, Richard Bell 887
Bullard, Robert L. 268, 283, 448, 509
Butler, Smedley Darlington 299, 318, 324, 337, 352, 406–10, 428, 432, 434, 463–64, 479, 604, 611, 642, 659, 667, 679
Brown, Preston 202
Brown, John Clifford 201

C

Central America 1321–22, 1390
Civil War (1861–65) 1807
Chaffee, Adna Romanza, 272, 298, 301 355
Chemical warfare 129, 136, 1104
China, 1901–41 general 356–61; army operations 362–71; marine operations 358–59, 403–37; naval operations 358–59, 361, 372–402
Civic action 1117–33
Clark, J. Reuben 465, 498
Colombia 1816
Columbus, New Mexico 519–25, 531, 541, 545–46
Chinese Civil War 1821
Cholera 232, 267
Costa Rica general 95
Counterinsurgency Use Small Wars/ Counterinsurgency
Crockett, Cary I. 286
Cuba 91, 95,
Cuba, 1906–9 general 438–50, 457; army operations 450–51, 453–55, 459 causes 443–45, 455; civic action 447, 460; economic aspects 457–58; marine operations 453–54; military advisory activities 452, 456; provisional government 446, 460
Cuba, 1917–22 general 772–76; causes 73, 773; naval operations 774, 777

524

Subject index 525

D

Dagu forts, China 303
Daly, Daniel Joseph 346
Darmond, James A.
Decorations 144
Dissertations and theses 46
Dominican Republic, 1916–24 general 95, 727–52; aerial operations 753–54; causes 113; atrocities 735; civic action 113, 758, 760–63; military advisory activities 755–57, 771; military government 736, 759, 762; military operations 132, 728, 735, 738, 743–52, 770; transportation 124, 126; withdrawal 764–69
Dominican Republic, 1965–66 general 91, 113, 616, 996, 1001, 1037, 1046, 1206–34, 1712; aerial operations 1257; army operations 1236–37, 1247–56; civil affairs 1008; elections 1217; marine operations 1242–46; naval operations 1239–41
Donoghue, M.J. 810

E

Edson, Merritt 866, 874, 881, 890
Egypt 1187–88
Eisenhower Doctrine 1163, 1165–67, 1169–70, 1174, 11744, 1180, 1187–88, 1190, 1192
El Salvador 980–81, 1131, 1217, 1526, 1760
Ethiopia 1642

F

Fagan, David 248
Finley, John P. 265, 267
Fletcher, Frank Friday 614, 623
France 808
French Revolution 1034
Fulbright, J. William 1213
Funston 199, 448, 578

G

Gardner, Cornelius 209
Gates, John M. 194
Germany 525, 732, 962, 964
Graves, William S. 817, 821,823, 825, 827, 831, 836
Grayson, William W. 204
Great Britain 798, 799, 808, 812, 1187, 1195
Grenada general 996, 1001, 1037, 1261, 1263, 1269, 1429–44; aerial operations 1460; African-American perspectives 1484–85; causes 1445–54; civil affairs 1008, 1468; Coast Guard operations 1476; documents 1434–35; legal aspects 1459; logistics 1464, 1466, 1470, 1473–74; media 1275–76, 1482–83; military operations 1429, 1455–70, 1473–78; naval operations 1472; political aspects 1430–33, 1436–37, 1440–43, 1445–54, 1479–81;
Greece 1019, 1029, 1517
Guatemala 91, 1019, 1131, 1526

H

Haiti, 1915–34 general 95, 113, 637–62, 692–702, 712–15, 1701; aerial operations 681; causes 663–65; civic action 703–11; military advisory activities 682–91; military operations 666–80; transportation 126; withdrawal 716–26
Haiti, 1991–96 general 113, 639, 660, 1261, 1238, 1512, 1515, 1701–9; civil affairs 1717; legal aspects 1714; military operations 1710–20
Hanneken, Herman H. 670
Hewitt, H. Kent 777
Helicopters 962
Histories (general) 71–116
Honduras 1131
Hoover, Herbert 95
Horton, William 344
Howell, Glenn F. 379
Hull, Cordell 726
Hunter, Robert Swart 889

I

Indexes (newspapers and periodicals) 16–44
Indochina 983, 1029, 1032, 1040, 1049, 1771
Intervention 47, 67, 91, 98–101, 104–5
Iraq War general 163, 169, 195, 1014, 1177, 1768, 1780, 1784, 1786, 1792–93, 1796, 1799, 1804, 1811, 1817, 1827, 1831–32, 1838, 1849, 1971–83; causes 1984–2006; general military studies 2007–21, 2024–59, 2066–80; aerial operations 2022–23; civil affairs/civic action 2108–18; ground combat 2024–59; Baghdad, battle 2060–61; Fallujah, battle 2062–65; information programs/psychological operations 2119–21; insurgency and counterinsurgency 2066–80; intelligence 2088–89; logistics 2105–7; media and the war 2121–23; military advisory activities 2096–2014; naval operations 2081–87; prisons and prisoners 2090–95; terminating the war 2124–31
Iran 1190
Israel 1190

J

Jikiri 278
Johnson, Benjamin O. 829

Johnson, James Weldon 652, 657
Johnson, Lyndon B. 1208, 1212–13, 1223–26, 1234
Johnson Doctrine 1226
Joint Readiness Training Center 1306
Jordan 1163, 1165, 1194–95
Jungle warfare 131, 140

K

Kennedy, John Fitzgerald 1054–61
Kenya 1029
Kimball 472
Knox, Philander 473
Korean War 960–61, 976, 1014, 1046, 1056, 1117, 1266, 1345, 1517
Kosovo general 1510, 1664–65, 1721–26, 1732; aerial operations 1722, 1730–33; marine operations 1743; military operations 1722, 1727–29, 1734–46

L

Lake Onega 814
Lansdale, Edward G. 1288
Laos 994, 1019, 1059
Lebanon, 1958, general 996, 1019, 1037, 1046, 1060, 1162–77, 1405; army operations 1178–82; documents 1162; influence 1203–5; logistics 1201–2; marine operations 1183–86, 1189, 1193, 1200–201; naval operations 1193, 1196–97; political aspects 1187–95;
Lebanon, 1982–84 general 1167, 1262–63, 1402–22; army operations 148 withdrawal 1423–27
Lee, Harry 770
Lee, William A. 875, 895
Libya 1050, 1263, 1882
Lukban, Vincente 207
LeRoy, James A. 172
Latin America 1357

M

MacArthur, Arthur 158, 176, 198
McCoy, Frank Ross 261, 936
Magoon, Charles E. 449
Malaya 976, 1013, 1029, 1032, 1049, 1106, 1171
Marts, Albert C. 420–21
Mexican Border, 1911–19; *see also* Punitive Expedition, 1916–17 and Vera Cruz, 1914 general 95, 490–500, 506–7, 511, 516; Columbus, New Mexico, battle, 519–25; medical and sanitary aspects 501, 514; military operations 501–2, 504–5, 508–10, 512, 514, 517, 518; National Guard mobilization 585–602; refugees 503
Mexican Revolution of 1910 general 1034

Mexico 842
Miles, Nelson A. 202
Military police 1364
Monroe Doctrine 104
Moro campaigns, 1902-c. 1913 general 145–52, 163, 180 249–93; maritime operations 276; transportation 285

N

Native Americans 177, 182, 185, 188, 196, 1003
National Associations for the Advancement of Colored People 657
National Guard; *see also* Mexican Border, 1911–19 196, 208, 211, 585–602, 1071, 1306, 1471, 1668, 1735, 1920
Nicaragua 95, 113, 1131, 1265, 1321, 1447–48
Nicaragua, 1909–25 general 113, 462, 466–67, 471, 473–75, 477–82, 484, 486, 867; causes 463, 465, 473–74, 478; legation guard 483–89; military operations 461, 463–64, 468–70, 475–76, 479, 481, 483, 485, 489; naval operations 472; public opinion 480
Nicaragua, 1927–33 general 841–64, 941, 944; aerial operations 860, 903–20; battles and campaigns 888–902; causes 841–44, 853; civic action 936–46; communications 930–31, 935; elections 936–40, 942, 946 military advisory activities 921–28 military operations 847, 849–51, 854, 856–58, 865–87; naval operations 933; political aspects 947–52; transportation 929; withdrawal 861, 953–59
Nigeria 1642
Nixon, Richard M. 1354
Nixon Doctrine 1354
North China, 1945–49, general 1134–50, 1161; diplomatic and political aspects 1146–50; military operations 1134, 1138–39, 1143, 1153–59; naval operations 1141; repatriation of Japanese personnel 1151–52; supply 1158
North Russia general 798–807; causes 807, 815–16; military operations 799–806, 808–12; naval operations 813–14

O

Otis, Elwell S. 157, 166
Olympia (U.S. Warship) 813
Oral histories 7

P

Pakistan 1190, 1816, 1834, 1862, 1868, 1880,
Palmer, Bruce 1215, 1236
Panama 95, 1261, 1269, 1355

Panama general 140, 1086, 1089, 1096, 1001, 1265–66, 1269, 1355, 1486–90, 1798; causes 1489–90; chronology 1495; civil affairs 1008, 1502; legal aspects 1498–99; marine operations 1501 media 1052, 1276; military government 1496, 1498–99, 1502; military operations 1487, 1491–1504; special military operations 1500
Panay Incident 392–95
Patton, George E. 579
Pendleton, Joseph 475
Pershing, John Joseph 250, 252, 255, 257–59, 273, 275, 288–90, 547, 553, 577
Persian Gulf War general 1014, 1262, 1269, 1510, 1515, 1534–54; aerial operations 1563–79; bibliography 1528; causes 1545, 1547, 1550, 1555–62; chemical warfare 1605; chronologies 1531–33; civic action 982, 1609; civil affairs 1608, 1610; dictionaries and encyclopedias 1529–30; legal aspects 1616–19; logistics 1606–7; marine operations 1588–98; media and public opinion 1052, 1275, 1620–33; military operations 1515, 1536–37, 1540–44, 1546–49, 1551–1153 1580–87; naval operations 162, 1599–1603; political aspects 1611–15; special military operations 1604
Philippine-American War general 145–52, 156, 157–98, 613, 1357, 1794, 1796, 1800, 1807; atrocities 174, 182, 185–86, 188–89, 194, 202, 205–6, 216, 220, 224; bibliographies 153–55; civic action 173, 181; documents 157–62; encyclopedias 74, 156; gender aspects 247; intelligence 179; medical aspects 231–32; military advisory activities 146, 148, 150–52, 178, 214; military operations 131, 199–230; naval operations 233–34; political aspects 235–45; racial aspects 168, 177, 182, 225–26, 246, 248
Philippine Constabulary 145–46, 148, 150–51, 178, 215, 271, 274, 280–81, 293
Philippine Scouts 152
Philippines 118, 980, 983, 1049, 1106, 1122, 1321, 1343, 1526, 1793, 1820
Pingree, Hazen S. 209
Plan of San Diego 495, 497
Platt Amendment 438–40, 454, 775
Port au Prince, Haiti 139
Powell Doctrine 1508, 1556
Puller, Lewis B. ("Chesty") 650, 875, 877, 895
Punitive Expedition, 1916–17 attack on Columbus, New Mexico 519–25; general 526–55; aerial operations 580–81; battles 556–60, 572–77; horses 583–84; intelligence 539; leaders 578–79; logistics 567–71; medical aspects 582; weapons and equipment 561–66;
Presidential war powers 79, 87, 102, 106–7, 111, 115, 833, 844, 1039, 1263, 1265–67, 1269–70, 1408, 1421, 1481, 1512, 1637, 1683, 1702, 1705, 1723
Price-Mars, Jean 662

R

Rhineland 732
Roman Empire 1807
Rowell, Ross E. 914–17
Rules of engagement 1244, 1247–48, 1251, 1253
Russia, 1918 general 778–87; diplomatic and political aspects 788–92; documents 782; military operations 780, 783, 794–97
Rapid Deployment Force 1352

S

Samoa 967
Sandino, Augusto Cesar 846, 852, 860, 864, 872–73, 899
Santo Domingo City 139
Sayre, John Nevin 947
School of the Americas 979
Scott, Hugh S. 546
Semenoff, Gregori 824
Semour expedition in the Boxer Uprising 322
Shoup, David M. 431
Siberia, 1918–20 general 817–18, 820–26; causes 822; diplomatic and political aspects 827–35; marine operations 840; military operations 817, 820, 836–37; naval operations 839; railways 819, 829, 830
Small wars/counterinsurgency general 71–116, 117–23, 127–28, 205, 882, 960–72, 988–1050, 1054–61, 1066, 1069, 1072–77, 1079, 1083–96, 1100–108, 1111–16, 1258–70, 1277–1358, 1505–27, 1765–1820; aerial operations 141, 1285, 1290, 1355, 1363, 1751, 1755–56, 1828–38; artillery 1102, 1287, 1355; bibliography 48, 51, 55, 58–59, 68; civic action 142, 1117–33, 1393–94, 1397–1401, 1839–42; civil affairs 1008, 1391–92, 1395–96; decorations 144, 1276; doctrine 117–23, 973–87, 1002, 1005, 1258–59, 1295–96, 1301, 1329, 1331, 1344, 1346, 1351, 1747–64; evacuation operations 1380–82; force structure 1359–79, 1843–51; intelligence 1109–10, 1386–90, 1822–27; legal aspects

1383–85, 1843–45; logistics 1821; marine operations 1371, 1377–79, 1396, 1773, 1849–50; media 1051–53, 1271–76; medical aspects 1394; naval operations 1016, 1081–82, 1299, 1369–70, 1372–76, 1851; systems analysis 1097–99, 1767; supply 125; tactics 129–40; transportation 124, 126, 143; weapons 1107–8

Small Wars Manual (1940) 117, 122–23, 1277
Smith, Jacob Hurd 206
Smith, Roy C. 384
Somalia general 981, 1262, 1510, 1512, 1515, 1634–44; intelligence 1654–56; legal aspects 1676–78, logistics 1657–58; marine operations 1655, 1665; media and public opinion 1053, 1276, 1640, 1679–80; military operations 1515, 1635, 1645–78; Mogadishu, battle, 1668–75
Special Air Warfare Center (USAF) 1290
Special Forces and similar elements 971, 1062–80, 1091, 1093, 1260, 1313, 1333, 1359, 1362, 1368, 1513, 1527, 1661, 1669, 1871, 1873, 1890, 1716, 1846
Special Operations School 1367
Special warfare 968, 971, 1010, 1062–80, 1091, 1093–94, 1260
Stevens, John F. 819
Stevenson, Adlai E. 1229
Stillwell, Joseph 369
Stimson, Henry L. 95, 841, 843
Strategic Air Command 1111

T

Taft, William Howard 172, 198, 238, 240, 498–500, 1537
Taylor, John R.M., 159, 170, 172
Texas Rangers, 504, 517
Thailand 994, 1124,
Thomas, Earl Albert 919
Tinio, Manuel 219
Transportation 143

U

United Nations 1521, 1641; bibliography 8–9
United States Air Force 1751; bibliography 3
United States Army 57, 7, 118–21, 131; bibliography 4, 5, 7, 52, 62–63
United States Coast Guard 1476

United States Congress; *see also* Presidential war powers 79, 84, 87, 102, 106, 107, 111, 115; bibliography 10–11, 15, 36, 49;
United States Marine Corps (general) 82, 88, 96,108–9, 117, 121–30, 132, 134–39, 269, 320–21, 327, 358–60, 1084, 1090, 1095, 1371, 1375, 1377–78, 1380–82; bibliography 6, 50, 55, 60,
United States Navy (general) 97, 110; bibliography 64–65;
United States Volunteers 224

V

Veracruz 139
Veracruz, 1914; *see also* Mexican Border, 1911–19 general 139,491, 496, 506, 603–20, 1216; aerial operations 635; civic action 625; military government 606, 625–27, 629–33; military operations 621–23, 628, 634, 636; political aspects 614–20; publicopinion 609; refugees 624; transportation 124
Vietnam 90, 111, 115, 122, 151, 163, 167, 174, 184–85, 209, 661, 974–76, 980, 982–84, 989, 994, 999, 1010, 1014, 1028, 1030, 1040, 1046, 1052, 1059–60, 1065, 1073, 1078, 1091, 1107–8, 1112, 1192, 1217, 1234, 1255, 1258, 1260, 1266, 1269, 1276, 1296, 1298, 1300–302, 1313, 1329, 1331, 1337, 1345, 1349, 1356, 1378, 1384–85, 1406, 1427, 1569, 1635, 1778–79, 1794, 1797, 1801, 1803–4, 1809–10, 1823, 1839, 1870, 1844, 1849
Voluntario troops 925

W

Wars powers Use Presidential war powers
Weinberger Doctrine Use Powell Doctrine
Wilson, Woodrow 81, 94
Wirkus, Faustin 707, 709–10
Women's International League for Peace and Freedom 640
Wood, Leonard 249–54, 258, 260, 266, 275, 289
World War II 962, 964, 1569

Y

Yangzi Patrol 356–58, 372–401
Yemen 1024